VISUAL INFORMATION REPRESENTATION, COMMUNICATION, AND IMAGE PROCESSING

OPTICAL ENGINEERING

Series Editor

Brian J. Thompson

*Distinguished University Professor
Professor of Optics
Provost Emeritus*

*University of Rochester
Rochester, New York*

1. Electron and Ion Microscopy and Microanalysis: Principles and Applications, *Lawrence E. Murr*
2. Acousto-Optic Signal Processing: Theory and Implementation, *edited by Norman J. Berg and John N. Lee*
3. Electro-Optic and Acousto-Optic Scanning and Deflection, *Milton Gottlieb, Clive L. M. Ireland, and John Martin Ley*
4. Single-Mode Fiber Optics: Principles and Applications, *Luc B. Jeunhomme*
5. Pulse Code Formats for Fiber Optical Data Communication: Basic Principles and Applications, *David J. Morris*
6. Optical Materials: An Introduction to Selection and Application, *Solomon Musikant*
7. Infrared Methods for Gaseous Measurements: Theory and Practice, *edited by Joda Wormhoudt*
8. Laser Beam Scanning: Opto-Mechanical Devices, Systems, and Data Storage Optics, *edited by Gerald F. Marshall*
9. Opto-Mechanical Systems Design, *Paul R. Yoder, Jr.*
10. Optical Fiber Splices and Connectors: Theory and Methods, *Calvin M. Miller with Stephen C. Mettler and Ian A. White*
11. Laser Spectroscopy and Its Applications, *edited by Leon J. Radziemski, Richard W. Solarz, and Jeffrey A. Paisner*
12. Infrared Optoelectronics: Devices and Applications, *William Nunley and J. Scott Bechtel*
13. Integrated Optical Circuits and Components: Design and Applications, *edited by Lynn D. Hutcheson*
14. Handbook of Molecular Lasers, *edited by Peter K. Cheo*
15. Handbook of Optical Fibers and Cables, *Hiroshi Murata*
16. Acousto-Optics, *Adrian Korpel*
17. Procedures in Applied Optics, *John Strong*
18. Handbook of Solid-State Lasers, *edited by Peter K. Cheo*
19. Optical Computing: Digital and Symbolic, *edited by Raymond Arrathoon*
20. Laser Applications in Physical Chemistry, *edited by D. K. Evans*
21. Laser-Induced Plasmas and Applications, *edited by Leon J. Radziemski and David A. Cremers*
22. Infrared Technology Fundamentals, *Irving J. Spiro and Monroe Schlessinger*
23. Single-Mode Fiber Optics: Principles and Applications, Second Edition, Revised and Expanded, *Luc B. Jeunhomme*
24. Image Analysis Applications, *edited by Rangachar Kasturi and Mohan M. Trivedi*
25. Photoconductivity: Art, Science, and Technology, *N. V. Joshi*
26. Principles of Optical Circuit Engineering, *Mark A. Mentzer*
27. Lens Design, *Milton Laikin*
28. Optical Components, Systems, and Measurement Techniques, *Rajpal S. Sirohi and M. P. Kothiyal*

29. Electron and Ion Microscopy and Microanalysis: Principles and Applications, Second Edition, Revised and Expanded, *Lawrence E. Murr*
30. Handbook of Infrared Optical Materials, *edited by Paul Klocek*
31. Optical Scanning, *edited by Gerald F. Marshall*
32. Polymers for Lightwave and Integrated Optics: Technology and Applications, *edited by Lawrence A. Hornak*
33. Electro-Optical Displays, *edited by Mohammad A. Karim*
34. Mathematical Morphology in Image Processing, *edited by Edward R. Dougherty*
35. Opto-Mechanical Systems Design: Second Edition, Revised and Expanded, *Paul R. Yoder, Jr.*
36. Polarized Light: Fundamentals and Applications, *Edward Collett*
37. Rare Earth Doped Fiber Lasers and Amplifiers, *edited by Michel J. F. Digonnet*
38. Speckle Metrology, *edited by Rajpal S. Sirohi*
39. Organic Photoreceptors for Imaging Systems, *Paul M. Borsenberger and David S. Weiss*
40. Photonic Switching and Interconnects, *edited by Abdellatif Marrakchi*
41. Design and Fabrication of Acousto-Optic Devices, *edited by Akis P. Goutzoulis and Dennis R. Pape*
42. Digital Image Processing Methods, *edited by Edward R. Dougherty*
43. Visual Science and Engineering: Models and Applications, *edited by D. H. Kelly*
44. Handbook of Lens Design, *Daniel Malacara and Zacarias Malacara*
45. Photonic Devices and Systems, *edited by Robert G. Hunsberger*
46. Infrared Technology Fundamentals: Second Edition, Revised and Expanded, *edited by Monroe Schlessinger*
47. Spatial Light Modulator Technology: Materials, Devices, and Applications, *edited by Uzi Efron*
48. Lens Design: Second Edition, Revised and Expanded, *Milton Laikin*
49. Thin Films for Optical Systems, *edited by Francoise R. Flory*
50. Tunable Laser Applications, *edited by F. J. Duarte*
51. Acousto-Optic Signal Processing: Theory and Implementation, Second Edition, *edited by Norman J. Berg and John M. Pellegrino*
52. Handbook of Nonlinear Optics, *Richard L. Sutherland*
53. Handbook of Optical Fibers and Cables: Second Edition, *Hiroshi Murata*
54. Optical Storage and Retrieval: Memory, Neural Networks, and Fractals, *edited by Francis T. S. Yu and Suganda Jutamulia*
55. Devices for Optoelectronics, *Wallace B. Leigh*
56. Practical Design and Production of Optical Thin Films, *Ronald R. Willey*
57. Acousto-Optics: Second Edition, *Adrian Korpel*
58. Diffraction Gratings and Applications, *Erwin G. Loewen and Evgeny Popov*
59. Organic Photoreceptors for Xerography, *Paul M. Borsenberger and David S. Weiss*
60. Characterization Techniques and Tabulations for Organic Nonlinear Optical Materials, *edited by Mark G. Kuzyk and Carl W. Dirk*
61. Interferogram Analysis for Optical Testing, *Daniel Malacara, Manuel Servin, and Zacarias Malacara*
62. Computational Modeling of Vision: The Role of Combination, *William R. Uttal, Ramakrishna Kakarala, Spiram Dayanand, Thomas Shepherd, Jagadeesh Kalki, Charles F. Lunskis, Jr., and Ning Liu*
63. Microoptics Technology: Fabrication and Applications of Lens Arrays and Devices, *Nicholas Borrelli*
64. Visual Information Representation, Communication, and Image Processing, *edited by Chang Wen Chen and Ya-Qin Zhang*
65. Optical Methods of Measurement, *Rajpal S. Sirohi and F. S. Chau*

Additional Volumes in Preparation

Integrated Optical Circuits and Components: Design and Applications, *edited by Edmund J. Murphy*

Computational Methods for Electromagnetic and Optical Systems, *John M. Jarem and Partha P. Banerjee*

Adaptive Optics Engineering Handbook, *edited by Robert K. Tyson*

VISUAL INFORMATION REPRESENTATION, COMMUNICATION, AND IMAGE PROCESSING

EDITED BY

CHANG WEN CHEN
University of Missouri–Columbia
Columbia, Missouri

YA-QIN ZHANG
Sarnoff Corporation
Princeton, New Jersey

MARCEL DEKKER, INC.　　　　NEW YORK · BASEL

ISBN: 0-8247-1928-X

This book is printed on acid-free paper.

Headquarters
Marcel Dekker, Inc.
270 Madison Avenue, New York, NY 10016
tel: 212-696-9000; fax: 212-685-4540

Eastern Hemisphere Distribution
Marcel Dekker AG
Hutgasse 4, Postfach 812, CH-4001 Basel, Switzerland
tel: 41-61-261-8482; fax: 41-61-261-8896

World Wide Web
http://www.dekker.com

The publisher offers discounts on this book when ordered in bulk quantities. For more information, write to Special Sales/Professional Marketing at the headquarters address above.

Copyright © 1999 by Marcel Dekker, Inc. All Rights Reserved.

Neither this book nor any part may be reproduced or transmitted in any form or by any means, electronic or mechanical, including photocopying, microfilming, and recording, or by any information storage and retrieval system, without permission in writing from the publisher.

Current printing (last digit):
10 9 8 7 6 5 4 3 2 1

PRINTED IN THE UNITED STATES OF AMERICA

About the Series

This volume in our Optical Engineering Series addresses the all-important field of visual information: how such visual information is represented, manipulated and transmitted, and then understood by the human observer or by a machine vision system.

The field of visual information representation has had an explosive growth in recent years with new results being applied, new technology being developed and new systems on-line. It is appropriate at this time to pause and assemble the current knowledge into an integrated volume. Thus, the various interconnected aspects of this field have been brought together by a series of authors who have been involved in the rapid growth and implementation of *Visual Information, Representation, Communication, and Image Processing*.

Brian J. Thompson
University of Rochester
Rochester, New York

Foreword

With the explosive growth of Internet traffic and popularity of the World Wide Web, access to information has never been easier. Information now appears in several forms, including text, graphics, speech, audio, images, and video. These various types of information formats, along with the increasing demand for such information, drive the need for more effective navigation techniques, data management functions, and higher-throughput networking.

Providing a high quality of service and a high degree of satisfaction presents a number of challenges, particularly for visual information. Visual data lacks adequate representation because a large number of bits are required to digitally represent visual information. Fortunately, the high degree of redundancy within the pixel domain means that effective compression can significantly reduce the number of bits or the required bit rate.

Research and development in the area of image and video compression has been ongoing since the mid-1980s, resulting in a great deal of progress. Many image and video coding standards have been established, each driven by current business needs and addressed by current technologies. One example is the MPEG-1 standards, which were developed to support VCR quality CD-ROM and VOD applications. The MPEG-1 format and resolutions were fundamentally driven by:

- CD-ROM technology for video information storage
- ADSL (Asymmetrical Digital Subscriber Line)
- Technology for delivery of VOD services over the local loop
- State-of-the-art compression techniques available at the time

Similar drivers can be listed for the H.261 video coding standards, which were developed for videophone and videoconferencing applications.

Today's new applications for visual information require not only highly efficient image and video coding techniques but also new and advanced representations of visual content. In addition, new technologies related to the creation, processing, storage, retrieval, networking, and consumption of visual information must also be considered.

Visual Information Representation, Communication, and Image Processing is a collection of topics dealing with new concepts, techniques, applications, and standards relating to the area of visual representation and communications. The editors, Professor Chang Wen Chen and Dr. Ya-Qin Zhang, have invited a host of world authorities to address several special topics of great importance. I am deeply honored by their invitation to write the Foreword and honestly believe that this book, the first of its kind, will have a profound impact on the future development of this area.

Ming-Lei Liou
Professor, Department of Electrical and Electronic Engineering
Director, Hong Kong Telecom Institute of Information Technology
The Hong Kong University of Science and Technology

Preface

A revolution has begun to reshape the video and multimedia industry, compelled by the influence of several factors including advances in compression algorithms, progress in semiconductor technology, explosive growth of the Internet, and convergence of international standards. This revolution is profoundly changing every aspect of image and video, from production to display, from storage to transmission, from representation to manipulation.

Research and development in visual information representation, communication, and image processing are moving forward at an extremely fast pace. The revolutionary technologies derived from this research and development have reached virtually every facet of daily life. Multimedia computer, videophone, video over the Internet, HDTV, digital satellite TV, and interactive games are just a few examples.

Visual Information Representation, Communication, and Image Processing brings together a wide spectrum of innovative ideas and perspectives related to the advances in these areas. We invited scholars and experts from diverse areas of research to contribute. Each chapter represents state-of-the-art contributions in the author's own field. The first chapter was written by the editors to provide an overview of the subsequent chapters. The rest of the contributions have been organized into two parts. The first part includes nine chapters that address recent researches in video coding and communication; the second includes six chapters that discuss various techniques in image manipulation and processing. Although the technologies of video coding and communication and of image manipulation and processing are closely coupled with each other, an integrated treatment of these two areas in a single volume is quite unique.

This book is intended to educate and intrigue those who are in a position not only to learn the basics of video communication and image processing but also to enhance the quality of future generations of video communication and image processing systems. The comprehensive coverage of a wide variety of topics makes this book useful to various professionals, including electrical engineers, computer scientists, imaging technologists, and vision scientists.

We would like to thank all the contributors for their timely response to the invitation to share their knowledge and experience at the forefront of the research and development in visual communication and image processing. We also thank the reviewers for their willingness to review the chapters on tight schedules and to offer invaluable comments and suggestions to improve the quality of each chapter. Without the contributions of the authors and reviewers, this book would not have been possible. We are also deeply indebted to Professor Ming Liou, who is the funding editor in chief of IEEE Transactions on Circuits and Systems for Video Technology, for kindly finding time in his busy schedule to write the Foreword and to Ms. Rita Lazazzaro and Ms. Kathleen Baldonado at Marcel Dekker, Inc., for their continued involvement and help since the beginning of this project.

We are greatly indebted to our families for their love and support throughout this project. Chang Wen Chen would like to thank his wife, Angel, for sharing every step toward the completion of this book with boundless love, patience, and support, and his children, Kenneth and Joanna, for bringing him joy and relaxation after every hard day. Ya-Qin Zhang is profoundly grateful to his wife, Jenny Wang, and daughter, Sophie Zhang, for their constant support, affection, and inspiration.

Chang Wen Chen
Ya-Qin Zhang

Contents

About the Series Brian J. Thompson	*iii*
Foreword Ming-Lei Liou	*v*
Preface	*vii*
Contributors	*xi*
Reviewers	*xv*

1. Recent Advances in Visual Information Representation, Communication, and Image Processing
   Chang Wen Chen and Ya-Qin Zhang — 1

2. Video Compression Using Residual Vector Quantization
   Mahesh Venkatraman, Heesung Kwon, and Nasser M. Nasrabadi — 7

3. Adaptive Quantization in Wavelet-Based Image and Video Coding
   Jiebo Luo and Chang Wen Chen — 39

4. Statistically Adaptive Wavelet Image Coding
   Bing-Bing Chai, Jozsef Vass, and Xinhua Zhuang — 73

5. Three-Dimensional Model-Based Image Communication
   Thomas S. Huang, Ricardo Lopez, Hai Tao, and Antonio Colmenarez — 97

6. Content-Based Video Compression and Manipulation Using Two-Dimensional Mesh Modeling
   A. Murat Tekalp and Peter van Beek — 129

7. Error Control and Concealment for Video Communication
   Qin-Fan Zhu and Yao Wang — 163

8. Multipoint Videoconferencing
   I-Ming Pao and Ming-Ting Sun — 205

9. Video Shot Detection and Analysis: Content-Based Approaches
   Qi Tian and HongJiang Zhang — 227

10.	MPEG-4: An Object-Based Standard for Multimedia Coding *Atul Puri and Alexandros Eleftheriadis*	255
11.	Video Coding Standards for Multimedia Communication: H.261, H.263, and Beyond *Tsuhan Chen*	317
12.	AM–FM Image Modeling and Gabor Analysis *Joseph P. Havlicek, Alan Bovik, and Dapang Chen*	343
13.	Electronic Digital Image Stabilization and Mosaicking *Carlos Morimoto and Rama Chellappa*	387
14.	Highly Robust Statistical Estimates Based on Minimum-Error Bayesian Classification *Xinhua Zhuang, Kannappan Palaniappan, and Robert M. Haralick*	415
15.	Learning in Computer Vision and Beyond: Development *Juyang Weng*	431
16.	Principles of Halftoning with Stochastic Screens *Qing Yu and Kevin J. Parker*	489
17.	An Image-Algebra-Based SIMD Image-Processing Environment *Joseph N. Wilson, E. Jason Riedy, Gerhard X. Ritter, and Hongchi Shi*	523

Index   *543*

Contributors

Alan Bovik, Ph.D. Department of Electrical and Computer Engineering, University of Texas at Austin, Austin, Texas

Bing-Bing Chai, Ph.D. Multimedia Technology Laboratory, Sarnoff Corporation, Princeton, New Jersey

Rama Chellappa, Ph.D. Computer Vision Laboratory, Center for Automation Research, University of Maryland, College Park, Maryland

Chang Wen Chen, Ph.D. Department of Electrical Engineering, University of Missouri–Columbia, Columbia, Missouri

Dapang Chen, Ph.D. Research and Development Department, National Instruments, Austin, Texas

Tsuhan Chen, Ph.D. Department of Electrical and Computer Engineering, Carnegie Mellon University, Pittsburgh, Pennsylvania

Antonio Colmenarez, M.S. Department of Electrical and Computer Engineering and The Beckman Institute, University of Illinois at Urbana–Champaign, Urbana, Illinois

Alexandros Eleftheriadis, Ph.D. Department of Electrical Engineering, Columbia University, New York, New York

Robert M. Haralick, Ph.D. Department of Electrical Engineering, University of Washington, Seattle, Washington

Joseph P. Havlicek, Ph.D. School of Electrical and Computer Engineering, University of Oklahoma, Norman, Oklahoma

Thomas S. Huang, Sc.D. Department of Electrical and Computer Engineering and The Beckman Institute, University of Illinois at Urbana–Champaign, Urbana, Illinois

Heesung Kwon, Ph.D. Department of Electrical Engineering, State University of New York at Buffalo, Buffalo, New York

Ricardo Lopez, Ph.D.   Evans and Sutherland, Salt Lake City, Utah

Jiebo Luo, Ph.D.   Imaging Science Technology Laboratory, Eastman Kodak Company, Rochester, New York

Carlos Morimoto, Ph.D.   Department of Computer Science, IBM Almaden Research Center, San Jose, California

Nasser M. Nasrabadi, Ph.D.   Department of the Army, U.S. Army Research Laboratory, Adelphi, Maryland

Kannappan Palaniappan, Ph.D.   Department of Computer Engineering and Computer Science, University of Missouri–Columbia, Columbia, Missouri

I-Ming Pao, M.S.   Department of Electrical Engineering, University of Washington, Seattle, Washington

Kevin J. Parker, Ph.D.   Department of Electrical and Computer Engineering and Center for Electronic Imaging Systems, University of Rochester, Rochester, New York

Atul Puri, Ph.D.   Department of Image Processing Research, AT&T Labs–Research, Red Bank, New Jersey

E. Jason Riedy, M.S.   Department of Computer and Information Science and Engineering, University of Florida, Gainesville, Florida

Gerhard X. Ritter, Ph.D.   Department of Computer and Information Science and Engineering, University of Florida, Gainesville, Florida

Hongchi Shi, Ph.D.   Department of Computer Engineering and Computer Science, University of Missouri–Columbia, Columbia, Missouri

Ming-Ting Sun, Ph.D.   Department of Electrical Engineering, University of Washington, Seattle, Washington

Hai Tao, Ph.D.   David Sarnoff Laboratory, Princeton, New Jersey

A. Murat Tekalp, Ph.D.   Department of Electrical and Computer Engineering, University of Rochester, Rochester, New York

Qi Tian, Ph.D.   Kent Ridge Digital Labs, National University of Singapore, Kent Ridge, Singapore

Peter van Beek, Ph.D.   Sharp Laboratories of America, Camas, Washington

Jozsef Vass, M.Sc.   Department of Computer Engineering and Computer Science, University of Missouri–Columbia, Columbia, Missouri

Mahesh Venkatraman, Ph.D. Department of Electrical Engineering, State University of New York at Buffalo, Buffalo, New York

Yao Wang, Ph.D. Department of Electrical Engineering, Polytechnic University, Brooklyn, New York

Juyang Weng, Ph.D. Department of Computer Science and Engineering, Michigan State University, East Lansing, Michigan

Joseph N. Wilson, Ph.D. Department of Computer and Information Science and Engineering, University of Florida, Gainesville, Florida

Qing Yu, Ph.D.[*] Department of Electrical and Computer Engineering and Center for Electronic Imaging Systems, University of Rochester, Rochester, New York

HongJiang Zhang, Ph.D. Hewlett-Packard Laboratories, Hewlett-Packard Company, Palo Alto, California

Ya-Qin Zhang, Ph.D. Multimedia Technology Laboratory, Sarnoff Corporation, Princeton, New Jersey

Qin-Fan Zhu, Ph.D. Systems Products Division, PictureTel Corporation, Andover, Massachusetts

Xinhua Zhuang, Ph.D. Department of Computer Engineering and Computer Science, University of Missouri–Columbia, Columbia, Missouri

---

[*] *Current Affiliation*: Kodak Research Laboratories, Eastman Kodak Company, Rochester, New York

Reviewers

Prof. Kiyoharu Aizawa Department of Electrical Engineering, University of Tokyo, Tokyo, Japan

Dr. David Beymer Autodesk, Mountain View, California

Prof. Shih-Fu Chang Department of Electrical Engineering, Center for Telecommunications Research, Columbia University, New York, New York

Dr. Tihao Chiang Multimedia Technology Laboratory, Sarnoff Corporation, Princeton, New Jersey

Prof. Jennifer Davidson Department of Electrical and Computer Engineering, Iowa State University, Ames, Iowa

Prof. Roger L. Easton, Jr. Chester F. Carlson Center for Imaging Science, Rochester Institute of Technology, Rochester, New York

Prof. Allen Hanson Department of Computer Science, University of Massachusetts, Lederle Graduate Research Center, Amherst, Massachusetts

Prof. Aggelos Katsagellos Department of Electrical and Computer Engineering, Northwestern University, Evanston, Illinois

Dr. Weiping Li OptiVision Inc., Palo Alto, California

Prof. Bede Liu Department of Electrical Engineering, Princeton University, Princeton, New Jersey

Dr. Amir Said Hewlett-Packard Laboratories, Palo Alto, California

Dr. Iraj Sodagar Multimedia Technology Laboratory, Sarnoff Corporation, Princeton, New Jersey

Dr. Gary Sullivan PictureTel Corporation, Peabody, Massachusetts

Dr. Huifang Sun Advanced Television Laboratory, Mitsubishi Electric ITA, New Providence, New Jersey

Prof. Yao Wang Department of Electrical Engineering, Polytechnic University, Brooklyn, New York

1
Recent Advances in Visual Information Representation, Communication, and Image Processing

Chang Wen Chen
University of Missouri–Columbia, Columbia, Missouri

Ya-Qin Zhang
Sarnoff Corporation, Princeton, New Jersey

1. INTRODUCTION

This book presents a wide range of topics in visual information representation, communication, and image processing with an emphasis on the most recent developments. The authors of this edited volume are leading researchers from all over the world in various fields of visual information processing and communication. The topics presented in this book range from several wavelet-based coding and compression schemes for storage and transmission to visual information processing for printing service, and from recent standards in video coding and multimedia communication to novel applications of multipoint video conferencing and 3D model-based image communication.

    The book consists of 17 chapters. These chapters can be approximately classified into two major parts. Chapters 2–11 cover the state-of-the-art research and standardization activities in image and video coding. Image and video signals have now become a primary source for modern visual communication because of the tremendous development in compression technology and image and video coding standards, as well as the advances in computer and communication applications. The last six chapters describe several novel visual information representation and processing techniques and their implementations. The diversity of these visual representation and processing techniques demonstrates that there are numerous applications in which the visual signal processing can play an important role. The authors of all these chapters are known for their important contributions in their respective fields. This introductory chapter will summarize major technical contributions in each chapter and offer our own perspectives on their potential impact on the general field of visual information representation and communication.

2. VIDEO CODING AND COMMUNICATION

The first three chapters address several different aspects of wavelet-based image and video coding. Chapter 2, by Venkatraman, Kwon, and Nasrabadi, describes a wavelet-based video coding algorithm using vector-based techniques. This algorithm used a lattice vector to quantize the vector bands created by a vector wavelet transform, resulting in improved coding efficiency and picture quality. Variable-rate entropy-constrained residual vector quantization with a variable block size has been applied for video coding. The entropy-constrained residual vector quantization has demonstrated the superiority of their algorithm in terms of rate-distortion performance, storage space, and complexity. Consequently, the proposed video coding scheme using residual vector quantization has become a part of the MPEG-4 video coding standard.

Chapter 3, by Luo and Chen, explores the characteristics of the subband coefficients in wavelet-based image and video compression with respect to both spectral and spatial localities. The goal of this approach is to solve a common problem with many existing quantization methods in which the inherent image structures are severely distorted when quantization becomes coarse at low bit rates. Recognizing that subband coefficients with the same magnitude generally often do not have the same perceptual importance, they propose a novel quantization scheme that is able to preserve clustered scene structures in the process of quantization. The adaptive quantization is implemented as a maximum *a posterior* (MAP) estimation-based clustering process in which subband coefficients are quantized to their cluster means, subject to local spatial constraints modeled by the Gibbs random fields. As a result, the available bits are allocated to visually important scene structures so that the information loss is least perceptible. Such adaptive quantization is very much desired in some low-bit-rate image and video coding applications in which scene structural information needs to be appropriately preserved, such as video phone and tactical applications.

Chapter 4, by Chai, Vass, and Zhuang, presents a novel strategy for data organization and representation for wavelet-based image coding. The scheme is called significance-linked connected component analysis (SLCCA) and adaptively exploits the statistical properties of wavelet-transformed images at each stage of the coding process. SLCCA exploits not only within-subband clustering of significant coefficients but also cross-subband dependency in the significant fields. The cross-subband dependency is effectively exploited by using the so-called significance link between a parent cluster and a child cluster. The key components of SLCCA include multiresolution discrete wavelet image decomposition, connected component analysis with subbands, significance-link registration across subbands, and bit-plane encoding of magnitudes of significant coefficients by adaptive arithmetic coding. The performance of SLCCA is evaluated against other state-of-the-art wavelet coders to demonstrate its superior coding efficiency.

The next two chapters investigate the modeling of image and video scenes on based realistic 3D models and flexible 2D meshes. Chapter 5, by Huang, Lopez, Tao, and Colmenarez, addresses image and video communication using a 3D model-based approach. The driving motivation in model-based coding is to create a more intelligent representation of the content in scenes so that a more efficient and flexible communications system can be designed. Although the model-based approach is powerful enough to be used in many scenarios, this chapter deals almost exclusively with the head-and-shoulders type of image and video. In the case of 3D model-based video communication, three issues need to be studied: (1) modeling, (2) synthesis, and (3) analysis. The chapter covers each of the three issues in detail and examines the uses of a 3D model-based approach in such areas as compression, virtual environments, and the developing MPEG-4 standard.

Chapter 6, by Tekalp and van Beek, discusses a 2D mesh representation of natural and synthetic video objects by means of mesh-based geometry and motion modeling and texture mapping. Such a 2D mesh representation enables the development of functionalities specified by both the MPEG-4 video coding standard, in which independent encoding of different visual objects is required, and the computer graphics standard VRML (Virtual Reality Modeling Language), in which modeling and interacting of synthetic scenes are accomplished through manipulation of elementary geometric objects and rendering of moving objects and moving texture mapping. This chapter not only describes 2D object-based mesh design and tracking algorithms but also 2D mesh-based video coding algorithms, including encoder/decoder architectures and compression algorithms for efficient transmission of mesh geometry and motion. Furthermore, 2D mesh-based video manipulation, including object transfiguration and augmented reality, is also discussed. The results of tracking, compression, and manipulation using 2D mesh modeling demonstrate that such a 2D mesh-based approach is efficient and flexible in content-based video compression and manipulation.

Chapter 7, by Zhu and Wang, covers error-control and error-concealment schemes in video communication, an important topic in visual communication when the transmission media is imperfect. Detection and correction of transmission errors are crucial in video communication, as most physical channels are actually imperfect. However, the task of error control and error concealment is very challenging, as most state-of-the-art video compression techniques use variable-length coding techniques in which one bit error can desynchronize the decoder and cause the loss of the entire block of coded visual information until the next synchronization codeword appears. This chapter presents various error-detection schemes that are currently adopted in various visual communication applications. The chapter also describes in detail the error concealment methods in three categories: forward, passive, and interactive. This chapter indeed provides an excellent coverage of error control and error concealment for video communication and can be invaluable to practical visual communication engineers.

The next two chapters describe two innovative applications in visual communication. Chapter 8, by Sun and Pao, presents the development of multipoint videoconferencing system extended from the more traditional point-to-point videoconferencing. The issues in multipoint videoconferencing involves networking, combining, and presentation of multiple coded video signals. This chapters first discusses the network configurations for supporting multipoint videoconferencing and reviews some existing schemes in continuous presence video bridges. It then provides a comparison between coded-domain video combining and pixel-domain video combining. The initial results show that the application of multipoint videoconferencing is very promising and techniques developed in multipoint videoconferencing can also be applied to other types of visual communication tasks, such as distance learning, remote collaboration, and video surveillance involving multiple sites.

Chapter 9, by Tian and Zhang, presents issues related to shot-based video content analysis and representation, and content-based video retrieval, browsing, and compression. This is an application-oriented research area of fast-growing interests to the visual communication community because of the need of such technology in interactive video and digital libraries. The authors first describe processing steps of video content analysis by drawing an analogy between text document image analysis and video data analysis, and they conclude that shots are a generic component for video content analysis. Then, they give a detailed description of shot-detection methods and algorithms, shot content attributes, and their application in content-based video retrieval and compression. The goal of video shot detection and analysis is to develop algorithms, tools, and systems that are able to extract and analyze basic elements and structures (both syntactic and semantic) of video, to make access, compression, and interaction of video more effective and content based. Many fine examples shown in this chapter illustrate that video shot detection and analysis through a content-based approach are able to provide well-structured visual information for video data analysis and recognition.

The next two chapters present two current video coding standards. Chapter 10, by Puri and Eleftheriadis, provides an overview of the current status of MPEG-4, an object-based standard for multimedia coding. Originally, MPEG-4 was conceived to be a standard for coding of limited complexity audio-visual scenes at very low bit rates. However, in July 1994, its scope was expanded to include coding of scenes as a collection of individual audio-visual objects and enabling a range of advanced functionalities not supported by other standards. After a brief overview of the status of related International Telecommunication Union–Telecommunication Standardization Sector (ITU-T) standards, the authors present an overview of the MPEG-4 in terms of requirements, test, video, audio, and systems. They also discuss the respective coding methods of MPEG-4 visual, audio, and systems standards. The issue of profiles and plans for verification tests are also presented. The authors conclude that the MPEG-4 standard is thus being designed to provide solutions for audio coding, video coding, systems multiplex/demultiplex, and scene composition in a truly flexible manner.

Chapter 11, by Chen, covers the video coding standards developed by ITU-T, formerly called the Consultative Committee of the International Telephone and Telegraph (CCITT). These standards include H.261, H.263, and a recent effort, informally known as H.263+, to provide a new version of H.263 (i.e., H.263 Version 2) in 1998. These video codec standards form important components of the ITU-T H-Series recommendations that standardize audiovisual terminals in a variety of network environments. The author presents, in detail, the techniques used in a historically very important video coding standard, H.261. He also discusses H.263, a video coding standard that has a similar framework to that of H.261, but with superior coding efficiency. Recent activities in H.263+ that resulted in a new version of H.263 with several enhancements are also covered.

3. VISUAL REPRESENTATION AND IMPLEMENTATION

Chapters 12–17 deals primarily with visual information representation and processing techniques, and their implementations. Chapter 12, by Havlicek, Bovik, and Chen, describes recent techniques for modeling and analyzing sophisticated images characterized by the presence of significant nonstationary multipartite structures. The authors show that sophisticated images can be effectively modeled using multicomponent multidimensional AM–FM functions, and computed representations based on such models can be obtained by the Gabor analysis. The individual components of a multicomponent AM–FM image model generalize the 2D Fourier transform kernel by admitting arbitrarily varying amplitude and phase modulations. Each component is capable of capturing essential nonstationary yet locally coherent structure. Such structure often contributes significantly to visual perception and interpretation. Computation of AM–FM image representations are useful in a variety of applications, including conjointly localized spatio-spectral analysis, image texture modeling, analysis, and segmentation, optical flow, 3D shape from texture, and phase-based computational stereopsis. The authors also demonstrate that biologically motivated multiband Gabor filterbanks can be designed to capture the essential and perceptually significant structure of an image using only a few components and will lead to significant future application in AM–FM based image and video coding schemes.

Chapter 13, by Morimoto and Chellappa, addresses an important issue in visual communication, the electronic digital image stabilization, and mosaicking. Image stabilization is defined as the process of generating a compensated video sequence, where the unwanted components of camera motion are smoothed or completely removed. This is vital in the capture of an excellent quality of visual signal without camera motion, which would otherwise present great difficulty in the subsequent processing, coding and manipulation of the video signals. The authors recognize that it is necessary to recover the motion of the camera, described by a global parametric transformation in order to generate a stabilized video sequence. Such parameters can also be used to align the input image frames, creating a panoramic view of the scene, also known as a mosaic. By comparing different motion models and estimation algorithms, particularly those suited for real-time implementation, the authors are able to report the performance of each motion model according to a group of measures designed to evaluate image stabilization systems.

Chapter 14, by Zhuang, Palaniappan, and Haralick, addresses another important issue in visual signal understanding. Aiming at providing a solution to the problem of lacking sufficient robustness in most existing computer vision or image understanding algorithms, the authors describe highly robust statistical estimates based on minimum-error Bayesian classification. The authors analyze the cause of low robustness with classical robust statistical methods such as the popular M-estimator and offers a solution to remedy the problem by applying the minimum-error Bayesian classification rule to delineate inliers from outliers. The minimum-error Bayesian classification naturally leads to a set of "partial" density models, each of which models the underlying unknown density function only partially. With extensive computer experiments, the authors conclude that such partial density modeling results in a highly robust estimator called the MF-estimator.

Chapter 15, by Weng, investigates some fundamental issues related to machine learning for image analysis. It introduces what is called the developmental approach to computer vision, in particular, and artificial intelligence, in general. The current technology for learning requires humans to collect images, store images, segment images, and train computer systems to use these images. The author argues that it is unlikely that such a manual labor process can meet the demands of many challenging recognition tasks that are critical for generating intelligent behavior, such as face recognition, object recognition, and speech recognition. The major goal of the developmental approach described in this chapter is to realize automation of general-purpose learning that enables machines to perform developmental learning over a long period. The machines must learn directly from continuous sensory input streams while interacting with the environment, including human teachers. The proposed learning model does not aim at developing intelligent programs for various tasks through programming and feeding data. Rather, it aims at learning through interactions with the machines, including demonstration, communication, action imposition, granting rewards, and executing punishments. This chapter discusses these fundamental issues and proposes a fundamentally new way of addressing the learning problem, one that unifies learning and performance phases and requires a systematic self-organization capability. Once successfully carried out, this learning research will have a major impact on many visual perception and image understanding tasks.

Chapter 16, by Yu and Parker, addresses the problem of visual representation when images are to be printed or displayed using a reduced number of quantization levels because of the limitations of their output levels. The goal of such visual representation is to reduce the quantization levels per pixel in a digital image while maintaining the gray or color appearance of the image at normal viewing distance. Because the human visual perception exhibits low-pass characteristics, the human vision system tends to average a region around a pixel and thus create the illusion of many gray and color levels even though the actual image is rendered with only black and white dots, or cyan, magenta, yellow, and black dots for the color case. This chapter begins by introducing the half-toning process that governs the distribution of these dots with the constraint that the perceived gray or color level is preserved. The chapter then focus on active research areas in stochastic screening, the techniques that exhibit relatively unstructured but visually pleasing half-tones. The authors consider the design of stochastic screens and their application in black-and-white half-toning, multilevel half-toning, and color half-toning. The design topics range from prototype stochastic screen, the Blue Noise Mask, the optimality of blue-noise binary patterns, to various modifications of stochastic screens to meet special application requirements, such as screens with dot-gain compensation, screens for fax encoding, and screens for multilevel output devices. The evaluation of these designs is accomplished based on the subjective human visual perception.

The last chapter, Chapter 17 by Wilson, Riedy, Shi, and Ritter, studies the implementation of image-processing algorithms using parallel computers. An efficient implementation is very much needed as many image-processing tasks are computationally intensive. This chapter considers the single-input multiple data (SIMD) mode of parallel processing based on image algebra principles. It is well recognized that SIMD parallel computers have been leading the way in improving processing speed for image- and video-related applications. However, current parallel programming technologies have not kept pace with the performance growth and cost decline of parallel hardware. The authors present a computing environment that integrates a SIMD mesh architecture with image algebra for high-performance image-processing applications. The environment describes parallel programs through a machine-independent, retargetable image algebra object library that supports SIMD execution on the Lockheed Martin, PAL-1 parallel computer. Program performance on this machine is improved through on-the-fly execution analysis and scheduling. The chapter describes the relevant elements of the system structure, outlines the scheme for execution analysis, and provides examples of the current cost model and scheduling system. The implementation on this specific parallel computer serves as an excellent guide for an implementation on other types of parallel computer for appropriate image-processing applications.

4. SUMMARY

In summary, we have compiled a collection of chapters representing the most recent advances in visual information representation, communication, and image processing. Although we attempt to produce an edited volume with coherent and fully consistent chapters, we also wish to maintain a full diversity of topics and new developments in these important research areas. We sincerely hope the diversity of these chapters provides the readers the opportunity to appreciate many facets of the research and development in visual information representation and communication.

2
Video Compression Using Residual Vector Quantization

Mahesh Venkatraman and Heesung Kwon
State University of New York at Buffalo, Buffalo, New York

Nasser M. Nasrabadi
U.S. Army Research Laboratory, Adelphi, Maryland

1. INTRODUCTION

Images are represented in the digital domain by a matrix/array of intensity values, and video sequences are represented by a series of matrices. These matrices are often large and require large amounts of storage space and/or transmission bandwidths. In the realm of limited resources (limited storage spaces and limited bandwidths), reducing the amount of data necessary to represent the digital imagery is quintessential. The process of reducing the amount of data, known as data compression, can be either with no loss in data (lossless) or with some degradation/distortion of the data (lossy). In lossless compression, redundancy in the data is removed to achieve a smaller representation, but the ratio of compression that can be achieved is small. On the other hand, lossy compression techniques trade off the compression ratio against the tolerated distortion. Most of the current image and video compression techniques are usually lossy because human visual perception can tolerate a limited amount of distortion in the presented visual data. Lossy compression can be achieved using quantization, a technique in which data are represented at a lesser numerical precision than the original representation.

In the early stage of video codec development, a simple interframe coding system was introduced by Mounts (1). This system was based on frame replenishment in which only the intensity of pixels that changed significantly between successive frames were transmitted. Only portions of a picture that contained high activity were encoded and transmitted; because small changes in the gray values of pixels were being ignored, poor reconstruction resulted. Haskell et al. (2) proposed a technique superior to frame replenishment, one that tried to predict the moving areas of a frame by using the information of a previous frame.

Video coding using vector quantization was introduced by Murakami et al. (3). In their algorithm, vector quantization was used to quantize blocks normalized by mean and variance. Goldburg and Sun (4) introduced several interframe coding systems, where fixed-block-size vector quantization, which incorporated label replenishment and codebook replenishment in their system, was applied on successive frames without using motion-compensation techniques.

Because successive frames in a video sequence are highly correlated, interframe predictors based on motion-compensated prediction significantly improve coding efficiency (5). Hybrid coding methods, combining differential pulse code modulation (or motion-compensated prediction between successive frames) and two-dimensional transform coding, were introduced to code residual signals in the transform domain (6,7). The motion-compensated frame difference signals, however, normally contain both high-activity regions and homogeneous regions representing motionless areas. It turns out that hybrid coding methods are not suitable for the motion-compensated frame difference signals containing high activity because of the low correlation between the pixels (8). Due to the poor performance of transform coding for active regions, there have been several research efforts to provide better coding for such regions by using methods such as quadtree decomposition (9–11). In quadtree-decomposition-based coding, quadtree representations of images are achieved by the successive subdivision of the image into four equal quadrants so that the content of each quadrant is approximately homogeneous, with little variation. Consequently, highly detailed regions are partitioned into smaller blocks and vector quantized so that these regions are reconstructed accurately. Different coding methods were applied to different size blocks. Transform vector quantization for larger blocks and spatial vector quantization for smaller blocks were used to obtain a better coding efficiency in Ref. 10. Interframe hierarchical vector quantization, introduced by Nasrabadi et al. (9), encoded small blocks representing high-contrast moving boundaries using vector quantization and large blocks representing smooth regions using the sample mean of the corresponding regions. The optimum quadtree decomposition, for images and video, for a given a set of quantizers was introduced by Sullivan and Baker (11). The optimum quadtree is determined by minimizing both the rate and the distortion to obtain the best reconstruction for a given fixed bit rate. With this optimum quadtree decomposition, one can actually achieve all the points on the operational rate-distortion curves.

Decomposition methods based on variable-size and variable-shape blocks and subsequent vector quantization on the motion-compensated frame difference signal have also been used recently to encode video (12). In this method, a classified vector quantizer was applied to encode small-dimensional vectors containing high activity to obtain high accuracy at the cost of a higher bit rate.

Use of finite-state vector quantization to compress image sequences was first introduced by Baker and Shen (13). Chen et al. (14) introduced an adaptive finite-state vector quantizer where the conditional replenishment of a supercodebook was used to reflect the local statistics of current frame.

Recently, a video coding algorithm was developed, as a part of the MPEG-4 standard, using vector-based techniques (15). This algorithm used a lattice vector to quantize the vector bands created by a vector wavelet transform, resulting in improved coding efficiency and picture quality. Variable-rate entropy-constrained residual vector quantization with variable block size has been recently applied for video coding (16). Previous work on entropy-constrained residual vector quantization (17,18) for still-image coding demonstrated the superiority of their algorithm in terms of rate-distortion performance, storage space, and complexity.

This chapter is organized as follows: Section 2 describes the basic principles of quantization and video compression. In Sec. 3, vector quantization and its application to video compression are briefly discussed. Section 4 describes the structure and design of residual vector quantization (RVQ), including a pruned variable-block-size RVQ, and the video compression using RVQ codec is introduced along with simulation results. Section 5 introduces the optimal quadtree-based vector quantization and its application to video compression with simulation results.

2. VIDEO COMPRESSION

2.1. Quantization

The quantization Q of a random variable $X \in \mathcal{R}$ (the real number) is a mapping from $\mathcal{R}$ to C, a finite subset of $\mathcal{R}$:

$$Q: \mathcal{R} \mapsto C, \quad C \subset \mathcal{R} \tag{1}$$

The cardinality, N_C, of the set C gives the number of quantization levels. The mapping Q is generally a staircase function, as shown in Fig. 1, where $\mathcal{R}$ is divided into N_C segments $[b_{i-1}, b_i)$, $i = 1, \ldots, N$. Each $X_n \in [b_{i-1}, b_i)$ is mapped to $c_i \in C$, where c_i is the reconstruction value.

A sequence of random variables X_n can be quantized using two methods. The first method involves each individual member of the sequence being quantized separately using the quantizer Q defined in Eq. (1). This method is called *scalar quantization*. In the second method, the sequence is grouped into blocks of adjacent members and the each block (a vector) is quantized using a *vector quantizer*. Vector quantization (VQ) and its application to video compression, the focus of this chapter, is explained in Sec. 3.

2.2. Video Compression

A video sequence is a three-dimensional signal of intensity with two spatial dimensions and a temporal dimension. A digital video sequence is a three-dimensional signal suitably sampled in all the three dimensions and is in the form of a three-dimensional matrix of intensity values. A typical video sequence has a significant amount of correlation between

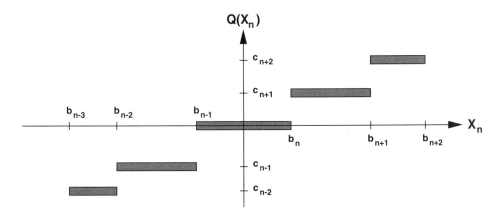

Fig. 1 Scalar quantizer Q of a random variable X_n.

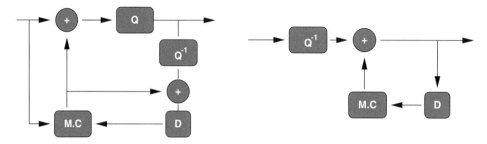

Fig. 2 Motion compensation for video coding.

neighbors in all three dimensions. The type of correlation in the temporal dimension is significantly different from the correlation in the spatial dimensions.

Many different approaches to video compression exist, and some international compression standards have been established. Among the different approaches are a class of algorithms that first attempt to remove correlations in the temporal domain and then deal with removing correlations in the spatial dimensions. Motion compensation is a popular technique for removing the correlations in the temporal domain. Motion compensation results in a residue sequence which is then quantized using two-dimensional quantization techniques similar to ones used for compressing still images.

A video scene usually contains some motion of objects, occlusion/exposure of areas due to movement of objects, and some deformation. The rate of this change is typically much smaller than the frame rate (i.e., the rate of sampling in the temporal dimension). Therefore, there is very little change between two adjacent frames. A motion-compensation algorithm uses this property to approximate the current frame using pieces from the previous frame. This results in a reasonable approximation of the current frame from the previous frame, with some side information in the form of motion vectors. The difference between the approximation of the current frame and the actual frame is quantized by a set of scalar or vector quantizers. Figure 2 shows the block diagram of the encoder and decoder. This difference between the approximation and the original is quantized by the quantizer Q. The encoder is a closed-loop system, as shown in the block diagram. It contains both a quantizer Q and an inverse quantizer Q^{-1}. Because the encoder uses the previous frame to approximate the current frame, the decoder needs the previous frame to generate the current frame. The decoder has only the quantized version of the previous frame and not the original frame. The inverse quantizer Q^{-1} in the encoder duplicates the decoder states at the encoder and gives the encoder access to a quantized version of the previous frame. The encoder uses this quantized version of the previous frame to generate an approximation of the current frame. This ensures that the approximation of the current frame generated from the previous frame is the same both at the encoder and decoder. Figure 3 shows the entropy of a sequence after decorrelation by (a) frame difference and (b) block motion estimation. It can be seen that the entropy in this case is reduced by a factor of 2 when compared to the original frame entropy. Between frames 105 and 120, the entropy of the residue due to motion estimation is significantly less than the entropy of the residue from frame differencing. This is due to significant motion of objects in the images during that period.

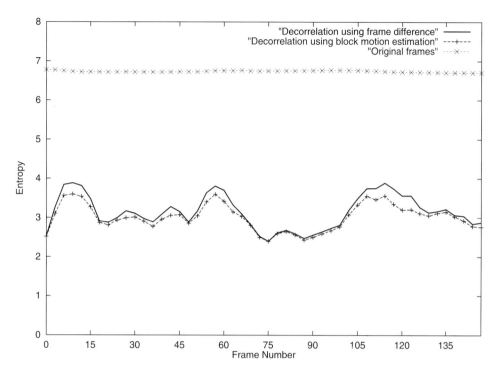

Fig. 3 Entropy of a sequence after decorrelation in the temporal dimension.

3. VECTOR QUANTIZATION

3.1. Introduction

A vector quantizer Q is a mapping from a point in k-dimensional Euclidean space $\mathcal{R}^k$ into a finite subset C of $\mathcal{R}^k$ containing N reproduction points or vectors:

$$Q: \mathcal{R}^k \mapsto C \qquad (2)$$

The N reproduction vectors are called *code vectors* and the set C is called the *codebook*: $C = (\mathbf{c}_1, \mathbf{c}_2, \ldots, \mathbf{c}_n)$ and $\mathbf{c}_i \in \mathcal{R}^k$ for each $i \in T = \{1, 2, \ldots, N\}$. The codebook C has N distinct members. The rate of the vector quantizer $r = \log_2(N)$ measures the number of bits required to index a member of the codebook.

A vector quantizer partitions the space $\mathcal{R}^k$ into cells $\mathcal{R}_i$,

$$\mathcal{R}_i = \{\mathbf{X} \in \mathcal{R}^k : Q(\mathbf{X}) = \mathbf{c}_i\} \quad \forall i \in T \qquad (3)$$

They represent the preimage of the points $\mathbf{c}_i$ under the mapping Q (i.e., $\mathcal{R}_i = Q^{-1}(\mathbf{c}_i)$). These cells have the following properties:

$$\cup_i \mathcal{R}_i = \mathcal{R}^k \qquad (4)$$

$$\mathcal{R}_i \cap \mathcal{R}_j = \{\emptyset\} \quad \forall i \neq j \qquad (5)$$

These equations imply that the cells are disjoint and that they cover the entire space $\mathcal{R}^k$.

The vector quantizers dealt with here have the following additional properties:

- They are regular. The cells of a regular vector quantizer $\mathcal{R}^k$ are convex and $\mathbf{c}_i \in \mathcal{R}_i$.
- They are polytopal. The cells of a polytopal vector quantizer are polytopal. Polytopes are geometric regions bounded by hyperplane surfaces. A polytopal region is the intersection of a finite number of subspaces.
- They are bounded. A vector quantizer is bounded if it is defined on a bounded domain $B \subset \mathcal{R}^k$ (i.e., every input vector $\mathbf{X}$ lies in B).

A vector quantizer consists of two operators: an encoder and a decoder. The encoder, $\mathbf{e}$, associates every input vector $\mathbf{X}$ to i, some member of the index set T,

$$\mathbf{e}: \mathcal{R}^k \mapsto T \tag{6}$$

The decoder, $\mathbf{d}$, associates i, some member of the index set T, to $\mathbf{c}_i$, some member of the reproduction set C,

$$\mathbf{d}: T \mapsto \mathcal{R}^k \tag{7}$$

The vector quantization is a combination of the encoder and decoder operations,

$$Q(\mathbf{X}) = \mathbf{d}(\mathbf{e}(\mathbf{X})). \tag{8}$$

The block diagram of the vector quantizer is shown in Fig. 4. The encoding operation is completely determined by specifying the partition of the input space. The encoder identifies the cell to which a given input vector belongs. The decoding operation is determined by specifying the codebook. Given the identity of the cell to which the input vector belongs, the decoder determines the reproduction vector that best represents the input vector. The decoder is very often of the form of a simple lookup table. Given the index, it returns the vector entry in the table corresponding to the index.

3.2. Quantization Error of Vector Quantizers

The performance of a vector quantizer can be evaluated by the average distortion introduced by encoding a set of training input vectors. Ideally, this distortion should be zero. The output of the decoder should be a close representation of the input vector. The expected value of the distortion measure represents the performance of the quantizer

$$\mathcal{D} = \mathbf{E}(\mathbf{d}(\mathbf{X}, Q(\mathbf{X}))) \tag{9}$$

where $\mathbf{d}(\mathbf{X}, Q(\mathbf{X}))$ represents the distortion introduced by the quantizer for the input vector $\mathbf{X}$.

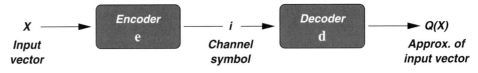

Fig. 4 The encoder/decoder model of a vector quantizer.

An important distortion measure is the squared-error distortion measure (Euclidean distortion/L_2 distortion). This distortion measure is especially relevant to image-coding problems, where the mean squared error is widely used as a quantitative measure of the performance of coding:

$$\mathbf{d}_{\mathrm{mse}}(\mathbf{X}, Q(\mathbf{X})) = \|\mathbf{X} - Q(\mathbf{X})\|^2 \tag{10}$$

$$\mathcal{D}_{\mathrm{mse}} = \mathbf{E}(\|\mathbf{X} - Q(\mathbf{X})\|^2) \tag{11}$$

Other distortion measures include weighed squared-error distortion measures, the Mahalanobis distortion measure, and the Itakura–Saito distortion measure.

3.3. Optimality Conditions for Vector Quantizers

An optimal vector quantizer is one that minimizes the overall distortion measure for any vector **X** with a probability distribution $P(\mathbf{X})$. A vector quantizer has to satisfy two optimality conditions to achieve this minimum distortion:

- For a given fixed decoder **d**, the encoder **e** should be the one that minimizes the overall distortion.
- For a given fixed encoder **e**, the decoder **d** should be the best possible decoder.

3.3.1. Nearest-Neighbor Condition

Given a decoder, it is necessary to find the best possible encoder. The decoder contains a finite set of vectors C, one of which is used to represent the input vector. For a given vector **X**, the vector $\mathbf{c}_i$ is the nearest neighbor if

$$\mathbf{d}(\mathbf{X}, \mathbf{c}_i) \leq \mathbf{d}(\mathbf{X}, \mathbf{c}_j) \quad \forall \mathbf{c}_j \in C$$

The overall distortion for a given fixed codebook C is given by

$$\mathcal{D} = \mathbf{E}(\mathbf{d}(\mathbf{X}, Q(\mathbf{X}))) = \int \mathbf{d}(\mathbf{X}, Q(\mathbf{X})) P(\mathbf{X}) \, d\mathbf{X}$$

Clearly,

$$\int \mathbf{d}(\mathbf{X}, Q(\mathbf{X})) P(\mathbf{X}) \, d\mathbf{X} \geq \int \mathbf{d}(\mathbf{X}, \mathbf{c}_i) P(\mathbf{X}) \, d\mathbf{X}$$

where $\mathbf{c}_i$ is the nearest neighbor of **X**. Therefore, the best possible encoder for a given decoder is the nearest-neighbor encoder.

3.3.2. Centroid Condition

For a fixed encoder, it is necessary to find the reproduction codebook which minimizes the overall distortion. For a given cell $\mathcal{R}_i$, the centroid $\mathbf{c}_i$ is defined as the point at which

$$\mathcal{D}(\mathbf{X}, \mathbf{c}_i) \leq \mathcal{D}(\mathbf{X}, \mathbf{c}) \quad \forall \mathbf{X}, \mathbf{c} \in \mathcal{R}_i, \mathbf{c}_i \in \mathcal{R}_i$$

For a given probability distribution and for a given encoder, the overall distortion is given by

$$\mathcal{D} = \int \mathbf{d}(\mathbf{X}, Q(\mathbf{X})) P(\mathbf{X}) \, d\mathbf{X} = \sum_i \int_{\mathcal{R}_i} \mathbf{d}(\mathbf{X}, \mathbf{c}) P(\mathbf{X}) \, d\mathbf{X}$$

Clearly,

$$\sum_i \int_{\mathcal{R}_i} \mathbf{d}(\mathbf{X}, \mathbf{c}) P(\mathbf{X}) \, d\mathbf{X} \geq \sum_i \int_{\mathcal{R}_i} \mathbf{d}(\mathbf{X}, \mathbf{c}_i) P(\mathbf{X}) \, d\mathbf{X}$$

Therefore, for a given encoder, the optimum decoder is the centroid of the nearest-neighbor partitions.

Consider A, a collection of all possible partitions of the input vectors, and C^o, a collection of all possible reproduction sets. The optimum vector quantizer is the pair

$$(\{\mathcal{R}_i\}, C), \quad \{\mathcal{R}_i\} \in A \text{ and } C \in C^o \tag{12}$$

such that $\mathcal{R}_i$ is the nearest-neighbor partition of C which contains the centroids of the partitions in $\mathcal{R}_i$. These two conditions are generalizations of the Lloyd–Max conditions for scalar quantizers.

3.4. Design of Vector Quantizers

The design of vector quantizers is a very difficult problem. For a given probability distribution $P(\mathbf{X})$, it is necessary to find the encoder and decoder that simultaneously satisfy both the nearest-neignbor condition and the centroid condition. Unfortunately, there exist no closed-form solutions for even simple distributions.

A number of design methods have been proposed for the design of vector quantizers. These are all iterative methods based on finding the best vector quantizer for a training set.

3.4.1. Generalized Lloyd's Algorithm

The generalized Lloyd's algorithm (GLA), also known as the LBG algorithm (19), is an iterative algorithm. This algorithm, which is similar to the k-means clustering algorithm, consists of two basic steps:

- For a given codebook C_t, find the best partition $\{\mathcal{R}_i\}_t$ of the training set that satisfies the nearest-neighbor neighborhood condition.
- For the new partition $\{\mathcal{R}_i\}_t$, find the best reproduction codebook C_{t+1} satisfying the centroid condition.

These two steps are repeated until the required codebook is obtained. The training algorithm begins with an initial codebook that is refined using the Lloyd's iterations until an acceptable codebook is obtained. A codebook is considered acceptable if the difference in the average error between the present and the previous codebooks is less than a threshold.

3.4.2. Kohonen's Self-Organizing Feature Map

The Kohonen's self-organizing feature maps (KSOFMs) can be used to design vector quantizers with optimal codebooks (20,21). In this method of codebook design, an error energy is formulated and minimized iteratively. This design procedure is sequential, as opposed to GLA, which uses the batch method of training.

3.5. Entropy-Constrained Vector Quantizer

For transmission over a binary channel, the index of the reproduction vector from a codebook C of size N is represented by a binary string of length $\lceil \log_2 N \rceil$ bits. Often it is possible

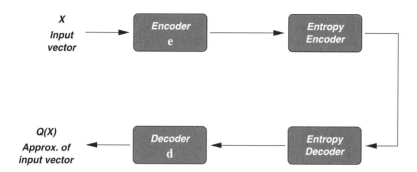

Fig. 5 Entropy coding of VQ indices.

to further reduce the number of bits required to represent the indices using entropy coding of the indices, as shown in Fig. 5. Entropy coding reduces the transmission entropy rate from $\lceil \log_2 N \rceil$ per block to almost the entropy rate of the index sequence. Because typical codebook design algorithms do not take into consideration the possible entropy rates of the index sequences, they do not combine with an entropy coder in an optimal way. The design of entropy-constrained vector quantizers (ECVQs) was studied among others by Chou et al. (22), who used a Lagrangian formulation with a gradient-based algorithm similar to the Lloyd's algorithm to design the codebooks.

Consider a vector $\mathbf{X} \in \mathcal{R}^k$ quantized by a vector quantizer with a codebook $C = \{\mathbf{c}_j : j = 1, \ldots, N\}$. Let $l(i)$ represent the length of the binary string used to represent the index i of the reproduction vector $\mathbf{c}_i$ of $\mathbf{X}$. Then, the functional that is minimized to design an ECVQ is given by (22)

$$J(\mathbf{e}, \mathbf{d}) = \mathbf{E}[d(\mathbf{x}_i, Q(\mathbf{x}_i))] + \lambda \mathbf{E}[l(i)] \tag{13}$$

Here, **e** and **d** are the VQ encoder and decoder, respectively. The index entropy $\log(1/p(i))$ is used in the algorithm to represent the length of the binary string required to represent the index i. The codebook is then designed in a manner similar to Lloyd's algorithm iteratively by choosing an encoder and decoder that decrease the functional (13) at every iteration. Experimental results have shown that the ECVQ design algorithm described gives an encoder–decoder pair which has a superior numerical performance.

3.6. Video Compression Using Vector Quantization

The residual signal obtained after motion compensation can be compressed using vector quantization. The two-dimensional signal is divided into blocks of equal size, as shown in Fig. 6. The VQ encoder that is used to compress the residual signal is a nearest-neighbor encoder. It has a reference lookup table that contains the centroids of the vector-quantizer partitions. The encoder compares each block, in some predefined scanning order, with each member of the lookup table to find the closest match in terms of the defined distortion measure (usually the mean squared error). The index of the closest matching code vector in the lookup table is then transmitted/stored as the compressed representation of the corresponding vector (block).

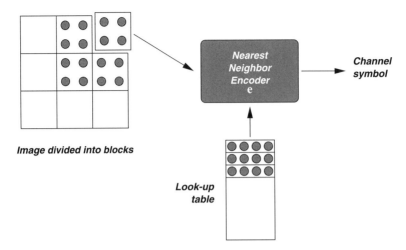

Fig. 6 VQ encoder for two-dimensional arrays.

The decoder is a simple lookup table decoder, as shown in the Fig. 7. The decoder uses the index symbol generated by the encoder as a reference to an entry in a lookup table present in the decoder. The lookup table present in the decoder is usually identical to the one in the encoder. This lookup table contains the possible approximations for the blocks in the reconstructed image. Based on the index, the approximate representation of the current block is determined. The decoder places this at the position corresponding to the scanning order to generate the reconstructed two-dimensional array. This array is used along with the motion-compensation algorithm to reproduce the compressed video sequence.

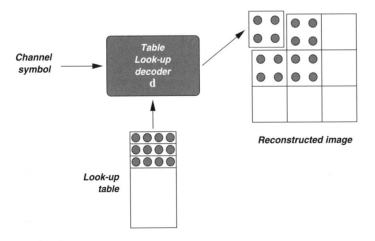

Fig. 7 VQ decoder for two-dimensional arrays.

4. RESIDUAL VECTOR QUANTIZATION

4.1. Introduction

Residual vector quantization (RVQ) is a structured vector quantization scheme proposed mainly to overcome the search and storage complexities of regular vector quantizers (23,24). Residual vector quantizers, also known as multistage vector quantizers, consist of a number of cascaded vector quantizers. Each stage has a vector quantizer with a small codebook; this vector quantizer refines the error due to the previous stage. Residual quantizers are successive refinement quantizers, in which the information to be transmitted/stored is first approximated coarsely and then refined in the successive stages.

4.2. Residual Quantization

Consider a random variable X with a probability density function (p.d.f.) $P(X)$. Let $Q^1(X)$ be a N^1-level quantizer and let its associated bit rate be $\log_2(N^1)$ bits. The quantization error of this quantizer is given by

$$R^1 = X^1 - Q^1(x^1) \tag{14}$$

The expected value of the distortion is given by

$$\mathcal{D}^1 = \int \mathbf{d}(X^1, Q^1(X^1)) P(X^1) \, dX^1 \tag{15}$$

If the random variable needs to be represented more precisely (i.e., if the expected value of the distortion needs to be smaller), the first quantizer residual can be quantized again by a second quantizer, Q^2. The quantizer Q^2 approximates the random variable R^1. Let $Q^2(R^1)$ be an N^2-level quantizer and let its associated bit rate be $\log_2(N^2)$ bits. The quantization error of this quantizer is given by

$$R^2 = R^1 - {}^{\prime}Q^2(R^1) \tag{16}$$

and the expected value of distortion is now

$$\mathcal{D}^2 = \int \mathbf{d}[X, (Q^1(X) + Q^2(R^1))] P(X) \, dX \tag{17}$$

This process can be thought of as a cascade of two quantizers, as shown in Fig. 8. The total bit rate of the quantization scheme is $\log_2(N^1) + \log_2(N^2)$. This scheme can be extended to any number of quantizers. A K-stage residual quantizer consists of K

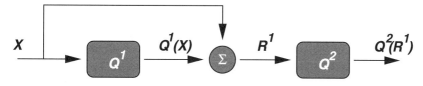

Fig. 8 Residual quantizer—cascade of two quantizers.

quantizers $\{Q^k: k = 1, \ldots, K\}$. Each quantizer Q^k quantizes the residue of the previous stage $R^{(k-1)}$. The total bit rate of the quantization scheme is given by

$$B = \sum_{k=0}^{K} \log_2(N^k) \tag{18}$$

where N^k is the number of quantization levels of the quantizer Q^k.

4.3. Residual Vector Quantizer

A residual vector quantizer is a vector generalization of the residual quantizer outlined above. A K-stage residual vector quantizer is composed of K vector quantizers $\{Q^k: k = 1, \ldots, K\}$. Each vector quantizer consists of its own codebook C^k of size N^k. The kth-stage vector quantizer operates on the residue $\mathbf{R}^{(k-1)}$ from the previous stage. The residue due to the first stage is given by

$$\mathbf{R}^1 = \mathbf{X} - Q^1(\mathbf{X}) \tag{19}$$

The final quantized value of the vector $\mathbf{X}$ is given by

$$Q(\mathbf{X}) = Q^1(\mathbf{X}) + Q^2(\mathbf{R}^1) + \cdots + Q^k(\mathbf{R}^{(k-1)}) + \cdots + Q^K(\mathbf{R}^{(K-1)}) \tag{20}$$

4.4. Search Techniques for Residual Vector Quantizers

The structure of a residual vector quantizer inherently lends itself to a number of possible encoding schemes. Two of the main characteristics of encoding in residual quantizers are as follows:

- Overall optimality—the least overall distortion at the end of the last stage of encoding
- Stagewise optimality—the least distortion possible at the end of each stage

4.4.1. Exhaustive Search

Exhaustive searches in residual quantizers aim at achieving the least overall distortion. In exhaustive search schemes, all possible combinations of all the stage quantizations are searched, and the combination giving rise to the least distortion is chosen. This search gives the best possible performance in the residual quantization scheme. However, this search scheme is computationally very expensive and is the same as that of a unstructured vector quantizer. The search complexity for a K-stage vector quantizer with codebook sizes $\{N_1, N_2, \ldots N_K\}$ is of the order $O(N_1 \times N_2 \times \cdots \times N_K)$. Exhaustive search schemes are not particularly conducive for progressive transmission schemes (successive refinement).

4.4.2. Sequential Search

Sequential searches in residual quantizers make full use of the structural constraint of the quantizer. The search process is stage by stage, where the quantization value that minimizes the distortion up to that stage is chosen. This search scheme, the least expensive of all search schemes, is inherently inferior to exhaustive schemes and usually leads to suboptimal overall distortion performance. The search complexity for a K-stage quantizer with

Video Compression

codebook sizes $\{N_1, N_2, \ldots N_K\}$ is of the order $O(N_1 + N_2 + \cdots + N_K)$. The search scheme is particularly well suited for progressive transmission schemes.

4.4.3. M-Search

A hybrid search scheme has been proposed (25), whose search complexity is less than that of a full search scheme but greater than that of a sequential search scheme. This scheme produces an overall distortion performance that is better than sequential search schemes and close to the exhaustive search scheme. In this scheme, a subset of the quantization values at each stage is chosen based on the least distortion, and these subsets are then searched in an exhaustive fashion to get the quantized value.

4.5. Structure of Residual Vector Quantizers

A quantizer partitions the input space into a finite number of polytopal regions (Fig. 9). The centroid of each polytope approximates all the input symbols that belong to that particular region. The process of finding the residue of the signal is equivalent to shifting the coordinate system to the centroid of the polytope. This process is repeated for all the polytopes. Therefore, we have a finite set of spaces, each corresponding to a polytope. These spaces are bounded by the underlying polytope of the partition (i.e., each of these spaces contain members around the origin which are limited in location by the polytope to which they belong).

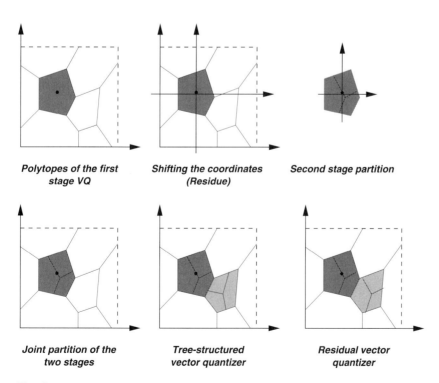

Fig. 9 Structure of a residual vector quantizer.

Now consider the quantization of each of these spaces. If the optimal quantizer is found for each of these spaces, their structure (i.e., the partition of the input space) may be totally different. Such a quantizer is called a tree-structured quantizer. If a constraint is imposed such that the same partition structure is used for all the subspaces, then the method of quantization is the residual quantization scheme.

4.6. Optimality Conditions for Residual Quantizers

4.6.1. Overall optimality

Consider a K-stage residual quantizer with a set of quantizers Q,

$$Q = \{Q^1, Q^2, \ldots, Q^k, \ldots, Q^n\}$$

and codebooks **C**

$$\mathbf{C} = \{C^1, C^2, \ldots, C^k, \ldots, C^n\}$$

with stage indexes $\mathbf{K} = \{k: k = 1, \ldots, K\}$. Each stage codebook C^k contains N^k code vectors $C^k = \{\mathbf{c}_1^k, \mathbf{c}_2^k, \ldots, \mathbf{c}_{N_k}^k\}$. As in the case of vector quantizers, we derive two optimality conditions. For the first condition, given the encoder, we find the best possible decoder. For the second condition, we find the best possible encoder for a given decoder. We derive the conditions for a particular stage, assuming that all other stages have fixed encoders and decoders.

4.6.1.1. Centroid Condition

The index vector $\mathbf{I}$ belongs to the index space $\mathcal{I} = \{\mathbf{I}: \mathbf{I} = (i^1, i^2, \ldots, i^k, \ldots, i^K), i^k = 1, \ldots, N^k\}$. Let the partition of the input space be $\mathcal{P}_\mathbf{I} = \mathcal{P}_{\mathbf{c}_{i^1}^1, \mathbf{c}_{i^2}^2, \ldots, \mathbf{c}_{i^k}^k, \ldots, \mathbf{c}_{i^K}^K}$ based on the different stage quantizers. This partition is based on a fixed encoder. To find the best decoder for the stage κ, let the decoders of stages $\{\mathbf{K}|\kappa\}$ be fixed. Overall distortion is given by

$$\mathcal{D}(\mathbf{X}, Q(\mathbf{X})) = \sum_{\mathbf{I} \in \mathcal{I}} \sum_{\mathcal{P}_\mathbf{I}} (\mathbf{X} - \mathbf{c}_{i^1}^1 - \mathbf{c}_{i^2}^2 - \cdots - \mathbf{c}_{i^{\kappa-1}}^{\kappa-1} - \mathbf{c}_{\iota^\kappa}^\kappa - \mathbf{c}_{i^{\kappa+1}}^{\kappa+1} - \cdots - \mathbf{c}_{i^K}^K)^2 P(\mathbf{X})$$

(21)

For the code vector $\mathbf{c}_{\iota^\kappa}^\kappa$ of the κth stage to be optimal, the following has to be true:

$$\frac{\delta \mathcal{D}(\mathbf{X}, Q(\mathbf{X}))}{\delta \mathbf{c}_{\iota^\kappa}^\kappa} = 0 \qquad (22)$$

that is,

$$\sum_{\mathbf{I} \in I_{\iota^\kappa}} \sum_{\mathcal{P}_\mathbf{I}} (\mathbf{X} - \mathbf{c}_{i^1}^1 - \mathbf{c}_{i^2}^2 - \cdots - \mathbf{c}_{i^{\kappa-1}}^{\kappa-1} - \mathbf{c}_{\iota^\kappa}^\kappa - \mathbf{c}_{i^{\kappa+1}}^{\kappa+1} - \ldots - \mathbf{c}_{i^K}^K) P(\mathbf{X}) = 0 \qquad (23)$$

where $I_{\iota_k} = \{\mathbf{I} \in \mathcal{I}: i_k = \iota_k\}$. Solving for $\mathbf{c}_{\iota^\kappa}^\kappa$, we get

$$\mathbf{c}_{\iota^\kappa}^\kappa = \frac{\sum_{\mathbf{I} \in I_{\iota^\kappa}} \sum_{\mathcal{P}_\mathbf{I}} (\mathbf{X} - \mathbf{c}_{i^1}^1 - \mathbf{c}_{i^2}^2 - \cdots - \mathbf{c}_{i^{\kappa-1}}^{\kappa-1} - \mathbf{c}_{i^{\kappa+1}}^{\kappa+1} - \cdots - \mathbf{c}_{i^K}^K) P(\mathbf{X})}{\sum_{\mathbf{I} \in I_{\iota^\kappa}} \sum_{\mathcal{P}_\mathbf{I}} P(\mathbf{X})}$$

(24)

Video Compression

A similar equation has been derived by Barnes and Frost (25). This has been described as the centroid of the grafted residue.

4.6.1.2. Nearest-Neighbor Condition

For fixed decoders and fixed encoders for stages $\mathbf{K}|\kappa$, the optimal stage-κ encoder is one that minimizes either the overall distortion or the distortion for that stage. In either case, the mapping that produces the least distortion is the nearest-neighbor mapping. For the case of exhaustive search decoders, the best encoder is the nearest-neighbor mapping for the direct sum codebook.

4.6.2. Causal Stages Optimality

For the encoder to be optimal in terms of quantizers up to the present stage, the optimality conditions are as follows.

4.6.2.1. Centroid Condition

Let the partition of the input space be $\mathcal{P}'_\mathbf{I} = \mathcal{P}_{c_{i^2}^2, c_{i^1}^1, \ldots, c_{i^k}^k, \ldots, c_{i^K}^K}$ based on the causal stage quantizers. For a fixed encoder, the optimal κth-stage code is given by

$$\mathbf{c}_{i^\kappa}^\kappa = \frac{\sum_{\mathbf{I} \in I_{i^\kappa}} \sum_{\mathcal{P}'_\mathbf{I}} (\mathbf{X} - \mathbf{c}_{i^1}^1 - \mathbf{c}_{i^2}^2 - \cdots - \mathbf{c}_{i^{\kappa-1}}^{\kappa-1} - \mathbf{c}_{i^{\kappa+1}}^{\kappa+1} - \cdots - \mathbf{c}_{i^K}^K) P(\mathbf{X})}{\sum_{\mathbf{I} \in I_{i^\kappa}} \sum_{\mathcal{P}'_\mathbf{I}} P(\mathbf{X})}$$

(25)

where $I_{i^\kappa} = \{\mathbf{I}: \mathbf{I} = (i^1, i^2, \ldots, i^k, \ldots i^K), i^k = 1, \ldots, N^k\}$ is the index vector of the causal stages including the present stage. This is the centroid of the direct partition up to the stage.

4.6.2.2. Nearest-Neighbor Condition

For fixed decoders and fixed encoders for stages $\mathbf{K}|\kappa$, the optimal κth-stage encoder is the nearest-neighbor mapping encoder.

4.6.3. Simultaneous Causal and Overall Optimality

Consider the κth-stage encoder of a $\mathbf{K}$-stage residual vector quantizer. Let us assume that it is optimal in terms of both causal as well as overall distortion. Then it satisfies the following two equations simultaneously:

$$\mathbf{c}_{i^\kappa}^\kappa = \frac{\sum_{\mathbf{I} \in I_{i^\kappa}} \sum_{\mathcal{P}_\mathbf{I}} (\mathbf{X} - \mathbf{c}_{i^1}^1 - \mathbf{c}_{i^2}^2 - \cdots - \mathbf{c}_{i^{\kappa-1}}^{\kappa-1} - \mathbf{c}_{i^{\kappa+1}}^{\kappa+1} - \cdots - \mathbf{c}_{i^K}^K) P(\mathbf{X})}{\sum_{\mathbf{I} \in I_{i^\kappa}} \sum_{\mathcal{P}_\mathbf{I}} P(\mathbf{X})}$$

(26)

$$\mathbf{c}_{i^\kappa}^\kappa = \frac{\sum_{\mathbf{I} \in I_{i^\kappa}} \sum_{\mathcal{P}'_\mathbf{I}} (\mathbf{X} - \mathbf{c}_{i^1}^1 - \mathbf{c}_{i^2}^2 - \cdots - \mathbf{c}_{i^{\kappa-1}}^{\kappa-1}) P(\mathbf{X})}{\sum_{\mathbf{I} \in I_{i^\kappa}} \sum_{\mathcal{P}'_\mathbf{I}} P(\mathbf{X})}$$

(27)

Because the two denominators are equal,

$$\sum_{\mathbf{I} \in I_{i^\kappa}} \sum_{\mathcal{P}_\mathbf{I}} (\mathbf{X} - \mathbf{c}_{i^1}^1 - \mathbf{c}_{i^2}^2 - \cdots - \mathbf{c}_{i^{\kappa-1}}^{\kappa-1} - \mathbf{c}_{i^{\kappa+1}}^{\kappa+1} - \cdots - \mathbf{c}_{i^K}^K) P(\mathbf{X})$$
$$= \sum_{\mathbf{I} \in I_{i^\kappa}} \sum_{\mathcal{P}'_\mathbf{I}} (\mathbf{X} - \mathbf{c}_{i^1}^1 - \mathbf{c}_{i^2}^2 - \cdots - \mathbf{c}_{i^{\kappa-1}}^{\kappa-1}) P(\mathbf{X})$$

(28)

Equation (28) gives

$$\sum_{\mathbf{I} \in I_{\iota^\kappa}} \sum_{\mathcal{P}_\mathbf{I}} (\mathbf{c}_{i^{\kappa+1}}^{\kappa+1} - \cdots - \mathbf{c}_{i^K}^K) P(\mathbf{X}) = 0 \qquad (29)$$

Basically, to have simultaneous global and stagewise optimality at any given stage κ, the sum of the code vectors of stages $\kappa + 1, \ldots, \mathbf{K}$ must equal zero. This suggests that the successive refinement residual vector quantizer is not optimal. A rigorous treatment of successive approximation is given by Equitz and Cover (26).

4.7. Design of Residual Vector Quantizers

The design of residual vector quantizers is based on a number of trade-offs, and different training methods are used for different coding schemes (27). Sequential search quantizer codebooks, in general, are different from exhaustive search quantizer codebooks. A number of design methods have been proposed. Chan et al. (28) have proposed a joint codebook design. The codebook is designed by fixing all but one stage and adapting one particular stage codebook to minimize overall distortion. During the next step of the iteration, another codebook is adapted, this being repeated in a cyclic manner until the required convergence is obtained. Barnes and Frost (25) use a similar algorithm for the codebook design. Rizvi and Nasrabadi (29) have proposed a design algorithm based on the Kohonen network, where an energy is iteratively minimized to reduce the overall distortion.

4.8. Residual Vector Quantization with Variable Block Size

The residual vector quantizer outlined thus far quantizes blocks (vectors) of the same size at each stage. When an input sequence to be quantized contains nonstationary artifacts, it is often difficult to compress using fixed-block-size quantizers. Blocks containing discontinuities are quantized rather poorly by all the stages or they require a large number of residual vector quantizer stages. One way to solve the problem is to use smaller block sizes at the later stages of the residual vector quantizer. A variable-block-size residual vector quantizer is shown in Fig. 10. In this figure, the first-stage quantizer uses blocks of size 4, the second-stage quantizer uses blocks of size 2, and the third-stage quantizer uses blocks of size 1 (the third-stage quantizer is a scalar quantizer in this example).

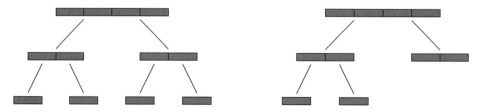

Fig. 10 Tree structure of the variable-block-size residual vector quantizer. Pruning of this tree corresponds to the variable-rate variable-block-size residual vector quantizer.

4.9. Pruned Variable-Block-Size Residual Vector Quantizer

Often, not every part of a digital signal requires to be quantized by all the stages of the residual vector quantizer. Sections of the signal containing no information or very little information can easily be represented by just one or two stages. Restricting the number of stages for a particular section of the signal is equivalent to pruning the tree structure as shown in Fig. 10. There are a number of different ways of pruning the tree. One significant characteristic of a pruned variable-block-size residual vector quantizer is that a small amount of side information needs to be stored/transmitted. This side information determines the particular tree structure used for every block. The tree structure is required by the decoder, as it needs to know the number of stages used for every part of the sequence.

4.9.1. Top-Down Pruning Using a Predefined Threshold

Deciding the number of stages to be used for a particular block can be done by examining the error after every stage. If the error at the end of a particular stage is less than a threshold, then the quantization can be stopped at that stage. The error is measured using a significance measure such as the L_2 norm (mean squared error). If the L_2 after stage κ is less than a threshold τ_κ, then quantization is stopped at that stage. The choice of the threshold determines the performance of this algorithm. When a single global threshold $\tau = \tau_1 = \cdots = \tau_k = \cdots = \tau_K$ is used, it directly controls the output bitrate of the quantizer.

4.9.2. Optimal Pruning in the Rate-Distortion Sense

This method is a bottom-up pruning technique in which a given block is quantized by all the stages. The quantization error is measured after every stage and stored. The tree is pruned by trading off the number of bits required to encode to a particular depth against the distortion. Let $\mathcal{T}$ represent the set of all possible tree structures and let $Q_T(\mathbf{X})$ be the quantized value of $\mathbf{X}$ corresponding to the tree structure T. Let $L(T)$ represent the number of bits required to represent a particular tree structure. For a particular tree structure T, let $\mathcal{T}_T$ represent the set of indices after quantization based on the tree structure and $L(\mathcal{T}_T)$, the number of bits required to encode the indices. A Lagrangian formulation can be made and the tree can be pruned based on this objective function. This is equivalent to finding a particular tree structure that minimizes the following cost function:

$$d(\mathbf{X}, Q(\mathbf{X})) + \lambda[L(\mathcal{T}_T) + L(T)] \tag{30}$$

The value of the Lagrangian multiplier λ controls the output bit rate of the quantizer. It controls the slope of the tangent to the $\mathbf{R}$–$\mathbf{D}$ curve of the quantizer at different operating points, as shown in Fig. 11.

4.10. Transform-Domain Vector Quantization for Large Blocks

Direct vector quantization of large blocks is computationally expensive and the design of the codebooks for large-block vector quantizers is difficult. The complexity of vector quantizers can be reduced with the use of transform vector quantization (30). With this method, the data vector is first transformed, using a decorrelating transformation Φ such as the discrete cosine transform (DCT). A masking function M is then applied to the transformed data to reduce the dimensionality of the vector. In its simplest form, the

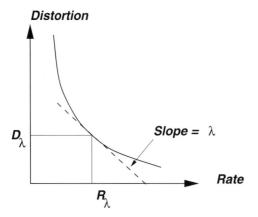

Fig. 11 Optimal pruning of the residual vector quantizer in the rate-distortion sense.

masking function is a binary vector, and it truncates the number of coefficients used. The masking function can also contain a normalizing factor for each coefficient based on its variance. The resulting vector is then quantized using a small-vector-dimension quantizer. Decoding is done by first applying the inverse of the mask function M^{-1} (usually in the form of padding with zeros) followed by the inverse transform Φ^{-1}. Use of a unitary transform like the DCT, which compacts the signal energy to a relatively small number of coefficients, leads to the requirement of a vector quantizer with much smaller dimensions. It is therefore possible to use transform vector quantizers in the initial stages of a variable-block-size residual vector quantizer.

4.11. Video Compression Using Residual Vector Quantization

The residual signal generated by the motion-compensation algorithm, as described in Sec. 2.2, can be compressed using a residual vector quantizer. This section describes a particular implementation of an encoder using residual vector quantization. In the first two stages of the encoder, quantization is performed in the transform domain, as shown in Fig. 12. The residual signal $r^0(i, j, t)$ is broken into blocks of dimension $m_1 \times n_1$, represented by $\mathbf{R}^0(I, J, t)$. The variance of each block is measured and compared to a threshold in order to determine if the block needs to be encoded. Often the background areas contain no information because the motion-estimation algorithm predicts the data perfectly. The vectors $\mathbf{R}^0(I, J, t)$ that require transmission are then transformed to obtain the transform-domain signal $\mathcal{R}^0(I, J, t)$, using a transform operator Φ:

$$\mathcal{R}^0(I, J, t) = \Phi[\mathbf{R}^0(I, J, t)]. \tag{31}$$

Next, a masking operator, M_1, is applied to the transformed vector to obtain a truncated vector, $\mathcal{R}_m^0(I, J, t)$, of reduced dimension, which is then quantized using the first-stage vector quantizer. Let $\mathbf{c}_{i^1}^1$ be the best matching code vector. Then, the approximation of

Video Compression

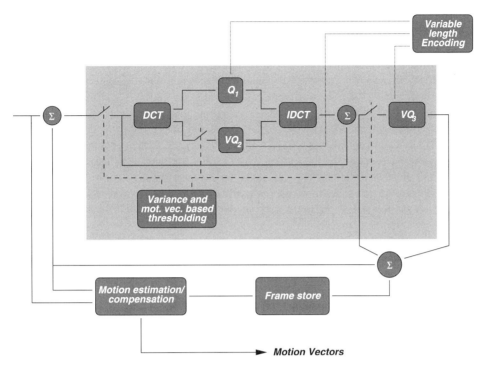

Fig. 12 Video encoder based on residual vector quantization.

$\mathbf{R}^0(I, J, t)$ is given by

$$Q_1[\mathbf{R}^0(I, J, t)] = \Phi^{-1}[M_1^{-1}(\mathbf{c}_{i'}^1)] \tag{32}$$

where M_1^{-1} and Φ^{-1} are the inverses of the masking function and the transform operator, respectively. The error after the first-stage quantization is given by

$$\mathbf{R}^1(I, J, t) = \mathbf{R}^0(I, J, t) - Q_1[\mathbf{R}^0(I, J, t)] \tag{33}$$

This residual vector is measured for significance, and if it requires further compression, a second mask, M_2, is applied to the transformed vector to obtain vector $\mathcal{R}_m^1(I, J, t)$. This vector is quantized by the second-stage VQ. If the best match code vector is $\mathbf{c}_{i'}^2$, the approximation of $\mathbf{R}^0(I, J, t)$ after the second stage is given by

$$Q_2[\mathbf{R}^1(I, J, t)] = \Phi^{-1}[M_1^{-1}(\mathbf{c}_{i'}^1) + M_2^{-1}(\mathbf{c}_{i'}^2)] \tag{34}$$

The masking functions M_1 and M_2 are binary templates which together select the first few perceptually significant coefficients. The masking function effectively creates a low-pass-filtered version of the signal by discarding the higher-frequency coefficients. The residue after the second stage is given by

$$\mathbf{R}^2(I, J, t) = \mathbf{R}^0(I, J, t) - Q_2[\mathbf{R}^1(I, J, t)] \tag{35}$$

This residual vector is then split into smaller blocks $\mathbf{R}_2'(I', J', t)$ of dimension $m_2 \times n_2$ for quantization by the subsequent stages. The vectors $\mathbf{R}_2'(I', J', t)$ are compared with a threshold to determine if they are significant enough to require transmission. The significant blocks that require transmission are quantized by using the third-stage vector quantizer, which gives an approximation $Q_3[\mathbf{R}^2(I_2, J_2, t)]$. The process of decomposition into smaller blocks and selective quantization is applied recursively in the later stages to obtain a good representation of the residual signal $r^0(i, j, t)$. The indices of the quantizers are entropy coded using adaptive arithmetic coding (31).

The block diagram in Fig. 13 shows that the operation of the decoder is not complex. The variable-length decoder recovers the bitmaps and the code vector indices from the arithmetic-encoded sequence. The decoders are constructed using lookup tables. Based on complexity requirements, the lookup tables of the first two stages can store either the transform-domain coefficients $c_{i^1}^1$ and $c_{i^2}^2$ or the reconstruction vectors $\Phi^{-1}[M_1^{-1}(c_{i^1}^1)]$ and $\Phi^{-1}[M_2^{-1}(c_{i^2}^2)]$. In the first case, two additional operations must be performed [a padding operation (M^{-1}) and an inverse transform operation (Φ^{-1})]. In the second case, memory requirements are significantly larger. If the later-stage vectors are required, a direct table lookup is performed, using the indices, and the low-dimensional vector is added to the first-stage reconstructed vector in the appropriate position. This process provides the reconstructed residual signal $\hat{\mathbf{R}}^0(I, J, t)$, which is then passed to the motion-compensation stage to obtain the reconstructed frame.

In the encoder, in order to achieve a true variable rate, decisions are made about the number of quantization stages required by each stage. These decisions have to be transmitted to the decoder in order to decode the encoded data correctly. The decisions are usually encoded as bitmaps for each stage. In the case of a three-stage encoder, three bitmaps are required for proper decoding. These bitmaps could require a significant portion of the bit budget if they are not intelligently encoded. The bitmap at the first stage is combined with the motion vector information to reduce the number of bits. If the motion vector of a block is nonzero, it is assumed that the block needs to be encoded by at least the first stage; thus, the overhead first-stage flag bit for blocks with nonzero motion vectors is eliminated. The bit budget for these bitmaps can be further reduced by using the correlation between the second- and third-stage bitmaps. The details of encoding the bitmaps are given in the next section.

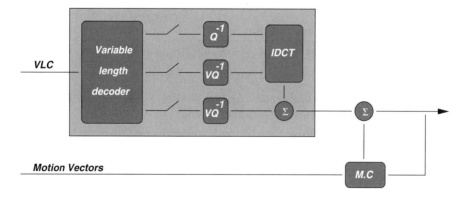

Fig. 13 Video decoder based on residual vector quantization.

4.11.1. Codec Architecture

The quantizer used to encode the residual signal is a three-stage quantizer. The first two stages work on transform domain coefficients and the third stage quantizer is a spatial domain quantizer. The first quantizer, a scalar quantizer, encodes the DC (direct current) coefficient. The quantization level is referred to in this chapter as DCQUANT. The second quantizer, a vector quantizer, encodes the low-frequency coefficients. The low-frequency DCT coefficients are zonally selected based on the perceptual significance, the first N coefficients (NSIGCOEFF), excluding the DC coefficient, are encoded. The second quantizer finds the best matching index (LFINDEX) from a codebook, which is then transmitted to the receiver. The first two quantizers work on blocks of size 8×8. The third quantizer is a vector quantizer with block size 4×4. This quantizer encodes the residue after the first two stages; the index is referred to as RESINDEX.

Each frame after motion estimation is divided into two regions based on the motion vectors. Blocks that have their motion vector equal to zero are classified as class-1 (C1); blocks and blocks with nonzero motion vectors are classified as class-2 (C2) blocks. Transmitting the classification information is not necessary, as it can be detected at the decoder based on the motion vectors.

The variance of each C1 block is compared with a threshold T_1. If the variance is less than the threshold, the block is not encoded; otherwise, the block is encoded by all the three stages and the bitstream then contains the three values: DCQUANT, LFINDEX, and RESINDEX. The decision about the number of stages, which is based on the threshold comparison, has to be transmitted to the decoder and is done using a flag bit (B1).

Each C2 block is encoded using the first-stage encoder, and the DCQUANT value is transmitted for every C2 block. The variance of the C2 block is compared with a threshold T_2 ($\leq T_1$) to determine its activity and hence make a decision about the number of stages required. If the block has low activity, only the DCQUANT value is transmitted. For high-activity blocks, all the three values, DCQUANT, LFINDEX, and RESINDEX, are transmitted to the receiver. The decision about the number of stages used for each block is transmitted in the form of a flag bit B2. The quantization scheme is summarized in Table 1. The parameters T_1 and T_2 are used to control the target bit rate. These parameters can also be used for rate control using feedback from an output buffer to obtain a signal with smaller bit-rate variation. The parameters to specify the encoder include the number of DC quantization levels (DCLVLS), the size of the second-stage codebook (LFCB_SZ), the size of the third-stage codebook (RESCB_SZ), and number of significant DCT coefficients used (NSIGCOEFF). The lookup tables required at both the encoder and decoder are the DC reconstruction values (DCQ), second-stage VQ codebook (LFCB), and the

Table 1 Quantization Scheme

Block type	B1	B2	DCQUANT	LFINDEX	RESINDEX
C1	0				
C1	1		x	x	x
C2		0	x		
C2		1	x	x	x

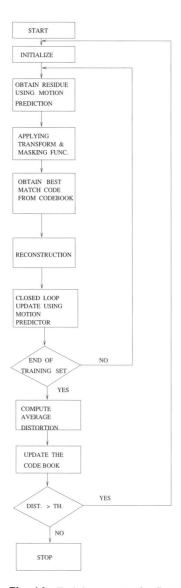

Fig. 14 Training process for first stage.

third-stage VQ codebook (RESCB). The algorithmic flowchart of the design of the codebooks of the codec are presented in Figs. 14 and 15.

4.11.2. Simulation Results

Simulation results are given for an outline encoder using a predefined threshold as well as an optimal pruning method in the rate-distortion sense (described earlier in this chapter) with the following parameters: The number of DC quantization levels DCLVLS was 8, the size of the second-stage codebook was 16 (LFCB_SZ = 16), and the size of the third-stage codebook was 128 (RESCB_SZ 128). The number of significant DCT coefficients used, NSIG-

Video Compression

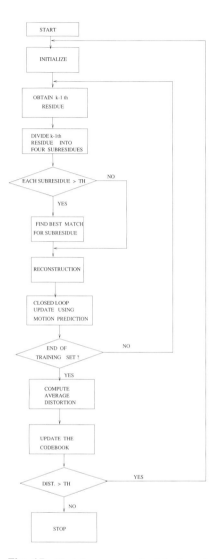

Fig. 15 Training process for kth stage (later stages).

COEFF, was 9. Simulation results are given for the encoder–decoder operating at three different very low bit rates. The results are compared with that of the H.263 codec (32,33). For the H.263 codec, all negotiable options were turned on except for PB frames to make a reasonable comparison with the RVQ codec (the PB frame mode can easily be incorporated in the RVQ codec). In the H.263 codec, the quantization parameter (QP) for the first frame was set to its minimum value to give the best performance for the intraframe. As we were interested in the steady-state characteristics and not the first few transient frames, the bits consumed in the first frame were not included in the bit-rate calculations. However, in order to understand how the coded first intraframe influences the motion-compensated residual signal, we also performed a coding experiment using coded intraframes, compressed at three different quality levels, with the results shown in Figs. 16 and 17. Two coding algorithms,

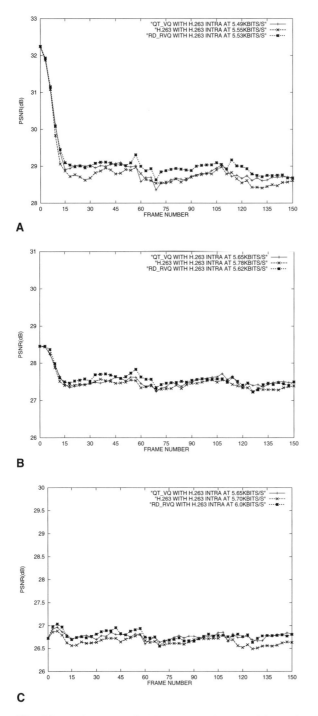

Fig. 16 Performance of video compression algorithms with the first intraframes encoded using H.263 intraframe coding algorithm. (A) PSNR results with the intraframe coded at 0.77 bpp; (B) PSNR results with the intraframe coded at 0.36 bpp; (C) PSNR results with the intraframe coded at 0.24 bpp.

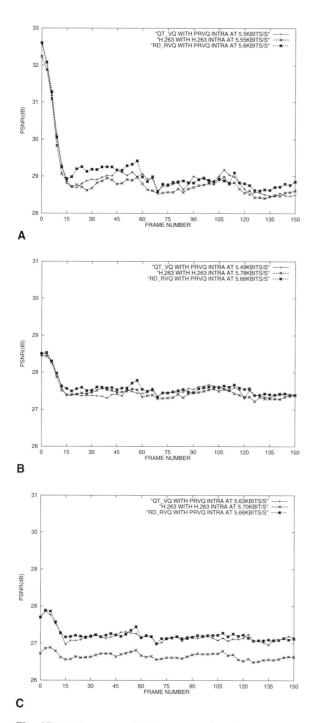

Fig. 17 Performance of video compression algorithms with different intraframe coding methods. (A) PSNR results with the intraframes coded at 0.77 bpp (H.263) and 0.83 bpp (PRVQ); (B) PSNR results with the intraframes coded at 0.36 bpp (H.263) and 0.35 bpp (PRVQ); (C) PSNR results with the intraframes coded at 0.24 bpp (H.263) and 0.27 bpp (PRVQ).

H.263 and predictive residual vector quantizer (PRVQ) developed by Rizvi et al. (18), were used to compress the intraframes for the RVQ and quadtree-based VQ codecs.

The sequences used were the popular test sequence—"salesman." Each of these sequences have 8-bit pixels, with frame size 144 × 176 and the frame rate was 10 frames/s. Performance was evaluated using the PSNR measurement to compare the two coding methods. The computation requirements for the RVQ codec include real-time DCT, which is the same as that of H.263. The transform-domain VQ has a codebook of size 16; therefore, its computational complexity is small. Only about 10% of the blocks (average over all the sequences tested) were encoded by the last VQ stage.

Forty motion-compensated difference frames extracted from four different original sequences (in which the test sequences were not included) were used to train the initial codebook of each stage (open-loop training). The initial codebooks were trained using a three-step process. The scalar quantizer was first designed using Lloyd's algorithm. The second- and third-stage initial codebooks were then generated using the k-means algorithm. In order to further train the initial codebooks in the presence of a motion-estimation predictor, we used the motion-compensated difference frame obtained from the current frame and the previously encoded frame using codebooks generated during the previous iteration of training

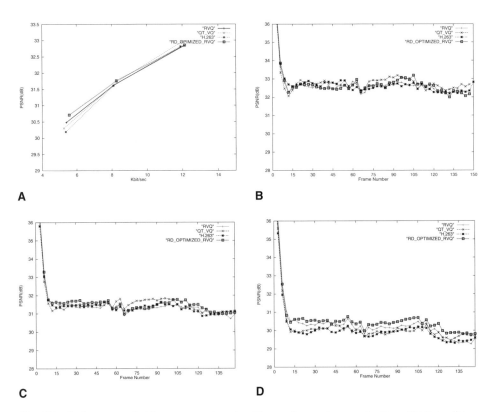

Fig. 18 Performance of video compression algorithms using vector quantization. (A) Rate-distortion performance; (B) PSNR results at 12 kbps; (C) PSNR results at 8.1 kbps; (D) PSNR results at 5.3 kbps.

process for closed-loop training. The three quantizers were then retrained using the entropy constraint in a closed-loop manner to improve the rate-distortion performance.

The rate-distortion performance of the two codecs are shown in Fig. 18a. It can be seen that both RVQ codecs (normal and rate-distortion-optimized RVQ) showed similar quality to H.263 at the higher bit rate and outperformed H.263 at the two lower bit rates. At very low bit rates, the PSNR for H.263 decreases drastically, whereas that of the RVQ codec decreases gradually. It can be seen that the performance on the H.263 codec for the salesman sequence decreases dramatically at very low bit rates; this is because of the rather large motion in this sequence. The RVQ codec handles such sequences very well at very low bit rates. Figures 18b, 18c, and 18d show the PSNR results of each reconstructed frame at three different bit rates. No rate control is used, and so a VBR bitstream with constant quality is generated. With the results shown in Figs. 16 and 17, the RVQ codec proved that it could handle the motion-compensated residues in wide statistical variation, caused by the coded first frames at different quality levels, better than H.263 codec. Figures 19b, 20b, and 21b show the reconstructed 30th frames for the sales-

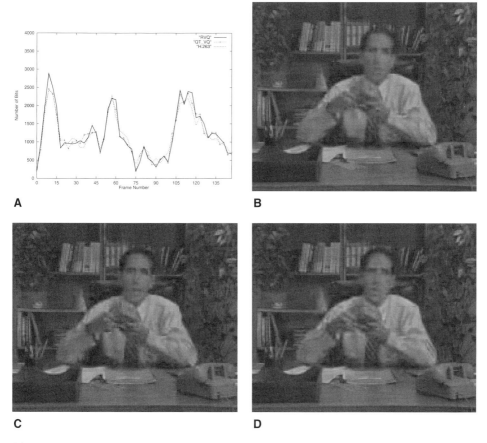

Fig. 19 Results for bit rate of approximately 12 kbps. (A) Frame bit rate for the sequence; (B) RVQ codec; (C) quadtree–VQ codec; (D) H.263 codec.

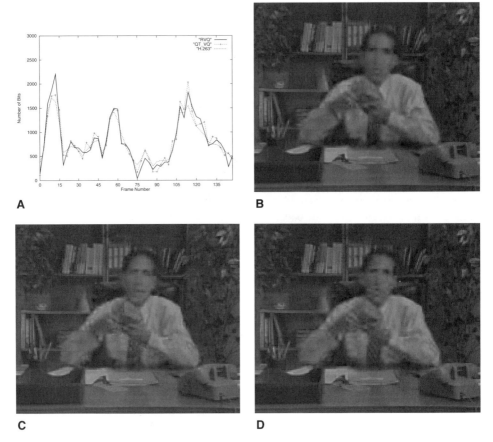

Fig. 20 Results for bit rate of approximately 8.1 kbps. (A) Frame bit rate for the sequence; (B) RVQ codec; (C) quadtree–VQ codec; (D) H.263 codec.

man sequence compressed at the three different bit rates using the RVQ encoder. Figures 19d, 20d, and 21d show the reconstructed frames compressed using the H.263 encoder at the same bit rates. It can be seen clearly especially at 5.4 kbps that H.263 suffers from blocking and smoothing, whereas the output of the RVQ codec is of much better visual quality.

5. QUADTREE-BASED VECTOR QUANTIZATION

5.1. Introduction

A quadtree is a hierarchical data structure used to represent regions, curves, surfaces, and volumes. Representations of regions by a quadtree is achieved by the successive subdivision of the image array into four equal quadrants. This process is known as a regular decomposition of an array. An image is thus decomposed into homogeneous regions with sides of lengths that are powers of 2. A tree of degree 4 (each nonleaf has four children)

Video Compression

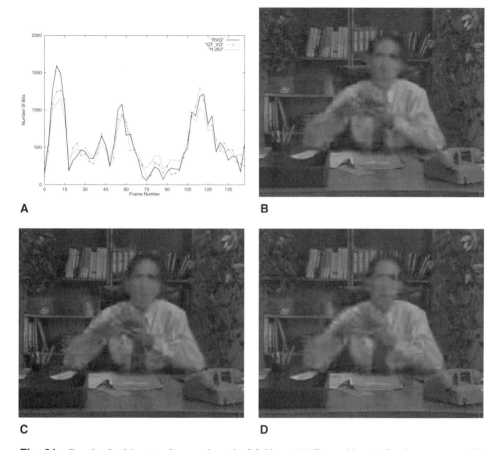

Fig. 21 Results for bit rate of approximately 5.3 kbps. (A) Frame bit rate for the sequence; (B) RVQ codec; (C) quadtree–VQ codec; (D) H.263 codec.

is generated to represent the image in terms of its homogeneous regions. The root node corresponds to the entire array, and each child of a node represents a quadrant of the region represented by that node. Leaf nodes of the tree correspond to those blocks for which no further subdivision is necessary. The above segmentation procedure is known as a top-down construction of the quadtree. Another possibility for constructing a quadtree is a bottom-up procedure, where small blocks are merged together recursively to form a larger block if they are homogeneous with respect to the merging criterion.

The regular decomposition method does not necessarily correspond to the segmentation of the image into maximal homogeneous regions. It is likely that unions of adjacent blocks form homogeneous regions. To obtain these maximal homogeneous regions, we must allow merging of the adjacent blocks. However, the resulting partition will no longer be represented by a quadtree; instead, the final representation is in the form of an adjacency graph. An alternative method to obtain maximal homogeneous regions is to use a decomposition technique that is not regular. This method will segment the image into rectangular blocks of arbitrary size, which will require a different coding procedure for each block size. Here, we use a regular decomposition method because the resulting blocks are squares, which will reduce the complexity of the encoder, the decoder, and the number of bits required to represent the binary quadtree.

5.2. Quadtree Decomposition

A quadtree decomposition results in an unbalanced tree structure with leaf nodes of different sizes. With a regular decomposition, the leaf nodes are restricted to square blocks. It is further possible to restrict the sides of the leaf nodes to a small range of values. This results in a tree structure with the leaf nodes being square blocks with a maximum of n different sizes. A signal decomposed using the above method can be compressed using vector quantization of the leaf blocks. This will require n different vector quantizers corresponding to the different block sizes. All leaf nodes of the same size are quantized by one vector quantizer, as shown in Fig. 22.

The criterion used in the quadtree decomposition is one of the main aspects in the design of a quadtree-based vector quantizer.

5.3. Optimal Quadtree in the Rate-Distortion Sense

Quadtree decomposition for vector quantization can take into account the distortion introduced by the vector quantizer. An optimal decomposition algorithm, in the rate-distortion sense, was introduced by Sullivan and Baker (11). The algorithm attempts to minimize a constrained error function defined as follows:

$$\mathcal{E}^q = \mathbf{d}(\mathbf{X}, Q^q(\mathbf{X})) + \lambda \mathbf{b}^q \qquad (36)$$

where $\mathbf{d}$ is the distortion introduced in quantizing the block k using a particular tree structure q, and $\mathbf{b}^q$ is the total number of bits used to represent the tree and the quantization indices. The Lagrangian multiplier, λ, controls the trade-off between the bit rate and distortion by determining the point on the $\mathbf{R}$–$\mathbf{D}$ curve as explained in Sec. 4.9.

Consider a block $\mathbf{X}^m$ of size $2^m \times 2^m$ and its descendents $\mathbf{X}_i^{m-1}$, $i = 1, \ldots, 4$. Let the distortion of quantizing the blocks $\mathbf{X}_i^{m-1}$ using the optimal quantizer be $\mathbf{d}_i^{m-1}$, and let the number of bits necessary to optimally quantize $\mathbf{X}_i^{m-1}$ be $\mathbf{b}_i^{m-1}$. Similarly, let the distortion of quantizing the block $\mathbf{X}^m$ be $\mathbf{d}^m$ and let the number of bits be $\mathbf{b}^m$. The four blocks $\mathbf{X}_i^{m-1}$ are merged and coded as a single block of size $2^m \times 2^m$ if the following condition is true:

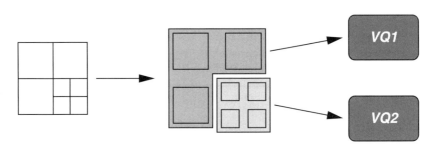

Fig. 22 Vector quantization of the quadtree leaf nodes.

Video Compression 37

$$\mathbf{d}^m + \lambda \mathbf{b}^m \le \sum_i \mathbf{d}_i^{m-1} + \lambda \sum_i \mathbf{b}_i^{m-1} \qquad (37)$$

The above criterion can be used to prune the tree in a bottom-up manner to obtain the **R–D** optimized hierarchical quantization scheme.

5.4. Video Compression Using Quadtree-Based Vector Quantization

The residual signal generated by the motion-compensation algorithm can be compressed using the quadtree vector quantizer (9). The residual signal $\mathbf{r}^0(i, j, t)$ is divided into blocks of size $2^m \times 2^m$. Each of these blocks is encoded using the **R–D** optimized quadtree-based vector quantizers. A **k**-stage hierarchical vector quantizer uses k vector quantizers that work blocks of size $2^n \times 2^n, n = m, \ldots, m - k$. The quadtree bitmap is encoded as shown in the Fig. 23.

5.4.1. Implementation Example and Simulation Results

Implementation of a video compression system using the quadtree VQ is outlined in this section. A simulation framework similar to that used in the RVQ video codec was used to evaluate the performance of the quadtree-VQ-based video compression algorithm. The residual signal after motion compensation was compressed using a three-stage quadtree VQ. The three quantizers used blocks of size 16×16, 8×8, and 4×4. Results are given for an encoder that uses scalar quantizers for blocks of size 16×16 and 8×8. Blocks of size 4×4 are compressed using a vector quantizer trained using the GLA algorithm. The quadtree was segmented using the **R–D** optimized algorithm. The performance was tested using the "salesman" sequence at three very low bit rates, as in the case of the motion-compensated RVQ video codec. Performance of the quadtree-VQ compression algorithm was numerically similar to the RVQ compression algorithm. Figures 18b, c, d show the PSNR results of each reconstructed frame at three different bit rates. Figures 19c, 20c, and 21c show the reconstructed 30th frames for the salesman sequence compressed at the three different bit rates using the quadtree-VQ encoder. It can be seen that the performance of the quadtree-VQ encoder is similar to that of the RVQ encoder at all the bit rates.

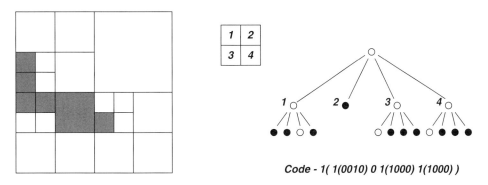

Fig. 23 Encoding the quadtree data structure.

REFERENCES

1. FW Mounts. Bell Syst Tech J. 48:2545–2554, 1969.
2. BG Haskell, FW Mounts, JC Candy. Proc IEEE, 60(7):792–800, 1972.
3. T Murakami, K Asai, E Yamazaki. Electron Lett 7:1005–1006, 1982.
4. M Goldburg, H Sun. IEEE Trans Commun C-B4(7):703–710, 1986.
5. AN Netravali, JD Robbins. Bell Syst Tech J. 58(3):631–670, 1979.
6. FA Kamangar, KR Rao. IEEE Trans Commun C-29(12):1740–1753, 1981.
7. AK Jain. Proc IEEE, 69(3):349–389, 1981.
8. M Kaneko, Y Hatori, A Koike. IEEE J Selected Areas Commun C-5(7):1068–1078, 1987.
9. NM Nasrbadi, S Lin, Y Feng. Optical Engineering 28(7):717–725, 1989.
10. J Vaisey, A Gersho. IEEE Trans Signal Process SP-40(8):2040–2060, 1992.
11. GJ Sullivan, RL Baker. IEEE Trans Image Process IP-3(3):326–331, 1994.
12. L Corte-Real, A Pimenta Alves. IEEE Trans Image Process IP-5(2):263–273, 1996.
13. RL Baker, H-H Shen. A finite-state vector quantizer for low-rate image sequence coding. Proceedings of IEEE International Conference on Acoustics, Speech and Signal Processing, 1987, pp 760–763.
14. W-T Chen, RF Chang, JS Wang. IEEE Trans Circuits Syst Video Technol CSVT-2(1):15–24, 1992.
15. W Li, HQ Cao, SA Ling, SA Segan, H Sun, JP Wus, YQ Zhang. IEEE Trans Circuits Syst Video Technol CSVT-7(1):146–157, 1997.
16. H Kwon, M Venkatraman, NM Nasrabadi. IEEE J Selected Areas Commun 15(9):1714–1725, 1997.
17. F Kossentini, MJT Smith. IEEE Trans Image Process IP-4(5):1349–1357, 1995.
18. SA Rizvi, NM Nasrabadi, W Lin-Cheng. Opt Eng 35(1):187–197, 1996.
19. Y Linde, A Buzo, RM Gray. IEEE Trans Commun C-28(1):84–95, 1980.
20. T Kohonen. Self-Organization and Associative Memory. New York: Springer-Verlag, 1984.
21. NM Nasrabadi, Y Feng. Int Conf Neural Networks I:101–105, 1988.
22. PA Chou, T Lookabaugh, RM Gray. IEEE Trans Acoustics Speech Signal Process ASSP-37(1):31–42, 1989.
23. S Roucos, J Makhoul, H Gish. Proc IEEE 73(11):1551–1587, 1985.
24. BH Juang, AH Gray. IEEE ICASSP, 1:597–600, 1982.
25. CF Barnes, RL Frost. IEEE Trans Inform Theory 39(2):565–580, 1993.
26. WHR Equitz, TM Cover. IEEE Trans Inform Theory 37(2):269–275, 1991.
27. CF Barnes, SA Rizvi, NM Nasrabadi. IEEE Trans Image Process IP-5(2):226–262, 1996.
28. WY Chan, S Gupta, A Gersho. IEEE Trans Commun 40(11):1693–1697, 1992.
29. SA Rizvi, NM Nasrabadi. IEEE J Selected Areas Commun 12(9):1452–1459, 1994.
30. RA King, NM Nasrabadi. Pattern Recogn Lett 1:323–329, 1983.
31. IH Witten, RM Neal, JG Cleary. Commun ACM, 30(6):520–540, 1987.
32. KN Ngan, D Chai, A Millin. IEEE Trans Circuits Systems Video Technol CSVT-6(3):308–312, 1996.
33. Telenor R&D. Tmn (h.263) encoder/decoder, ftp://bonde.nta.no/pub/tmn. TMN codec, May 1996.

3
Adaptive Quantization in Wavelet-Based Image and Video Coding

Jiebo Luo
Eastman Kodak Company, Rochester, New York

Chang Wen Chen
University of Missouri–Columbia, Columbia, Missouri

1. INTRODUCTION

The rapid development of high-performance computing and communication has opened up tremendous opportunities for various computer-based applications with image and video communication capability. However, the data required to represent the image and video signal in digital form continues to overwhelm the capacity of many communication and storage systems. Therefore, a well-designed compression element is often the most important component in such visual communication systems.

Over the years, various frameworks have been proposed to deal with image and video compression at different bit rates. JPEG is a Discrete Cosine Transform (DCT)-based coding standard for still images (1). Video coding standards, such as H.261 (2) and MPEG (3), are also DCT-based coding schemes with block-based motion-estimation and motion-compensation capabilities. However, at low bit rates, such block DCT-based standard coding schemes generally suffer from the visually annoying "blocking effect" originating from the simple but unnatural rectangular block partition. They are also limited by the performance and the complexity of motion estimation and motion compensation in video coding. Therefore, an alternative coding scheme free of the "blocking artifact," without or with less demanding motion-estimation and motion-compensation requirements for video coding, is desired at low bit rates. A multidimensional subband coding scheme has been proposed (4,5), which generally employs no estimation or compensation of interframe motions. Subband or wavelet coding is especially advantageous at low bit rates when the "blocking artifacts" resulting from DCT-based coding or vector quantization have become noticeable and visually objectionable. Two-dimensional subband image coding has been investigated with much success (6). The application of three-dimensional (3-D) subband decomposition to video coding has also recently been attempted with initial success (5,7,8). In addition, the architecture for real-time implementation of 3-D subband video coding has been proposed with comparable computational complexity as the motion-compensated DCT scheme and more complicated data storage and movement procedures (9).

Image and video coding schemes based on subband decomposition exploit the difference in perceptual response so that the compression strategies can be adjusted to each individual subband. Pyramid subband coding is equivalent to wavelet transform coding (10). Wavelet transform coding resembles the human visual system (HVS) in that an image is decomposed into multiscale representations. Moreover, wavelets have good localization properties both in space and frequency domains (11). These two features provide excellent opportunities for incorporating the properties of the HVS and devising appropriate coding strategies to achieve high-performance image and video compression. In general, for a target bit rate, a higher compression ratio in high-frequency subbands, where the distortion becomes less visible, allows the low-frequency subbands to be coded with high fidelity. Although this is not unique to subband schemes, prioritized coding is limited in a DCT-based scheme because of the sole use of frequency representation. Decomposed subbands provide a joint spatial-frequency representation of the signal. Therefore, one can devise a coding scheme to take advantage of both the frequency and spatial characteristics of the subbands. In other words, one can determine the perceptual importance of the subband coefficients based on not only the frequency content but also the spatial content, or scene structures. The combination of a high compression ratio for perceptually insignificant coefficients and high fidelity for perceptually significant coefficients provides a promising alternative to high-quality image and video coding at low bit rates.

For high-frequency subbands, where the correlation has already been reduced by subband decomposition, various scalar and vector quantization schemes have been proposed, including Pulse Code Modulation (PCM, Scalar quantization) (5), finite-state scalar quantization (12), vector quantization (VQ) (13), edge-based vector quantization technique (14), geometric vector quantization (GVQ) based on constrained sparse codebooks (8), a scalar quantization that utilizes a local activity measure in the base band to predict the amplitude range of the pixels in the upper bands (15), and so forth. All these schemes have been proposed to take advantage of the characteristics of the high-frequency subbands in order to increase the coding efficiency.

However, a common problem with many existing quantization methods is that the inherent image structures are severely distorted with coarse quantization. An apparent drawback of the conventional scalar quantization schemes is the inefficiency in approaching the entropy limit. Therefore, image fidelity cannot be properly maintained when the quantization becomes very coarse at low bit rates. Vector quantization, on the other hand, would generally achieve better coding efficiency. In general, VQ is performed by approximating the signal to be coded by a vector from a codebook generated from a set of training images based on minimizing the MSE (mean square error) (13). In the case of GVQ, the structure and sparseness of the high-frequency data is exploited by constraining the number of quantization levels for a given block size. The number of levels and block size determine the bit rate, and the levels and shape adapt for each block (8,16,17). However, in general, the creation of a universal codebook for any image is impossible. The performance of vector quantization applied to a particular image largely depends on a codebook generated in advance and is not adaptive to a given signal. This inability of signal-dependent adaptation will limit the exploitation of the individualized correlation in an arbitrarily given image. Moreover, to form vectors, rectangular block partitioning of images is usually adopted. At low bit rates, such block partitioning often destroys the inherent scene structure of a given image, because the approximation of a given block by a vector from the codebook could alter the position, orientation, and the strength of the structures within the block, such as edge segments. As a result, at low bit rates, vector quantization often produces visible blocking artifacts which severely degrade the image quality. The blocking artifacts become less of an issue when VQ is used to encode the subband images due to the inherent filtering in the subband synthesis. Moreover, the codebook generation and the searching against the codebook in vector quantization are usually computationally expensive. Some suboptimal implementations are often adopted in practice primarily to reduce the computational complexity (18). These problems would adversely affect the coding performance.

The proposed adaptive quantization with spatial constraints is intended to resolve the aforementioned problems. The incorporation of Gibbs random fields as spatial constraints in a clustering process enables the quantization to be both signal adaptive and scene adaptive. Such a quantization constitutes the major distinction of this scheme from the existing ones because it is designed to exploit both the spectral *and* spatial localization *simultaneously*. In this scheme, an adaptive clustering with spatial constraints is applied to the sparse and highly structured high-frequency bands to accomplish the quantization. The incorporation of localized spatial constraints is justified and facilitated by the existence of good spatial localization in the subbands decomposed using wavelets. Upon clustering, the representation of each pixel by its cluster mean is equivalent to a quantization process. However, such quantization enables us to preserve the important scene structures and eliminate most isolated nonprominent impulsive noises which have negligible perceptual significance. The compression ratio of these quantized high-frequency subbands can be greatly increased because the entropy has been reduced due to the smoother spatial distribution of each cluster within these subbands. In addition, the reconstructed images from these quantized high-frequency subbands can also be enhanced in the postprocessing stage using an enhancement algorithm based again on a Gibbs random field so that the reconstruction noise can be suppressed while the image details are well preserved.

We implement the clustering as a Bayesian estimation through optimal modeling of the intensity distributions and efficient enforcement of various spatial constraints in different subbands. We use the terminology of ''scene adaptive'' and ''signal adaptive'' to emphasize two different aspects of our algorithm. First, the signal-adaptive property refers to the modeling and exploitation of the intensity distribution of the coefficients. This is accomplished by using a Laplacian model to model the intensity distribution of each cluster in a Bayesian estimation framework. Second, the scene-adaptive property refers to the modeling and exploitation of the spatial redundancies in a given high-frequency subband. This is accomplished by using Gibbs random fields tuned according to the orientation and the resolution of each subband. We argue that the scene adaptivity and signal adaptivity are generally related to the exploitation of the psychovisual redundancies within the framework of wavelet decomposition.

This chapter is organized as follows. Section 2 briefly describes the subband analysis and synthesis scheme for image and video coding. In particular, we discuss the spatiotemporal decomposition of video signals and the characteristics of each subband and the corresponding coding strategies. Section 3 briefly summarizes the conventional scalar quantization and vector quantization schemes, and their variations to incorporate the HVS. Section 4 introduces the adaptive quantization algorithm and its implementations. In particular, detailed discussions are devoted to optimal Laplacian modeling of the cluster distributions, the effective enforcement of various spatial constraints using GRFs, and an efficient noniterative implementation of the clustering-based quantization. In Sec. 5, we discuss issues beyond quantization, including an enhancement technique for the postprocessing of the reconstructed images from the quantized subbands. Experimental results are presented in Sec. 6. Section 7 concludes with some discussions.

2. SUBBAND SCHEMES FOR IMAGE AND VIDEO CODING

Subband coding was initially developed for speech coding by Crochiere et al. (19) and has since proved to be a powerful technique for both speech and image compression. The extension of the subband coding to multidimensional signal processing was introduced in Ref. 4 and the application to image and video compression has been attempted with much success (6,20,21). In image compression, the subband decomposition is accomplished by passing the image data through a bank of analysis filters. Because the bandwidth of each filtered subband image is reduced, they can be subsampled at its new Nyquist frequency, resulting in a series of reduced-size subband images. These subbands are more tractable than the original signal in that each subband image may be coded separately, transmitted over the communication system, and decoded at the destination. These received subband images are then upsampled to form images of original size and passed through the corresponding bank of synthesis filters, where they are interpolated and added to obtain the reconstructed image.

Three-dimensional subband coding was originally proposed in Ref. 5 as a promising technique for video compression. It has shown comparable performance to other methods, such as transform coding and vector quantization. The video signal is decomposed into temporal- and spatial-frequency subbands using temporal and spatial analysis filterbanks. In the following, the procedure for spatio-temporal decomposition and reconstruction of the video signal using the 3-D subband scheme is described. The two-dimensional scheme for image signal is a special case in that only spatial analysis and synthesis are involved. To recognize the difference between the temporal-frequency response and the spatial-frequency response of the HVS, the filterbanks used for temporal decomposition are often different from those for spatial decompositions. After subband decomposition, each subband would exhibit certain distinct features corresponding to the characteristics of the filterbanks. These features are utilized in the design of compression strategies in order to fully exploit the redundancy in the decomposed subbands.

2.1. Three-Dimensional Subband Spatio-Temporal Decomposition

To minimize the computational burden of the temporal filtering in decomposing the video signal, temporal decomposition is based on the 2-tap Haar filterbank (5,8,17). This also minimizes the number of frames that need to be stored and the delay caused by the analysis and synthesis procedures. The temporal decomposition results in two subbands: the high-pass temporal (HPT) band [i.e., frame difference (FD)] and the low-pass temporal (LPT) band.

In the case of spatial analysis and synthesis, longer-length filters can be applied, as these filters can be operated in parallel and the storage requirements are not affected by the filter length. Therefore, spatial decomposition, both horizontal and vertical, is often based on multitap filterbanks. With separable filters, multidimensional analysis and synthesis can be carried out in stages of directional filtering. To achieve high compression, the lowest-frequency band can be further decomposed in a tree-structure fashion. The high-frequency subbands contain structures approximately aligned along the horizontal, vertical,

Adaptive Quantization

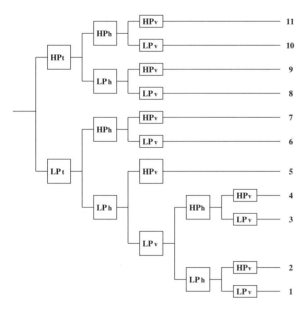

Fig. 1 The 11-band tree-structured decomposition for video signal.

or diagonal direction Fig. 1 shows an 11-band tree-structured decomposition scheme for video signals. The template for displaying the decomposed 11-band subband images is shown in Fig. 2.

In this research, wavelet filterbanks, namely the Daubechies 9/7 biorthogonal wavelets (13), are employed to decompose and reconstruct the signal. The regularity and orthogonality of the wavelet filterbanks ensure the reconstruction of image and video signals with high perceptual quality. Moreover, it has been shown (10,13,22) that the wavelet transform corresponds well to the human psychovisual mechanism because of its localization characteristics in both space and frequency domains. Note that the choice of wavelets also corresponds well to the proposed quantization scheme. First, the good localization of wavelet decomposition in frequency domain offers good frequency separation that

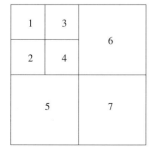 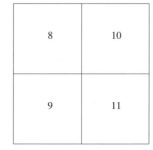

Fig. 2 Template for displaying the 11-band decomposition scheme.

facilitates efficient compression; second, and more importantly, the good localization of wavelet decomposition in the spatial domain justifies and facilitates the incorporation of spatial constraints in the quantization. Appropriate spatial constraints can then be efficiently enforced to identify and preserve perceptually important components in the process of quantization.

2.2. Characteristics of Subbands and Corresponding Coding Strategy

After the spatio-temporal decomposition, the resultant subbands exhibit quite different characteristics from one to another and have different perceptual responses. The quantization strategies need to be designed to suit individualized subbands to achieve optimal representation with minimum visible distortions. For the 11-band decomposition of the video signal, the characteristics of each band are summarized in the following. For the decomposition of 2 D still images, similar characteristics exist.

- Band 1 is a low-resolution representation of the original image and has similar histogram characteristics, but with a much smoother spatial distribution. It can be efficiently coded using differential pulse code modulation (DPCM).
- Bands 2–7 contain spatial high-frequency components of the LPT band. They consist of different amount of "edges" and "impulses," corresponding to different directions and resolution levels. Because the signal power and the perceptual importance, in general, decreases as the resolution level increases, the bit allocation should be adjusted accordingly.
- Band 8 is the low-pass spatial band of the HPT band and contains most motion energy. It needs more bits or finer quantization when the motion activity is high.
- Bands 9–11 represent spatial high-frequency components of the FD. They usually contain very low signal energy and are of low perceptual sensitivity.

The quantization and coding algorithm should be developed based on these characteristics. In general, the strategies are summarized as follows: Subbands at the lower-resolution levels (with smaller index in Fig. 2) contain most signal energy and are of higher visual significance. They require higher-quality coding and hence finer quantization; subbands on the higher levels are quantized coarsely or may be discarded.

3. CONVENTIONAL AND VARIATIONAL SCALAR AND VECTOR QUANTIZATION

In this section, we will briefly review the conventional scalar quantization and vector quantization techniques and some of their variations to accommodate perceptual factors. It will be shown that both conventional scalar and vector quantization schemes have very limited adaptivity with respect to the scene structure and the characteristics of an individual decomposed subband. It is also difficult for VQ schemes to accommodate the properties of the HVS, which is vital in evaluating the reconstructed images, in particular at low bit rates. In other words, these schemes have limitations in exploiting the unique spatial and spectral localities of wavelet decomposition as well as the psychovisual redundancies in the subbands.

3.1. Scalar Quantization

Scalar quantization is straightforwardly applicable to the quantization of decomposed subbands. In many reported subband or wavelet coding schemes, uniform quantization is often adopted. This simplicity in the quantization process leaves the burden of achieving efficient coding to the subsequent symbol coding and entropy coding processes (5,23–26). Nonuniform scalar quantizer can also be designed based on the unique probability density function (p.d.f.) of the subband coefficients using the Lloyd–Max algorithm (18).

The single most used trick when applying uniform quantization to the high-frequency subbands is to allocate a relatively large quantization interval around zero, known as the "dead zone" (23) (shown in Fig. 6). This is used to take advantage of the zero-mean, near-Laplacian p.d.f. of a high-frequency subband and the fact that those coefficients with low magnitudes generally correspond to noise. As a result, the entropy coder can effectively take advantage of the very peaked p.d.f. of the quantized subband coefficients, as shown in Fig. 3.

Simplicity is the greatest advantage of scalar quantization. That is the major reason why it is used even in the well-known EZW (embedded zerotree wavelet) coding algorithm (24). On the other hand, low coding efficiency is its greatest disadvantage. However, if combined with some very elegant symbol coding algorithms and some most sophisticated entropy coding algorithms (adaptive arithmetic coding (27), trellis coding (25,26)), scalar

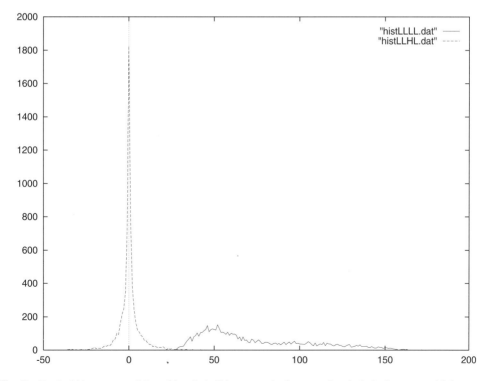

Fig. 3 Typical histograms of the subbands (solid curve—the low-pass band; dashed curve—a high-pass band). The horizontal axis is the intensity axis and the vertical axis is the histogram count axis.

quantization, uniform (24,26) or nonuniform (25,28), is also able to yield a very competitive coding performance.

3.2. Vector Quantization

In general, scalar quantization is inadequate in removing the statistical dependency between coefficients in a local neighborhood. According to Shannon's rate-distortion theory, higher coding efficiency can always be obtained when vectors rather than scalar quantities are coded (18). Such coding gain is dependent on the statistical correlation, described by the joint probability distribution functions, between samples that constitute the vectors. After the wavelet decomposition, the correlation between coefficients, in particular those not belonging to edge structures, is statistically significant. This makes VQ an attractive choice for coding the subband images.

The standard vector quantization, known as the Linde–Buzo–Gray (LBG) algorithm (29), has proven to be a powerful tool for image compression on its own right (30). The principle is to encode a sequence of samples, which is called a vector, rather than encoding each sample individually. Encoding is performed by approximating the sequence to be coded by a vector belonging to a predetermined catalog, usually known as a codebook. The codebook is generated and optimized using the LBG algorithm with a mean square error (MSE) criterion. This process is usually referred to as training in which a classification is performed upon a training set composed of vectors from, preferably, representative, images, and the codebook is updated based on the classification until it converges to a local optimum. Each of the vectors in the codebook is indexed. At the encoder side, only the index of the vector in the codebook that best approximates (in terms of MSE) a given input sample vector needs to be encoded. At the decoder side, the reconstruction is simply a table lookup process, provided that the decoder has the same codebook as the encoder.

The wavelet decomposition naturally lends itself to the generation of a multiresolution codebook for each resolution level and preferential direction. As addressed in Ref. 10, a multiresolution codebook design provides several advantages over a global codebook. First, a global design results in edge smoothing, whereas the multiresolution design preserves edges. Second, multiresolution design facilitates a more efficient search for the best coding vector because only the appropriate subcodebook (resolution and direction) is checked. Consequently, not only is the computation reduced, but the searching is also less prone to local minima. It is also hoped that the spatial and frequency characteristics of the image are accounted for by the wavelet decomposition, and this frees the VQ from using more sophisticated distortion measures such as a HVS-weighted MSE, which would undoubtfully increase the already intensive computation.

3.2.1. VQ for Fixed-Length Coding

The original LBG algorithm (29) assumes fixed-length coding. After the bit allocation, the size of the codebook is determined based on the available bit budget and then fixed. The codebook resulting from the LBG algorithm is unstructured and represents a Voronoi partition of the vector space. The training algorithm is iterative and may converge to a local minimum. The quantization process requires a full search in the codebook and intensive computation.

3.2.2. Entropy-Constrained VQ

Entropy-constrained vector quantization (ECVQ) is designed to minimize the distortion for a fixed-target entropy of the quantization output rather than a fixed number of code vectors (31). Among all known vector quantizers, ECVQ gives the best coding efficiency when followed by entropy coding. In principle (31), a Lagrange multiplier formulation is used to enforce the entropy constraint in the codebook design. The initial reconstruction vectors can use the LBG codebook as if the bit-rate constraint (i.e., the Lagrange multiplier factor λ) equals zero. The number of initial reconstruction vectors must be sufficiently large. Otherwise, the resulting codebook will not achieve the theoretical minimum distortion for a given bit rate. Redundant vectors are automatically eliminated in the iterative training process. The training process for ECVQ is similar to LBG because it also requires a full search. Due to the increased size of the codebook (corresponding to a finer partition of the vector space), the convergence (speed and optimality) of ECVQ can be a serious issue.

3.2.3. Lattice VQ

Lattice vector quantization (LVQ) has been proposed to reduce the computation (32). LVQ is an extension of uniform quantization to the vector space. It has been shown (33) that, given a smooth one-dimensional p.d.f. and a sufficiently fine quantizer, a uniform scalar quantizer when followed by entropy coding approaches the minimum MSE for a given entropy. In multidimensional space, this corresponds to dense lattice quantizers. For example, a hexagonal partition of a 2-D space is called an $A2$-lattice, which is known to yield the lowest MSE among various partitions. The orthogonal lattice in 2-D space is called the Z^2-lattice. Similarly, the $A3$-lattice, Z^4-lattice, and so on can be defined. However, the reconstruction vector of each cell is the centroid of the region rather than the corresponding lattice point itself. By exploiting the structure inherent to a lattice quantizer, the codebook generation and the quantization process can be made very fast. The computational cost for these lattice VQ schemes is typically orders of magnitude less than their unstructured counterparts. The size of a lattice VQ codebook is theoretically infinitely large, but, in practice, the limited range of the input vectors only requires the codebook size to be finite. In general, the lattice codebook size is typically an order of magnitude larger for a given bit rate.

3.3. Comparison of Vector/Scalar Quantization

Senoo and Girod (34) did a comparative study of the performance of vector quantizers for subband images assuming entropy coding of the quantization output. Using MSE as a distortion measure, they concluded that the lattice VQ gave the best practical performance considering the distortion and the computation load needed to arrive at the distortion. In most cases, VQ optimized for variable-length coding (VLC) can significantly improve the coding efficiency over that designed for fixed-length coding (FLC). The authors found it satisfying that the simple lattice quantizers were able to yield near-optimal performance when combined with the wavelet/subband decomposition. Optimal bit allocation can be accomplished using the Lagrange multiplier approach. ECVQ performed the best but was limited by implementation issues (the codebook size) at very high bit rates. The $A2$-lattice quantizer was a close second, especially at very high bit rates. Fixed-word-length VQ (LBG), and the scalar quantizer (uniform) followed with significantly lower performance.

A somewhat different conclusion was drawn by Antonini et al. in an independent study. They found that the PSNR could be improved by about 3 dB using ECVQ instead of the LBG algorithm, but the computation time became prohibitively expensive. The difference in these two studies may be due to the implementation issues. As stated before, the number of initial vectors must be sufficiently large for ECVQ to perform to its capability. One thing that has to be kept in mind is, as in the scalar quantization case, lattice quantizers were developed to yield minimum MSE at *high bit rates* (33).

3.4. Perceptually Tuned Scalar and Vector Quantization

There is one thing in common for the scalar quantization and vector quantization described above. Like all traditional image compression techniques, including DCT and DPCM, they have been designed to exploit the statistical redundancy present in the data representing images, which is essentially visual information. Removing statistical redundancy can only give a limited amount of compression; to achieve higher compression, some of the statistically nonredundant information has to be removed. In doing so, these statistical coding techniques introduce ungraceful visual degradation due to the lack of a mechanism to evaluate the perceptual consequence of the information loss. Image degradation becomes visually annoying when coding errors are produced in visually important parts of the image, such as edges.

Many researchers have recognized that image is more than the numbers used to represent it. A natural evolution started. By using methods of image decomposition that closely mimic the human visual system, image compression can then take into account the importance of each individual coefficient and code it accordingly.

What are the important characteristics of the HVS as an information-processing system? At a low level, objects are composed of structures made up of surfaces of the same color, albedo, or texture bounded by edges. Although the appearance, color, and texture of an object can be greatly affected by the orientation of the surface and the illumination from the light source, edges are usually the most important cues for visual recognition. Therefore, a good image compression method should minimize the edge distortion. In visual psychophysics, it has been discovered (35) that the HVS filters the image into a number of bands, each approximate one octave wide in frequency. An image is considered to be composed of information at a number of scales (35) in the spatial domain. Moreover, spatial localization is another important characteristic because the physical phenomena that give rise to the intensity changes in the image are spatially localized. The octave wavelet decomposition and the resulting multiscale spatial representation correspond well to the first two characteristics. Therefore, the focus in wavelet-based image compression remains on how to exploit the spatial localized structures within the subbands.

Visual psychophysics studies have also revealed that there are a number of factors affecting the frequency and spatial sensitivities of the human eyes to the noises. The error caused by the quantization which represents the image with fewer levels can be considered as a noise source. These factors include the background luminance, the proximity to edges, the frequency band, and texture masking (22).

- Background luminance is related to noise sensitivity by Web's law. The eyes are less sensitive to noise with brighter background luminance. In wavelet-based coding, the baseband coefficients provide such background luminance information.
- Edge proximity relates noise sensitivity to distance from and magnitude of the edge. Because most of the coefficients that we intend to send are assumed to be part of an edge, the spatial locality of the subband octave provides distance information, and the energy of the lower frequency bands indicate edge magni-

tude. The noise sensitivity decreases for increasing edge magnitude and decreasing distance from the edge.
- Frequency-band sensitivity is fixed for each octave, scale, orientation, and luminance–chrominance channel. There are some empirical models based on psychovisual experiments, such as the well-known HVS MTF (modulation transfer function) for modeling perceptual frequency response, the CCIR 451-2 frequency-weighting curve (36), and some other measurements (22,37).
- Texture masking reduces the sensitivity to noise if there is high activity in the locality of the coefficients. The energy of lower-frequency coefficients can indicate the texture activity level.

Therefore, instead of quantizing the subbands based solely on the frequency response background luminance, edge proximity, frequency-band sensitivity, and texture-masking information can be used to construct a perceptual quantization for *each* subband coefficient. Safranek and Johnston (37) combined the above information to determine a perceptual threshold for each subband coefficient. They then used the minimum threshold value to quantize the entire subband. Lewis and Knowles (22) used a similar calculation to estimate the perceptual threshold of each 2×2 block, and then quantized each pixel in the block with this threshold using a linear mid-step quantizer. For example, in Eq. (1), the quantization stepsize is determined as follows:

$$qstep(r, s, x, y) = (q_0)[\text{frequency}(r, s)][\text{luminance}(r, x, y)][\text{texture}(r, x, y)^{0.034}] \quad (1)$$

where q_0 is a normalization constant and r, s, x, and y denotes octave, orientation, and position (horizontal and vertical), respectively. They also used this threshold to normalize the coefficient value before edge detection. After the number of quantization levels have been determined and the coefficients have been quantized, the near-Laplacian distribution of the coefficients is still approximately true. Compression is then achieved using entropy coding. After the quantization, tree structuring and block coding have also been used to exploit interband and intraband correlation in Ref. 22, whereas in Ref. 37, DPCM is used.

4. ADAPTIVE QUANTIZATION OF HIGH-FREQUENCY SUBBANDS

Many attempts in low-bit-rate subband coding have been concentrated in the study of characteristics of the high-frequency subbands so that the features of these subbands can be incorporated in the design of coding algorithms (8,14–16,22). One characteristic of the high-frequency subbands is their less significant perceptual responses. They can often afford coarse representations that would result in fewer bits needed to code the image without introducing much visible distortion in the reconstructed images. Another important characteristic of the high-frequency subbands is the spatial structures in these subbands. These structures appear as sparse ''edges'' and ''impulses'' that correspond mainly to a few strong intensity discontinuities in temporal or spatial domains. In general, strong and clustered ''edges'' and ''impulses'' are of significant visual importance and need to be preserved in the quantization. On the other hand, there are some nonstructural weak impulses corresponding to the noises that have much less visual importance but would need a considerable amount of bits to code. Removal of these noises would lead to significant coding gain with perceptually negligible distortion in the reconstructed image. In addition, these sparse "edges" and "impulses" exhibit well-defined directional arrangement in accordance with the filtering direction in the subband analysis.

To achieve the desired simultaneous scene adaptivity and signal adaptivity, we propose a novel quantization scheme for high-frequency subbands based on the concept of adaptive clustering with spatial constraints. This scheme utilizes Gibbs random fields to enforce neighborhood constraints in order to remove those isolated "impulses" and weak local variations whose contributions to the reconstruction are negligible. The smoothing of the perceptually insignificant pixels is accomplished in a scale-dependent way similar to the perception of the HVS. As visual psychophysics states, the HVS is sensitive to not only the frequency contents but also several spatially localized characteristics, including the background luminance and contrast, the proximity to edges, texture masking, and scale (22,35,37). As will be shown, the entropy of the subband images after the proposed adaptive quantization is reduced without significant perceptual distortions in the reconstructed images. It is the principle of scene-adaptive and signal-adaptive quantization as the result of the exploitation of the HVS, and the spatial and spectral localities of wavelet transform, that constitutes the fundamental difference between this quantization scheme and the existing ones.

4.1. Relationship to Perceptual Grouping

In the perceptual literature, the Gestalt psychologists of the 1920s and 1930s investigated questions of how the human visual system groups together simple visual patterns. More recently in computer vision literature (38), these Gestalt investigations have inspired work in perceptual grouping, an area championed by Lowe (39) and Witkin and Tenenbaum (40). In particular, Lowe (39) defines perceptual grouping as a basic capability of the human visual system to derive relevant grouping and spatial structures from an image *without* prior knowledge of its contents. As will be shown later, the adaptive quantization is able to group together the subband coefficients likely to have come from intrinsic objects in the original scene, without requiring specific object models (41). The quantization depends on the local scene structure and is therefore *scene adaptive*. Upon the completion of such an adaptive clustering and quantization, the high-pass subbands contain mainly refined "edges" or "clumps" over a much cleaned background. Because the "noises" are largely removed and the "edges" are redefined using only a few levels, the images are significantly less busy with greatly reduced entropy.

4.2. Adaptive Quantization Algorithm

The proposed adaptive quantization of high-frequency subbands is accomplished through an adaptive clustering process. In this clustering-based quantization, each pixel is quantized to its cluster mean according to its intensity and its neighborhood constraints modeled by a Gibbs random field. Such a clustering process results in an adaptive quantization in two aspects. First, the quantization is *signal adaptive* because the number of quantization levels needed and the value of these quantization levels are determined according to the statistical characteristics and the perceptual frequency response of each subband. Second, through enforcing spatial constraints, isolated pixels or pixels representing local noisy variations are quantized to the mean of the cluster to which majority of their neighbors belong and therefore are absorbed by the neighborhood. With such a constrained clustering, the spatial distribution of the subband, especially the noisy background, becomes rather smooth. However, the prominent structures and details with significant perceptual importance are preserved mimicking the HVS perception.

We have tailored the clustering algorithm proposed in Refs. 42 and 43 to develop an enhanced adaptive clustering algorithm. It has been shown in Refs. 42 and 44–46 that images can be modeled by a Gibbs random field and image clustering can be accomplished through a maximum a posteriori probability (MAP) estimation. Using Bayes' theorem and the log-likelihood function, the Bayesian estimation that yields the MAP of the clustering x given the image y can be expressed as

$$\hat{x} = \arg\max_{x} p(x|y) \tag{2}$$

$$= \arg\max_{x}\{\log p(y|x) + \log p(x)\}$$

where $p(x)$ is the a priori probability of the clustering x and $p(y|x)$ represents the conditional probability of the image data y given the clustering x. There are two components in the overall probability function. The conditional probability corresponds to the adaptive capability that forces the clustering to be consistent with intensity distribution of the corresponding cluster. The prior probability corresponds to the spatial smoothness constraints which will be characterized by a Gibbs random field (GRF) There are several distinctions between our adaptive quantization algorithm and the GRF-based clustering algorithms in (42,43). First, we have different models for the a priori probability $p(x)$. The GRF (i.e., the parameter β) is adjusted according to the orientation and the resolution of each subband in our algorithm. Second, we have a different model for the conditional probability $p(y|x)$. We use a Laplacian model, as opposed to a Gaussian model, to model the intensity distribution of each cluster. Finally, we develop a noniterative implementation suitable for quantization, where the means are not obtained iteratively, but obtained in advance using a Lloyd–Max quantizer. These aspects of differences will be elaborated in detail in the following subsections.

4.2.1. Modeling of the Spatial Constraints

Gibbs random fields, the practical equivalence of Markov random fields, have been widely used to represent various types of spatial dependency in images (42,44). A Gibbs random field can be characterized by a neighborhood system and a Gibbs potential function. A Gibbs distribution can then be defined as

$$p(x) \propto \exp\left\{-\sum_{c} V_c(x)\right\} \tag{3}$$

where $V_c(x)$ is called the clique potential. Associated with the neighborhood system are cliques and their potentials. A clique c is a set of sites where all elements are neighbors (47). In this study, we consider that a 2-D image is defined on the Cartesian grid and the neighborhood of a pixel consists of its four nearest pixels.

Image clustering constrained by Gibbs random fields is accomplished by assigning labels to each pixel in the given image according to its own intensity value and the properties of its neighbors. A label $x_s = i$ indicates that the pixel s belongs to the ith class of the K classes. According to the essential property of an Markov random field, the conditional probability $p(y|x)$ and thus the clustering depends only on the local neigh-

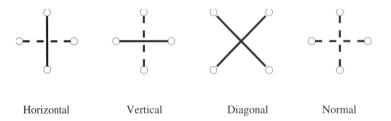

Fig. 4 Cliques for subbands with different preferential directions.

borhood constraints. A two-point clique potential function suitable for clustering can be defined as

$$V_c(i,j) = \begin{cases} -\beta & \text{if } x_i = x_j \text{ and } s, t \in C \\ +\beta & \text{if } x_i \neq x_j \text{ and } s, t \in C \end{cases} \quad (4)$$

Note that the maximization of the overall posterior probability implies the pursuit of the lowest potential state. Therefore, by penalizing inhomogeneous clustering with positive potential (β) and rewarding homogeneous clustering with negative potential ($-\beta$) within local neighborhoods, this potential function can be used to enforce desired spatial constraints to achieve homogeneous clustering if an appropriate neighborhood system c and a proper parameter β are selected.

We have developed four types of clique for the parameterization of the Gibbs random field according to the characteristics of the high-frequency subbands. In Fig. 4 solid lines indicate strong connections with a large β along the nonpreferential directions to enforce strong smoothness constraints; the dashed lines represent weak constraints with a small β along the preferential directions. The preferential direction of a subband is defined as the direction along which the structures are aligned, and it is perpendicular to the filtering direction. Within each subband, the image details along the dashed-line direction can be preserved and the smoothing is done mainly along the nonpreferential direction. The cliques shown in Fig. 4, from left to right, are suitable for the horizontal, vertical, and diagonal high-frequency subbands, and the lowest-frequency subband, respectively. Note that the proposed adaptive quantization may not be suitable for the lowest-frequency subband in the cases where the latter needs to be coded with high fidelity to ensure overall high-quality reconstruction. In this band of the lowest resolution, each pixel corresponds to manifold pixels in the original image and the small quantization error will be magnified in the reconstruction. Only when it is necessary to apply the adaptive quantization to the baseband at a very low bit rate, relatively weaker spatial constraints can be enforced using a normal clique as illustrated in Fig. 4. Furthermore, β can be adjusted to the resolution level of each subband to reflect different neighborhood constraints on the grid at different scales. In general, a larger β is used for the subbands on the higher-resolution levels in accordance with the increase in resolution and scale. For example, β can be doubled, each time moving to the next higher resolution level. β can also be related to bit allocation in progressive coding in that larger β is used for the subbands on higher levels to reduce the bitstream when bits run out. Such flexible parameterization of the Gibbs random field allows us to preserve the most significant structures in a given subband under the bit-rate constraints.

4.2.2. Modeling of the Cluster Intensity Distribution

It has been shown that the overall distribution of a high-frequency subband, as shown in Fig. 3, can be optimally modeled by a Laplacian with zero mean. Such modeling yields the best coding performance under optimal bit allocation (10). Within each high-frequency subband, nonzero coefficients are basically clustered into "edges" (i.e., oscillating positive or negative "strips" over the fairly uniform zero background) or appear as isolated "impulses." For a quantization scheme that is scene adaptive, it needs to preserve those critical positive, negative, and zero values which are of perpetual significance in the reconstruction. PCM was first introduced to quantize these subbands and a "dead zone" technique (23) was proposed to suppress visually insignificant noise around zero by setting a relatively larger quantization interval around zero. This technique allows finer quantization of the tails of the Laplacian distribution because the pixels of larger amplitude are often of greater visual importance (5,8,23). However, the noise suppression using this technique is limited to smoothing only the noise close to the zero background and leaves noises in the rest range of the intensity distribution unaffected.

There are several possible models for the individual intensity distribution $p(y|x)$ of each cluster x in Eq. (2), including Gaussian, generalized Gaussian, and Laplacian probability density functions (p.d.f.). In the case of clustering (43), the conditional density is typically modeled as a Gaussian process with mean μ and variance σ:

$$p(y|x) \propto \frac{1}{\sqrt{2\pi\sigma}} \exp\left\{-\sum_s \frac{1}{2\sigma^2}(y-\mu)^2\right\} \qquad (5)$$

With a Gaussian model, we can derive the overall probability density as

$$p(x|y) \propto \sum_s \left\{\ln \frac{1}{\sqrt{2\pi\sigma}} - \frac{1}{2\sigma^2}(y-\mu)^2\right\} - \sum_c V_c(x). \qquad (6)$$

However, for a clustering-based quantization, such an assumption would not lead to an optimal modeling. It is natural to model the individual cluster conditional density as a Laplacian process considering that the overall intensity distribution can be optimally approximated by a Laplacian source. For a given cluster, if we assume

$$p(y|x) \propto \frac{1}{\sqrt{2\sigma}} \exp\left\{-\sum_s \frac{\sqrt{2}}{\sigma}|y-\mu|\right\}, \qquad (7)$$

then the overall probability density becomes

$$p(x|y) \propto \sum_s \left\{\ln \frac{1}{\sqrt{2\sigma}} - \frac{\sqrt{2}}{\sigma}|y-\mu|\right\} - \sum_c V_c(x). \qquad (8)$$

To examine the validity of the modeling of the cluster distribution $p(y|x)$, we construct the overall distribution based on cluster distributions such that the actual distribution of the coefficients in a given subband is modeled as the superposition of individual cluster distributions whose statistical parameters are obtained from the optimal clustering. As clearly shown in Fig. 5, superposition of multiple Gaussian distributions is unable to yield a satisfactory approximation to the overall histogram. Not only can individual Gaussian modes be identified, but the characteristics of the distribution (e.g., first-order and second-order derivatives of the distribution) are also quite different. This is due to the fact that the

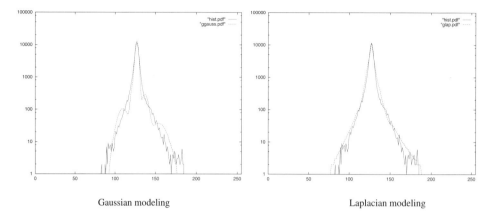

Fig. 5 Modeling of the intensity distribution in high-frequency subbands (blowup in log scale). The horizontal axis is the intensity axis (shifted by 127); the vertical axis is the histogram count axis.

exponent term in a Gaussian distribution is quadratic, whereas in a Laplacian distribution, it is essentially linear. On the other hand, the composite distribution of multiple Laplacian distributions is very consistent with the overall Laplacian distribution, especially in the tail parts where perceptually important information usually resides. Goodness-of-fit tests can also show the superiority of the multimodal Laplacian modeling of the cluster distribution with less fitting error (48). In the case of clustering-based adaptive quantization, as a result of the optimal modeling of the cluster distribution, we are able to obtain optimal quantization and therefore achieve optimal reconstruction from the quantized high-frequency subbands. Comparison of the quantized subbands using a different modeling is given in Fig. 13.

Note that the reconstruction levels are global for the entire subband and therefore only labels need to be coded and transmitted. However, this quantization scheme is indeed *adaptive* for the following reasons. The local neighborhood of each pixel site s changes from one location to another; therefore, two coefficients with the same intensity value are not necessarily quantized to the same level. Depending on the quantization (or clustering) of the local neighboring coefficients, a coefficient is quantized according to a local Bayesian estimation based on (a) its own intensity value, (b) its neighboring coefficients, and (c) the orientation and the resolution of the subband it belongs to, as if spatially adaptive ''local quantization tables'' were utilized. Virtually, there exists a ''local quantization table'' according to the local spatial configuration of each site. Therefore, using just one set of cluster means, we are able to achieve a spatially adaptive quantization under the framework of Bayesian estimation.

With the GRF-based spatial constraints, how a pixel is quantized is not only determined by its intensity but also by its neighborhood spatial constraints. It is noteworthy that this adaptive quantization scheme cannot be achieved by the combination of scalar quantization and noise filtering, such as median filtering. Seemingly, median filtering can be used to remove impulsive noise while preserving edges. However, median filtering is appropriate for normal images containing regions. It is the existence of regions that generates the necessary majority votes so that edges of the region can be preserved. The subband images are essentially composed of thin ''edges'' and isolated ''impulses,'' with literally *no* regions, over the zero background. Although median filtering can remove "impulses," it would also remove those thin and long structures, such as meandering edge segments. A prominent spike of large amplitude would also be removed by median filtering, but it can be preserved by the adaptive quantization because the first energy term in Eq. (8) would be large enough so that it is not absorbed by the neighborhood. Furthermore, the spatial information within a median filtering window is not preserved in the median filtering. The reason being, that median filters (and other order-statistics filters) seek to obtain only one good representative among the N neighboring pixels and this median can be any of the N values. Therefore, the spatial localization of the thin edges can be altered, although within a local window, during the median filtering process.

The incorporation of Gibbs random fields in the MAP estimation allows us to achieve a similar but better effect of "dead zone" which was originally proposed in Ref. 23. Unlike the original approach which generates a "dead zone" simply by intensity thresholding, we achieved an improved "dead zone" which suppresses noises according to both the intensity and local spatial constraints. Moreover, the adaptive quantization is capable of suppressing noise in the entire range of the intensity distribution, instead of being limited to the zone around zero. As illustrated in Fig. 6, without spatial constraint the partition of clusters is such that the zones of clusters are separated. With the incorporation of the spatial constraints, the zones of clusters are actually overlapped with each other. This overlapped partition allows us to achieve an overlapped quantization, which is fundamentally different from all existing quantization schemes. Therefore, it enables the suppression of noises in the entire range of the distribution. The actual quantization intervals are essentially enlarged, not just for the central zone around zero but for all the quantization intervals.

4.2.3. Implementations

The original adaptive K-mean clustering based on Eq. (8) can be implemented using a local optimization technique called the *iterative conditional mode* (ICM) (49). The ICM is efficient to enforce local spatial constraints (44). At first, an initial clustering x is obtained through the simple K-mean algorithm. In this study, an odd K is chosen for the total number of levels for each subband, as the histograms of the high-frequency subbands are approximately symmetric around zero. K can be assigned according to the perceptual importance of each subband (i.e., the characteristics of the HVS (22)) and the principle of optimal bit allocation. The subbands of lower resolution often have larger dynamic

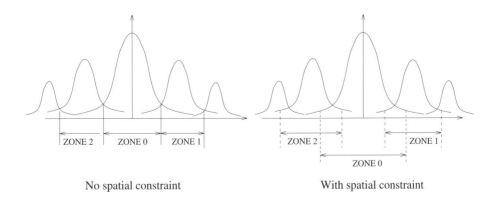

Fig. 6 Dead zone effect.

range. Therefore, they are assigned more levels because they carry more perceptually important information. Then, the overall probability function is maximized in a site-by-site fashion, with the mean μ and the variance σ of each cluster being updated after each iteration. The optimization is accomplished through alternating between the MAP estimation of the clustering and the iterative update of the cluster means and variances. Such a alternating process is repeated until no pixel change class. The result is the adaptive clustering of the given high-frequency subband. Finally, the quantized subband is obtained by replacing each pixel with its cluster mean.

There are still some problems with the ICM implementation of this clustering-based adaptive quantization. First, the intensity distribution and the spatial constraints are coupled in an iterative process in the ICM process. Even a small parameter β can impose very strong constraints of the Gibbs random field over large distances through clique interactions in successive iterative processes. Therefore, some edge-enhancing effect can occur, which is not desired in the case of quantization if image fidelity is the concern. Second, the iterative implementation is still considered time-consuming, although the ICM is one of the computationally least expensive optimization techniques (42). In the case of video communication where large amount of subbands are generated in the spatio-temporal decomposition, it cannot afford an expensive computation because real-time processing is often required.

For the clustering-based adaptive quantization, we developed a two-step noniterative implementation. At first, a Lloyd–Max scalar quantizer is found whose optimal reconstruction levels are used as the means of clusters. MAP estimation of the clustering is then accomplished in virtually one iteration because the cluster means and variances have been predetermined. The spatial constraints are only used to eliminate those nonprominent impulsive pixels while preserving the important structures. In our experiments, the cluster means (i.e., the reconstruction levels in quantization) obtained using iterative implementation turned out to be very close to those obtained using a Lloyd–Max quantizer. This observation is not surprising because both implementations optimize similar objective functions. However, the noniterative implementation not only is computationally efficient but, more importantly, produces better reconstruction results because the local spatial constraints are more appropriately enforced. Meanwhile, the saving in computation is on the order of 10-fold.

5. BEYOND QUANTIZATION

5.1. Coding of the Quantized High-Frequency Subbands

Coding of an image generally includes two distinct operations: quantization and symbol coding. The adaptive quantization with spatial constraints is capable of removing the "noises" of low perceptual significance, which would otherwise need considerable bits to code. The quantized high-frequency subbands are then coded by a symbol coder, which generally includes an entropy coder. With the reduction of entropy upon the adaptive quantization, a lower bit rate is expected from the entropy coding. The entropy coder consists of a variable-word-length coder to code the labels of the nonzero values of the clustered subbands and a run-length coder to code their corresponding locations (20). Different scanning schemes can be used for the individual subband to increase the run length, as these clustered high-frequency subbands are composed of well-defined "edges" whose directions correspond to the direction of the high-pass filtering used to obtain the decomposition. Because of the smoother background in the quantized subbands, a Hilbert–Peano scan (50) can also be very effective. Another scheme for increasing the run length is to partition the subbands into nonoverlapping blocks (23). Through such partitioning, the local area of zero values can be better exploited to improve the run-length coding efficiency.

In our experiments, we will use the directional scan schemes followed by a run-length coding. The horizontal and vertical subbands are scanned accordingly. We also use the horizontal scan for diagonal subbands for simplicity, as we found that the gain margin is rather small using a diagonal zigzag scan. Recently, zerotree-based coding algorithms have achieved great success in wavelet-based coding (24,28) due to the efficient symbol coding techniques which exploits the intrinsic parent–descendent dependencies in the wavelet decomposition. We will adopt the zerotree coding technique proposed in Ref. 24 to code the quantization level indices in the experiments where more levels of wavelet decomposition are selected. Upon adaptive quantization, the subband contains very few ''clustered edges,'' which consist of nonzero coefficients, over a very clean zero background. Therefore, the zerotree coding can be very efficient.

5.2. Enhancement Algorithm

There are some artifacts in the reconstructed image due to the quantization in different frequency subbands. These artifacts generally appear as a ringing effect around sharp edges, loss of fine details, and blotchiness in the slowly varying regions. Although the loss of fine details is difficult to recover, the other typical artifacts in wavelet-based coding are not as visually annoying as the blocking effect, and some of them can be removed or reduced. A Gibbs random field is again applicable as a spatial constraint to remove these artifacts and enhance the reconstructed image. The enhancement is also formed as an MAP estimation.

The conditional probability of the quantization y given the original data x can be written as

$$p(y|x) = \begin{cases} 1, & y = \mathbf{Q}[\mathbf{W}(x)] \\ 0, & y \neq \mathbf{Q}[\mathbf{W}(x)] \end{cases} \tag{9}$$

where $\mathbf{Q}$ stands for the quantization and $\mathbf{W}$ denotes the wavelet transform. The conditional probability states that the estimated image should conform to the quantized data. This constraint can be enforced by projecting the estimated image back to the transform domain (i.e., decomposing the image in the same way as before) and adjusting the pixels so that same quantized subband image is maintained.

We use a specific Gibbs random field, the Huber–Markov random field model, to model the a priori probability. Its potential function $V_{c,T}(x)$ is in the form of Eq. (10). The Huber minimax function has been successfully applied to the removal of blocking effect in low-bit-rate transform coding (51,52). It can be written as

$$V_{c,T}(x) = \begin{cases} x^2, & |x| \leq T \\ T^2 + 2T(|x| - T), & |x| > T \end{cases} \tag{10}$$

The desirable property of this minimax function is its ability to smooth the artifacts in the image while still preserving the image detail, such as edges and regions of textures. If we define the gray-level differences between the current pixel $x_{m,n}$ and the pixels within its neighborhood $N_{m,n}$ as

$$\{x_{m,n} - x_{k,l}\}_{k,l \in N_{m,n}}, \quad 1 \leq m, n \leq N \tag{11}$$

then these differences can be used as the argument of Huber minimax function. The quadratic segment of the minimax function imposes least mean square smoothing of the artifacts when the local variation is below T. On the other hand, the linear segment of the Huber minimax function enables the preservation of the image detail by allowing large discontinuities in the image with a much lighter penalty. The overall enhanced image is given by

$$\hat{x} = \arg\min_{x \in \mathcal{X}} \sum_{k,l \in N_{m,n}} V_{c,T}(x_{m,n} - x_{k,l}), \quad 1 \leq m, n \leq N \tag{12}$$

Because the projection to the constraint space $\mathcal{X} = \{x: y = \mathbf{Q}[Hx]\}$ requires a full cycle of subband analysis and synthesis, a suboptimal solution with the least computation would be the unconstrained noniterative estimation of Eq. (12). The Huber–Markov random field model also results in a very low computational complexity. To compute the derivative of the function for performing local gradient–descent in an ICM-like scheme, only linear operations are involved.

6. EXPERIMENT RESULTS

Experimental results have been obtained using the test image "lena" and the test video sequence "salesman." As discussed previously, the temporal filterbank is the 2-tap Haar filterbank. The Daubechies wavelet 9/7 biorthogonal filterbank (10) is selected for the spatial analysis and synthesis. The decomposition, quantization, reconstruction and enhancement of "Lena" and a typical frame of "salesman" sequence are shown in Figs. 7–13. To examine the quantization results, the quantized subbands are displayed with a mid-gray cluster corresponding to the zero value, darker clusters to the negative values, and brighter clusters to the positive values, as shown in Figs. 7 and 11. The spatial distribution of the quantized subband is made much smoother because of the incorporation of spatial constraints. Using the adaptive quantization, we remove those perceptually negligible noisy contents and only preserve those visually important components in the high-frequency subbands (see Fig. 7). To boost the contrast and emphasize the effect of the adaptive quantization for display purposes, histogram equalization has been performed on those subband images. The numerical results on entropy reduction are presented in Tables 1 and 2.

In terms of the modeling of the intensity distribution, multiple Laplacian modeling is able to produce the most coherent quantization. In terms of the implementation, the noniterative quantization enforces the spatial constraints more rigorously than the quantization through the ICM. This may need some explanation. It is clear that the interaction between

Fig. 7 A 4-band decomposition of the "Lena" image. *Left*: original subbands; *right*: quantized high-frequency subbands.

Adaptive Quantization

Fig. 8 Reconstruction of "Lena" using the EZW algorithm. *Left*: the original "official" "Lena" image; *right*: the reconstructed image.

Fig. 9 Reconstruction of "Lena" with the adaptive quantization and the EZW algorithm. *Left*: the reconstructed image; *right*: the enhanced image.

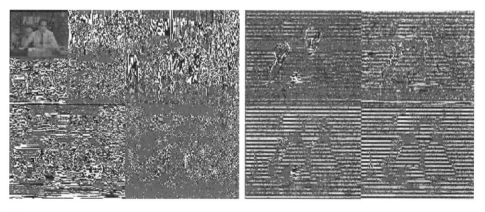

Fig. 10 An 11-band decomposition of the "salesman" sequence. *Left*: low-pass temporal (LPT) bands; *right*: high-pass temporal (HPT) bands.

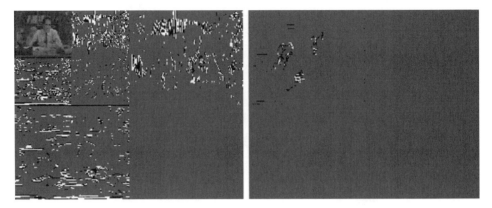

Fig. 11 The quantized subbands of the "salesman" sequence. *Left*: low-pass temporal (LPT) bands; *right*: high-pass temporal (HPT) bands.

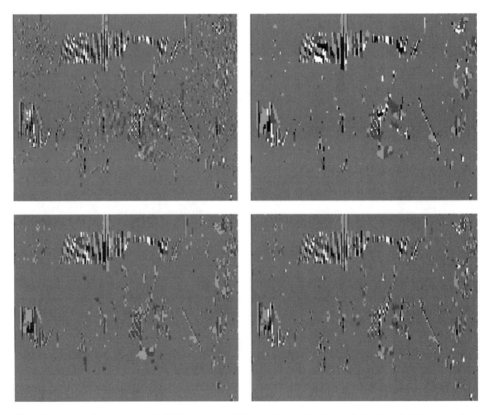

Fig. 12 Quantization of a high-frequency subband (blowup). *Upper-left*: Lloyd–Max quantizer without spatial constraints; *upper-right*: adaptive quantization with Gaussian modeling; *lower-left*: adaptive quantization with Gaussian modeling and ICM; *lower-right*: adaptive quantization with Laplacian modeling and NICM. The quantized subbands of the "salesman" frame.

Adaptive Quantization

Fig. 13 Coding of the "salesman" sequence. *Left*: original frame; *right*: reconstructed frame.

Table 1 PSNR of the Reconstruction and Overall Entropy Reduction in High-Frequency (HF) Subband

Quantization scheme	PSNR (dB)	PSNR after enhancement (dB)	Average HF entropy	Average HF entropy after quantization
"lena," Gaussian modeling, ICM	35.53	35.57	3.46	0.320
"lena," Gaussian modeling, NICM	35.54	35.62	3.46	0.318
"lena," Laplacian modeling, NICM	36.29	36.32	3.46	0.316
"salesman," Gaussian modeling, ICM	32.01	32.16	2.58	0.132
"salesman," Gaussian modeling, NICM	31.92	32.07	2.58	0.131
"salesman," Laplacian modeling, NICM	32.97	33.15	2.58	0.129

Table 2 Entropy Reduction After Quantization for the "Salesman" Sequence

Subbands		Before quantization	After quantization	Quantization levels
LPT	LLLL	6.66	2.85	Stepsize $\Delta = 8$
	LLHL	3.98	0.70	7
	LLLH	4.18	0.66	7
	LLHH	2.83	0.16	5
	HL	3.65	0.29	3
	LH	3.61	0.27	3
HPT	LL	1.70	0.07	3

the spatial constraints and the image force is desired for a image segmentation problem. However, in the adaptive quantization, we consider that both the elimination of nonstructural coefficients and the preservation of original image scene structures are important. We found that the adaptive quantization through the ICM sometimes altered the original image structure, or created some nonexistent structures. The reason is that the high-frequency subbands contains mostly thin *edgelike structures* rather than *regions* contained in normal images. Overall, the noniterative implementation (NICM) with Laplacian modeling outperforms the other two combinations. This is clear from Fig. 12, where the adaptive quantization eliminates the noises without altering those important scene structures. Such a combination also yields the highest PSNR, as is shown in Table 1. In Table 2, the entropy of the "salesman" subband images before and after the quantization shows significant entropy reduction in the high-frequency bands. Because we use directional scan and run-length coding, the entropy of these high-frequency subbands is calculated using the first-order entropy of appropriate direction instead of the zero-order entropy, which is based solely on the histogram. In general, the first-order entropy is smaller than the zero-order entropy. Those insignificant subbands are discarded and therefore are not listed in this table. The entropy of the lowest-frequency subband is obtained using DPCM and is included in the overall entropy.

To demonstrate the effectiveness of the proposed adaptive quantization, Shapiro's state-of-the-art embedded zerotree wavelet (EZW) coding algorithm (24) is adopted. Compression results are obtained using six-level wavelet decomposition for the following cases: (a) using the original EZW algorithm and (b) cascading the adaptive quantization with the EZW algorithm. The comparison is done using the 512×512 "official" "Lena" image (24). Higher peak signal to noise ratio (PSNR) and better visual quality is obtained at a low bit rate of 0.25 bits per pixel (bpp) by combining the proposed quantization with the zerotree coding. Noticeably, the rim of the hat, the shoulder, and the face are reproduced much better in Fig. 9 than in Fig. 8. The reason being that the available bits are concentrated on the scene structures in high-frequency subbands that correspond to these important edges in the original image. If the enhancement technique is applied, visually aesthetic reconstruction can be produced with slight PSNR improvements. The remaining minor ringing artifacts and blotchiness are completely removed while the image details are preserved. The PSNR of the reconstructed image in Fig. 9 obtained using the proposed adaptive quantization is 33.52 dB, compared to 33.17 dB by the EZW reported in Fig. 8. The final PSNR after the enhancement is slightly higher at 33.91 dB because the improvements occur at only a small portion of the image pixels, such as around sharp edges. However, corresponding visual improvements are of significant importance.

For the "salesman" sequence, we achieved the 40:1 compression required for videoconferencing. The compression ratio of 40:1 for Common Image Format (CIF) sequence means the luminance signal is coded at 304 kbps, which leaves 64 kbps for the chrominance signal and 16 kbps for the audio in a 384-kbps video conferencing application, similar to the scheme adopted in Refs. 8 and 17. The PSNR of our results is 33.97 dB and is lower than H.261. However, the perceptual quality of our coded video is better. Note that we only use the 2-tap Haar filter for temporal decomposition, and only decompose the LPT band into two levels (therefore we cannot take advantage of the efficient zerotree coding). Nevertheless, the good visual quality of these reconstructed images suggests that the proposed quantization approach is very promising in image and video compression because it is capable of preserving those visually significant components at low bit rates through its signal-adaptive and scene-adaptive quantization. Recently, Pearlman's group reported a 3-D subband video coding scheme using improved zerotree coding (53), which is able to match the PSNR performance of the motion-compensation-based schemes, such as H.261 and H.263.

Adaptive Quantization

Experiments have also been conducted for medical images. The magnetic resonance (MR) images are 256 × 256 slices from a volume of a 3-D brain scan, provided by the Department of Neurology, University of Rochester Medical Center. The ultrasound images are 564 × 412 frames from a sequence of an abdominal scan, provided by the Department of Radiology, University of Rochester Medical Center. For the ultrasound images, the two-level decomposition is applied. For the MR images, the three-level decomposition is applied. In the ultrasound images, the horizontal stripes and the appearance of the tissue are critical for diagnostic purpose. In the MR image, the boundaries between different tissues, such as those between the gray matter and the white matter, are extremely important. These structures need to be preserved even at relatively low bit rates. For comparison, JPEG-compressed images using the coder developed by the Independent JPEG Group are also obtained. At the same bit rate, the JPEG-compressed MR image (see Fig. 14c) is apparently useless for diagnostics. To achieve a comparable quality of the compressed

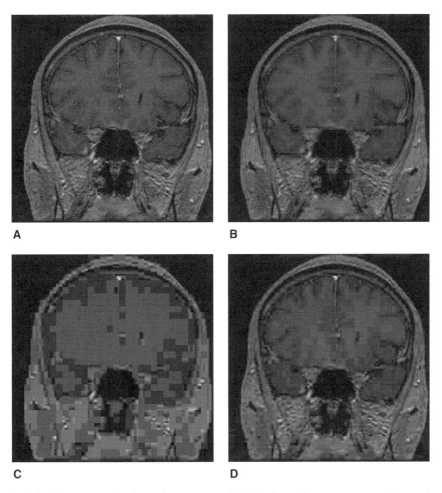

Fig. 14 Compression of MRI images: (A) original, (B) wavelet compression with adaptive quantization (0.194 bpp, PSNR = 27.41 dB); (C) JPEG with the same bit rate (PSNR = 23.25 dB); (D) JPEG with same PSNR (0.35 bpp).

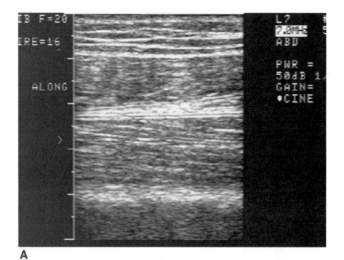

A

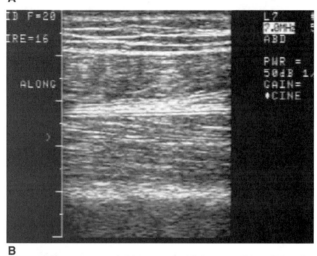

B

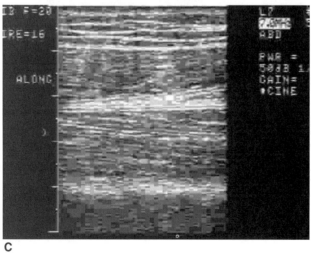

C

image (see Fig. 14b) obtained using wavelet-based coding with the adaptive quantization (note that zerotree coding is not applicable in this case), the bit rate for Fig. 14d is increased by more than 75%. Still, the blocking artifacts are visible enough to cause inaccuracy in diagnosis. The same is true for the ultrasound image. At low bit rates, the wavelet coder is able to preserve clinically useful anatomical information significantly better than the standard JPEG coder, as shown by Figs. 15b and 14b.

7. DISCUSSIONS AND CONCLUSIONS

It is well known (35) that the HVS tends to be attentive to the major structured discontinuities within an image, rather than intensity changes of individual pixels. Therefore, a desired property for a quantization scheme is the capability of the high-fidelity representation of major scene structures. Unlike the DCT-based schemes in which spatial information is lost after the transform, the wavelet transform preserves both spatial and frequency information in the decomposed subbands.

7.1. Beyond Simple Statistical Models

Because the nature of image scene structures is nonstationary and varies for each individual image, a simple statistical model, as adopted by many existing quantization schemes, is often inadequate for individual scene representation. The combination of a scene structure model and a conventional statistical model will be more appropriate to characterize both the random and deterministic scene distributions within an image. Because scene structures of objects can often be represented by edges, a primitive candidates for the scene structure description will be the location, strength, and orientation of edges. In wavelet coding, such edge information is already available in the high-frequency subbands. The issue is how to combine such information with statistical models to achieve a scene-adaptive and signal-adaptive quantization.

7.2. Binding Scene Structures Using a Dynamic Neighborhood System

The proposed quantization scheme has provided us an effective way of distinguishing perceptually more important structures from less important ones. Within the high-frequency subbands, the strong and clustered edges correspond to important scene structures and are retained, whereas the weak and isolated impulses correspond to perceptually negligible components and are discarded. To identify these clustered edges, neighborhood coefficients need to be bound together to determine the presence of scene structures. The binding of scene structures is accomplished by the introduction of naturally defined Gibbs neighborhood systems in the proposed adaptive quantization, whereas in vector quantization, it is accomplished by artificial block partition, which is often inconsistent with the

Fig. 15 Compression of ultrasound images: (A) original; (B) wavelet compression with adaptive quantization (0.217 bpp, PSNR = 23.50 dB); and (C) JPEG (0.215 bpp, PSNR = 22.0 dB).

natural boundaries of objects. It is noteworthy that a Gibbs neighborhood system is of a dynamic nature because the neighbors of each individual coefficient are different from one location to another. Such a dynamic, individualized neighborhood system is consistent with the natural representation of spatial dependencies and is, therefore, able to overcome the potential scene distortions caused by any artificial partitioning.

In summary, this novel *scene*-adaptive and *signal*-adaptive quantization scheme is able to resolve the common problems with some existing quantization methods designed for wavelet-based compression. The novelty of the proposed quantization lies in the way we exploit both the spatial and frequency redundancies in the subbands, which are generally related to the psychovisual redundancy of the HVS. The principle of the scene-adaptive and signal-adaptive quantization is fundamentally different from existing scalar or vector quantization schemes in that we combine both a scene structure model with a conventional statistical model. This quantization scheme has the individuality of scalar quantization in that each coefficient is inspected with regard to its perceptual importance, but in a more efficient way. It also exploits the local spatial correlation as vector quantization, but in an essentially different way such that it is able to preserve inherent image structures even at low bit rates. Both algorithmic analysis and experimental results have shown that the proposed adaptive quantization provides a promising way of achieving efficient image and video compression at low bit rates. In addition, such adaptive quantization has many impacts on the subsequent coding and transmission in such aspects as coding efficiency, coding artifacts reduction, transmission loss concealment, and transmission noise reduction which are currently under investigation (54).

7.3. Injecting Vision into Coding

The ultimate judgment of the image and video compression is done by the viewer, the human being. Traditionally, image and video coding has been treated as a data compression problem and the coding approaches have been developed based on information theory and statistical signal processing. Recently, a trend toward vision-based coding has started. Some initial successes have been reported (22,55,56) and this trend will play very important roles in the so-called ''second-generation'' or even ''third-generation'' coding. However, the vision-based coding cannot come from nowhere. Rather, it has added a new dimension to the well-established traditional coding. In the following, we attempt to give an overview of the image and video coding from the perspective of vision and point out the coordinates of the adaptive quantization within such a universe.

In Fig. 16, this research can be regarded as working jointly at both the sub-mid-level and sub-high-level of the vision-based coding. Traditional image coding schemes, represented by DPCM, DCT, and VQ, treat image and video signals as realizations of certain random processes and take advantage of the statistical redundancies in signal samples to achieve data compression. These schemes utilize low-level pixel-based vision features and are considered low-level coding. At the other end, high-level coding approaches, represented by model-based coding (57), describe the image in such terms of object and action, and aim to achieve a very high coding efficiency through high-level description and image synthesis. In between these two ends is the mid-level coding. The proposed adaptive quantization exploits such visual features as clusters, edges and, structures, whereas the face-location-based prioritized quantization exploits regions and semantics to some degree. High-level coding is promising, yet very difficult. By investing the efforts at two sublevels that are of higher potential and more tangible at the same time,

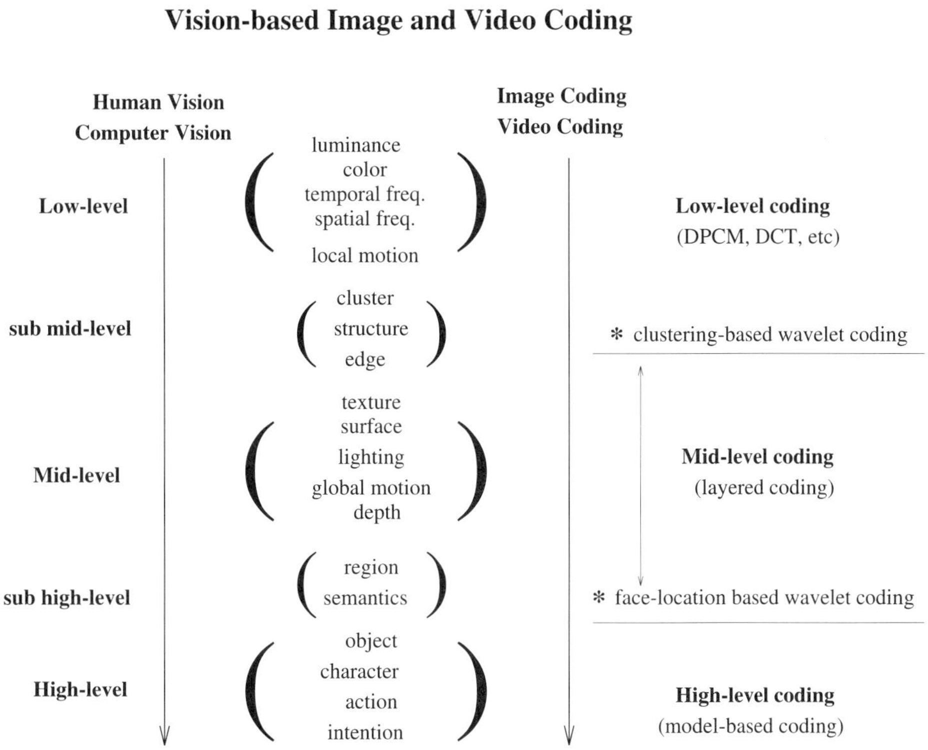

Vision-based Image and Video Coding

Human Vision / Computer Vision		Image Coding / Video Coding
Low-level	(luminance, color, temporal freq., spatial freq., local motion)	Low-level coding (DPCM, DCT, etc)
sub mid-level	(cluster, structure, edge)	* clustering-based wavelet coding
Mid-level	(texture, surface, lighting, global motion, depth)	Mid-level coding (layered coding)
sub high-level	(region, semantics)	* face-location based wavelet coding
High-level	(object, character, action, intention)	High-level coding (model-based coding)

Fig. 16 The framework of the vision-based image and video coding and the coordinates of the adaptive quantization.

this research is expected to achieve high performance in image and video coding at low bit rates. Recently, some researchers have seen the complementarity of waveform-based lower-level schemes and model-based high-level schemes (58). On the one hand, model-based coding can be employed to improve the performance of waveform-based coding schemes by efficiently representing model compliance objects. On the other hand, possibly as important, if not more, waveform-based coding can be employed to benefit semantic-based coding by reducing image analysis complexity and image synthesis complexity and handling unmodeled objects. How to efficiently combine these two different but complementary coding schemes, or, more generally, how to efficiently combine low-level and high-level vision-based coding, is surely worth further investigation.

7.3.1. Incorporation of Human Vision: Sensitivities of the HVS

Psychovisual redundancy is an important dimension to explore in image and video coding. Ideally, we can use our knowledge of the HVS to determine the least amount of information required in terms of human perception (55). For example, the error caused by the quantization which represents the image with fewer levels can be considered as a noise source. However, visual psychophysics discovers that there are a number of factors affecting the frequency and spatial sensitivities of the human eye to the noises. As discussed previously, these factors include the background luminance, the proximity to edges, the frequency band, and texture masking (22).

Therefore, instead of quantizing the subbands based solely on the frequency response, these perceptual factors can be used to construct a perceptual quantization for *each* subband coefficient because wavelet transform offers good spatial and spectral localities (22,37). In the future, such psychovisual factors can be accounted for, along with Gibbs random fields, in the adaptive quantization scheme to improve the coding performance.

7.3.2. Incorporation of Computer Vision: ROI Versus Perceptual Grouping

Here, we make a distinction between human vision issues and computer vision issues for image compression purpose. The human vision issues discussed above refer to low-level, subconscious, innate aspects of the HVS, which can be described by those measures and used in the quantization design. The high-level recognition-oriented aspects of the HVS are referred to as the computer vision issues which can only be facilitated by computer vision techniques. For example, a new face-location technique based on the extraction of the facial contour is proposed in Ref. 59. It has been combined with the wavelet-based image and video compression scheme for improved videoconferencing (59). An approximate face region mask is generated and used for a prioritized quantization of the decomposed subbands. Through the incorporation of computer vision techniques, coarser quantization of the background enables the face region to be quantized finer and coded with higher quality at the same bit rate. In particular, the adaptive quantization can be applied in such a matter that the face region is assigned more quantization levels and weaker spatial homogeneity constraints. Experimental results (59) have shown that this approach is promising in videoconferencing applications because the perceptual image quality of the face region can be greatly improved. Although a contour-based face-location technique is used in Ref. 59, motion information can be incorporated for robust location and tracking of the face region in a more complex background using full 3-D information. Semantic information, such as the detection of the mouth and the eyes, shall also be incorporated to aid the face detection (60).

We believe that computer vision techniques can be combined with compression schemes to improve the performance of the overall coding system. Although face location can be employed to improve the visual quality of the video in videoconferencing, such a prioritized compression scheme can be readily extended to the location and tracking of any region of interest (ROI) or object of interest in the video scene. For example, the product that the ''salesman'' is demonstrating should be of particular interest in that setting. Therefore, in addition to the face, the product in his hand can also be located, tracked, and coded with high fidelity. A completely automatic robust object-location technique for arbitrary complex scenes may be very difficult to design. However, it is possible to control the image acquisition condition to simplify the task of computer vision or initialize the location process with some human–machine interaction.

In the case when such *explicit* segmentation of ROI is unattainable due to technical difficulties, or undefinable for all possible viewing purposes (different viewers, or the same viewer given different visual inspection tasks), we can devise and rely on some mechanisms for detecting *generic* salient features in the image (i.e., features that generally catch the viewer's attention) (61). Such perceptual grouping will then direct the coding strategy to concentrate the available bit budget on the selected regions to maximize the perceived quality of the reconstructed image. The scene-adaptive aspect of the quantization

Adaptive Quantization

is aimed to address such a generic perceptual grouping without resorting to higher-level image understanding. Of course, motion is a very important cue, in addition to spatial features, for determining saliency and attention in video coding. One way of capturing such temporal saliency is described in the 3-D spatio-temporal subband decomposition scheme. Through adaptive quantization in the DFD image, the temporal saliency can be located and coded without resorting to explicit motion estimation or object tracking.

7.4. Spectrum of the Redundancies in Wavelet-Based Coding

To conclude this chapter, quantization issues in wavelet-based compression are reviewed from the perspective of the coding redundancies and the corresponding exploiting strategies. In Table 3, the spectrum of the coding redundancies is categorized under the intraband redundancy, the cross-band redundancy, the psychovisual redundancy, and the symbol coding redundancy. The specifications of these redundancies and the corresponding coding methods are also given to the best of the current knowledge.

The goal is to fully exploit the coding redundancies in order to achieve high-performance coding. As can be seen from Table 3, the adaptive quantization is capable of exploiting, to a large extent, the redundancies with a wavelet-based coding scheme via the MAP formulation and properly tuned Gibbs random fields. Although the adaptive quantization is separate from the subsequent cross-band coding and symbol coding, it is clearly amenable to the best known such coding techniques. In our opinion, this adaptive quantization represents an effort toward the development of a low-bit-rate compression

Table 3 The Spectrum of Coding Redundancies and the Corresponding Coding Strategies Devised to Exploit Such Redundancies (SASA-Q—Signal-Adaptive and Scene-Adaptive Quantization)

Redundancy	Description	Strategy
Intraband redundancy	Directional correlation	• Directional scan • Hilbert–Peano scan • Run-length coding • SASA-Q (directionally tuned GRF)
	Near-Laplacian spectral distribution	• Nonuniform scalar quantization • SASA-Q (MAP)
	"Clumps" + "speckles" spatial distribution	• SASA-Q (MAP) • GVQ
Cross-band redundancy	Tree-structured insignificance map	• Zerotree coding
Psychovisual redundancy	Spatial sensitivity spatial selectivity	• HVS spatial-feature-weighted quantization • SASA-Q (MAP)
	Frequency sentitivity	• HVS MTF-weighted quantization • SASA-Q (quantization level allocation)
Symbol coding redundancy	First-order and higher-order statistics	• Adaptive arithmetic coding • Trellis coding • Scalar–vector quantization

algorithm through the exploitation of the coding redundancies in terms of both the *information theory* and the *human visual system*.

ACKNOWLEDGMENTS

This work is supported by NSF Grant EEC-92-09615 and a New York State Science and Technology Foundation Grant to the Center for Electronic Imaging Systems at the University of Rochester.

REFERENCES

1. WB Pennebaker, JL Mitchell. JPEG Still Image Data Compression Standard. New York: Van Nostrand Reinhold, 1993.
2. M Liou. Commun ACM 34:59–63, April 1991.
3. DL Gall. Commun ACM 34:46–58, April 1991.
4. M Vetterli. Signal Process 6:97–112, 1984.
5. G Karlsson, M Vetterli. Three dimensional sub-band coding of video. Proceedings International Conference Acoustics, Speech and Signal Processing, 1988, pp 1110–1113.
6. JW Woods, SD O'Nell. IEEE Trans Acoustics Speech Signal Process ASSP-34:1278–1288, 1986.
7. RH Bamberger. New subband decomp coders for image and video compression. Proceedings International Conference Acoustics, Speech and Signal Processing, San Francisco, CA, pp III-217–III-220, 1992.
8. C Podilchuk, A Jacquin. In: Human Vision, Visual Processing, and Digital Display III. SPIE, San Jose, CA, 1992, vol 1666, pp 241–252.
9. J Hartung. Architecture for the real-time implementation of three-dimensional subband video coding. Proceedings International Conference Acoustics, Speech and Signal Processing, San Francisco, 1992, pp III-225–III-228.
10. M Antonini, M Barlaud, P Mathieu, I Daubechies, Image coding using vector quantization in the wavelet transform domain. Proceedings International Conference Acoustics, Speech and Signal Processing, San Francisco, 1990, pp 2297–2300.
11. I Daubechies. Commun Pure Appl Math 41:901–996, 1988.
12. T Naveen, JW Woods. Subband finite state scalar quantization. Proceedings International Conference Acoustics, Speech and Signal Processing, Minneapolis, MN, 1993, pp V-613–V-616.
13. M Antonini, M Barlaud, P Mathieu, I Daubechies. IEEE Trans. Image Process IP-1:205–220, 1992.
14. N Mohsenian, NM Nasrabadi. Subband coding of video using edge-based vector quantization technique for compression of the upper bands. Proceedings International Conference Acoustics, Speech and Signal Processing, San Francisco, 1992, pp III-233–III-236.
15. O Johnsen, OV Shentov, SK Mitra. A technique for the efficient coding of the upper bands in subband coding of images. Proceedings International Conference Acoustics, Speech and Signal Processing, Albuquerque, NM, 1990, pp 2097–2100.
16. C Podilchuk, NS Jayant, P Noll. Sparse codebooks for the quantization of non-dominant subbands in image coding. Proceedings International Conference Acoustics, Speech and Signal Processing, Albuquerque, NM, 1990, pp 2101–2104.
17. C Podilchuk, NS Jayant, N Farvardin. IEEE Trans Image Process IP-2, Feb. 1995, 11:125–139.
18. A Gersho, R Gray. Vector Quantization and Signal Compression. Boston: Kluwer Academic Publishers, 1992.
19. RE Crochiere, SA Webber, JL Flanagan. Bell Syst Tech J 55:1069–1985, 1976.

20. H Gharavi, A Tabatabai. IEEE Trans Circuits Syst. CS-35: 207–214, 1988.
21. PH Westerink, J Biemond, DE Boekee, JW Woods. IEEE Trans Commun C-36:713–719, 1988.
22. AS Lewis, G Knowles. IEEE Trans Image Process IP-1: 244–250, April 1992.
23. H Gharavi. In JW Woods, ed. Subband Image Coding. Boston: Kluwer Academic Publishers, 1991, pp 229–272.
24. JM Shapiro. IEEE Trans Signal Process SP-41:3445–3462, December 1993.
25. P Sriram, MW Marcellin. IEEE Trans Image Process IP-3, September 1994, pp. 725–733.
26. TL Joshi, VL Crump, TR Fisher. IEEE Trans Circuits Syst Video Technol CSVT-5:515–523, December 1995.
27. IH Witten, JC Clearly. Commun ACM 30:520–540, June 1987.
28. A Said, WA Pearlman. IEEE Trans Circuits Syst Video Technol CSVT-6(6); 1996, pp. 243–250.
29. Y Linde, A Buzo, RM Gray. IEEE Trans Commun C-28:702–710, January 1980.
30. NM Nasrabadi, RA King. IEEE Trans Commun C-36, August 1988, pp. 957–971.
31. PA Chou, T Lookabaugh, RM Gray. IEEE Trans Acoustics Speech Signal Process ASSP 37: 31–42, 1989.
32. JH Conway, JA Sloan. IEEE Trans Inform Theory, IT-28:227–232, March 1982.
33. A Gersho. IEEE Trans Inform Theory IT-25:373–380, July 1982.
34. T Senoo, B Girod. IEEE Trans Image Process IP-1:526–533, October 1992.
35. D Marr. Vision. New York: Freeman, 1982.
36. M Kunt, O Johnsen. Proc IEEE 68:770–786, July 1986.
37. RJ Safranek, JD Johnston. A perceptually tuned sub-band image coder with image dependent quantization and post-quantization data compression. Proceedings International Conference Acoustics, Speech and Signal Processing, 1989, pp 1945–1948.
38. WEL Grimson, Object Recognition by Computer: The Role of Geometric Constraints. Cambridge, MA: MIT Press, 1990.
39. DG Lowe. Perceptual Organization and Visual Recognition. Boston: Kluwer Academic Publishers, 1985.
40. AP Witkin, M Tenenbaum. In: A Rosenfeld and J Beck, eds. Human & Machine Vision. New York: Academic Press, 1983.
41. A Sha'ashua, S Ullman. Structural saliency: The detection of globally salient structures using a locally connected network. Proceedings 2nd International Conference on Computer Vision, Tampa, FL, 1988, pp 321–327.
42. S Geman, D Geman. IEEE Trans Pattern Anal. Machine Intell. PAMI-6:721–741, 1984.
43. T Pappas. IEEE Trans Signal Process SP-40:901–914, 1992.
44. J Luo, CW Chen, KJ Parker. J Electron Imaging 4(2):189–198, 1995.
45. H Derin, H Elliot. IEEE Trans Pattern Anal. Machine Intell PAMI-9:39–55, 1987.
46. S Lakshmanan, H Derin. IEEE Trans Pattern Anal. Machine Intell PAMI-11:799–813, 1989.
47. J Besag. J Roy Statist Soc 36:192–326, 1974.
48. J Luo. Wavelet-based low bit rate image and video compression with adaptive quantization, coding and postprocessing. PhD thesis, University of Rochester, Rochester, NY, 1995.
49. J Besag. J Roy Statist Soc 48:259–302, 1986.
50. A Perez, S Kamata, E Kawagushi. In: AG Tescher, ed. Proc. SPIE Conference Application of Digital Image Processing XIV, San Diego, CA: SPIE, 1991, pp 354–361.
51. RL Stevenson. Proceedings International Conference on Acoustics, Speech and Signal Processing, Minneapolis, MN, 1993, pp V-401–V-404.
52. J Luo, CW Chen, KJ Parker, TS Huang. A new method for block effect removal in low bit rate image compression. Proceedings International Conference on Acoustics, Speech and Signal Processing, Australia, 1994, pp V341–V344.
53. Y Chen, WA Pearlman. Three-dimensional subband coding of video using the zerotree method. Proceedings SPIE Symposium on Visual Communication and Image Processing, Orlando, FL, 1996, pp 1302–1312.

54. Z Sun, J Luo, CW Chen, KJ Parker. Wavelet-based compression technique for wireless image communication. Proceedings SPIE Aerospace/Defense Sensing and Controls Symposium, Orlando, FL, 1996.
55. N Jayant, J Johnson, R Safranek. Proc IEEE 81:1385–1422, 1993.
56. JYA Wang, EH Adelson. IEEE Trans Image Process 3:625–638, September 1994.
57. K Aizawa, TS Huang. Proc IEEE, 83:175–193, 1995.
58. H Li, A Lundmark, R Forchheimer. IEEE Trans Image Process IP-3:589–609, September 1994.
59. J Luo, CW Chen, KJ Parker. Face location in wavelet-based video compression for high perceptual quality videoconferencing. Proceedings International Conference Image Processing, Washington, DC, 1995, pp II 583–II 586.
60. G Yang, TS Huang. Pattern Recogn 27:53–63, January 1994.
61. RD Rimey, CM Brown. Int J Comput Vision 12:173–207, 1994.

4
Statistically Adaptive Wavelet Image Coding

Bing-Bing Chai
Sarnoff Corporation, Princeton, New Jersey

Jozsef Vass and Xinhua Zhuang
University of Missouri–Columbia, Columbia, Missouri

1. INTRODUCTION

In the mid-1980s, wavelet theory was developed in applied mathematics (1–3). Soon, subband coding (4), which has been a very active research area for image and video compression, was identified as the wavelet's discrete cousin. Furthermore, a fundamental insight into the structure of subband filters was developed from wavelet theory that led to a more productive approach to designing the filters (1,5,6). Thus, subband and wavelet are often used interchangeably in the literature.

Two types of subband decomposition are commonly used in image compression: uniform and pyramidal decomposition. Uniform decomposition (7) divides an image into equal-sized subbands (Fig. 1a). By contrast, pyramidal decomposition represents an octave-band (dyadic) decomposition, offering a multiresolution representation of the image as illustrated in Fig. 1b. Most of the subband image coders published recently are based on pyramidal decomposition.

Conventional wavelet or subband image coders (5,8) mainly exploit the energy compaction property of subband decomposition by using optimal bit-allocation strategies. The drawback is apparent in that all zero-valued wavelet coefficients, which convey little information, must be represented and encoded, biting away a significant portion of the bit budget. Although this type of wavelet coder provides superior visual quality by eliminating the blocking effect in comparison to block-based image coders such as JPEG (9), their objective performance measured by the peak signal-to-noise ratio [PSNR, Eq. (1) in Sec. 4] increases only moderately.

A fundamental issue in wavelet coding is: If the image admits a stochastic model such as a Markov random field, then what are the statistical characteristics of the corresponding wavelet-transformed image? We will give a brief description of our initial exploration in this chapter. Empirically, it has been observed that a wavelet-transformed image has the following statistical properties:

1. Spatial-frequency localization
2. Energy compaction

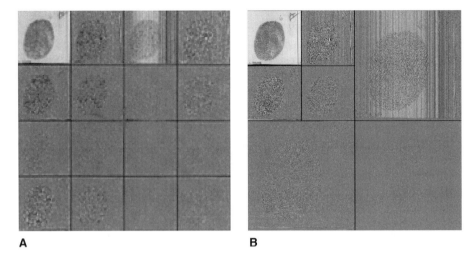

Fig. 1 (A) Uniform wavelet decomposition; (B) pyramidal wavelet decomposition.

3. Within-subband clustering of significant coefficients
4. Cross-subband similarity
5. Decaying of coefficients magnitudes across subbands

In recent years, we have seen an impressive advance in wavelet image coding. The success is primarily attributed to innovative strategies for data organization and representation of wavelet-transformed images which exploit the above statistical properties one way or the other. Three such top-ranked wavelet image coders have been published, namely Shapiro's embedded zerotree wavelet coder (EZW) (10), Servetto et al.'s morphological representation of wavelet data (MRWD) (11), and Said and Pearlman's set partitioning in hierarchical trees (SPIHT) (12). Both EZW and SPIHT exploit cross-subband dependency of *insignificant* wavelet coefficients, whereas MRWD does within-subband clustering of *significant* wavelet coefficients. As a result, the PSNR of reconstructed images is consistently raised by 1–3 dB over block-based transform coders.

In this chapter, we present a novel strategy for data organization and representation for wavelet image coding termed significance-linked connected component analysis (SLCCA). SLCCA adaptively exploits the statistical properties of wavelet-transformed images at each stage of the coding process. SLCCA strengthens MRWD by exploiting not only within-subband clustering of significant coefficients but also cross-subband dependency in the significant fields. The cross-subband dependency is effectively exploited by using the so-called significance link between a parent cluster and a child cluster. The key components of SLCCA include multiresolution discrete wavelet image decomposition, connected component analysis within subbands, significance link registration across subbands, and bit-plane encoding of magnitudes of significant coefficients by adaptive arithmetic coding.

The remainder of the chapter is organized as follows. In Sec. 2, the aforementioned statistical properties of wavelet-transformed images are discussed. Then, the data organization and representation strategies used in EZW, MRWD, and SPIHT are analyzed. Our wavelet image coding algorithm, SLCCA, is presented in Sec. 3. In Sec. 4, the performance of SLCCA is evaluated against other wavelet coders. The last section concludes the chapter.

2. STATISTICAL PROPERTIES OF WAVELET-TRANSFORMED IMAGE AND THEIR APPLICATION IN IMAGE COMPRESSION

The wavelet transform cuts up signals into different frequency bands and then analyzes each band with a resolution matched to its scale. As an adaptive alternative to the classical short-time Fourier transform (STFT) (13), wavelet transform is of primary interest for the analysis of nonstationary signals. In contrast to STFT, which uses a single analysis window, the wavelet transform provides constant relative bandwidth frequency analysis (namely it uses short time windows at high frequencies and long time windows at low frequencies). When wavelet transform is applied to images, the spatial domain naturally replaces the time domain.

2.1. Statistical Properties of Wavelet-Transformed Images

Discrete-wavelet-transformed images demonstrate the following statistical properties and their exploitation continually proves to be important for image compression.

2.1.1. Spatial-Frequency Localization

Although both Finite Impulse Response (FIR) and Infinite Impulse Response (IIR) filters can be used for subband decomposition, FIR subband filters are most commonly used. Thus, our discussion here will be based on FIR filters. When FIR filters are used, each wavelet coefficient contains only features from a local segment of an input image. Because subband coding decomposes an image into a few frequency bands with almost no overlap, each subband is frequency localized with nearly independent frequency content. In brief, each wavelet coefficient represents information in a certain frequency range at a certain spatial location.

2.1.2. Energy Compaction

A natural image is typically composed of a large portion of homogeneous and textured regions and a rather small portion of edges including perceptually important object boundaries. Homogeneous regions have the least variation and primarily consist of low-frequency components; textured regions have moderate variation and consist of a mixture of low- and high-frequency components; and edges show the most variation and are mainly composed of high-frequency components. Accordingly, wavelet transform compacts the most energy distributed over homogeneous and textured regions into the low-pass subband. Each time a low-pass subband at a fine resolution is decomposed into four subbands at a coarser resolution, critical sampling is applied that allows the newly generated low-pass subband to be represented by only one-fourth of the size of the original low-pass subband. By applying this decomposition process repeatedly for a few times on an image, the energy will be effectively compacted into a few wavelet coefficients. Figure 2 shows the one- and two-scale wavelet decompositions of "Lena" 's feather as well as the corresponding energy distribution surfaces. After the two-scale decomposition, most energy is well compacted into one-sixteenth of the total wavelet coefficients.

2.1.3. Within-Subband Clustering of Significant Coefficients

A wavelet coefficient c is called *significant* with respect to a predefined threshold T if $|c| \geq T$; otherwise, it is deemed *insignificant*. An insignificant coefficient is also known as

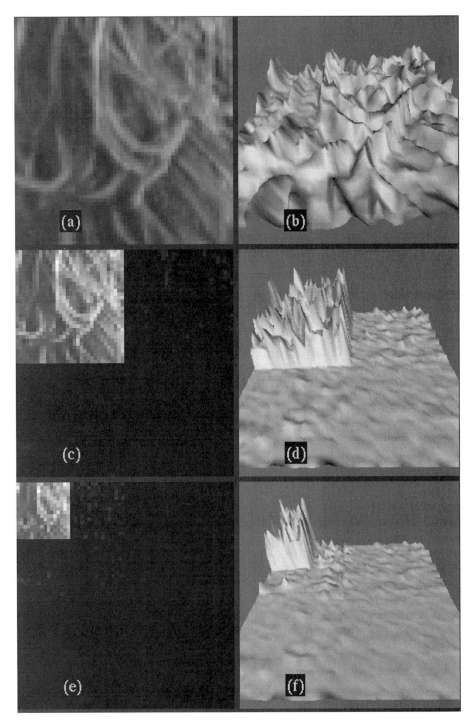

Fig. 2 Energy compaction property of wavelet decomposition. (a) Part of the ''Lena'' image and (b) its surface plot. (c) One-scale and (e) two-scale wavelet decomposition; the corresponding surface plots are shown in (d) and (f), respectively.

a zero coefficient. Empirically, it has been found that significant coefficients within subbands are more clustered than a two-dimensional (2-D) Poisson distribution which shares the same marginal probability (11). Due to the absence of high-frequency components in homogeneous regions and the presence of high-frequency components in textured regions and around edges, significant coefficients in high-pass subbands usually appear at the spatial locations of edges or textures of high energy. In other words, they are indicative of prominent "discontinuity" or prominent "changes," a phenomenon which tends to be clustered. The within-subband clustering of the "Lena" image is shown in Fig. 3a. Note how the clusters of significant coefficients in high-pass subbands delineate the contour of "Lena."

2.1.4. Cross-Subband Similarity

According to the definitions in Refs. 10 and 14, relative to a given wavelet coefficient, all coefficients at finer scales which correspond to the same spatial location are called its *descendents*; accordingly, the given coefficient is called their *ancestor*. Specifically, the coefficient at the coarse scale is called the *parent* and all four coefficients corresponding to the same spatial location at the next finer scale of similar orientation are called *children* (Fig. 4). Although the linear correlation between the values of parent and child wavelet coefficients has been empirically found to be extremely small as expected, there is likely additional dependency between the magnitudes of parent and children. Experiments showed that the correlation coefficient between the squared magnitude of a child and the squared magnitude of its parent tends to be between 0.2 and 0.6, with a strong concentration around 0.35 (10).

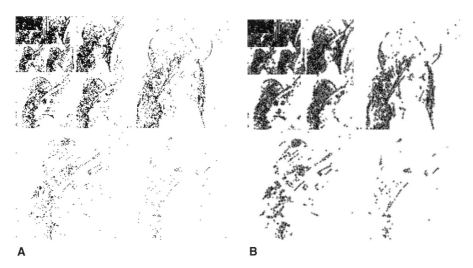

Fig. 3 Significance map for six-scale wavelet decomposition, $q = 11$. (A) Significance map after quantization: white pixels denote insignificant coefficients and black pixels denote significant coefficients. (B) The transmitted significance map (after removing clusters having only one significant coefficient): white pixels denote insignificant coefficients that are not encoded, black and gray pixels denote encoded significant and insignificant wavelet coefficients, respectively.

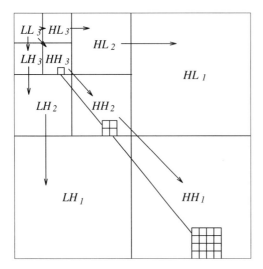

Fig. 4 Illustration of parent–child relationship between subbands at different scales. The pixel drawn in HH_3 is the parent of the four pixels in HH_2.

2.1.5. Decaying of Coefficient Magnitudes Across Subbands

Although it appears to be difficult to characterize and make a full use of the cross-subband magnitude similarity, a reasonable conjecture based on experience with real-world images is that the magnitude of a child is generally smaller than the magnitude of its parent. By assuming the Markov random field as the image model, we are able to prove that, statistically, the magnitude of wavelet coefficients decays exponentially from the parent to its children (15,16), providing a strong theoretical support for EZW, SPIHT, and SLCCA.

2.2. Overview of Data Organization and Representation Strategies

Although all wavelet-based image coding algorithms take advantage of energy compaction and spatial-frequency localization properties of wavelet-transformed images, the greatest contribution to improvement in coding efficiency is the exploitation of the within-subband and cross-subband statistical properties. There exist two efficient approaches to the organization and representation of wavelet coefficients in the literature. Whereas EZW and SPIHT use the regular tree structure to approximate the spatial similarity in insignificant fields across subbands, MRWD finds irregular clusters of significant fields within subbands. Among the top three wavelet image coders, SPIHT performs the best, in general.

It is commonly accepted from the source coding theory that, in general, an image compression technique grows computationally more complex as it becomes more efficient. EZW interrupts this tendency by achieving outstanding performance with very low computational complexity. It efficiently identifies and approximates arbitrary-shaped regions with wavelet coefficients equal to zero (i.e, zero regions) across subbands by the union of highly constrained tree-structured zero regions called *zerotrees* (the structure formed by the pixels in HH_3, HH_2, and HH_1 in Fig. 4). Meanwhile, it defines the significant fields outside these zero regions by progressively refining the magnitudes of the coefficients. It is apparent that each *zerotree* can be effectively represented by its root symbol. On the other hand, there may still be many zero coefficients that cannot be included in the highly structured zerotrees. These isolated zeros remain expensive to represent.

The SPIHT seeks to enhance EZW by partitioning the cross-subband tree structure into three parts: tree root, children of the root, and nonchild descendents of the root, which are illustrated by the pixels shown in HH_3, HH_2, and HH_1 in Fig. 4, respectively. It is obvious that the nonchild descendents comprise a majority of the population in the tree structure. When a child coefficient is found significant, EZW represents and encodes all four grandchild coefficients separately, even if all nonchild descendents are insignificant. In contrast, SPIHT treats the insignificant nonchild descendents as a union and employs a single symbol to represent and encode it. This fine set partitioning strategy leads to an impressive increase in PSNR by 0.86–1.11 dB over EZW on the ''Lena'' image (see Table 4), indicating that SPIHT exploits cross-subband dependency more efficiently than EZW.

Different from EZW and SPIHT, MRWD directly forms irregular-shaped clusters of significant coefficients within subbands. The clusters within a subband are progressively delineated by insignificant boundary zeros through morphological conditioned dilation operation, which utilizes a structuring element to control the shape and size of clusters, as well as the formation of boundaries. For most structuring elements, a formed cluster could be neither 4-connected nor 8-connected. With MRWD, the boundary zeros of each cluster still need to be coded, but the expensive cost of representing and encoding isolated zeros in EZW is largely avoided. As a result, MRWD constantly outperforms EZW. For instance, it gains 0.32–0.62 dB over EZW on the ''Lena'' image, as shown in Table 4. Nevertheless, MRWD *does* need to specify a seed (a pixel from which a cluster is originated) for each cluster and encode its positioning information as overhead. As a large number of clusters are involved, the overall overhead may take up a significant portion of the bit budget.

3. SIGNIFICANCE-LINKED CONNECTED COMPONENT ANALYSIS

The SLCCA attempts to exploit all the aforementioned statistical properties of wavelet-transformed images. In this section, the key features of our wavelet coder SLCCA are first described. Then a complete algorithm is presented.

3.1. Formation of Connected Components Within Subbands

It has been seen that a rather large portion of wavelet coefficients are usually insignificant; significant coefficients within subbands are more clustered than points from a 2-D Poisson distribution having the same marginal probability (11) (Fig. 3a). Therefore, organizing and representing each subband as irregular-shaped clusters of significant coefficients provides an efficient way for encoding. Clusters are progressively constructed by using conditioned dilation (17), resulting in an effective segmentation of the within-subband significant regions. This idea was sketched in Ref. 11. In the following, we discuss the issue with regard to the selection of structuring elements.

Suppose A is a binary image, B a binary structuring element, and $M \subset A$ a marker image. Then, the *conditioned dilation* is defined as

$$D^1(M, A) = (M \oplus B) \cap A$$

where $\oplus$ denotes the morphological dilation (18,19) and $\cap$ denotes the intersection. Let

$$D^n(M, A) = D^1(D^{n-1}(M, A), A)$$

Then, $D^\infty(M, A)$ defines a cluster in A. For a digital image, the cluster is formed in finite iterations when $D^n(M, A) = D^{n-1}(M, A)$.

In the case of clustering in the significance field, the binary image A represents the significance map:

$$A[x, y] = \begin{cases} 1 & \text{if the wavelet coefficient } c \text{ at location } [x, y] \text{ is significant} \\ 0 & \text{otherwise} \end{cases}$$

The marker $M \subset A$ represents the seeds of each cluster.

Traditionally, a connected component is defined based on one of the three types of connectivity: 4-connected, 8-connected, and 6-connected, each requiring a geometric adjacency of two neighboring pixels. Because the significant coefficients in the wavelet field are only loosely clustered, the conventional definition of a connected component will produce too many components, affecting the coding efficiency. Thus, we may use symmetric structuring elements with a size larger than 3×3 square, but we still call the segments generated by conditioned dilation *connected components* even if they are not geometrically connected. Some of the structuring elements tested in our experiments are shown in Fig. 5; the ones in Figs. 5a and 5b generate 4- and 8-connectivity, respectively. The structuring elements in Figs. 5c and 5d represent a diamond of size 13 and a 5×5 square, respectively. The latter two may not preserve geometric connectivity within a component, but may perform better than the former in terms of coding efficiency.

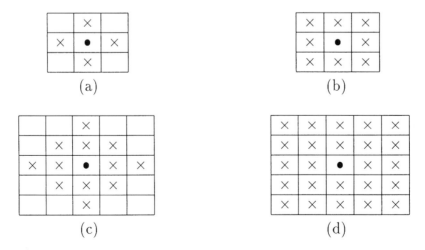

Fig. 5 Structuring elements used in conditioned dilation.

Table 1 Performance Comparison of Different Area Thresholds on "Lena" Image at 0.25 bpp, Using 5 × 5 Structuring Element (Fig. 5d)

Area threshold	No. of clusters	Significant coefficients	Transmitted insignificant coefficients	Performance PSNR (dB)
0	452	11,200	38,690	34.26
1	253	11,545	36,290	34.35
2	188	11,638	35,028	34.34
3	152	11,703	34,301	34.32

To effectively delineate a significant cluster, all zero coefficients within the neighborhood B of each significant coefficient in the cluster need to be marked as the boundary of the cluster. By increasing the size of the structuring element, the number of connected components decreases. On the other hand, a larger structuring element results in more boundary zero coefficients. The optimal choice of the size of the structuring element is determined by the cost of encoding boundary zeros versus that of encoding the positional information of connected components. Because the significance-link technique will substantially reduce the positioning cost, relatively smaller structuring elements can be selected for connected component analysis.

As extremely small clusters usually do not produce discernible visual effects, and those clusters render a higher insignificant-to-significant coefficient ratio than large clusters, they are eliminated to avoid more expensive coding. As the area threshold increases, the number of clusters decreases, which results in the reduction of the required cluster positioning information. As illustrated in Table 1, the zero area threshold has the worst performance. All other area thresholds have a similar performance.

In the article by Luo et al. (20), the authors also propose wavelet coefficients clustering for image compression. They use clustering as a tool for quantization; that is, wavelet coefficients are clustered together and quantized to the mean value of the given cluster. The wavelet coefficients are then coded by either using traditional run-length coding or Shapiro's EZW algorithm. In SLCCA, we use clustering to register and transmit the significance map; that is, clustering is our tool for data organization. The two algorithms justify each other in the sense that clusters having perceptually less significant information can be removed to raise the coding gain without compromising the objective and subjective quality.

The connected component analysis is illustrated in Fig. 3. The significance map obtained by quantizing all wavelet coefficients with a uniform scalar quantizer with step size $q = 11$ is shown in Fig. 3a. The 22748 significant wavelet coefficients form 1654 clusters using the structuring element shown in Fig. 5c. After removing connected components having only one significant coefficient, the number of clusters is reduced to 689. The final encoded significance map is shown in Fig. 3b. It is clear that only a small fraction of zero coefficients is encoded.

3.2. Significance Link in the Wavelet Pyramid

The cross-subband similarity among *insignificant coefficients* in the wavelet pyramid has been exploited in EZW and SPIHT that greatly improves the coding efficiency. On the other hand, it is found that the spatial similarity in wavelet pyramid is not strictly satisfied (i.e., an insignificant parent does not warrant all four children insignificant). The "isolated zero" symbol used in EZW indicates the failure of such a dependency. The similarity described by the zerotree in EZW and the similarity described by both zerotree and insignificant all-nonchild descendents in SPIHT are more of a reality when a large threshold is used. As was stated in Refs. 10 and 21, when the threshold decreases (for embedding) to a certain point, the tree structure or set-partitioned-tree structure are no longer efficient.

In the proposed algorithm, as opposed to EZW and SPIHT, we attempt to exploit the spatial similarity among *significant coefficients*. However, we do not seek a very strong parent–child dependency for each and every significant coefficient. Instead, we try to predict the existence of clusters at finer scales. As pointed out earlier, statistically the magnitudes of wavelet coefficients decay from a *parent* to its *children*. It implies that in a cluster formed within a fine subband, there likely exists a significant child whose parent at the coarser subband is also significant. In other words, a significant child can likely be traced back to its *parent* through this *significance linkage*. It is crucial to note that this significance linkage relies on a much looser spatial similarity.

Two connected components or clusters are called *significance-linked* if the significant parent belongs to one component, and at least one of its children is significant and lies in another component (Fig. 6). If the positional information of the significant parent in the first component is available, the positional information for the second component can be inferred through marking the parent as having a *significance link*. As there are generally many significant coefficients in connected components, the likelihood of finding a significance link between two components is fairly high. Apparently, marking the significance link costs much less than directly encoding the position, and a significant savings on encoding cluster positions is thus achieved. The savings from using the significance link increases as the bit rate increases, ranging from 527 bytes at 0.25 bit per pixel (bpp)

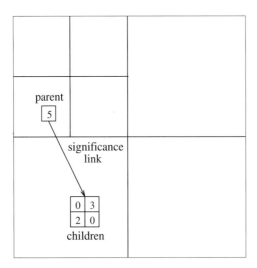

Fig. 6 Illustration of significance link. The values are the magnitudes of quantized coefficients. Nonzero values denote significant coefficients.

to 3103 bytes at 1 bpp for the "Lena" image. Among all, using the significance link makes a major difference between SLCCA and MRWD.

3.3. Bit-Plane Organization and Adaptive Arithmetic Coding

As in most image compression algorithms, the last step of SLCCA involves entropy coding for which adaptive arithmetic coding (22) is employed. Adaptive arithmetic coder updates the corresponding conditional probability estimation every time the coder visits a particular context. Thus, the local probability distributions is well exploited, resulting in a higher compression than that achieved by fixed-model arithmetic coder.

In order to exploit the full strength of an adaptive arithmetic coder, it is preferable to organize outcomes of a nonstationary Markov source into such a stream that local probability distribution is in favor of one source symbol in a certain segment of the data stream. This is the basic idea behind the well-known lossless bit-plane coding, in which an original image is divided into bit planes with each bit plane being encoded separately. Because more significant bit planes generally contain large uniform areas, the entropy coding techniques can be more efficient.

This idea is employed by SLCCA to encode the magnitudes of significant coefficients in each subband. The magnitude of each significant coefficient is converted into a binary representation with a fixed length determined by the maximum magnitude in the subband. From the energy compaction property of the wavelet-transformed image, in high-pass subbands there are more coefficients with small magnitude than those with large magnitudes, as shown in Fig. 7. This implies that more significant bit planes would contain a

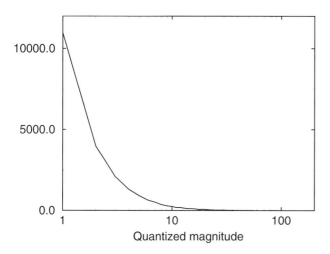

Fig. 7 Distribution of significant coefficients of the "Lena" image when quantized for the final rate, 0.5 bpp.

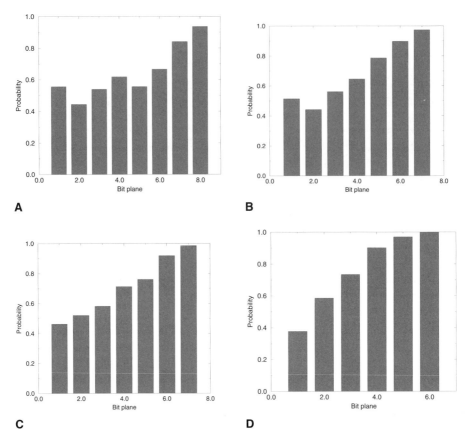

Fig. 8 Zero symbol probabilities in four subbands for ''Lena'': (A) 3rd subband; (B) 5th subband; (C) 7th subband; and (D) 12th subband.

lot more 0's than 1's, as shown in Fig. 8. Accordingly, the adaptive arithmetic coder would generate more accurate local probability distributions in which the conditional probabilities for ''0'' symbols are close to 1 for the more significant bit planes. The context used to determine the conditional probability model of the significant coefficient at $[x, y]$ is related to the significance status of its parent and its eight neighbors. Let $K_p[x, y]$ denote the significance status of the parent; that is, $K_p[x, y] = 1$ if the parent is significant, otherwise $K_p[x, y] = 0$. Let $K_n[x, y]$ denotes the number of significant coefficients in a 3×3 *causal* neighborhood of the current pixel $[x, y]$. The adaptive context $K[x, y]$ is selected by $K[x, y] = K_n[x, y] + 9K_p[x, y]$, which yields a total of 18 possible models.

Another method of encoding the coefficients magnitudes is to encode different magnitude values as different source symbols in entropy coding. Experiments have been conducted to compare this method to the bit-plane coding scheme and the results are shown in Table 2. In all cases, bit-plane coding saves 271–1264 bytes over direct magnitude encoding. The reason is that for direct magnitude coding, the large number of source symbols demands much more updating steps for the coder to acquire a good estimate of the source distribution compared to bit-plane coding where only a two-symbol alphabet

Statistically Adaptive Wavelet Image Coding

Table 2 Compressed File Size (Bytes) Comparison Between Bit-Plane Coding and Direct Magnitude Coding on the 512 × 512 "Lena" Image

	PSNR dB				
Algorithm	34.33	35.14	36.42	37.38	40.44
Bit-plane coding	8,193	9,828	13,107	16,383	32,769
Direct magnitude coding	8,464	10,151	13,575	16,992	34,033

is used. As a result, bit-plane magnitude coding exploits the local source distribution more effectively than direct magnitude coding, resulting in a higher compression ratio.

The idea of bit-plane encoding is also used in both EZW and SPIHT, but in a different manner. In EZW, the idea is realized through progressive transmission of magnitudes, with the "0" bits before the first "1" bit being encoded as either "zerotree" or "isolated zero." Similar to EZW, in SPIHT, initial "0" bits are represented as part of the insignificant set until the occurrence of the first "1" symbol. Then the magnitudes are coded by progressive transmission.

3.4. Description of SLCCA

In the following, we present the encoding algorithm of SLCCA.

> **Step 1.** Form a subband pyramid and quantize all wavelet coefficients with a uniform scalar quantizer.
> **Step 2.** Perform connected component analysis of significant coefficients within each subband and remove extremely small connected components.
> **Step 3.** Form a scan list containing all the coefficient positions in the subband pyramid as follows. Starting from the coarsest subband, scan subbands according to the order *LL, LH, HL, HH* (Fig. 4). Within each subband, scan the coefficients from left to right, top to bottom. Go to the next finer scale if all coefficients in the current scale have been scanned.
> **Step 4.** Let c be the first coefficient in the scan list.
> **Step 5.** If c is found significant and has not been encoded, go to **Step 6;** or else, let c be the next coefficient in the scan list and repeat **Step 5.**
> **Step 6.** Encode the position $[x, y]$ of c.
> **Step 7.** Encode the sign (POS or NEG) of c.
> **Step 8.** If c is the parent of a child cluster that has not been linked to any other coefficient, then
> > **8.1** encode a special symbol (LINK);
> > **8.2** move the child position in a first-in–first-out (FIFO) queue to store the information that the child cluster has been linked.
>
> **Step 9.** For every $[\Delta x, \Delta y]$ in a predefined neighborhood, do
> > If $c[x + \Delta x, y + \Delta y]$ is significant and has not been encoded,
> > > then $x = x + \Delta x$, $y = y + \Delta y$, let c be the coefficient at $[x, y]$ and go to **Step 7.**
> >
> > If $c[x + \Delta x, y + \Delta y]$ is insignificant,
> > > then encode a ZERO symbol.

Step 10. If the FIFO queue is not empty, take the next child position out of the queue, and go to **Step 7,** otherwise, let c be the next coefficient in the scan list and go to **Step 5.**

Step 11. Encode the magnitude of significant coefficients in bit-plane order using the adaptive arithmetic coder.

The decoding algorithm is straightforward and can be obtained by reversing the encoding process.

4. PERFORMANCE EVALUATION

4.1. Natural Image Compression

The SLCCA is evaluated on 8 natural 512×512 gray-scale images shown in Fig. 9. The performance is compared with the best wavelet coders EZW, MRWD, and SPIHT. Each original image is decomposed into a six-scale subband pyramid using 10/18 filters (Table 3) obtained from ftp.cs.dartmouth.edu. No optimal bit allocation is carried out in SLCCA. Instead, all wavelet coefficients are quantized with the same uniform scalar quantizer. As usual, the distortion is measured by the peak signal-to-noise ratio (PSNR) defined as

$$\text{PSNR (dB)} = 20 \log_{10}\left(\frac{255}{\text{RMSE}}\right) \qquad (1)$$

where RMSE is the root mean squared error between the original and reconstructed images. All the reported bit rates are computed from the actual file sizes.

Table 4 shows the comparison among four wavelet coders on the ''Lena'' image at different bit rates. SLCCA consistently outperforms EZW, MRWD, and SPIHT. Compared to EZW, SLCCA gains 1.08 dB in PSNR on average. When compared to MRWD, SLCCA is superior by 0.27–1.07 dB. Compared to SPIHT, SLCCA gains 0.18 dB on average.

Table 5 compares the performance of SLCCA, EZW, and SPIHT on the ''Barbara'' image. There exist many versions of the ''Barbara'' image. The one used in this experiment was obtained from ftp.cs.dartmouth.edu, which is the same as that used by EZW and SPIHT. On average, SLCCA is superior to EZW by 1.67 dB, and to SPIHT by 0.79 dB. The original ''Barbara'' image, and the reconstructed images at 0.25 bpp, 0.5 bpp, and 1.0 bpp are shown in Figs. 10a–10d respectively.

The comparison between SLCCA and SPIHT on the rest of the test images is shown in Table 6. SLCCA consistently outperforms SPIHT. It appears that SLCCA performs significantly better than SPIHT for images which are rich in texture; see, for instance, the results of ''Barbara,'' ''Baboon,'' ''Boat,'' and ''Tank.'' For images which are relatively smooth, the difference in performance between SLCCA and SPIHT is small, as indicated by the results of ''Goldhill,'' ''Couple,'' and ''Man.''

4.2. Texture Image Compression

To further verify the above observation, we compare the performance of SLCCA and SPIHT on 8 typical 256×256 gray-scale texture images shown in Fig. 11. The results at 0.4 bpp are summarized in Table 7, indicating that SLCCA constantly outperforms SPIHT by 0.32–0.70 dB. An explanation is as follows. When textured images are encoded, the wavelet transform is unlikely to yield many large zero regions for lack of homogeneous

Statistically Adaptive Wavelet Image Coding

Fig. 9 The 512 × 512 test images. From left to right, top to bottom: "Lena," "Barbara," "Baboon," "Couple," "Man," "Boat," "Tank," and "Goldhill."

Table 3 Biorthogonal Filterbank Used in SLCCA

Tap	Value	Tap	Value
	(a) Analysis		(b) Synthesis filters
1	2.885256E−02	1	9.544158E−04
2	8.244478E−05	2	−2.727196E−06
3	−1.575264E−01	3	−9.452462E−03
4	7.679048E−02	4	−2.528037E−03
5	7.589077E−01	5	3.083373E−02
6	7.589077E−01	6	−1.376513E−02
7	7.679048E−02	7	−8.566118E−02
8	−1.575264E−01	8	1.633685E−01
9	8.244478E−05	9	6.233596E−01
10	2.885256E−02	10	6.233596E−01
		11	1.633685E−01
		12	−8.566118E−02
		13	−1.376513E−02
		14	3.083373E−02
		15	−2.528037E−03
		16	−9.452462E−03
		17	−2.727196E−06
		18	9.544158E−04

Table 4 Performance Comparison [PSNR (dB)] on the 512 × 512 "Lena" Image

Algorithm	Rate (bpp)					
	0.125	0.25	0.30	0.40	0.50	1.00
EZW	30.23	33.17	—	—	36.28	39.55
MRWD	—	—	34.07	—	36.60	40.17
SPIHT	31.09	34.11	34.95	36.24	37.21	40.41
SLCCA	31.38	34.33	35.14	36.42	37.38	40.44

Table 5 Performance Comparison [PSNR (dB)] on the 512 × 512 "Barbara" Image

Algorithm	Rate (bpp)					
	0.125	0.25	0.30	0.40	0.50	1.00
EZW	24.03	26.77	—	—	30.53	35.14
SPIHT	24.86	27.58	28.56	30.10	31.39	36.41
SLCCA	25.45	28.43	29.39	30.93	32.28	37.15

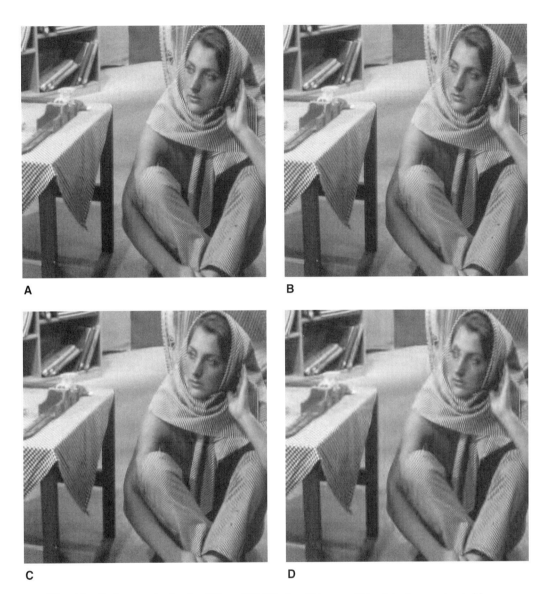

Fig. 10 Coding results for the 512 × 512 "Barbara" image: (A) original reconstructed image; at: (B) 1.0 bpp, PSNR = 37.15 dB; (C) 0.5 bpp, PSNR = 32.28 dB; (D) 0.25 bpp, PSNR = 28.43 dB.

Table 6 Performance Comparison [PSNR (dB)] of SPIHT and SLCCA on Different 512 × 512 Natural Images

Rate (bpp)	Algorithm	Image					
		"Baboon"	"Couple"	"Man"	"Boat"	"Tank"	"Goldhill"
0.25	SPIHT	23.27	29.25	30.01	30.97	29.36	30.56
	SLCCA	23.44	29.38	30.07	31.09	29.44	30.64
0.3	SPIHT	23.76	30.07	30.74	31.77	29.77	31.15
	SLCCA	23.99	30.10	30.81	31.94	29.89	31.24
0.4	SPIHT	24.66	31.29	31.94	33.16	30.52	32.18
	SLCCA	24.96	31.43	32.04	33.44	30.65	32.32
0.5	SPIHT	25.64	32.45	33.08	34.45	31.16	33.13
	SLCCA	25.83	32.55	33.12	34.68	31.28	33.24
1.0	SPIHT	29.17	36.58	37.34	39.12	33.78	36.55
	SLCCA	29.33	36.58	37.33	39.28	33.99	36.66

regions. Thus, the advantage of using an insignificant tree as in EZW, or an insignificant part-of-tree structure as in SPIHT, is weakened. On the other hand, SLCCA uses significance-based clustering and significance-based between-cluster linkage, which are not affected by the existence of textures.

4.3. Comparison with the Latest Wavelet Coding Algorithms

Recently, we learned about two new wavelet image coding algorithms: Xiong et al.'s wavelet-based spatial-frequency quantization (23) (SFQ), and a latest version of Servetto et al.'s MRWD (24).

In SFQ, the zerotree structure is optimized for a given target bit rate using the Lagrange multiplier method in the operational rate-distortion sense. The optimization procedure yields remarkable performance, but at the price of much higher computational complexity.

In the latest version of MRWD (24), all wavelet coefficients are coded regardless of their significance. Raster scan is the basic scan order to encode the coefficients in a subband. When a significant coefficient (the seed of a cluster) is encountered, a special symbol is encoded followed by the entire cluster before continuing the raster scan. The context of the adaptive arithmetic model is primarily based on the significance of the parent coefficient. Because a single uniform quantizer is used for bit-rate control, the complexity of both the encoding and decoding process is comparable to SLCCA.

The performance comparison of the latest MRWD, SFQ, and SLCCA is given in Table 8. In the case of the "Lena" image, SFQ and SLCCA yield comparable performances. For the other two test images, SLCCA is superior on "Barbara," whereas SFQ is superior on "Goldhill." The differences in PSNR for both images are about 0.1 dB. Finally, both SLCCA and SFQ outperform MRWD.

Statistically Adaptive Wavelet Image Coding

Fig. 11 The 256 × 256 texture images. From left to right, top to bottom: "fingerprint," "sweater," "grass," "pig skin," "raffia," "sand," "water," and "wool."

Table 7 Performance Comparison [PSNR (dB)] of SPIHT and SLCCA on 256 × 256 Texture Images at 0.4 bpp

Image	SLCCA	SPIHT
"Fingerprint"	27.61	27.07
"Sweater"	41.83	41.48
"Grass"	25.45	24.82
"Pig skin"	26.82	26.50
"Raffia"	20.93	20.30
"Sand"	24.18	23.63
"Water"	29.76	29.19
"Wool"	26.40	25.70

4.4. SLCCA for Intraframe Coding in Video Compression

In most video coding algorithms, the frames in a video sequence are divided into two categories: intraframe and interframe. Intraframes are coded using still-image compression algorithms, whereas interframes are predicted from intraframes or other interframes. In the existing international standards for video compression, such as H.261, H.263, MPEG-1, and MPEG-2, intraframes are encoded using block-based DCT coding algorithms similar to JPEG. Table 9 compare the results of encoding the intraframe by both H.263 and SLCCA. Figures 12 and 13 show the reconstructed frames from both algorithms for the "Foreman" and "Hall Monitor" sequences, respectively. It is clear that for intraframe compression, SLCCA outperforms H.263 in both objective and subjective measure and thus could be a very good candidate for intraframe compression.

Table 8 Performance Comparison of the Latest MRWD, SFQ, and SLCCA on "Lena," "Barbara," and "Goldhill"

Image	Algorithm	Rate (bpp)				
		0.25	0.30	0.40	0.50	1.00
"Lena"	MRWD[a]	34.12	34.93	36.20	37.18	40.33
	SFQ	34.33	35.07	36.43	37.36	40.52
	SLCCA	34.33	35.14	36.42	37.38	40.44
"Barbara"	MRWD[a]	27.86	28.71	30.22	31.44	36.24
	SFQ	28.29	29.21	30.77	32.15	37.03
	SLCCA	28.43	29.39	30.93	32.28	37.14
"Goldhill"	MRWD[a]	30.53	31.14	32.19	33.15	36.56
	SFQ	30.71	31.34	32.45	33.37	36.70
	SLCCA	30.64	31.24	32.32	33.25	36.66

[a] Data from Ref. 24.

Statistically Adaptive Wavelet Image Coding

Table 9 Performance Comparisons of I-Frame Coding Results from H.263 and SLCCA

	Image			
	"Foreman"		"Hall Monitor"	
Rate (bits) Algorithm	14122	27132	14164	28280
H.263	31.62	36.82	31.42	37.57
SLCCA	32.69	37.57	32.04	38.34

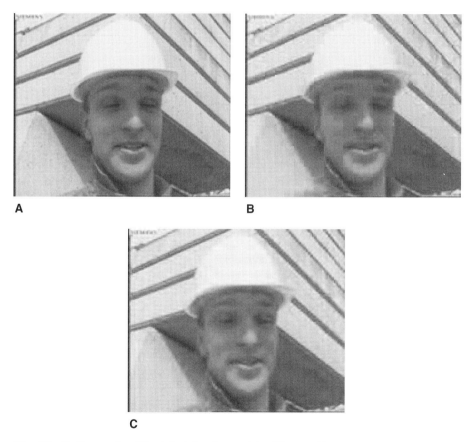

Fig. 12 Coding results: (A) original first frame of the "Foreman" sequence. Reconstructed images from (B) H.263 and (C) SLCCA at 14 kb/frame.

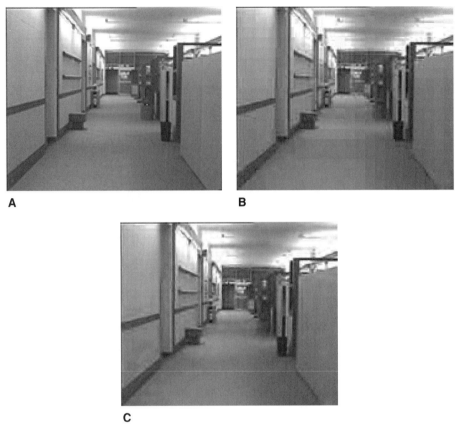

Fig. 13 Coding results: (A) original first frame of the "Hall Monitor" sequence. Reconstructed images from (B) H.263 and (C) SLCCA at 14 kb/frame.

5. CONCLUSIONS

A new image coding algorithm termed significance-linked connected component analysis is proposed. The algorithm takes advantage of two properties of the wavelet decomposition: the within-subband clustering of significant coefficients, and the cross-subband dependency in significant fields. The significance link is employed to represent the positional information for clusters at finer scales, which greatly reduces the positional information overhead. The magnitudes of significant coefficients are coded in the bit-plane order so that the local statistic in the bitstream matches the probability model in adaptive arithmetic coding to achieve further savings in the bit rate. Extensive computer experiments show that SLCCA is among the best image coding algorithms reported in the literature.

ACKNOWLEDGMENTS

The authors would like to thank the reviewer for his comments and suggestions. The surface plots of Fig. 2 were generated by Interactive Image SpreadSheet (IISS) (25) developed at NASA Goddard Space Flight Center.

REFERENCES

1. I Daubechies. Ten Lectures on Wavelets. Philadelphia: Society for Industrial and Applied Mathematics, 1992.
2. O Rioul, M Vetterli. IEEE Signal Process Mag 8(4):14–38, October 1991.
3. M Vetterli, J Kovačević. Wavelets and Subband Coding. Englewood Cliffs, NJ: Prentice-Hall, 1995.
4. JW Woods, ed. Subband Image Coding. Norwell, MA: Kluwer Academic Publishers, 1991.
5. M Antonini, M Barlaud, P Mathieu, I Daubechies. IEEE Trans Image Proces IP-1(2):205–220, 1992.
6. SG Mallat. IEEE Trans Pattern Anal Machine Intell PAMI-11:674–693, July 1989.
7. JW Woods, SD O'Neil. IEEE Trans Acoustics Speech Signal Process ASSP-32(5):1278–1288, 1988.
8. N Farvardin, N Tanabe. In: Proceedings of SPIE Image Processing Algorithms and Techniques. SPIE, Santa Clara, CA, 1990, vol 1244, pp 240–254.
9. GK Wallace. Commun ACM, 34(4):30–44, 1991.
10. JM Shapiro. IEEE Trans Signal Process SP-41(12):3445–3462, 1993.
11. S Servetto, K Ramchandran, MT Orchard. Wavelet based image coding via morphological prediction of significance. Proceedings of IEEE International Conference on Image Processing, 1995, pp 530–533.
12. A Said, WA Pearlman. IEEE Trans Circuits Syst Video Technol CSVT-6(3):243–250, 1996.
13. JB Allen, L R Rabiner. Proc IEEE 65:1558–1564, 1977.
14. AS Lewis, G Knowles. A 64 Kb/s video codec using the 2-D wavelet transform. Proceedings of Data Compression Conference, Snowbird, UT, 1991.
15. BB Chai, J Vass, X Zhuang. IEEE Trans Image Process (in press), 1999.
16. X Li, X Zhuang. The decay and correlation properties in wavelet transform. Technical Report, University of Missouri-Columbia, 1997.
17. L Vincent. IEEE Trans Image Process IP-2(2):176–201, 1993.
18. RM Haralick, SR Sternberg, X Zhuang. IEEE Trans Pattern Anal Machine Intell PAMI-9(4): 532–550, 1987.
19. RM Haralick, LG Shapiro. Computer and Robot Vision. Reading, MA: Addison-Wesley, 1992.
20. J Luo, CW Chen, KJ Parker, TS Huang. IEEE Trans Circuits Syst Video Technol CSVT-7(2): 343–357, 1997.
21. A Said, WA Pearlman. IEEE Trans Image Process IP-5(9):1303–1310, 1996.
22. IH Witten, M Neal, JG Cleary. Commun ACM 30(6):520–540, 1987.
23. Z Xiong, K Ramchandran, MT Orchard. IEEE Trans Image Process IP-6(5):677–693, 1997.
24. S Servetto, K Ramchandran, MT Orchard. IEEE Trans Image Process (in press), 1999.
25. AF Hasler, K Palaniappan, M Manyin, J Dodge. Computers Phys 8(3):325–342, 1994.

5
Three-Dimensional Model-Based Image Communication*

Thomas S. Huang and Antonio Colmenarez
University of Illinois at Urbana-Champaign, Urbana, Illinois

Ricardo Lopez
Evans and Sutherland, Salt Lake City, Utah

Hai Tao
David Sarnoff Laboratory, Princeton, New Jersey

1. INTRODUCTION

The chapter is organized into four main sections. Section 1 begins with a general introduction to the model-based paradigm and the different frameworks within which it can be used. Sections 2 and 3 deal with three of the main steps in the model-based approach: analysis, synthesis, and modeling. Finally, we will conclude with a description of a real-time model-based coding system being developed at the Beckman Institute and present some encouraging results. Throughout the chapter, we will discuss some of the previous approaches to the different problems in model-based coding and then present some of the current work being done by the authors. For the purpose of this chapter and, in fact, for most model-based coding research, we concentrate our efforts on head-and-shoulders type of images. To that end, we consider issues such as rigid and nonrigid tracking, object modeling, and computer graphics synthesis in the context of human head and faces. This is not to say that the model-based approach is not of use in other scenarios, however, because it can be extended to deal with many types of objects, both synthetic and real.

1.1. The Model-Based Approach

The problem of extracting information from a scene is a difficult task and is often ill-posed, impractical, or unrealizable. Extra information or constraints are usually required, but closed-form solutions are rarely available. In most cases, it is easier to model the scene and adjust the parameters of such model until it matches the observations. Those parameters

---

* This work was supported in part by Army Research Laboratory under Cooperative Agreement No. DAAL01-96-2-0003, in part by Joint Services Electronics Program Grant ONR N00014-96-1-0129, and in part by an AT & T Fellowship.

would provide the desired information from the observed scene. Additionally, such a set of parameters is expected to provide a compact representation of the nonredundant information of the scene and, therefore, provide the key for advanced, more efficient, and flexible coding techniques. This is the basic idea of model-based approaches in computer vision and visual communications.

1.2. Model-Based Analysis

Pattern recognition and image understanding techniques can be improved if knowledge (models) are made available to them. Figure 1 shows the overall scheme of a model-based analysis system. The input image is preprocessed, possibly with the help of the knowledge provided by the model, to obtain the raw information that is matched with the model output. The model is then tuned iteratively until the modeling error is minimized. The parameters that produce the best match with the observation are expected to accurately describe the ground truth of the observed scene.

One example of a sucscessful implementation of an analysis system driven by a model is reported in Refs 1 and 2. In this system, the global head pose and the individual feature position are tracked over long video sequences using a three-dimensional model of the head; the feature positions are located with the help of templates that are synthetized using the model at an approximated head pose. This implementation is discussed in more detail in the later sections of the chapter.

Expression recognition can be improved by using motion parameters obtained via model-based analysis. Such motion parameters are normalized with the global head pose and, therefore, reduce the limited scenario in which current applications succeed. On the other hand, a parametric description of the facial motion obtained with model-based approaches might be better to describe expressions, as it could be closely related to an actual physical system that causes such gesture patterns.

1.3. Model-Based Coding

In the past decade or so, research in the application of the model-based approach to image and video coding has been strong. The main motivation has been the drastic reduction in bandwidth for transmission as well as obtaining a higher level of representation for video streams. In this context, we assume that both the transmitting and receiving parties have knowledge of the three-dimensional (3-D) model. By analyzing the input video stream,

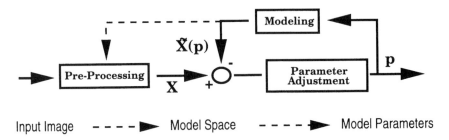

Fig. 1 Model-based analysis: Information is extracted from the scene by tuning the parameters until the modeling error is minimized.

3-D Model-Based Image Communication

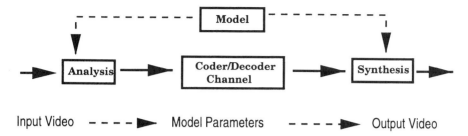

Fig. 2 Model-based coding: Model parameters are encoded and sent through the channel; then, the receiver synthetizes the output.

high-level information regarding the activity in the scene is extracted based on the models available. This information is sent to the receiving end where it is applied to the local model, Fig. 2. There are several advantages to this approach for coding video:

- Model parameters provide a compact representation of the geometry of the scene as well as its motion. Most of the underlying redundancy is covered with the global object position and its articulated local motion.
- Model-based coding allows operations such as rendering synthetic images at different views, virtual environments, and so forth.

1.4. Virtual Agent

Model-based approaches find an important application in human–computer interface research, where the ultimate goal is to have the computer behave like an agent (Fig. 3). This virtual agent should not only recognize and understand the user, but also look, move, and respond accordingly.

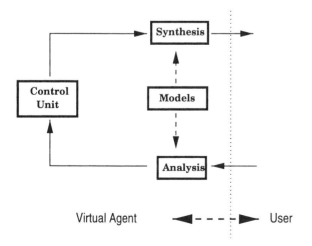

Fig. 3 Virtual agent: A fully automatic, synthetic agent that allows a natural interaction between humans and computers.

Modeling information should include realistic gestures and motion. Off-line analysis of human behavior should be carried out to capture the motion patterns that the virtual agents will adopt (e.g., gestures, lip motion, etc.). On-line analysis would provide the system with the ability to perceive the user.

2. MODELING AND SYNTHESIS

Modeling and animation of human faces has been an important research issue with many practical applications in the fields of computer animation, model-based video compression, and human–computer interaction. The objective is to generate photo-realistic facial images. To achieve this goal, several issues including the geometric face model, the face articulation model, and the synthesis techniques have to be extensively investigated.

A geometric model includes the geometric surface mesh, the texture information, and the rendering environment description. The deformation of this model is described by a facial articulation model, which is usually consistent with the physical rules to make the rendered face images visually convincing. Depending on the nature of an application, many techniques are applicable to the human face modeling process. One approach to simplify the procedure is to divide a model into four layers and build these layers one upon another (3). The bottom layer is the geometric model. The next three layers are the parameter layer, the expression and viseme layer, and the script level. Each higher level is an abstraction of the lower levels.

2.1. Geometric Face Modeling

The surface representation plays an important role in a geometric face model. There are many ways of describing a 3-D surface. The polygonal mesh surface model, free-form parametric surface model, and implicit geometric surface model are most commonly used. It is still very challenging to model detailed facial features such as wrinkles and marks using these methods. Texture mapping technique is considered to be a solution for this problem. However, because of the technical limitations, the texture mapping technique was not applied in most early face models. In recent years, graphics workstations are capable of rendering texture mapping in real time and this solution becomes more appealing.

In addition to the geometric model, an environment model, which describes the relationship between the facial model and its environment variables such as the lighting sources and the background description, should also be carefully considered for achieving realistic visual effects in a video compression system.

2.1.1. 3-D Surface Modeling

Most of the early human face surface models are composed of irregular triangular meshes. Precise approximation is achieved when the shape is sampled sufficiently. The advantage of using this configuration lies in the fact that triangular meshes are very flexible in modeling complicated objects. Also, they can be rendered efficiently.

For describing irregular shapes such as a human face, a free-form surface model is preferred to a solid surface model. Two main types of free-form surface are parametric surfaces and implicit surfaces. Parametric surfaces such as Bézier surfaces and B-spline surfaces have been investigated intensively in the areas of approximation theory and computer-aided geometry design for many years (4,5). Unlike polygonal surface models which only achieve zero-order continuity (C^0), this approach is capable of constructing surfaces with higher order of continuity. A drawback of this method, however, is that it is computational expensive. Fortunately, this may not be a problem in the future.

A straightforward approach to improve the face surface model is to derive a free-form model directly from an existing polygonal model by applying interpolation techniques. Algorithms for interpolating triangular or rectangular meshes are both available. At first look, a triangular mesh interpolation approach is very attractive. However, after careful analysis and experiments we found that many problems in triangular patch interpolation are not yet solved (6–9). One problem is the choice of normal vectors on vertices. They are important parameters which will affect the shapes significantly. So far, no algorithm guarantees optimal solutions in all situations. Another problem is the computational cost. The triangular-patch interpolation is much slower than the rectangular-mesh interpolation. The reason is that for rectangular mesh, a 2-D interpolation process is decomposed into two 1-D interpolation processes. A disadvantage of rectangular mesh is its lack of flexibilities in modeling complicated shapes. To compensate for this drawback, hierarchical rectangular meshes are adopted in our system.

2.1.2. Generic Face Model from MRI Data

To build a generic face model, the first step is to acquire the positions of sample points on the surface of a real human face. Several techniques are presented to accomplish this task. Using a Cyberware 3D color scanner data is a handy approach. However, the problem is that the internal structures of a human face, such as the bone structure and the muscles, are not observable. An alternative method is to analyze magnetic resonance imaging (MRI) data, which give us the information about both the surface and the inner structures (10). The process of modeling face surface using MRI data includes the following three steps:

1. Contour fitting in each interested MRI data slice. In this stage, a fixed number (25 points in our model) of sample points on the surface contour in each interested MRI data slice is manually extracted. By assuming that a face is symmetric, only 13 points are needed in each slice. These sample points are adjusted so that the interpolated B-spline curve from these points fits the contour. Forty-one data slices are sampled in our system for the face surface.
2. 2-D Interpolation. From Step 1, a 25×41 rectangular mesh is obtained. The bi-cubic B-spline interpolation scheme is applied to this mesh to calculate a bi-cubic B-spline surface model.
3. Refine the interested regions. This step is accomplished by repeating Steps 1 and 2 on local features such as nose and ears. For example, in our model, the nose is refined to a 10×8 mesh; the ears are refined to 12×15 meshes. A picture of the derived generic face model is shown in Fig. 4a.

2.1.3. Face Geometric Model Fitting

After a generic geometric facial model has been derived from a particular data set, the next step is to fit the model to a specific person. This process is also called *model fitting*. The model fitting problem has been investigated in various research fields such as scattered-data interpolation (11), free-form surface model deformation (12,13), and elastic object deformation (14). The common goal of these processes is to transform shapes in three dimensions smoothly. In the facial model fitting case, an additional constraint is imposed: the resulting model should be consistent with the geometry of a real human face.

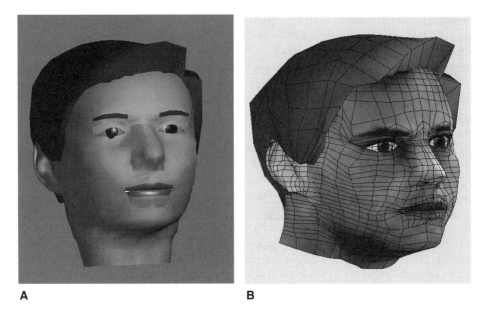

Fig. 4 (A) A bi-cubic B-spline facial surface model; (B) texture-mapped version of a fitted model.

2.1.3.1. Problem Formation

The 3-D model fitting problem is generally stated as follows: Given 3-D surface models A and B and some points p_i, $i = 0, 1, \ldots, n - 1$ on surface A and their corresponding points q_i, $i = 0, 1, \ldots, n - 1$ on surface B, find a C^1 continuous mapping function F: $A \rightarrow B$ that satisfies $F(p_i) = q_i$ and $F(A)$ is a reasonable facial surface model.

The above statements also address a 3-D scattered data interpolation problem. One of the special cases in which all p_i and q_i are coplanar has been intensively investigated for image warping. Most image warping algorithms (15,16) can be extended conveniently for 3-D model fitting.

2.1.3.2. Image Warping Methods

To understand the image warping process, imagine that there is a rubber sheet with an image printed on it. Image warping has the same effect on the image as stretching the rubber sheet at interested points. Suppose the 2-D displacements at some image feature points are given, then an appropriate displacement interpolation scheme is necessary to derive the displacements of all other image points. Some algorithms using this interpretation have been proposed. Three major categories are triangulation-based methods, inverse distance methods, and radial basis function methods (17). In triangulation-based methods, a Delaunay triangulation is first constructed from image feature points, then polynomial interpolation of displacement values is performed in the image space and the warping function is found. Recently, it has been proved that this approach is optimal in terms of minimizing roughness, although the triangulation process involves no knowledge of the displacement values (18). In inverse distance methods or radial function methods, the basic idea is to interpolate the displacement values using weighted averages. The weights are derived from inverse distances or radial function values. The farther an image point is away from a feature point, the less it is affected. A common problem of these methods is that they treat all feature points equally. When the warping problem is a multilevel mapping by nature, like human expression synthesis, they often fail to produce expected results. To overcome this problem, a local, bounded, radial-basis method is introduced. The key idea is to define an effective range of a radial function. This one more dimension of freedom produces pleasanter results if the feature points and function bounds are carefully chosen. Ruprecht and Müller (17) wrote an excellent survey on these image warping algorithms.

2.1.3.3. Face Model Fitting Using Voronoi-Weighted Diagram

In our approach, each feature point is assigned a weight according to its influence. A weighted Voronoi diagram is constructed based on this information. Then, interpolation methods are applied to this Voronoi diagram. In this process, only the feature points with the largest weights generate Voronoi cells. This also means that only those points with large weights are mapped to their final positions. Once the feature points with large weights have been mapped correctly, their weights are reduced to the maximum weight of those unmapped feature points and their displacement values are set to zero. This procedure is performed iteratively until all feature points are correctly mapped. The advantage of this approach is that the underlying triangulation is changing between each iteration to fit the displacement scale. Both global rigid motions and local nonrigid deformations are modeled appropriately using this approach. More details are given in Ref. 19. A fitted face model is shown in Fig. 4b.

2.1.4. Face Geometric Model Compression

For quadrilateral meshes, scalable compression schemes are easy to derive (20). Here, the term *scalable* means that a layered coding scheme can generate object-oriented wire frames of different resolutions from a single source. As a result, a face model has multiple representations. When the model is rendered on a high-performance platform, a high-resolution version is used. Otherwise, a simplified version of the same model is adopted. Working in concert with existing texture-coding standards, this type of wire-frame compression technique provides scalability functionality to many computer facial animation systems. The proposed wire-frame compression scheme consists of the following three steps:

1. Coordinate system transformation. The first step is to transform the wire-frame data to a cylindrical coordinate system. The three resulting coordinates are radius, angle, and height (r, q, h). For a single rectangular mesh, each is represented as a matrix. The transformed coordinates are smoother than the original data and are more efficiently encoded (Fig. 5a).
2. Intramode and intermode coding. The wire-frame data are coded in either intramode or intermode. In the intramode, the entire wire-frame structure is coded into multiple layers of bitstreams and transmitted to the decoder. In the intermode, only the prediction errors of the wireframes are transmitted. Both coding modes are used for the purpose of downloading a new face model. However, the intermode exploits the predictive coding further when the decoder and the encoder have the same base surface model.
3. Pyramid progressive coding. The resulting wire-frame data is progressively coded, as shown in Fig. 5b. The down-sampling operation computes the average position of four neighboring points, which is then quantized and forms the next layer of data (lower spatial resolution) to be coded. The residual errors between

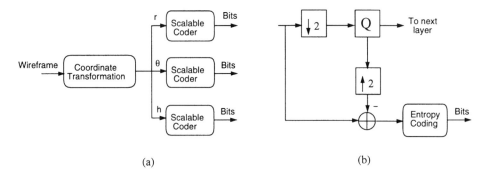

Fig. 5 (a) Block diagram of the wire-frame compression scheme; (b) pyramid progressive coding method.

the original data and the quantized average data are compressed using an entropy-coding scheme and transmitted to the decoder. The advantage of this scheme is that the computations are simple, local, and can be performed in parallel.

2.2. Facial Articulation Modeling

2.2.1. Articulation Parameters

For the animation of the facial model, using an accurate facial articulation model is crucial. One of the early methods for understanding facial motions is the *facial-action-coding system* (FACS) developed by Ekman and Friesen (21). The FACS system defined a minimum set of facial deformations for driving a face model.

Recently, some physical-based models have been developed (22,23). In these models, the skin and the tissues are modeled as elastic materials. Muscle activities are considered to be the stimulation of this mechanical system. The state of this system with minimum energy is computed using finite-element method. Although these models are close to the real facial motion model, physically, they often fail to generate realistic results. The reason is that the real facial muscle system is very complicated and most existing systems are only very rough approximations.

Another type of dynamic facial model describes the facial articulations in terms of surface deformations. An example is the facial-animation parameter (FAP) set defined by the MPEG-4 synthetic and natural hybrid coding (SNHC) subgroup (24). By defining this set of parameters, communications between distributed applications such as talking head, teleconferencing, and intelligent human agents are allowed.

There are a total of 68 parameters in FAP. Sixty-six of them are articulation parameters, or, more precisely, the position offsets of facial feature points relative to their neutral positions. For example, open-jaw depicts the downward movement of the jaw. Most FAP parameters can be obtained by computer vision techniques such as point-tracking techniques. In our system, the FAP stream is used to articulate the face model.

2.2.2. Model Deformation

Based on the FACS system, some geometric articulation models have been developed (25,26). In these models, the geometric deformations of a single or a group of muscle actions are described. The rendered results are the combination of these deformations. Magenant-Thalmann et al. (25) proposed a facial articulation model called the *abstract-muscle-action procedures* (AMAP), in which the facial movements are described by the displacements of the surface vertices. These displacements are implemented in many procedures. Each procedure is defined individually. This approach is similar to a performance-driven model in a sense that they both only describe the deformations in pure geometric terms. However, the AMAP is relatively tedious to develop. To overcome this problem, Kalra et al. (26) developed a free-form surface deformation system to emulate the facial motions. To some extend, this method simplifies the AMAP model. However, choosing the bounding boxes of the deformation parallel pipes is still a time-consuming task.

Pure geometric approaches usually are not successful in modeling nonlinearity in complex articulations. Physical models are introduced to handle these situations. Waters et al. (22,23) developed a three-layer muscle model. In this model, three connected spring layers are configured to model the outtermost facial skin, intermediate layer of soft tissues, and the underlying fixed bone structure. Muscles are modeled as elastic line segments with one end attached to the bone layer and the other end attached to the tissue layer. Without any muscle action, the facial expression stays neutral. When the contraction parameters of some muscles are provided, the dynamic system is no longer stable. Then, the finite-element method (Euler method) is applied to find the minimum energy state, which is the final facial expression. Platt and Badler (27) also developed a similar system for a high-resolution face surface model. A common problem of physical models is that the user has no control over the final results once the physical parameters are set. Usually, it is difficult to foresee the animation results from these data.

In some situations, because the goal is to make the synthesized face visually identical to a given image or a given video sequence, it is inefficient to articulate a complex model with the underlying structures of a face. With this motivation, some so-called *performance-driven* facial-articulation models were developed (28,29). The kernels in these models usually are tracking algorithms. Makers are put on the a face object to make the tracking process easier. Texture-mapping techniques are also exploited to create photo-realistic face images (30). In our implementation, because FAP parameters are employed, this type of articulation model is particularly appropriate. Based on this approach, we also integrated the physical muscle model and pure geometric articulation model into our system. Figure 6 shows a synthesized expression of "surprise."

2.2.3. Articulation Parameter Stream Compression

In our scheme, the principal component analysis (PCA) technique is applied to exploit the correlations among FAP parameters and to represent them in an optimal coordinate system in which the energy of the original signal concentrates in a smaller subspace. For the new representation, the correlations among different components are zero.

Suppose that for each time instance, or frame, the original FAP parameter is a 68-dimension vector v_i and that the ensemble average of v_i is $\bar{v}$. Then, the covariance matrix is $C = [1/(k-1)]\sum_{i=1}^{k}(v_i - \bar{v})(v_i - \bar{v})'$, where k is the number of FAP vectors. The eigenvectors of C are orthogonal to each other and span a new coordinate system. The eigenvalues of C indicate the energy distribution over each coordinate.

Then, the most significant eigenvalues are extracted and a subspace is formed by eigenvectors corresponding to these eigenvalues. This subspace contains most of the original signal energy. In other words, the projection of the original FAP vector in this subspace is a good approximation of that vector. From MPEG-4 test sequences that have been investigated, it was observed that the mean square error is less than 2% of the original

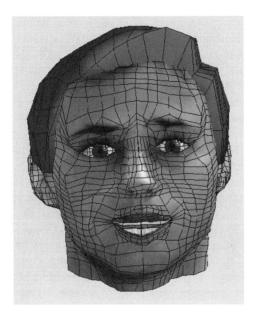

Fig. 6 Synthesized facial expression of "surprise."

signal energy if eight most significant eigenvectors are used. As a result, the representation dimension is reduced dramatically from 68 to 8. This projection process is also known as Karhunen–Loeve transform (KLT). After appropriate quantization, these eight components are differentially encoded. A block diagram of this method is shown in Fig. 7.

The source of compression in the PCA scheme is the correlations among FAP param-

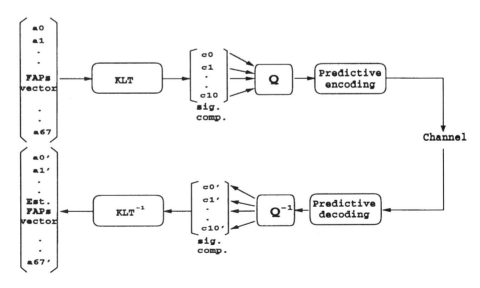

Fig. 7 Block diagram of a FAP compression scheme.

eters. These correlations reflect the fact that each expression involves many physically related muscle movements. Principal component analysis is a powerful tool to take advantage of this property.

2.3. Synthesis

Fast rendering of a realistic articulated face model is demanded by real-time applications. Several issues including environment model and texture-mapping techniques have to be carefully considered.

2.3.1. Environment Model

To get realistic animation results, environment variables such as the lighting sources and the background have to be considered. If it is assumed that the positions and the properties of lighting sources are known, the following procedure will incorporate the texture-mapping process and the lighting model into face model synthesizer. First, the facial surface model is rendered without texture mapping. Then, the texture is attached to the surface using α blending techniques. The simplest blending function is a liner interpolation, which is written as

$$c = \alpha c_t + (1 - \alpha)c_l, \quad \alpha \in [0, 1],$$

where c_l is the color of the surface with no texture attached to it, c_t is the color of the texture, and α is the transparency property of the texture. This process is automatically performed on some SGI workstations.

A more complicated problem is to combine a face model with its background. A possible solution for generating a simple background is to produce models for all background objects such as walls and windows. An even simpler approach is to put a background image behind the 3-D face model. If the background is not complicated and the view projection is stable, and the facial model is always in front of the background objects, then this approach often gives satisfactory results.

2.3.2. Texture Mapping

The texture-mapping technique is applied to generate the fine structures of a human face, like wrinkles and scars. With the help of dedicated graphics chips, workstations are able to perform texture mapping of high-resolution images in real time. The key issue in texture mapping is to establish the one-to-one correspondence between the face image (texture) and the 3-D model surface. When the face surface model is represented in planar polygons (triangles, for example), the mapping process is straightforward. First, the polygons in the 3-D model are projected to the 2-D face image, which produces a 2-D triangle. Then, the texture in the 2-D triangle is mapped to the 3-D triangle surface. Actually, only the coordinates of the vertices in 2-D need to be fed to the graphics synthesizer. The texture-mapping function is then trivially generated.

For the B-spline surface model, as discussed in previous sections, the problem is how to find the corresponding area of a bi-cubic Bézier patch in a face image. A property of the 3-D Bézier curve is that its projection on any 2-D plane is a 2-D Bézier curve. As a result of this property, in a face image plane, the corresponding area of a 3-D Bézier patch is also a Bézier patch. From this conclusion, a texture-mapping scheme is easily derived. First, all control points of a Bézier patch are projected to the image plane. Then, the mapping function is stated as

$$C(s, t) = I(s, t), \quad s, t \in [0, 1],$$

where $C(s, t)$ is the 3-D Bézier patch and $I(s, t)$ is the Bézier patch in a face image plane.

A problem with above method is that when a single face image is used, for example, only the front-view image is considered; a 3-D Bézier patch with a large area may be projected to a Bézier patch with a very small area in a face image. When this patch is rendered, a low-resolution texture is observed in that region. For facial model, patches on cheeks are vulnerable to this problem. A solution is to acquire both front-view and side-view texture images and blend them later in a rendering stage.

3. ANALYSIS

The analysis stage, in the context of model-based image communication, consists of extracting higher-level information from the input video sequence. We discuss two main types of information, both pertaining to motion estimation: rigid head motion tracking and nonrigid facial motion estimation.

3.1. Review of Past and Current Work

This section describes recent and current work in the field of analysis for model-based video and image communication. We first list some excellent reviews in the field of Model-Based Video Coding (MBVC) and then examine in detail one of the major research areas: head and facial feature tracking.

3.1.1. MBVC systems

The majority of research in MBVC has been driven by two goals:

- *Realistic* reproduction of the original input sequence, using the model-based approach to drive the bit rate below those achievable with conventional coders.
- *Synthetic* reproduction of the original input sequence using analysis of the 2-D scene to extract higher-level parameters to drive an artificial head model at the decoder.

In general, most approaches have been geared toward one of the above goals; however there are systems proposed to handle both cases effectively (31). Some excellent reviews on MBVC and its recent progress can be found in Refs. 32–34. We next discuss in more detail the recent work done in the areas related to head and feature tracking.

3.1.2. Rigid Motion

In this section, we review recent work in 3-D head tracking which involves the recovery of the 3-D rigid motion of a head in a 2-D video sequence. Most of the recent head-tracking (or more generally, 3-D motion estimation) research has taken one of two approaches: *motion from feature tracking* and *motion from optic flow or primitive vectors*. We consider both approaches in the following review.

In Ref. 30, the authors use optical flow equations to estimate both the global head motion and local facial expressions. Similiarly, in the work by Li et al., they estimate rotation parameters by enforcing the optical flow equations (35,36). Depth information is assumed in later work by using the CANDIDE face model and small motion from frame to frame is assumed. Nakaya and Harashima use the distribution of 2-D motion vectors on the head (computed on a block basis) to compute 3-D rotation and translation (31). The least square method is used and depth values are taken from the wire-frame head model. A straightforward template-matching technique is used by Kokuer and Clark to track features in two dimensions. A cylindrical head shape is assumed to compute the corresponding 3-D motion for each axis independently (37). Basu et al. use a 3-D ellipsoidal model of the head to interpret optical flow in terms of possible rotations of the head (38,39). This method seems to work well; however, the computation of optical flow is intensive and no 2-D tracking of facial features is performed. They test their method with real and synthetic sequences. Bozdaği et al. present an algorithm that uses an optical-flow-based framework to estimate 3-D global motion, local motion, and the adaptation of the wire-frame model simultaneously (40,41). The entire wire frame is flexible and can vary from frame to frame to minimize the error in the optic flow equations. To overcome the disadvantages of traditional optic-flow-based methods, they also include photometric effects. The approach seems to have a good theoretical basis but suffers from high computational load at each frame.

Horprasert et al. employ a parameterized tracking method to track five feature points on the face. Information regarding invariant cross-ratios from face symmetry and statistical modeling of face structures are used to compute three rotation angles (42). In this approach, head orientations close to frontal are very sensitive to tracking localization. In the research by Fukuhara and Murakami, a set of five facial features are tracked in two dimensions, giving motion vectors from frame to frame. These vectors are input to a three-layer neural net to determine the 3-D motion. The neural net is trained using many possible motion patterns of the head with an existing 3-D model (43). There are several drawbacks, including nonautomatic initial feature selection and a simple template matching to track the features. Also, the recovered 3-D motion is restricted to be one of a discrete number of possible motions determined by the training.

3.1.3. Nonrigid Motion

Another major research topic is the estimation of the nonrigid facial motion in a video sequence. This motion is the result of the many facial expressions humans make to communicate; therefore, research in the area of expression detection and recognition are relevant. The majority of work has concentrated on computing facial motion from frontal or near-frontal pose images. More recently, the incorporation of varying head pose has been examined.

Matsuno et al. use a deformable two-dimensional net, which they call a Potential Net, to detect expressions in input images. A training procedure is used to build a model of the net deformations for different facial expressions (44). In the work by Yacoob and Davis, an optic flow approach is used to analyze and represent facial dynamics from sequences. A midlevel symbolic representation is used to detect six expression as well as eye-blinking (45). Black and Yacoob use local parametric models to track both rigid and nonrigid motion. Six basic expressions are detected and some attempt is made to account for global, rigid, head motions (46). Essa and Pentland propose a new method of representing expression by building a database of facial expression characterized by muscle activations. They then use this database to recognize expression with an underlying physics-based model as well as by matching spatio-temporal motion–energy templates (47).

3.2. Tracking Rigid Global Motion

The overall pose computation module we propose is shown in Fig. 8. This module, one of the most crucial steps in the tracking algorithm, provides higher-level information on the 3-D movement of the head for use in synthesis, motion prediction, dynamic feature set changes, and so forth. As we see in Fig. 8, there are three main steps. First, an initial estimate of the scale and transformation matrix is obtained by using the 2-D–3-D feature

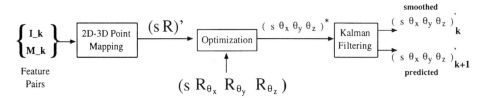

Fig. 8 The 3-D motion estimation/filter/prediction module.

point correspondences. Next, an optimization stage computes the best true rotation matrix using a gradient descent algorithm. Finally, the resulting angles and scale factor are optimally filtered to smooth the data and to predict the motion in subsequent frames.

3.2.1. 2-D-3-D Pose Estimation

The traditional pose estimation problem has a long history and many major issues have been encountered. We are faced with two main limitations that restrict the approaches we can use:

- A small number of facial features is available that is both *salient* and *rigid*.
- A nontrivial amount of *localization error* in the tracked points.

These two factors discourage the use of more sophisticated algorithms which require large numbers of feature pairs or are very sensitive to noise. Also, we do not make any assumptions on camera calibration, or the availability of any camera parameters. The general pose estimation problem in our scenario is assumed to be the following. The imaging model for true perspective projection converts model points to image points as follows (coordinates are in the camera coordinate system):

$$\begin{Bmatrix} x_i \\ y_i \end{Bmatrix} = f * \begin{Bmatrix} X_i/Z_i \\ Y_i/Z_i \end{Bmatrix}.$$

Because it is difficult to analyze systems with this model, a common approximation is a scaled orthographic projection system. In this model, object points are all assumed to have the same depth, $\overline{Z}$:

$$\begin{Bmatrix} x_i \\ y_i \end{Bmatrix} = f * \begin{Bmatrix} X_i/\overline{Z} \\ Y_i/\overline{Z} \end{Bmatrix} = \frac{f}{\overline{Z}} \cdot \begin{Bmatrix} X_i \\ Y_i \end{Bmatrix}. \tag{1}$$

$$= s \cdot \begin{Bmatrix} X_i \\ Y_i \end{Bmatrix}. \tag{2}$$

However, we do not have $\{X_i, Y_i, Z_i\}^T$ (the model points in the camera coordinate system). What we have is $\{U_i, V_i, W_i\}^T$ (model points in the model coordinate system). We express the transformation from the object coordinate system to the image coordinate system as

$$\begin{Bmatrix} X_i \\ Y_i \\ Z_i \end{Bmatrix} = \mathbf{R}_{oi} \cdot \begin{Bmatrix} U_i \\ V_i \\ W_i \end{Bmatrix} + T_{oi}.$$

If we use the first point as a reference,

$$\begin{Bmatrix} X_i - X_0 \\ Y_i - Y_0 \\ Z_i - Z_0 \end{Bmatrix} = \mathbf{R}_{oi} \cdot \begin{Bmatrix} U_i - U_0 \\ V_i - V_0 \\ W_i - W_0 \end{Bmatrix}, \tag{3}$$

where $\mathbf{R}_{oi} = [\mathbf{i}, \mathbf{j}, \mathbf{k}]^T$. Substituting Eq. (3) into Eq. (2) and using the corresponding image point as reference, we obtain

$$\begin{Bmatrix} x_i - x_0 \\ y_i - y_0 \end{Bmatrix} = s \cdot \begin{bmatrix} \mathbf{i} \\ \mathbf{j} \end{bmatrix} \cdot \begin{Bmatrix} U_i - U_0 \\ V_i - V_0 \\ W_i - W_0 \end{Bmatrix}. \tag{4}$$

Assuming we have at least four noncollinear and unique points, $i = 0$ to 3, we can create the following linear system:

$$\mathbf{Image} = s \cdot \begin{bmatrix} \mathbf{i} \\ \mathbf{j} \end{bmatrix} \cdot \mathbf{Model}.$$

After a simple matrix inversion (which exists because of the noncollinearity constraints),

$$s \cdot \begin{bmatrix} \mathbf{i} \\ \mathbf{j} \end{bmatrix} = \mathbf{Image} \cdot (\mathbf{Model})^{-1}. \tag{5}$$

We can then use the properties of orthonormal matrices to compute the desired parameters s, $\mathbf{i}$, $\mathbf{j}$, and $\mathbf{k}$:

$$\|s \cdot \mathbf{i}\| = \|s \cdot \mathbf{j}\| \Rightarrow s_{avg} = \frac{\|s \cdot \mathbf{i}\| + \|s \cdot \mathbf{j}\|}{2}$$

$$\hat{\mathbf{i}} = \frac{s \cdot \mathbf{i}}{\|s \cdot \mathbf{i}\|} \quad \text{and} \quad \hat{\mathbf{j}} = \frac{s \cdot \mathbf{j}}{\|s \cdot \mathbf{j}\|}.$$

$$\hat{\mathbf{k}} = \hat{\mathbf{i}} \times \hat{\mathbf{j}}$$

So, our final result is an estimate of the pose and scale of the object from the set of four feature point matches:

$$\hat{s} \cdot \hat{\mathbf{R}}_{oi} = s_{avg} \cdot \begin{bmatrix} \hat{\mathbf{i}} \\ \hat{\mathbf{j}} \\ \hat{\mathbf{k}} \end{bmatrix}. \tag{6}$$

Because we are assuming an orthographic + scale projection system, we cannot recover the translation in the Z axis. However, the 2-D translation, **T**, can be calculated as

$$\hat{\mathbf{T}} = \begin{bmatrix} x_0 \\ y_0 \end{bmatrix} - \hat{s} \cdot \begin{bmatrix} \hat{\mathbf{i}} \\ \hat{\mathbf{j}} \end{bmatrix} \cdot \begin{bmatrix} U_0 \\ V_0 \\ W_0 \end{bmatrix}. \tag{7}$$

The overhead caret symbol indicates that these are estimates to the true pose and scale due to the measurement noise in the image points. Ideally, this estimate would represent a true orthonormal rotation matrix. However, in calculating Eq. (5), we have forced the transformation to map model points to noisy image points and the result is not a true rotation.

3.2.2. Pose Optimization

The recovered pose from the previous section is only an estimate to the true pose (for the assumed projection model). Applying this nonorthonormal transform to the 3-D head model results in nonrigid deformations, which is not desired. Also, because we would like to apply filtering techniques to the recovered pose, it is necessary to convert the estimated transform to parameters which would make sense to filter, such as rotation angles about each axis. Our approach is the following.

- Express the pose as three consecutive rotations, one about each axis
- Apply optimization techniques to recover optimal angles and scale for the estimated pose
- Filter these angles to obtain optimal estimates for the current pose and to predict future poses

The desired rotation matrix can be represented in an infinite number of ways. We choose to represent it as rotations of (θ_y, θ_x, θ_z) about the respective axes:

$$s \cdot R = s \cdot R_{\theta_y} R_{\theta_x} R_{\theta_z}. \tag{8}$$

We can solve for the desired angles and scale by defining the following error measure:

$$\epsilon = f(\theta_x, \theta_y, \theta_z, s)$$
$$= \|\hat{s} \cdot \hat{\mathbf{R}}_{oi} - s \cdot R_{\theta_y} R_{\theta_x} R_{\theta_z}\|_F^2,$$
$$(\theta_x, \theta_y, \theta_z, s)^* = \min_{(\theta_x, \theta_y, \theta_z, s)} \epsilon.$$

This minimization can be solved by using a version of Powell's quadratically convergent method. The result, then, is a representation of the pose as three angles and a scale factor that minimize the error with respect to the computed transform.

3.2.3. Filtering and Prediction

Localization errors in the feature-tracking module propagate to the recovered angles and scale computed above. When used for synthesis, applying these pose computations to a head model results in jerky head movements, which is visually unacceptable. One way to overcome this is to use optimal filtering techniques to process the measurements of angles and scale for each frame. To do this, we implement one of the more well-known optimal filters, the discrete Kalman filter. The Kalman filter is a recursive procedure that consists of two stages: time updates (or prediction) and measurement updates (or correction). At each iteration, the filter provides an optimal estimate of the current state using the current input measurement and produces an estimate of the future state using the underlying state model.

The values which we want to smooth and predict are the three angles and scale that determine the 3-D pose: $(s, \theta_x, \theta_y, \theta_z)$. We can filter these independently of each other, using a discrete-time Newtonian physical model of rigid-body motion. In general, the linear difference equations for each process can be written as

$$\mathbf{x}_{k+1} = A_k \mathbf{x}_k + B\mathbf{u}_k + \mathbf{w}_k, \tag{9}$$

$$\mathbf{z}_k = H_k \mathbf{x}_k + \mathbf{v}_k, \tag{10}$$

where

$$\mathbf{x} = \begin{bmatrix} \theta \text{ (angle)} \\ \omega \text{ (veloc.)} \\ \alpha \text{ (accel.)} \end{bmatrix}, \quad A = \begin{bmatrix} 1 & 1 & 1 \\ 0 & 1 & 1 \\ 0 & 0 & 1 \end{bmatrix},$$

$$H = [1\ 0\ 0].$$

We can ignore the terms $B\mathbf{u}_k$ because we assume no external driving forces. The variables $\mathbf{w}$ and $\mathbf{v}$ represent the process and measurement noise, respectively, and are independent and white, with normal probability distribution

$$p(w) \approx N(0, Q), \quad p(v) \approx N(0, R).$$

With that said, the filtering algorithm is, for each incoming process measurement z_k,

$$K_k = P_k^- H_k^T (H_k P_k^- H_k^T + R_k)^{-1},$$

$$\hat{x}_k = \hat{x}_k^- + K(z_k - H_k \hat{x}_k^-),$$

$$P_k = (I - K_k H_k) P_k^-,$$

$$\hat{x}_{k+1}^- = A_k \hat{x}_k + B u_k,$$

$$P_{k+1}^-1 = A_k P_k A_k^T + Q_k.$$

The filtered estimate can be used to synthesize the 3-D motion of the head model, and the prediction can be used to produce templates for tracking the facial features in the next frame.

3.3. Tracking Nonrigid Local Motion

In contrast to the estimation of the global head pose, the estimation of the nonrigid motion from a single view is ill-posed; that is, given the two-dimensional displacement of a marking point in the face surface, the three-dimensional motion of the corresponding point in the head model cannot be computed uniquely.

One possibility is to consider stereo vision, in which the actual three-dimensional local motion of all the points that can be accurately matched from one view to the other can be easily obtained from geometrical constrains provided by the camera calibration parameters. This approach, however, might not be suitable for scenarios such as the one in low-rate video coding where only a single camera is most likely to be used or accurate camera calibration can not be achieved.

Another possibility is to impose some constraints to reduce the degree of freedom of the nonrigid local motion of the facial features so that it can be estimated from a single view. Finally, a model-based approach for nonrigid motion estimation of facial features can be used. Such an approach would require a head/face model in which nonrigid motion of the facial features is parameterized.

3.3.1. 2-D Motion + Constraints

One of the methods available for computing the 3-D motion vectors associated with nonrigid facial motion is the application of constraints to the 2-D motion of the image points. As mentioned earlier, there are two type of motion the points on the image plane can undergo: rigid and nonrigid motion. The location of the rigid points allows us to compute the 3-D rigid motion and then align the 3-D head model to the 2-D image. The nonrigid features will undergo the estimated rigid motion plus some nonrigid motion specific to that region of the face. Each image frame is composed of points from each of these sets; so for images at time t_i, we have

$$\text{Image}^i = \{I^i_{\text{rigid}}\ I^i_{\text{nonrigid}}\}$$

$$\text{Image}^i_{\text{rigid}} = \lfloor T_i(\circ \text{Model})_{\text{rigid}} \rfloor_{2d}$$

$$\text{Image}^i_{\text{nonrigid}} = \lfloor T_i(\circ \text{Model})_{\text{nonrigid}} \rfloor_{2d} + \Delta V_i.$$

The first of these equations states that the image at time t_i is composed of an image resulting from rigid motion and an image resulting from nonrigid motion. In the second equation, the image resulting from rigid motion is shown to be the result of transforming the 3-D model and projecting into the 2-D plane. The third equation states that the image resulting from nonrigid motion is composed of the rigid transformation, followed by projection to two dimensions and then some 2-D deformation. We can then use the location of the nonrigid features to back-project onto a 3-D surface and retrieve the new model coordinates for that particular image feature. One way to create this surface is to use a Cyberware head scan of an individual, with appropriate smoothing to reduce surface discontinuities (Fig. 9). The underlying assumption we make here, of course, is that points on the face travel along its surface in three dimensions. This is certainly a valid assumption for facial features such as the eyebrow corners. However, even for points around the mouth and eyes, this assumption is valid enough to be used in recovering good estimates of the 3-D motion vectors. The estimation of the new 3-D points can be used to create a modified 3-D model:

$$\text{Model}' = \text{Model} + \lceil \text{Image}^i_{\text{nonrigid}} \rceil_{3d}.$$

Finally, using the new 3-D model plus a reference pose (e.g., the pose of the head in the first frontal image frame), we can create a new image. The result is a pair of images that use the same frame of reference and can be used to compute the facial-action parameters that describe the rigid and nonrigid motion in the sequence. After computing the desired facial motion, we can apply this to the synthetic 3-D model along with the computed rigid head pose. This procedure is shown later in Fig. 22.

$$\text{Image}^{i'} = \lfloor T_0 \circ (\text{Model}') \rfloor_{2d}$$

$$\text{FAP} = f(I^0, I^{1'})$$

3-D Model-Based Image Communication

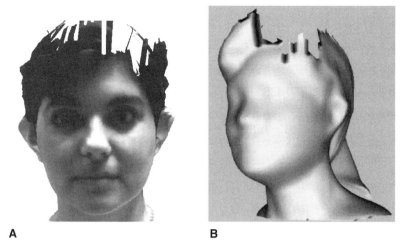

Fig. 9 (A) Original Cyberware head scan; (B) smooth surface approximation used to recover 3-D coordinates.

At this point, we should mention that the constraint that points move along surface of the model is by no means the only one we can impose. We can also recover 3-D-motion vectors for several facial features by enforcing the constraint that the motion is perpendicular to the normal vector of the surface of the face.

First, we need to obtain the normal vectors of the face surface given by the Cyberscan range data. Let S be a surface given by the range data as

$$S = \{\mathbf{X}(i,j) \in \Re^3, i = 0, 1, \ldots, N, j = 0, 1, \ldots, M\}$$

We compute the normal vector $\mathbf{n}(\mathbf{X}_o) = [-n_x - n_y\ 1]'$, by fitting the plane

$$(\mathbf{X} - \mathbf{X}_o) \cdot \mathbf{n}(\mathbf{X}_o) = 0$$

to the surface S using minimum least squares over a region around $\mathbf{X}_o$; that is, we find (n_x, n_y) that minimizes the error:

$$\sum_{(i,j) \in W_{\mathbf{X}_o}} [z(i,j) - z_o - n_x(x(i,j) - x_o) - n_y(y(i,j) - y_o)]^2,$$

where $\mathbf{X}(i,j) = [x(i,j)\ y(i,j)\ z(i,j)]'$ and $\mathbf{X}_o = [x_o\ y_o\ z_o]'$ are surface points and $W_{\mathbf{X}_o}$ is a region around $\mathbf{X}_o$.

Let $\mathbf{X}_o \in \Re^3$ be the position of a facial feature in the face surface of the head model, and let $\mathbf{T}$ be a 4×4 homogeneous transformation matrix representing the estimated head pose at a given frame such that

$$\mathbf{X}'_o = \mathbf{T} \cdot \mathbf{X}_o = \mathbf{R} \cdot \mathbf{X}_o + \mathbf{t},$$

where $\mathbf{X}'_o \in \Re^3$ is the position of the feature in the image space, $\mathbf{R}$ is a rotation matrix, and $\mathbf{t}$ is a translation vector. Note that although the z component of $\mathbf{X}'_o$ is not used to render a synthetic image with orthographic projection, it is well defined. On the other

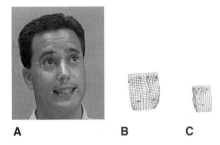

Fig. 10 Results of 3-D-motion vector computation. (A) Location of 2-D features in the video sequence; (B) the computed 3-D-motion vectors at the estimated pose; (C) the 3-D-motion vectors seen from another viewpoint.

hand, let $\mathbf{X}'_1$ be the position of facial feature in the image frame of the video sequence obtained by template matching.

First, we rotate the normal vector of the surface, $\mathbf{n}' = \mathbf{R} \cdot \mathbf{n}$, so that it can be used in the image space. Then, we enforce the motion to lay perpendicular to this normal vector; that is, we compute z'_1 from

$$z'_1 = z'_o - \frac{n'_x}{n'_z}(x'_1 - x'_o) - \frac{n'_y}{n'_z}(y'_1 - y'_o).$$

Finally, we compute the 3-D-motion vector $\mathbf{X}_1 - \mathbf{X}_o$ in the model space using the inverse rotation matrix:

$$(\mathbf{X}_1 - \mathbf{X}_o) = \mathbf{R}^{-1} \cdot (\mathbf{X}'_1 - \mathbf{X}'_o).$$

Shown in Fig. 10 is an example of this approach.

3.3.2. Model-Based Nonrigid Local-Motion Estimation

Another approach to recovering nonrigid local motion is to use a model-based approach with parametric models. Let us assume we have a face/head model in which each vertex position is obtained from

$$\mathbf{x}'_i = \mathbf{x}_i + \mathbf{v}_i(\alpha), \quad i = 1, \ldots, N, \tag{11}$$

where $\mathbf{x}_i, i = 1, \ldots, N$ provides the shape of the individual being modeled, $\mathbf{v}_i(\alpha)$ is a parameterized description of each vertex motion, and the vector α represents the nonrigid motion state.

Note that for rigid points such as the eye corners, nose, and so forth, the motion component is zero. Therefore, independently of the facial expression, the head head pose $\mathbf{T}$ can be estimated from the rigid points of the face.

All the facial features not occluded at a pose $\mathbf{T}$ are projected to the image plane at the position

$$\mathbf{X}'_i(\alpha) = \mathbf{T} \cdot (\mathbf{x}_i + \mathbf{v}_i(\alpha)), \quad i \in S, \tag{12}$$

where S is the list of visible points at the given view.

The problem of estimating the nonrigid motion can be solved by finding an α that minimizes the overall error:

3-D Model-Based Image Communication

$$\epsilon^2 = \sum_{i \in S} \|\mathbf{X}'_i(\alpha) - \mathbf{X}_i\|^2, \tag{13}$$

where $\mathbf{X}_i$ is the position of the vertex i obtained from the input video frame using a visual matching technique.

With this techniques, the quality of this analysis depends on the accuracy with which the model describe the face expressions. Simple models might not provide good results. However, the optimization algorithm might show convergence problems if the models are too complex or have too many degrees of freedom.

4. MBVC SYSTEM IMPLEMENTATIONS

Although progress in MBVC has been steady, it is still a very immature field compared to traditional coding techniques. One of the most lacking accomplishments has been the development of complete coders/encoders with which we can test the validity of the research done so far. We aim to provide such a test bed (Fig. 11) and to demonstrate its application in a realistic scenario. The coding system we assume is shown in Fig. 12. It takes a computer graphics approach and analyzes the video sequences to extract higher-level knowledge regarding motion, expressions, and so forth. These parameters are then sent along the channel to drive a head model on the receiving end. Although this method

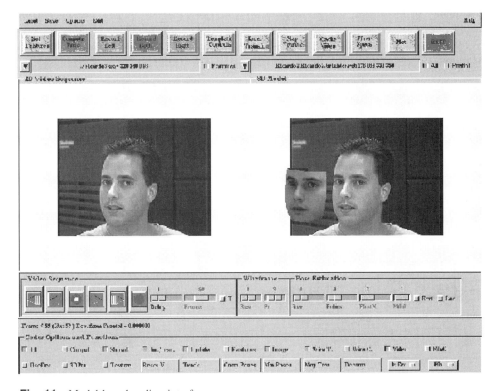

Fig. 11 Model-based coding interface.

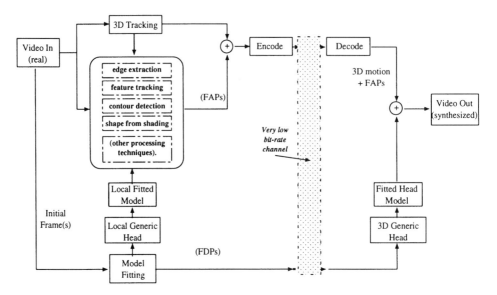

Fig. 12 Model-based video coding system.

has potentially lower bandwidth, the end result is a video sequence that is dependent on the quality of the underlying model, how well it is fitted to the subject, the underlying parameterization (muscle, tissue, etc.), rendering hardware available, and other factors. We now give a brief description of our basic approach. The research discussed in previous sections is used to implement the various stages of the system.

4.1. Tracking System

In Fig. 13, we can see the main tracking system for 3-D and 2-D motion. It consists of two main steps: the *initialization* and the main *tracking loop*.

4.1.1. Initialization

Initialization of the facial feature locations is done automatically using the first image frame and the texture map obtained from the Cyberware scanner (Fig. 14). After the initial

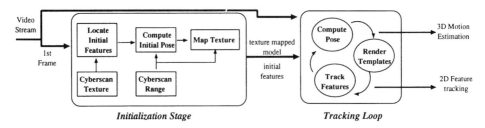

Fig. 13 Object-tracking system.

3-D Model-Based Image Communication

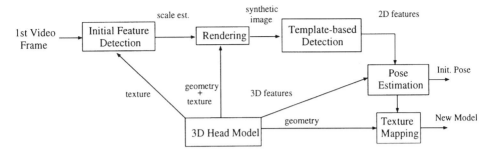

Fig. 14 Initialization for the tracking system.

features points are obtained, the 3-D pose is computed and applied to the head model (48). This aligns the model with the first video frame so that the initial texture mapping can be performed. This texture map, along with any texture updates, is then used to create templates for the subsequent frames.

4.1.2. Tracking Loop

The main tracking loop consists of three steps:

1. Compute pose from 2-D–3-D point pairs.
2. Render templates using pose and 3-D head model.
3. Locate features in current frame.

We can note at this stage that the system has three major outputs which can be of use in a MBVC system: *3-D motion estimates, 2-D feature tracking,* and *synthesized approximations to the original input sequence.* Also, using the methods discussed earlier, we can convert the 2-D feature tracking into the appropriate 3-D motion vectors to drive the facial expressions of a synthetic model. (See Fig. 15.) The current system will implement basic eyebrow, eye, mouth, and jaw movements.

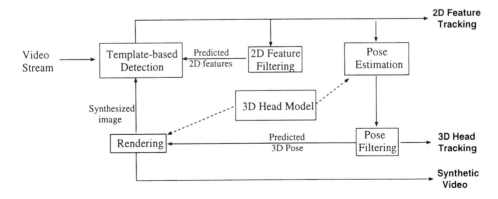

Fig. 15 Main tracking loop.

4.1.3. Head Modeling

Our proposed tracking system uses an underlying 3-D model of the object being tracked. We choose a straightforward method of obtaining head models by using a Cyberware 3-D range scanner. The 3-D coordinates of the features to be tracked are obtained directly from the range data. One factor, which is a concern for real-time rendering purposes, is the high resolution of the data. To remedy this, a subsampled version of the head scans were used, along with texture-mapping techniques, to create very accurately synthesized images.

4.2. Practical Implementation Issues

Currently, the system is implemented on an SGI Onyx Reality Engine. Using the highly optimized rendering pipeline, the rigid-head tracking algorithm runs at greater

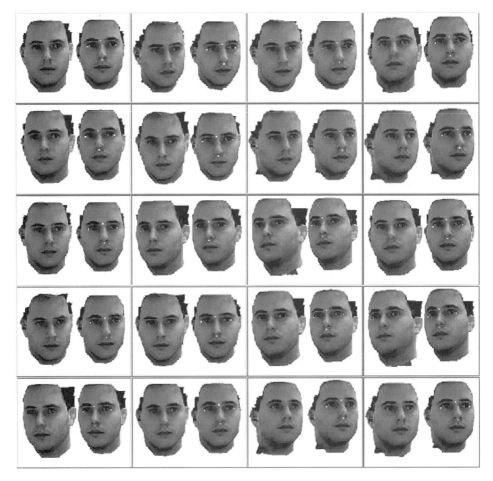

Fig. 16 Visual comparison of recovered pose; selected frames. *Left*: rendered model; *right*: original frame.

than 10 fps (frames per second). Input to the system is a monitor-mounted COHU gray-scale camera with video field size of 360×243 and image frame size of 720×486. The most computationally expensive step in the video analysis is the template-matching stage. This has been implemented in parallel using mutliple processors to increase performance and maintain frame rates when tracking large numbers of features. Video frames are processed as quickly as possible and no attempt is made to keep pace with the steady 30-fps video input signal. For all graphics operations, standard OpenGL libraries have been used so that the system is easily implemented on other platforms (Sun, HP, PC). The system has also been used in conjunction with a D1 digital-tape machine and VLAN interface to analyze long video sequences without loss of frames. This is useful in conjunction with other projects such as video analysis for gisting and face recognition using video sequences, as both research areas can make substantial use of head tracking in general.

4.3. Experiments and Results

Our proposed head- and feature-tracking system was tested on several real and synthetic video sequences. The synthetic sequences were created using a Cyberware Head Scan and a prespecified motion path, so that the ground truth for the angles, scale, and translation was known. In the real sequences, the frames were gray scale, 320×240, and captured at 30 fps. In Fig. 16, we see the original synthetic sequence side by side with the texture-mapped head model rendered at the computed pose. Because we know the motion in the synthetic case, we can compare the recovered angles and scale directly with the ground truth. Figure 17a shows plots of the recovered angles (optimal, filtered, and true values). It is also interesting to examine the accuracy of the Kalman filter predictions for the 3-D pose. Figure 17b show a plot of the predicted and actual angles for the Y axis using the Kalman filter model.

For the real video experiments, ground-truth values are not known. However, a visual comparison can be made to see how accurately the computed angles and scale follow the 3-D motion of the head. In Fig. 18, we see the original sequence side by side

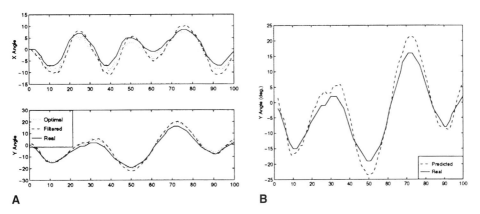

Fig. 17 Results of pose recovery for the synthetic sequence. (A) Optimal, filtered, and true measurements—*top*: θ_X; *bottom*: θ_Y. (B) Predicted and true measurements for the θ_Y.

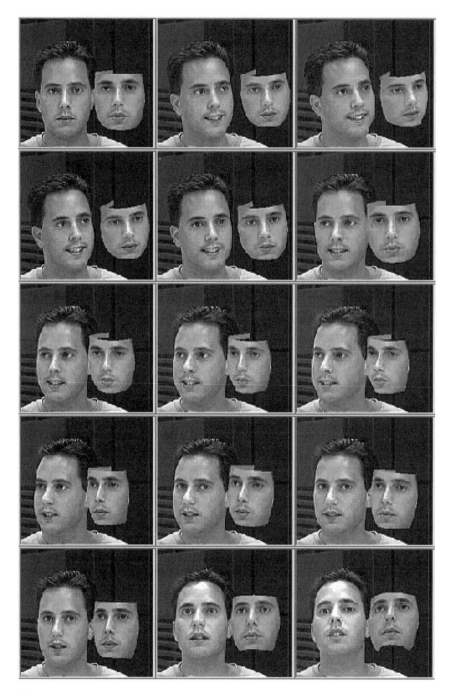

Fig. 18 Visual comparison of recovered pose; selected frames from sequence A. *Left*: original frame; *right*: rendered model.

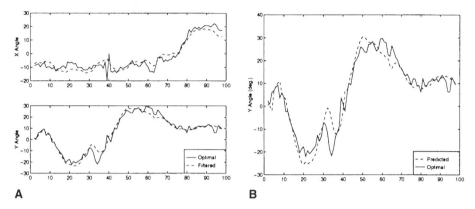

Fig. 19 Results of pose recovery for real sequence A. (A) Optimal and filtered measurements—*top*: θ_X; *bottom*: θ_Y. (B) Predicted and optimal measurements for Y angle.

with the texture-mapped head model rendered at the computed pose. Figure 19a shows plots of the recovered angles for one of the test sequences (optimal and filtered values). Figure 19b shows plots of the predicted pose and the actual computed pose angles for the same sequence. Finally, in Fig. 20, we can see the 3-D wire-frame model overlayed on the original sequence using the computed pose.

We can also evaluate the performance of our system as a facial feature tracker. We show the results of tracking features in two dimensions that undergo global motion as well as global + nonrigid local motion (Fig. 21). The rigid points were the outer eye corners, nose base, and middle of the nose bridge. Nonrigid points were the tips of the eyebrows and the mouth corners.

Finally, we examine the computation of the 3-D-motion vectors for nonrigid points. As explained earlier, these vectors are computed from the 2-D trajectories of

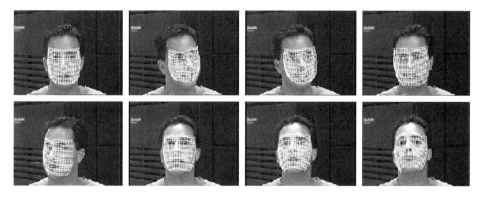

Fig. 20 Three-dimensional model wire-frame tracking. Wire-frame mesh is overlayed on the original images using the recovered pose.

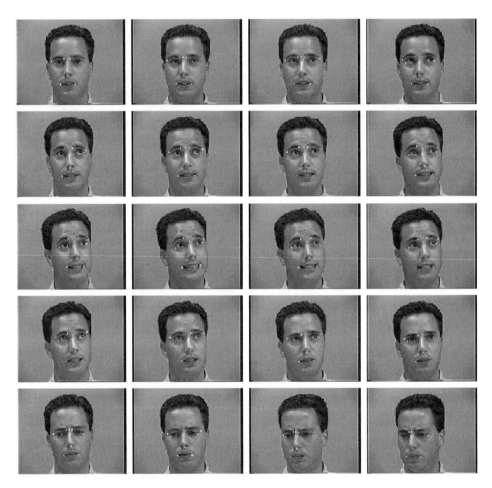

Fig. 21 Feature-tracking results; selected frames from sequence B.

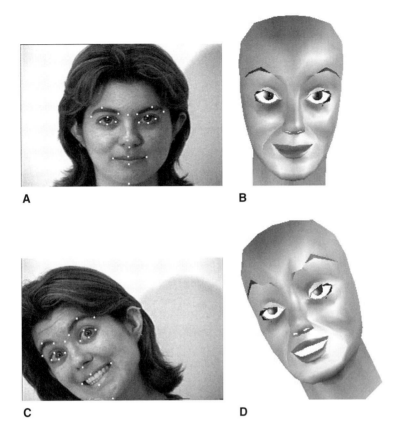

Fig. 22 (A) Reference frame from video sequence; (B) neutral head model; (C) expression at nonfrontal pose: (D) synthesized expression.

nonrigid features points during the tracking. Motion is constrained in three dimensions to lie on the surface of the head model using a locally planar surface assumption. Figure 22 shows the results for the eyebrow and mouth corners. The recovered 3-D vectors have been applied to the well-known Parke head model (49) to synthesize the nonrigid motion.

5. FUTURE WORK

Future work by the authors is planned in many aspects of model-based image communication. Improvements on automatic model fitting as well as more robust tracking methods are being examined. Both theoretical and practical research is being pursued and the feasibility of real-time systems is still a driving factor. Also, the establishment of the MPEG-4 SNHC standard will undoubtedly affect research directions and applications.

6. CONCLUSIONS

In this chapter, we have explored the area of image and video communication using model-based techniques. The three main areas of the model-based paradigm were discussed:

modeling, analysis, and *synthesis.* Past work in these areas were presented as well as ongoing work of the authors. We also presented a real-time model-based coding system which included rigid and nonrigid head tracking, modeling, and synthesis. The results obtained from this system are encouraging and helped to validate the model-based framework as a viable approach to the coding and representations of video and images.

REFERENCES

1. R Lopez, A Colmenarez, T Huang. Vision-based head and facial feature tracking. Displays Fed Lab Annual Symposium, Adelphi, MD, 1997.
2. A Colmenarez, R Lopez, T Huang. 3d Model-based head tracking. In: Visual Communications and Image Processing. SPIE Press, 1997.
3. N Magnenat-Thalmann, D Thalmann. Complex models for animating synthetic actors. IEEE Computer Graphics Applic 11(5):32–44, 1991.
4. W Boehm, G Farin, J Kahmann. A survey of curve and surface methods in CAGD. Computer Aided Geomet Design 1(1):1–60, 1984.
5. G Farin. Curves and Surfaces for Computer Aided Geometric Design. Orlando, FL: Academic Press, 1993.
6. G Farin. Triangular Bernstein–Bézier patches. Computer Aided Geometr Design 3(2):83–128, 1986.
7. G Nielson. A Transfinite, Visually Continuous, Triangular Interpolant. Philadelphia: SIAM, 1987, pp 235–246.
8. J Peters. Local smooth surface interpolation: a classification. Computer Aided Geometr Design 7(1–4):191–195, 1990.
9. B Piper. Visually Smooth Interpolation with Triangular Bezier Patches. Philadelphia: SIAM, 1987, pp 221–234.
10. H Tao. Face modeling. Technical report, University of Illinois at Urbana-Champaign, Dept. of Electrical and Computer Engineering, 1996.
11. N Arad, D Reisfeld. Image warping using few anchor points and radial functions. In: Computer Graphics Forum, Volume 14. London: Eurographics Basil Blackwell Ltd, 1995, pp 35–46.
12. C Frederick, EL Schwartz. Conformal image warping. IEEE Computer Graphics Applic 10(2): 54–61, 1990.
13. TW Sederberg, SR Parry. Free-form deformation of solid geometric models. In: DC Evans, RJ Athay, eds. Computer Graphics (SIGGRAPH '86 Proceedings), 1986, Vol. 20, pp 151–160.
14. IA Essa, S Sclaroff, A Pentland. A unified approach for physical and geometric modeling for graphics and animation. In: A Kilgour, L Kjelldahl, eds, Computer Graphics Forum (EUROGRAPHICS '92 Proceedings), 1992, Vol. 11, pp 129–138.
15. G Wolberg. Skeleton-based image warping. Visual Computer 5(1/2):95–108, 1989.
16. G Wolberg. Digital Image Warping. Los Alamitos, CA: IEEE Computer Society Press, 1990.
17. D Ruprecht, H Müller. Image warping with scattered data interpolation. IEEE Computer Graphics Applic 15:37–43, March 1995.
18. F Aurenhammer. Voronoi diagrams—A survey of a fundamental geometric data structure. ACM Comput Surveys, 23(3):345, 1991.
19. H Tao, T Huang. Multi-scale image warping using weighted voronoi diagram. IEEE International Conference Imaging Processing (ICIP'96), 1996.
20. H Tao, T Huang, H Chen, TPJ Shen, A Bayya. Technical description of uiuc/rockwell MPEG-4 snhc proposal. ISO/IEC JTC1/SC29/WG11 MPEG96/M1239. September 1996.
21. P Ekman, WV Friesen. Facial Action Coding. Palo Alto, CA: Consulting Psychologists Press Inc., 1978.
22. K Waters. A muscle model for animating three-dimensional facial expression. In: MC Stone, ed. Computer Graphics (SIGGRAPH '87 Proceedings), 1987, Vol. 21, pp 17–24.

23. D Terzopoulos, K Waters. Analysis and synthesis of facial image sequences using physical and anatomical models. IEEE Trans Pattern Anal Machine Intel PAMI-15, 1993.
24. MPEG-4 Video and SNHC. Text for CD 14496-2 visual. ISO/IEC JTC1/SC29/WG11 N1902. October 1997.
25. N Magnenat-Thalmann, E Primeau, D Thalmann. Abstract muscle action procedures for human face animation. Visual Computer 3(5):290–297, 1988.
26. P Kalra, A Mangili, N Magnenat-Thalmann, D Thalmann. Simulation of facial muscle actions based on rational free form deformations. In: A Kilgour and L Kjelldahl, eds., Computer Graphics Forum (EUROGRAPHICS '92 Proceedings), 1992, Vol. 11, pp 59–69.
27. SM Platt, NI Badler. Animated facial expressions. Computer Graphics 15(3):245–252, 1981.
28. L Williams. Performance-driven facial animation. In: F Baskett, ed. Computer Graphics (SIGGRAPH '90 Proceedings), 1990, Vol. 24, pp 235–242.
29. IA Essa. Analysis, interpretation and synthesis of facial expressions. Technical Report 303, MIT Media Lab, 1994.
30. CS Choi, K Aizawa, H Harashima, M Takebe. Analysis and synthesis of facial image sequences in model-based coding. IEEE Trans Circuits Syst Video Technol CSVT-4(3):257–275, 1994.
31. Y Nakaya, H Harashima. Model-based/waveform hybrid coding for low-rate transmission of facial images. IEICE Trans Commun E75-B(5):377–384, 1992.
32. K Aizawa, TS Huang. Model-based image coding: Advanced video coding techniques for very low bit-rate applications. Proc IEEE 83:259–271, February 1995.
33. DE Pearson. Developments in model-based video coding. IEEE 83:892–906, June 1995.
34. T Ebrahimi, E Reusens, W Li. New trends in very low bitrate video coding. Proc IEEE 83:877–891, June 1995.
35. H Li, P Roivainen, R Forchheimer. 3-d Motion estimation in model-based facial image coding. IEEE Trans Pattern Anal Machine Intell PAMI-15(6):545–555, 1993.
36. H Li, P Roivainen, R Forchheimer. Two-view facial movement estimation. IEEE Trans Circuits Sys Video Technol 4(3):276–287, 1994.
37. M Kokuer, AF Clark. Feature and model tracking for model-based coding. Proceedings of the IEEE Conference on Image Processing and its Applications, Maastricht, Netherlands, 1995, pp 135–138.
38. S Basu, I Essa, A Pentland. Motion regularization for model-based head tracking. Technical Report 362, MIT Media Laboratory Perceptual Computing Section, Cambridge, MA, 1996.
39. I Essa, T Darnell, A Pentland. Tracking facial motion. Proceedings of IEEE Nonrigid and Articulated Motion Workshop, Austin, TX, 1994, pp 36–42.
40. G Bozdagi, M Tekalp, L Onural. 3-d Motion estimation and wireframe adaptation including photometrics for model-based coding of facial image sequences. IEEE Trans Circuits Syst Video Technol CSVT-4(3):246–256, 1994.
41. G Bozdagi, M Tekalp, L Onural. An improvement to MBASIC algorithm for 3-d motion and depth estimation. IEEE Trans Image Process IP-3(5):711–716, 1994.
42. T Horprasert, Y Yacoob, LS Davis. Computing 3-d head orientation from a monocular image sequence. Proceedings of Conference on Computer Vision and Pattern Recognition, Seattle, WA, 1994, pp 242–247.
43. T Fukuhara, T Murakami. 3-d Motion estimation of human head for model-based image coding. IEE Proc 140:26–35, 1993.
44. K Matsuno, CW Lee, S Kimura, S Tsuji. Automatic recognition of human facial expressions. In IEEE International Conference on Computer Vision. Los Alamitos, CA: IEEE Computer Society Press, 1995, pp 352–358.
45. Y Yacoob, LS Davis. Recognizing human facial expressions from long image sequences using optical flow. IEEE Trans Pattern Anal Machine Intell PAMI-18(6):636–642, 1996.
46. MJ Black, Y Yacoob. Tracking and recognizing rigid and non-rigid facial motions using local parametric models of image motions. Technical Report CS-TR-3401, Xerox Palo Alto Research Center, 1995.

47. IA Essa, AP Pentland. Facial expression recognition using a dynamic model and motion energy. In IEEE International Conference on Computer Vision. Los Alamitos, CA: IEEE Computer Society Press, 1995, pp 360–367.
48. A Colmenarez, TS Huang. Maximum likelihood face detection. Proceedings International Conference on Automatic Face and Gesture Recognition, Killington, VT, 1996, pp 139–152.
49. F Parke. Parameterized models for facial animation. IEEE Computer Graphics Applic Mag 12:61–68, 1982.
50. A Azarbayejani, T Starner, B Horowitz, A Pentland. Visually controlled graphics. IEEE Trans Pattern Anal Machine Intell 15(6):602–605, 1993.
51. T Chen, HP Graf, K Wang. Lip synchronization using speech-assisted video processing. IEEE Signal Process Lett 2(4):57–59, 1995.
52. L Tang, T Huang. Analysis-based facial expression synthesis. Proceedings of the IEEE International Conference on Image Processing, Austin, TX, 1994, pp 98–102.
53. IA Essa. Analysis, interpretation and synthesis of facial expressions. PhD thesis, Massachusetts Insitute of Technology, Cambridge, MA, 1995.
54. MPEG Integration Group. MPEG-4 synthetic/natural hybrid coding: Call for papers. ISO/IEC JTC1/SC29/WG11 N1195, March 1996.
55. MPEG Integration Group. MPEG-4 synthetic/natural hybrid coding development of media model standards. ISO/IEC JTC1/SC29/WG11 N1199, March 1996.
56. L Tang, T Huang. Automatic construction of 3D human face models based on 2D images. Proceedings of the IEEE International Conference on Image Processing, Lausanne, Switzerland, 1996, pp 198–202.
57. K Waters, D Terzopolous. The computer synthesis of expressive faces. Phil Trans Roy Soc London Biol Sci 335(1273):87–93, 1992.
58. D Terzopolous, K Waters. Analysis and synthesis of facial image sequences using physical and anatomical models. IEEE Trans Pattern Anal Machine Intell PAMI-15(6):569–579, 1993.
59. T Huang, R Lopez. Computer vision in next generation image and video coding. Recent Developments in Computer Vision. New York: Springer-Verlag, 1996, pp 13–22.
60. R Lopez, T Huang. Head pose computation for very low bit-rate video coding. In: V Hlavac, R Sara, eds. Computer Analysis of Images and Patterns. Prague: Springer-Verlag, 1995.
61. R Lopez, T Huang. 3d head pose computation from 2d images: Templates versus features. IEEE International Conference in Image Processing, 1995, pp 220–224.
62. R Lopez, T Huang. Simplification of 3d scanned head data for use in real-time model based coding systems. In: Proceedings of Visual Communications and Image Processing, Philadelphia: SPIE Press, 1996, pp 218–222.
63. TS Huang, R Lopez. Computer vision in next generation image and video coding. Second Asian Conference on Computer Vision (ACCV), 1995.
64. K Aizawa, CS Choi, TS Huang. Human facial motion analysis and synthesis with application to model-based coding. In: MI Sezan, RJ Lagendijk, eds. Motion Analysis and Image Sequence Processing, Boston: Kluwer Academic, 1993.

6

Content-Based Video Compression and Manipulation Using Two-Dimensional Mesh Modeling

A. Murat Tekalp
University of Rochester, Rochester, New York

Peter van Beek
Sharp Laboratories of America, Camas, Washington

1. INTRODUCTION

Digital video on CD-ROM and digital television were made possible by the well-known MPEG-1 and MPEG-2 international standards for video compression, respectively. However, these standards do not allow content-based interactivity with video, because they compress and store it in a frame-by-frame format. The advent of interactive video over the World Wide Web has resulted in increased demand for composition of natural and synthetic visual content as well as reuse of archived video at the desktop. This, in turn, necessitates new content-based representations for video compression and storage, which allow for object-based search, authoring, production, and distribution of synthetic and natural content, as well as the means to synchronize and composite these object-based data at the user terminal.

Two groups, MPEG (Motion Picture Experts Group) and the VRML (Virtual Reality Modeling Language) Consortium, are currently working to standardize object-based representation of natural and synthetic scene content. MPEG aims to develop an object-based video compression standard, MPEG-4 (1,2), which allows for independent encoding of different visual objects, as an extension of previous MPEG standards. These objects are composited at the decoder to form display frames. The visual objects may have natural or synthetic content, including video objects, animated talking heads with texture mapping and synthesized speech, animated human bodies, text and graphics overlay, and so on. In MPEG-4 terminology (3), a Video Object (VO) refers to spatio-temporal data associated with a semantically meaningful part of a natural scene, and a Video Object Plane (VOP) is a two-dimensional (2-D) snapshot of a video object at a particular time instant. Video objects and video object planes are simply the object-based equivalents of video sequences and frames. Video objects are modeled by their shape, motion, and texture (color). An object's shape is represented by an alpha plane, which is compressed as a bitmap. The motion representation is based on the well-known translational block model, where the bounding box of the object is divided into macroblocks, and each macroblock may have up to four motion vectors. The texture or motion-compensation prediction error of each 8 × 8 block is DCT encoded, where blocks that are not entirely within the shape are appropriately padded. Face and body animation are accomplished by a set of face and body definition and animation parameters which may transform a general 3-D facial mesh into a personalized face or body model and render face and body expressions. The composition is modeled by a scene graph, where elementary objects (represented by graph nodes) are connected together to form composite audio-visual objects (AVO). These composite AVOs are then passed to a presenter for rendering.

The Virtual Reality Modeling Language, on the other hand, was introduced by the computer graphics community for modeling of and interacting with synthetic scenes. Whereas VRML 1.0 included only static scenes composed of elementary geometric objects, VRML 2.0 allows for rendering of moving objects and moving texture mapping (4). Object motions are specified by user-provided parametric motion trajectories, whereas moving texture maps must be downloaded in one of the accepted formats before the session starts. The geometry of synthetic objects can be represented in VRML by 3-D primitive models (such as Cones, Boxes, etc.) and by polygonal mesh models (called IndexedFaceSets). Work is currently in progress to support streaming (using real-time transmission) of moving texture maps and object synchronization in VRML to provide complete user interaction with natural and synthetic content. However, features supported by different VRML-compliant browsers may vary significantly from platform to platform. Convergence of MPEG-4 and VRML functionalities is likely in the future, as MPEG-4 introduces functionalities for composition of natural and synthetic content and VRML is working on inclusion of real-time moving textures. In fact, the binary scene description language of MPEG-4 is closely related to VRML.

This chapter introduces a 2-D mesh representation of natural *and* synthetic video objects by means of mesh-based geometry and motion modeling and texture mapping, extending user interaction available with synthetic objects to the domain of natural video content in both MPEG-4 and VRML domains. Section 2 provides an overview of 2-D object-based mesh modeling, including functionalities enabled by 2-D mesh modeling. Section 3 provides descriptions of 2-D object-based mesh design and tracking algorithms. Two-dimensional mesh-based video coding algorithms are discussed in Section 4, including encoder/decoder architectures and compression algorithms for efficient transmission of mesh geometry and motion. Section 5 describes 2-D mesh-based video manipulation, including object transfiguration and augmented reality. Section 6 contains selected results of tracking, compression, and manipulation.

2. MOTION AND SHAPE MODELING OF VIDEO OBJECTS

The translational-block motion model has proven to be satisfactory for video compression in terms of compression efficiency in international standards such as MPEG-1 and MPEG-2. Based on this observation, translational-block modeling has also been adopted by MPEG-4. Hence, MPEG-4 provides content-based functionalities within the framework of translational-block modeling. An alternative to block modeling that supports many desirable functionalities is mesh-based modeling. This section introduces mesh-based motion and shape modeling for object-based video coding. We start by comparing block-based and mesh-based modeling of frame-to-frame motion. Parametric motion modeling and mesh-based motion estimation are discussed next. The last subsection compares block-based and mesh-based modeling of arbitrary shaped video objects.

2.1. Block- Versus Mesh-Based Motion Modeling

In block-motion modeling (5), an independent motion model is estimated for each square block of pixels (from the current frame to a reference frame). Most block-motion estimation algorithms allow only a translational motion model, although generalized block-motion estimation methods have been proposed that can employ parametric motion models (see Sec. 2.2) (6). In block-based motion compensation, the best matching blocks in the reference frame may overlap (even though blocks in the current frame do not), thus violating spatial neighborhood relationships from frame to frame (see Fig. 1). Furthermore, certain pixels in the reference frame may not be used at all. This often causes the well-known blocking artifacts in low-bit-rate video compression. Overlapped block-motion compensation has been proposed to overcome the blocking artifacts (7). A 2-D *mesh* is a planar graph that tessellates (or partitions) a 2-D image region into polygonal patches. The vertices of the polygonal mesh elements are referred to as the *node points* of the mesh. Usually, the polygonal patches are triangles or quadrangles, corresponding to triangular or quadrilateral meshes, respectively. Recently, 2-D meshes have been investigated in video coding for motion modeling as an alternative to the block-motion model (8–12). Two-dimensional mesh-based motion modeling differs from block-based motion modeling in that the patches overlap neither in the reference frame nor in the current frame; see Fig. 1. Instead, polygonal patches in the current frame are deformed by the movements of the node points into polygonal patches in the reference frame, and the texture inside each patch in the reference frame is *warped* onto the current frame as a function of the node-point motion vectors. Thus, neighboring patches in the reference frame remain neighbors in the current frame. This implies that the original 2-D motion field can be compactly represented by the motion of the node points, from which a continuous, piecewise smooth motion field can be reconstructed.

An advantage of the mesh model over the translational-block model is its ability to represent more general types of motion. At the same time, mesh models constrain the movements of adjacent image patches. Therefore, they are well suited to represent mildly deformable but spatially continuous motion fields. An advantage of the block-based model

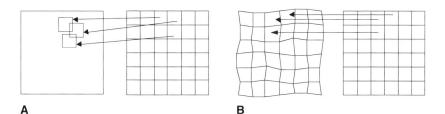

Fig. 1 Two-dimensional block-based (A) versus 2D mesh-based motion modeling; (B) Arrows show corresponding patches in the reference image (on the left) and current image (on the right). In both cases, patches that best match a given patch in the current frame are searched in the reference image, corresponding to backward motion estimation.

is its ability to handle discontinuities in the motion field; however, such discontinuities do not always coincide with block borders, as discussed in Ref. 8. Note that a mesh-based motion field can be described by approximately the same number of parameters as a translational block-based motion field.

2.2. Parametric Motion Modeling

In parametric modeling, the motion over a polygonal patch is described by a mapping H_m, where m denotes the index of the patch. The arrow denotes a vector. The mapping describes the geometric relation between patch m of the current image at time t' and the reference image at time t using a parametric transformation. Use of such parametric transformations for texture mapping is often referred to as *image warping*. Prediction of intensity samples in patch m of the image $I_{t'}(\vec{x}')$ at time t' can then be described by warping a corresponding patch of the reference image $I_t(\vec{x})$ at time t as

$$I_{t'}(\vec{x}') = I_t(\vec{x}) = I_t(H_m(\vec{x}')), \quad \vec{x}' \in D_m \tag{1}$$

where H_m is the geometric transform for patch m and D_m is the domain of patch m.

Common transforms include the *affine*, *bilinear*, and *perspective* transforms (5,13). It can be shown that the affine and perspective coordinate transformations correspond to the orthographic and perspective projections of a 3-D rigid motion of a planar object, respectively (5). The bilinear transformation is not related to a practical physical motion. The affine transform is very convenient for a number of reasons. Affine transforms can model translation, rotation, scaling, reflection, and shear, and preserve straight lines. An affine transform can describe the mapping between two arbitrary nondegenerate triangles while guaranteeing the continuity of the mapping across the boundaries of adjacent triangles. Finally, its linear form leads to a lower computational complexity. The affine transform between coordinates (x', y') at time t' and (x, y) at time t has the form

$$x = a_1 x' + a_2 y' + a_3 \tag{2}$$
$$y = a_4 x' + a_5 y' + a_6$$

where a_i are the affine parameters. The six degrees of freedom in the affine transform matches that of warping a triangle by the motion vectors of its three vertices. Therefore, the mapping can also be described as the interpolation across the triangle of the three motion vectors (14). Note that triangular polygon meshes are predominant in 3-D object modeling and graphics rendering. Similar equations can be used in 3-D computer graphics to describe texture mapping, a popular procedure for adding detail to rendered surfaces (15).

2.3. Motion Estimation

Motion estimation methods can be classified as *backward* or *forward*. The former, in the case of block-motion modeling, consists of searching for a block of pixels in a previous reference image that best matches a block of pixels in the current image (see Fig. 1). In the case of mesh modeling, backward estimation refers to searching in a previous reference image for the best locations of the node points of a mesh tessellating the current image, such that the polygonal image patches of the previous image match those in the current image as close as possible, under a piecewise transform. In backward mesh-motion estima-

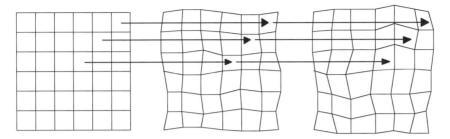

Fig. 2 Mesh-based motion modeling with forward estimation and tracking.

tion, one usually sets up a new *regular* mesh on each current frame to avoid sending mesh geometry (node locations) at every frame. Because a regular mesh can be set up at both the encoder and decoder without geometry overhead, it suffices to transmit only node motion vectors.

In forward mesh-motion estimation (see Fig. 2), one sets up a mesh in a previous reference image and searches for the best locations of the node points in the current image. In this case, there are two options. The first option is to set up a new mesh only in the first reference image and continue to search for node motion vectors in successive imaged using the most recently updated mesh, thus *tracking* image features through the entire sequence. This option is also known as forward tracking mesh and is illustrated in Fig. 2. Mesh tracking enables manipulation and animation of graphics and video content, as detailed in the following sections. The initial mesh may be regular or can be adapted to the image, in which case it is called a *content-based mesh*. The second option is to set up a nonuniform (adaptive) mesh on the reconstructed previous image using a standard algorithm at both the encoder and decoder.

Various techniques have been proposed for node motion vector estimation. The simplest method is to form blocks that are centered around the node points and then use a gradient-based technique (e.g., the method of Lucas and Kanade (16)) or block-matching to find motion vectors at the location of the nodes (9). Hexagonal matching (8) and closed-form matching (17) techniques find the locally optimal motion vector at each node under the parametric warping of all patches surrounding the node while enforcing mesh connectivity constraints. Another method is iterative gradient-based optimization of node-point locations, taking into account image features and mesh deformation criteria (11).

2.4. Block- Versus Mesh-Based Video Object Modeling

So far, we discussed block- versus mesh-based motion modeling and compensation for video frames. In content-based video compression and manipulation, the shape, motion, and texture of each arbitrary shaped video object need to be modeled and encoded independently. In the currently evolving MPEG-4 Video Verification Model (3), the shape of a VOP is represented by a bitmap, called an alpha plane, whereas the texture (color) of a VOP is represented by a texture plane, similar to the representation discussed in Ref. 18. Pixels in the alpha plane can take on multiple values to represent opacity or transparency. The values of pixels in the texture plane are defined if the corresponding alpha-plane pixel is nonzero. The motion of the VO is represented by a translational-block model. To this

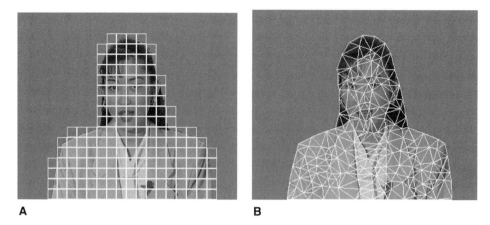

Fig. 3 Two-dimensional block-based modeling (A) versus 2D mesh modeling (B) of the "Akiyo" video object.

effect, the bounding box of the first VOP is tessellated into 16×16 macroblocks (MB), where each MB is allocated four or less motion vectors. Block modeling of an arbitrary shaped VOP is depicted in Fig. 3.

The mesh model offers a versatile alternative, in particular *content-based meshes*, which are designed to match and track specific scene content (19–21). Two-dimensional content-based mesh modeling corresponds to *nonuniform* sampling of the motion field at a number of salient feature points (node points) in a VOP. Additionally, the content-based mesh model includes a compact representation of the shape of each VOP by the polygonal boundary of the mesh. Thus, the proposed 2-D object-based mesh representation is able to model the shape and motion of a VOP in a unified framework, which is also extensible to the 3-D object modeling. Furthermore, the use of forward mesh-based motion estimation allows for tracking and enables more functionalities in terms of video manipulation than does the block-based motion model.

If the first content-based mesh is designed on the original VOP, the initial mesh geometry has to be transmitted, in addition to all node motion vectors. Note that the mesh geometry needs to be transmitted only once, as subsequent forward motion estimation is based on the most recently updated mesh. Alternatively, the initial mesh may be designed on a *reconstructed* VOP, in which case it does *not* have to be transmitted because the decoder can mimic the encoder in this case provided that the mesh design algorithm is fixed a priori.

In this work, we employ forward-tracking content-based meshes. The proposed object-based mesh representation provides the following functionalities:

Video Object Compression. Two-dimensional mesh modeling may improve compression efficiency in two ways:
- Mesh model provides better modeling of smooth motion fields (than translational-block model), which may result in less visually disturbing blocking artifacts at very low bit rates.
- Alternatively, we can choose to transmit texture maps only at selected key frames and animate these texture maps (without sending any prediction

error image) for the intermediate frames using mesh-based motion information.

Video Object Manipulation
- High-quality spatio-temporal interpolation: Assuming a content-based mesh representation, affine interpolation preserves high spatial frequencies better than linear interpolation and offers improved spatial object interpolation (zooming). Similarly, affine/parametric motion modeling provides better motion-compensated temporal interpolation (frame rate up-conversion), which may suggest that temporal scalability may be achieved by postprocessing at the decoder without a need for an enhancement layer in the bitstream. Motion-compensated temporal interpolation is different from self-transfiguration in that the frequency of key frames in self-transfiguration is dependent on the content of the clip, whereas the frame rate in temporal interpolation is fixed independent of content.
- Augmented reality: merging virtual (computer-generated) images with real moving images (video) to create enhanced display information. The computer-generated images must remain in perfect registration with the moving real images (hence, the need for tracking). Merging computer-generated and real moving images can be performed based on 2-D or 3-D mesh-based modeling of the said objects.
- Synthetic-object transfiguration/animation: replacing a natural video object in a video clip by another video object. The replacement video object may be extracted from another natural video clip or may be transfigured from a still-image object using the motion information of the object to be replaced (hence, the need for a temporally continuous motion representation in the bitstream).

Content-Based Video Indexing
- Mesh modeling provides accurate object trajectory information that can be used to retrieve visual objects with specific motion.
- Mesh modeling provides a vertex-based object shape representation which is a more efficient than the bitmap representation for shape-based object retrieval.

3. OBJECT-BASED 2-D MESH DESIGN AND TRACKING

This section describes algorithms for 2-D mesh design and tracking. Video object tracking is a very challenging problem in general, as one may need to take into account the mutual occlusion of scene objects, which leads to covering and uncovering of object surfaces projecting into the image. However, the complexity of the object-based tracking problem depends on the type of video source at hand. We consider two different types of video source. Type-1 sources are such that the shapes of each video object plane (VOP) are available as well as the intensities/colors at all pixels within each VOP. An example of a type-1 sequence is one where VOPs are shot by chroma-keying (blue-screening) techniques. In type-2 sources, pixel intensities in the covered parts of each VOP are not available. This case arises, for example, if the VOPs are extracted from a single camera shot (usually by user interaction). In type-1 sources, covered and uncovered VOP regions result mainly due to self-occlusion. In type-2 sequences, covered and uncovered VOP regions can also result from object-to-object interactions as they move independently of each other. Based on the knowledge of the (un)covered regions, one can update or refine each mesh so as to keep the motion representation accurate. In the following, we mainly discuss tracking of the VO mesh node points for type-1 sequences, where all VOP intensities, their alpha planes, and composition orders are known. Here, each VO sequence is processed and compressed independently. Note that the mesh-based representation can also be applied to type-2 sequences with or without alpha planes. The reader is referred to Refs. 22 and 23 for further discussion of this more difficult problem.

An overview of the video object tracking procedure can be given by the block diagram and illustration in Fig. 4. First, a 2-D mesh is designed for the initial video object

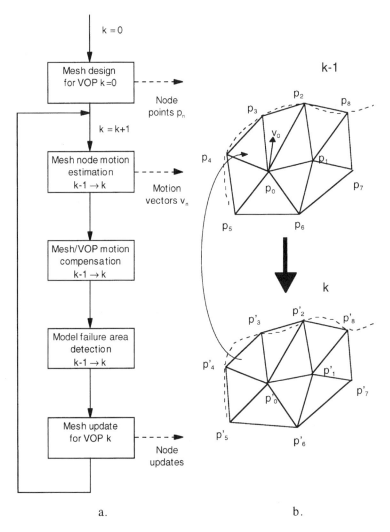

Fig. 4 (a) Outline of the object-based mesh tracking algorithm; (b) illustration of mesh-based warping and motion compensation for part of a mesh; the mesh itself is shown by solid lines, whereas the actual VOP boundary is shown by dashed lines; p_n are nodes in the mesh of VOP $k - 1$ and p'_n are nodes in the mesh of VOP k.

plane, which can be a regular mesh with uniform topology (e.g., in case of a rectangular VOP) or a content-based mesh (e.g., in case of an arbitrary shaped VOP). In addition, the mesh model may be hierarchical or nonhierarchical. We employ *Delaunay triangulation* to construct a content-based mesh from selected node points. The location of a mesh node is denoted by $\vec{p}_n$. In subsequent VOPs the mesh is tracked from the previous to the current VOP. The tracking algorithm implements the following steps: A forward-motion vector (between the previous and current frames/VOPs) is estimated for each node point, denoted by $\vec{v}_n$. These motion vectors are used to warp the texture of the mesh elements (patches) from the previous frame/VOP to the current frame/VOP. Mesh propagation is constrained to preserve mesh topology. From the error between the warped previous frame/VOP and actual current frame/VOP, model-failure regions are identified. In these regions, the mesh geometry may be updated by adding or removing certain nodes. Various steps of this procedure are described in detail in the following.

3.1. Object-Based 2-D Mesh Design

Mesh design takes place in three steps. First, the VOP boundary (extracted from the alpha plane) is approximated by a polygon. The vertices of this polygon are taken as the mesh boundary points. Next, interior nodes are selected. Finally, Delaunay triangulation is applied to define the mesh topology.

3.1.1. VOP Boundary Polygonization and Selection of Mesh Boundary Nodes

The first step for designing an object-based mesh is polygonization of the VOP shape boundary (assuming that the alpha plane for each VOP is available). The resulting polygon becomes the boundary of the object mesh. The vertices of the boundary polygon will serve as node points of the 2-D object mesh. We have used a fast sequential polygonal approximation algorithm (24).

3.1.2. Selection of Mesh Interior Nodes

Additional nodes, besides the vertices of the VOP boundary polygon, are selected in the interior of the VOP using one of the following approaches:

- A. *Uniform mesh.* A given number of node points is placed on a uniformly spaced grid over the object region. Interior nodes are connected to each other and to boundary nodes (which may result in nonuniform triangular elements near the object boundaries (21)). In any case, the mesh design can be replicated at the decoder. Therefore, no node positions need to be transmitted (except for object boundaries).
- B. *Content-based mesh.* The basic principle of this method is to place node points such that triangle edges (to a certain extent) align with image-intensity edges and the density of node points is proportional to the local motion activity (19,25). The former is attained by placing node points on spatial edges (pixels with a high spatial gradient). The latter is achieved by allocating node points in such a way that a predefined function of the displaced frame difference (DFD) within each triangular patch attains approximately the same value. An outline of the content-based node-point selection algorithm is as follows.
    1. Compute an image containing the displaced frame difference inside the VOP, named DFD(x, y). For instance, this can be computed using a forward

dense motion field from the current frame k to the next frame $k + 1$. Alternatively, this image can contain past quantized prediction error—available at the decoder. In any case, areas in this image with high pixel values signal that the motion cannot be estimated well locally. More nodes will be placed in these areas than in areas with low prediction error value, thus creating a finer motion representation in the former areas.

2. Compute a "cost function" image:

$$C(x, y) = |I_x(x, y)|^2 + |I_y(x, y)|^2$$

where $I_x(x, y)$ and $I_y(x, y)$ stand for the partial derivatives of the intensity with respect to x and y coordinates evaluated at the pixel (x, y). The cost function is related to the spatial-intensity gradient so that selected node points, hence the boundaries of the triangular patches, tend to coincide with spatial edges.

3. Initialize a *label* image to keep track of node positions and pixel labels. Label all pixels as *unmarked*. Denote the number of available nodes by N.

4. (Re)compute the *average* displaced frame difference value,

$$\text{DFD}_{\text{avg}} = \frac{1}{N} \sum_{(x,y)} [\text{DFD}(x, y)]^p$$

where $\text{DFD}(x, y)$ is the displaced frame difference or prediction error image computed in step 1 and the summation is over all *unmarked* pixels, N is the number of currently available nodes, and $p = 2$.

5. Find the *unmarked* pixel with the highest $C(x, y)$ and label this point as a node point. Note that marked pixels cannot be labeled as nodes. Decrement N by 1.

6. Grow a square or circular region about this node point until the sum Σ $[\text{DFD}(x, y)]^p$ over the *unmarked* pixels in this region is greater than DFD_{avg}. Continue growing until the radius of this region is greater or equal than some prespecified value. Label all pixels within the region (depicted in Fig. 5) as *marked*.

7. If $N > 0$, go to step 4; otherwise, the desired number of node points, N, is selected and the algorithm stops.

The growing of marked pixels in step 6 ensures that each selected node is not closer to any other previously selected nodes than a prespecified minimum distance. At the same time, it controls the node-point density in proportion to the local motion. In reference to Fig. 5, a small circle indicates a high temporal activity and a large circle indicates low temporal activity. The pixels within each circle are marked, so that another node point cannot be placed within a marked circle.

3.1.3. Constrained Delaunay Triangulation

In the case of a content-based mesh, constrained *Delaunay triangulation* (26,32) is employed after all node points are selected to construct a content-based triangular mesh within each VOP (see, e.g., Fig. 3). The edges of the VOP boundary polygon are used as constraints in the triangulation, to make sure that polygon edges become triangle edges and that all triangles are inside the polygon. Note that the same procedure should be followed by both encoder and decoder. Delaunay triangulation is a well-known technique in compu-

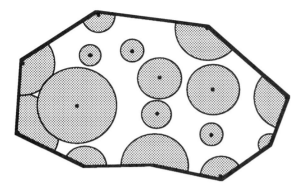

Fig. 5 Illustration of the node-point selection procedure. Inside the VOP polygonal boundary, regions are grown around node points and pixels inside these regions are marked, so that another node point cannot be placed within a marked circle. Each region grows until the sum of the DFD values inside the region attains a certain value. Circular regions with small radii correspond to regions with high temporal activity; regions with large radii correspond to regions with low temporal activity.

tational geometry and has a number of properties that make it well suited for representation of smoothly varying motion fields.

3.2. Node-Point Motion Estimation

Motion estimation is done in all VOPs except the first VOP, in order to propagate a mesh from the previous VOP to the current VOP. For all the mesh node points, a motion vector has to be computed, pointing from a node location in the previous VOP $k-1$ to a location in the current VOP k, using forward estimation.

Motion vectors of node points inside a VOP can be estimated in several ways, such as block-matching, generalized block-matching, gradient-based methods, hexagonal matching (8), or closed-form least squares matching (17,25). We have used either block-matching or the gradient-based method of Lucas and Kanade (16) to estimate the motion at locations of node points, and hexagonal matching (8) for motion vector refinement. For nodes which are close to the VOP boundary, only YUV data of that VOP is taken into account. Note also that prior to motion estimation, the previous and current VOPs are padded beyond their boundaries.

The motion estimation of node points at VOP boundaries is constrained (except for the background VOP). For example, in reference to Fig. 4b, the motion vectors of nodes at the boundary of the foreground VOP $k-1$ must point to a point on the boundary of VOP k. This can be achieved by restricting the search space during motion estimation or by projecting the boundary nodes onto the actual boundary after motion estimation.

After motion estimation, each node $\vec{p}_n$ has a motion vector $\vec{v}_n$. It is possible that the estimated node motion vectors are inconsistent in the sense that they lead to crossovers of edges in the 2-D mesh and do not preserve the 2-D topology of the mesh. This is

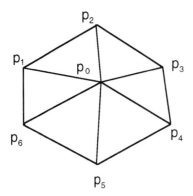 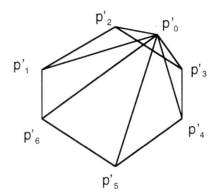

Fig. 6 Illustration of a combination of motion vectors corrupting the mesh structure. A node that effectively moves outside its enclosing polygon is found to have a badly estimated motion vector.

illustrated in Fig. 6, where node $\vec{p}_0$ at time $k - 1$ is connected to the nodes $\vec{p}_1, \vec{p}_2, \vec{p}_3,$ $\vec{p}_4, \vec{p}_5,$ and $\vec{p}_6$. The motion vector $\vec{v}_0$ at node $\vec{p}_0$ must be such that $\vec{p}'_0 = \vec{p}_0 + \vec{v}_0$ lies inside the polygon formed by the surrounding motion-compensated nodes $\vec{p}'_1, \vec{p}'_2, \vec{p}'_3,$ $\vec{p}'_4, \vec{p}'_5,$ and $\vec{p}'_6$ at time k. However, because the motion-estimation method used to compute the motion vectors does not employ such a constraint, this condition may be violated.

We employ a postprocessing algorithm to preserve the connectivity of the patches. This is achieved by ordering the sampled motion vectors according to a measure of confidence. In the case of a motion vector crossover, the motion vectors with the lowest confidence are interpolated using those with higher confidence to eliminate crossovers as follows:

1. Determine the ordering of the nodes for postprocessing as follows:
   - For each node n, find its enclosing polygon, defined by nodes connected to node n.
   - Calculate the criterion function

     $$O_n = \frac{\text{ADFD}_n}{\sigma_n^2}$$

     where ADFD_n and σ_n^2 denote the average displaced frame difference and intensity variance inside the polygon enclosing the node n, respectively.
   - The node with the highest O_n will be processed first. Such prioritization enables resolution of conflicts at nodes with lowest confidence motion vectors first in favor of those nodes with more reliable motion vectors.

2. Scan the nodes in the order determined in step 1 to detect nodes with inconsistent motion vectors as follows. At each node $\vec{p}_n$:
   - Find the all the nodes connected to node $\vec{p}_n$ and label them as $\vec{p}_l$, $l = 1, \ldots, L_n$, where L_n is the number of nodes connected to $\vec{p}_n$.
   - Find the motion compensated node locations $\vec{p}'_n$ and $\vec{p}'_l$, $l = 1, \ldots, L_n$, at time k using the motion vectors at the nodes. Form the polygon defined by $\vec{p}'_l$, $l = 1, \ldots, L_n$.

- If $\vec{p}'_n$ is inside the polygon, go to next node in order.
- If $\vec{p}'_n$ is outside the polygon, interpolate the motion of the node from the motion of its neighboring nodes as follows:

$$u = \frac{\sum_{l=1}^{L_n} u_l/d_l}{\sum_{l=1}^{L_n} 1/d_l} \quad \text{and} \quad v = \frac{\sum_{l=1}^{L_n} v_l/d_l}{\sum_{l=1}^{L_n} 1/d_l}$$

where (u_l, v_l) is the node motion vector at the node $\vec{p}_l$, and d_l is the distance between node $\vec{p}_l$ and node $\vec{p}_n$.

This postprocessing results in increased robustness to errors in motion estimation around occlusion boundaries.

3.3. Mesh-Based Motion Compensation

Motion compensation is performed by warping the triangular patches from the previous VOP to the current VOP using the estimated node motion vectors. Note that prior to the warping, the image containing the previous VOP is padded (repetitive padding), in case some area of triangles in the previous VOP falls outside the actual VOP region.

Each triangular patch in the VOP corresponding to a mesh triangle is predicted by warping it according to an affine transform using Eq. (2). Postprocessed node motion vectors establish a set of point correspondences between the previous and current VOPs, which are used to determine a set of backward affine spatial transformations from time k to $k - 1$ as given by Eq. (2). For each triangular patch, the three motion vectors for the three nodes provide three correspondences, as illustrated in Fig. 7. From these three correspondences, the set of affine parameters $a_1, \ldots, a_6$, can be computed using the six resulting linear equations. Then, all pixels (x', y') within the patch of the current VOP are motion compensated from the previous VOP by using an affine mapping as depicted in Fig. 7. Bilinear interpolation is used when the corresponding location (x, y) in the previous VOP is not a pixel location.

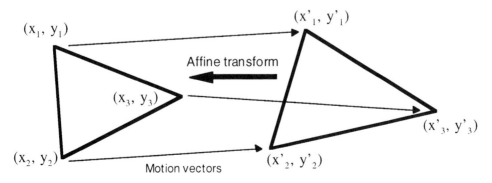

Fig. 7 Motion compensation using the backward affine transform, computed from node-point correspondences at time $k - 1$ and k of a triangular mesh element.

Note that some pixels in the current VOP may fall outside the mesh that models the current VOP, because the boundary of the mesh is only a polygonal approximation to the true boundary. These pixels exterior to the mesh but inside the VOP need to be motion compensated as well. Each of these pixels is motion compensated by computing a motion vector derived from the mesh node motion vectors. This motion vector is estimated by interpolating the motion vectors of the two nearest nodes on the polygonal mesh boundary.

3.4. Mesh Propagation and Refinement

In order to have a temporally continuous motion representation (which is important for functionalities that require object tracking), the 2-D mesh designed for a VOP at time $k - 1$ (to predict frame k) needs to be propagated to VOP k (to predict VOP $k + 1$) rather than designing a new mesh for VOP k.

The basic scheme for forward tracking of the mesh from time $k - 1$ to k is to propagate all nodes by their motion vectors, as explained in the previous subsections. Then, subsequent motion estimation and motion compensation is applied on the propagated mesh, rather than designing a new mesh for every VOP. This results in a spatio-temporally continuous motion representation, which is important for functionalities that require object tracking. However, the 2-D mesh model can fail in certain regions of the image, due to covering or uncovering of object parts, inaccuracies in node-point motion estimation, and/or other limitations of the parametric motion model. Therefore, the mesh may be updated optionally to refine the modeling locally. This is realized by detecting areas where the parametric motion model fails and inserting new nodes in the mesh locally.

Model failure (MF) regions are detected by thresholding the difference between the actual intensity image and the motion-compensated intensity image

$$\left| I_k(x, y) - \tilde{I}_k(x, y) \right| < T_{\text{MF}}$$

where $\tilde{I}_k(x, y)$ denotes the motion-compensated image based on the parametric (piecewise affine) motion field, and T_{MF} is the MF region detection threshold. To obtain "cleaner" MF regions, the resulting binary image is postprocessed using morphological smoothing

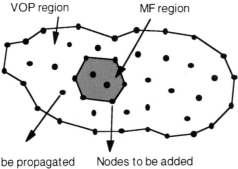

Fig. 8 Illustration of the mesh propagation and refinement procedure.

operators and by removing small pixel clusters. It is expected that with successful tracking, the MF regions will reduce to uncovered background regions [note that the uncovered background region(s) refer to pixels in the next image which are uncovered as a result of the motion].

New mesh nodes need to be added in uncovered background (UB) regions for purposes of subsequent tracking of newly appearing objects. The UB region is approximated by the MF region whose boundary is modeled by a polygon. Any nodes inside the MF boundary polygon are deleted. The vertices of the MF are accepted as nodes for mesh refinement. Additional nodes within the MF region may be selected by using the mesh design discussed earlier. The regions within which nodes are propagated and added are depicted in Fig. 8. Following the refinement step, Delaunay triangulation is reapplied to obtain a new mesh for the next VOP.

4. COMPRESSION OF VIDEO OBJECTS AND 2-D DYNAMIC MESHES

This section discusses methods for video object and 2-D dynamic mesh coding, where dynamic mesh coding refers to mesh geometry and motion compression. First, we discuss general encoder and decoder architectures for coding video objects with associated dynamic mesh data, which enable mesh-based manipulation of the video objects at the receiver. We then present novel methods for 2-D dynamic mesh coding. In the following, I-VOP and P-VOP refer to the object-based equivalents of I-frame (intraframe) and P-frame (predicted frame).

4.1. Encoder and Decoder Architectures

We present three multimedia encoder architectures to encode video objects and dynamic mesh data that are capable of rendering 2-D mesh-based functionalities. Two of them employ a stand-alone video object encoder, such as the basic MPEG-4 Video Verification Model (VM), as the base layer to compress a natural video object, and an auxiliary encoder to compress the associated 2-D mesh geometry and motion. The MPEG-4 video VM is based on translational-block prediction using overlapped block-motion compensation (OBMC) and block-based shape and texture compression. Both of these architectures allow skipping the enhancement layer coding in case the enhanced mesh-based functionalities are not needed. The third architecture is an *integrated* coder which employs mesh-based motion estimation and compensation for video object compression. All three architectures are depicted in Fig. 9. Figure 9a shows a *simulcast* architecture where mesh-motion vectors are spatially predicted without reference to the block-based video object encoder; hence, mesh compression is entirely independent of video object compression. Figure 9b depicts a functionality-*scalable* architecture that predicts mesh motion vectors from block-based VM motion vectors; hence, there is a clear dependency relation. In the integrated coder, which is depicted in Fig. 9c, mesh-based motion estimation and compensation replaces the block-based motioncompensation of the video VM. In the standard scalable and simulcast decoders, a demultiplexer separates the coded descriptions of the mesh geometry, node-point motion vectors, block-based motion vectors, shape, and video object texture. The mesh geometry and node motion bits are decoded by the auxiliary coder, whereas the video object texture is decoded by the standard video VM decoder. The mesh–object

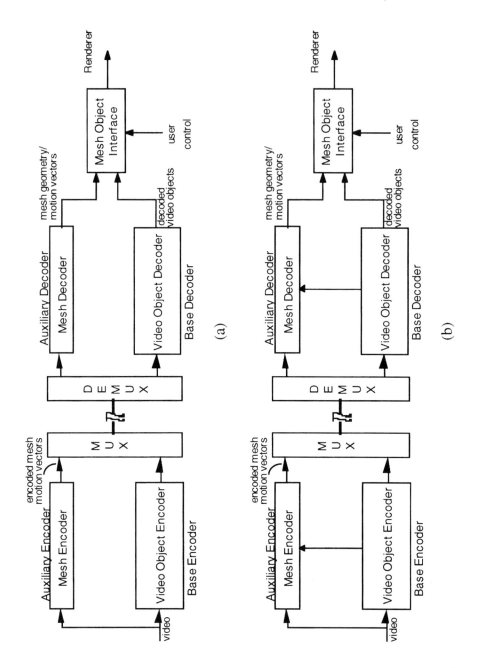

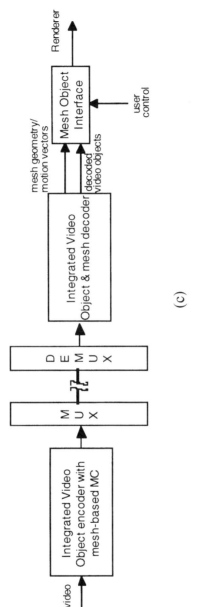

Fig. 9 Architectures for content-based encoders with 2-D mesh modeling: (a) mesh-simulcast architecture; (b) mesh-scalable architecture; (c) integrated coder architecture.

interface animates the 2-D mesh object using the node-point motion vectors and specified user-control parameters. The integrated decoder follows similar steps except that block-based VM motion vectors are not available.

4.2. 2-D Mesh Geometry and Motion Compression

The mesh geometry and node-point motion vectors need to be compressed for efficient transmission (27). Because we construct a 2-D triangular mesh by *Delaunay* triangulation of a set of node points at both the encoder and decoder, the mesh triangular topology (links between node points) need not be coded; only the node-point coordinates $\vec{p}_n = (x_n, y_n)$ are coded. Coding of node-point coordinates is necessary for the I-VOPs. For the next VOPs, rather than encoding the new mesh geometry again, we encode the node motion vectors for one VOP to the next and keep the topology fixed. Node-point coordinate coding is a simple spatial predictive compression scheme. There are two alternative methods for predictive coding of the mesh motion vectors; the first is based on using motion vectors of neighboring nodes as predictors; the second is based on using block-based motion vectors of the video VM as predictors. Note that the second method can only be applied if the video VM runs at the same temporal rate as the auxiliary mesh.

4.2.1. Coding of Node Positions

The position of all node points needs to be encoded only for I-VOPs of a video object. For all other VOPs, only the motion vectors of all node points as well as any mesh structure update information (such as insertion or deletion of nodes) need to be encoded.

We employ a simple spatial prediction scheme to encode the coordinates of node points in a I-VOP. A linear ordering of the node points is computed such that each node is visited only once. When a node is visited, its position is differentially encoded with respect to the previously encoded node; that is, the difference between the position of the present node and the reconstructed value of the previous node is encoded using an entropy coder. Because each node is encoded in reference to the previous node, the ordering information is not transmitted. The ordering is such that the boundary nodes are visited first. For coding efficiency reasons, the ordering of the other nodes should be such that it minimizes the sum of the bits spent. However, determining such an ordering is a computationally intractable problem for arbitrary sized meshes; hence, we propose a suboptimal algorithm, which employs a *nearest-neighbor* strategy to order the nodes.

This procedure is illustrated on an example 2-D mesh in Fig. 10. Starting from the top-left node $\vec{p}_0$ (coded without prediction), the boundary node $\vec{p}_1$ is found and the difference between $\vec{p}_0$ and $\vec{p}_1$ is encoded; then all other boundary nodes are encoded in a similar fashion. Then, the not previously encoded interior node that is nearest to the last boundary node is found and the difference between these is encoded. Then, the not previously encoded node nearest to the last encoded node is found and the difference is encoded, and so on. Every node point has an x and y coordinate, $\vec{p}_n = (x_n, y_n)$, each of which is subtracted from the corresponding coordinate of the previously coded node point. The two resulting difference values are coded using variable-length coding. The decoder simply decodes the node positions by adding each difference value just received to the position of the previously decoded node.

By sending the total number of nodes and the number of boundary nodes before the coordinates, the decoder is able to tell how many nodes will follow and how many of those nodes comprise the mesh boundary. Once all node locations are recovered for a given frame, Delaunay triangulation is used at the decoder to define the 2-D mesh topology.

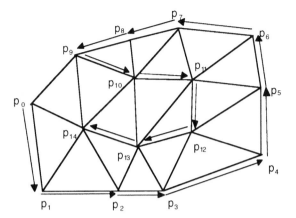

Fig. 10 Illustration of linear ordering of node points for spatial predictive coding of the geometry of a 2-D triangular example mesh and ordering of the node points to be encoded predictively. First, the boundary nodes are visited (according to connectivity); then, the interior nodes are visited (according to proximity; i.e., the next node is always the nearest node that is not already encoded).

To accommodate dynamic changes of the mesh geometry (streaming geometry compression), we allow incremental geometry updates (such as insertion of a new node, deletion of a node, putting a node into a hidden state, bringing a node back to visible state) to be transmitted. Insertions of new node points require transmission of their positions; this can be done using the technique discussed above, except that only a small set of node points need to be encoded, instead of all node points of a mesh. The other node updates do not require transmission of the node position, only the index of the node. Therefore, only the type of update and node index are encoded for each update. Finally, a termination codeword is transmitted to signal that the last node update has been sent.

4.2.2. Coding of Node Motion Vectors

4.2.2.1. Coding of Node Motion Vectors with Spatial Prediction

Self (spatial)-prediction of motion vectors $\vec{v}_n$ of node points $\vec{p}_n = (x_n, y_n)$ of a 2-D triangular mesh entails prediction of each motion vector in reference to already encoded motion vectors. Specifically, to encode the motion vector of a certain node that is the vertex of a triangle, we use the motion vectors of the two other vertices of that triangle to predict the motion vector; that is, the motion vector $\vec{v}_n$ at the node point $\vec{p}_n$ that is part of a triangle $t_k = \langle \vec{p}_l, \vec{p}_m, \vec{p}_n \rangle$ can be predicted in reference to the motion vectors $\vec{v}_l$ and $\vec{v}_m$ of the nodes $\vec{p}_l$ and $\vec{p}_m$.

To encode all motion vectors, one needs to find an ordering of the nodes, such that for every node motion vector to be encoded, one can find a triangle in the mesh that contains this particular node as well as two other nodes, of which the motion vectors have been encoded previously. To find this ordering, we make use of the fact that if there is a triangle t_k of which all three node motion vectors have been encoded, there must be at least one other neighboring triangle t_w that has two nodes in common with t_k (assuming all the triangles in the mesh form one connected patch). In general, there can be three such neighboring triangles, one for each side of t_k; however, on the boundaries of the mesh, there may be less than three. Because the motion vectors of the two nodes that t_k and t_w have in common have already been encoded, one can use these two motion vectors to predict the motion vector of the third node in t_w (if not already encoded). The prediction can be computed by inverse distance-weighted averaging of the two reference motion vectors.

Thus, given an initial triangle with encoded node motion vectors, one can always find neighboring triangles, each with a node motion vector that can be predicted from previously encoded node motion vectors. Then, the neighboring triangles of those triangles are found, and so on until all triangles in the mesh have been processed. This iterative process defines in effect a *breadth-first traversal* of the triangles in the mesh (as opposed to depth-first) (28). This traversal of the mesh can be performed by both the encoder and decoder in a unique manner, as the topology of the mesh is known a priori. The breadth-first traversal process is illustrated in the Fig. 11a for an example mesh. Start with a triangle t_0 on the boundary of the mesh. Encode the motion vector of its first node $\vec{p}_0$ without prediction and encode the motion vector of its second node $\vec{p}_1$ using the motion vector of $\vec{p}_0$ as a prediction. Node $\vec{p}_2$ is the first node for which *two* other node motion vectors (those of $\vec{p}_0$ and $\vec{p}_1$) can be used as predictors. Triangles t_1 and t_2 are both neighbors of t_0, having two nodes in common with t_0. Processed in that order, the motion vector of $\vec{p}_3$ is encoded next, and then the motion vector of $\vec{p}_4$ is encoded, each using previously encoded node motion vectors as predictors. The traversal of the mesh continues until all triangles $t_0, \ldots, t_8$ have been visited once. This traversal defines the ordering of the mesh nodes $\vec{p}_0, \ldots, \vec{p}_8$, each of which is visited at least once. Visited triangles are *marked* such that they are not visited more than once. Nodes may be visited more than once during the traversal of the triangles; nodes should be *marked* at the time of encoding, such that their motion vectors are not encoded again. The *breadth-first traversal* of all triangles can be performed by the decoder in a unique manner, as the topology of the mesh is known a priori.

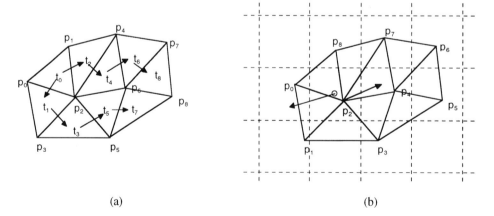

Fig. 11 (a) Breadth-first traversal of triangles of a 2-D triangular mesh and ordering of the node points to be coded in spatial motion vector prediction. (b) Relation between mesh nodes and block-motion vectors in node motion vector prediction from block motion vectors. The node motion vector of node $\vec{p}_2$ is defined from VOP $k - 1$ to k. The corresponding block in VOP k and its block motion vector are shaded.

4.2.2.2. Coding of Node Motion Vectors with Prediction from Block-Based Motion Vectors

Mesh tracking uses forward motion vectors $\vec{v}_n$ defined from frame $k - 1$ to k, for each node point $\vec{p}_n = (x_n, y_n)$. The MPEG-4 video VM, as well as all other block-based motion-compensated video coders, assigns a backward motion vector (from frame k to $k - 1$) to each 8×8 or 16×16 block B. If a single motion vector was assigned to a 16×16 macroblock (MB), then it is replicated four times to generate motion vectors for 8×8 blocks. For intracoded or skipped blocks or blocks that fall outside of arbitrary shaped video object planes (VOP), the corresponding motion vectors are set equal to zero.

Here, we present a method for predicting mesh node motion vectors from the 8×8 block motion vectors to encode the difference (prediction error) vectors. The method assumes some local spatial coherence of the motion field. To encode the forward motion vector $\vec{v}_n$ of the node point (x_n, y_n) from frame $k - 1$ to frame k, take the following steps, further illustrated in Fig. 11b:

1. Find the 8×8 block B in frame k which contains the node point (x_n, y_n). Find the four blocks adjacent to B, that is, above, below, to the left of, and to the right of B.
2. Form a prediction $\vec{w}_n$ for the motion vector v_n by computing the (component wise) median of the block motion vectors of the block B and its four neighboring blocks.
3. Encode the difference vector $\vec{v}_n - \vec{w}_n$ by variable-length coding.

The decoder simply has to find the same blocks in VOP k, compute the prediction, and add the decoded prediction error vector to it to obtain the original node motion vector.

4.3. The Integrated Encoder

The integrated coder uses mesh-based warping for motion compensation. Because the mesh boundary is an approximation to the actual VOP boundary, special attention has to be given to pixels near the boundary. First, the previous reference VOP is padded before motion compensation, because there may be pixels inside the region covered by the mesh in the reference VOP that have a zero alpha value. Furthermore, pixels inside the current VOP with a nonzero alpha plane that are outside the region covered by the mesh have to be motion compensated separately, as explained in Sec. 3.

To code the actual shape of the VOP and to code the motion-compensated texture differences, we have used the tools provided by the MPEG-4 video VM. The VM uses a tessellation of the VOP into 16×16 macroblocks. Each macroblock contains a 16×16 alpha plane block, four 8×8 luminance blocks, and two 8×8 chrominance blocks. The alpha plane block is encoded by a modified run-length encoding scheme. The six texture blocks are encoded using the DCT, zigzag scanning of the coefficients, uniform quantization, and variable-length coding. The luminance and chrominance blocks are either repetitively padded or zero padded before applying the DCT. For further details, we refer to Refs. 29 and 30.

5. 2-D MESH-BASED VIDEO MANIPULATION

A major functionality provided by the 2-D mesh representation of video objects is content-based video manipulation. The 2-D mesh representation of a video object is analogous to the 3-D polygon mesh representation used in computer graphics. In computer graphics, the animation parameters of a mesh model is often synthetically specified, whereas the animation parameters of our 2-D mesh model is derived from a natural video object by tracking. Thus, the proposed mesh modeling of natural video objects allows us to animate a still image of one object by motion parameters of another similar object and/or interactively combine natural and synthetic objects within a unified framework. In addition, the mesh structure facilitates high-quality spatio-temporal video interpolation. The details of 2-D mesh-based video manipulation are described in the following subsections.

5.1. Object Transfiguration

Object transfiguration refers to synthesizing an animated video object (natural or synthetic) from a still image of the object by texture mapping on a dynamic 2-D mesh representation. We classify object transfiguration as *self-transfiguration* or *synthetic transfiguration*. Self-transfiguration, which refers to animating a video object from its first VOP and its own dynamic mesh representation, has applications to video object compression. Synthetic transfiguration refers to synthesizing an animated video object from a still image (the replacement object) and a dynamic mesh representation of a similar object with the desired motion. It is used for special-effects editing (e.g., to replace an original video object by a synthesized object that is given the motion of the original video object) or clip art animation (e.g., animate a cartoon character by the motion of a natural video object). Synthetic transfiguration requires a preprocessing step to register the initial appearance of the replacement object with that of the object to be replaced. This may require resizing of the replacement object as well as prewarping for small adjustments in its initial pose.

Given the dynamic mesh model, there are two alternative ways to implement texture mapping for object transfiguration: the fixed-reference approach and moving-reference approach, which are depicted in Fig. 12. In the fixed-reference-frame method, all successive frames of the animated video object are obtained by warping a single reference texture map, as shown in Fig. 12a. This method is best suited when there are no covered/uncovered object regions and the mesh geometry is not updated during the life span of the animation/transfiguration. The moving-reference-frame method is depicted in Fig. 12b, where the previous frame serves as the reference frame for texture mapping (i.e., each frame is obtained by warping the previous frame).

The fixed-reference-frame method can be used for mesh-based rendering when the mesh structure is not updated between successive frames. Otherwise, the moving reference frame with backward motion estimation at the frames where the mesh structure is updated needs to be employed. The moving-reference method for rendering can also be employed with block-motion estimation and OBMC, although the resulting rendered images look significantly blurred compared with mesh-based rendering.

5.2. Augmented Reality

Augmented reality refers to blending synthetic (computer-generated) video, images, text, and/or graphics with natural moving images (video) to create enhanced display information. The synthetic content must remain in registration/synchronization with the moving real images. Blending synthetic or natural content with a moving object can be performed by first motion-tracking the object to be augmented, registering the augmentation object with the initial appearance of object to be augmented, and then 2-D mesh-based texture

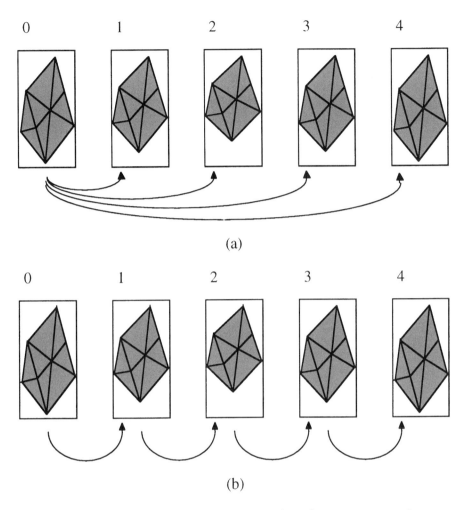

Fig. 12 (a) Fixed-reference texture mapping; (b) moving-reference texture mapping.

mapping of the augmentation object onto the object to be augmented with proper alpha blending. Registration of the augmentation object with the initial appearance of the object to be augmented is achieved by a piecewise affine mapping of the augmentation object. This affine mapping (also called prewarping) is specified by M point correspondences that are interactively registered by the user.

There are two ways to perform the rendering of the augmented object after the augmentation object is registered onto the initial appearance of the object to be augmented: (a) the first composited object can be rendered using the mesh data (similar to self-transfiguration); (b) only the prewarped (registered) augmentation object is rendered using the mesh data, and each instance of the rendered augmentation object is alpha-blended separately onto the originals of the object to be augmented. Note that, in each case, the rendering of the composited object or the augmentation object can be performed by the fixed-reference or moving-reference texture mapping. Two of these options are depicted in Figs.

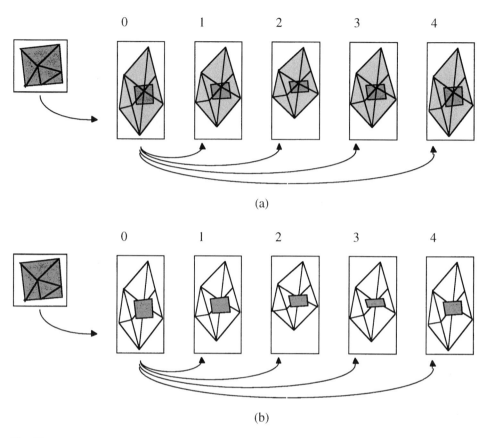

Fig. 13 Augmented reality using 2-D mesh warping. The augmentation VOP and its corresponding mesh which is used to register the augmentation object onto the reference VOP are illustrated on the left. The augmentation process is illustrated on the right for time instants 0 to 4. In (a), the augmentation VOP is mapped onto the reference VOP containing the original texture; the augmented initial VOP is then texture mapped to other time instances. In (b), only the augmentation VOP is texture mapped by the dynamic mesh, which can then be alpha-blended with the corresponding VOPs of the original object.

13a and 13b. Figure 13a shows the case where the augmentation object (on the left) is prewarped onto the initial appearance of the object to be augmented, and then the composite object is rendered using the fixed-reference texture mapping approach (on the right). Figure 13b shows the case when the augmentation object is separately rendered using the fixed-reference texture mapping approach and then overlaid onto the original frames of the object to be augmented. Successful augmentation requires accurate tracking of the motion of the original video object so that the augmentation object can be rendered in perfect registration with the original object. For more details of this procedure, see Refs. 22 and 31.

A number of variations of the procedure discussed above have been implemented which provide different kinds of animation results. For instance, a video object can be augmented by a *moving* video object instead of a still image as well. This can be done by repeating the mapping of video object planes into the reference VOP each time a new augmented VOP is generated. Finally, by *alpha-blending* the new video object plane with the original VO instead of mapping it on top, other effects can be generated. Thus, natural and synthetic video data can be manipulated interactively by the user to create a range of special effects.

5.3. Spatio-Temporal Interpolation

Another interesting functionality provided by 2-D mesh-based modeling is spatio-temporal interpolation. Mesh-based *spatial* interpolation (e.g., for *zooming* in or out on an object) is simply performed by first scaling the mesh and then applying texture mapping onto the scaled mesh, where the affine transform is defined by correspondences between the original and scaled meshes. *Temporal* motion-compensated interpolation (e.g., for *frame-rate up-conversion*) is handled by computing an intermediate mesh geometry in between the two given meshes and then applying forward and/or backward texture mapping from one or both of the given object texture maps. A unique intermediate mesh geometry can be computed by linearly interpolating the node-point trajectories between the two given meshes. Alternatively, higher-order spline interpolation can be employed by using more than two original meshes. The actual texture mapping can be forward, backward, or a weighted average of the two original VOPs, much like the bidirectional prediction (B-frame) mode in MPEG-1/2/4.

6. EXPERIMENTAL RESULTS

This section reports selected video object tracking, compression, and manipulation results. The experiments were conducted with the video objects and video object planes described in Table 1. The Akiyo, Sean, and Bream sequences are originally in 30-Hz YUV 4:2:0 format, 288 lines by 352 pixels, with two-level alpha planes and containing 300 frames. The meshes for Akiyo, Sean, and Bream were content-based meshes, designed on the first VOP of each sequence. Only one VOP is used of the Cyclamen sequence (480 lines by 720 pixels) which was used only to evaluate geometry coding.

During mesh design, the number of mesh nodes in the initial VOP was computed automatically to be on the order of the number of 16×16 macroblocks present in the first VOP, in the cases of Akiyo, Sean, and Bream.

Several motion-estimation techniques were used for tracking of the above video objects. In particular, block-matching was used to compute node motion vectors in the Akiyo and Sean sequences; a hierarchical implementation of the gradient-based technique

Table 1 Video Objects (and VO Planes) Used in the Experiments

Sequence	VO number	Frame rate	VOPs	Motion range
Akiyo	1 (woman)	10 Hz	0, 3, 6, ..., 297	16
Sean	1 (man)	10 Hz	0, 3, 6, ..., 297	16
Bream	1 (fish)	15 Hz	60, 62, 64, ..., 110	16
Cyclamen	1 (flower)	—	40	—

of Lucas and Kanade was used in the Bream sequence, as well as a generalized hexagonal matching technique (see Refs. 8 and 31).

6.1. Tracking Results

This section illustrates some tracking results obtained with the algorithms discussed in Sec. 3. In this example, we illustrate tracking of a content-based mesh of the Bream and Sean video objects.

The initial mesh of the Bream VO, which contained 165 nodes, was designed on VOP 60 and then tracked over 25 more VOPs of the VO (i.e., VOPs 62, 64, 66, ..., 108, 110). This is illustrated in Fig. 14 using three snapshots of the tracked mesh at VOPs 60, 100, and 108. This part of the Bream sequence contains a limited amount of global motion combined with deformable motion, increasing toward the end. Most of the mesh nodes were tracked satisfactory. The initial mesh of the Sean VO, which contained 193 nodes, was designed on VOP 0 and then tracked over 99 more VOPs of the VO (i.e., VOPs 3, 6, 9, ..., 294, 297). As an illustration, Fig. 14 provides three snapshots of the tracked mesh at VOPs 30, 90, and 180. The Sean sequence contains a limited amount of motion in the form of head movements, small body movements, and movements of the arms and hands. Most of the motion was tracked satisfactory, although small errors are observed in the vicinity of the neck due to uncovered regions.

6.2. Mesh Geometry and Motion Compression Results

In this section, we report results on compression of the mesh geometry and motion. We evaluated the performance of our mesh geometry compression algorithm by comparing it to the compression performance of `gzip` on four different 2-D meshes; see Table 2. We evaluated the performance of our mesh motion compression algorithm (based on node motion vector coding with spatial prediction) by comparing it to coding of a sequence of meshes one by one (using the mesh geometry compression algorithm for each mesh) on three different mesh sequences; see Table 3. The mesh data was obtained by automatic mesh design and tracking as described earlier.

These results show that our mesh geometry compression algorithm far outperforms a basic compression algorithm such as `gzip`, furthermore, the results show that our mesh motion vector compression algorithm outperforms compression of a sequence of mesh geometries.

6.3. Comparison of Block-Based and Mesh-Based Video Object Encoding

This section describes results on video object and mesh compression using the schemes outlined previously. The coding performance is reported in terms of bit rate and peak signal-to-noise ratio (PSNR). We compare the performance of the *integrated* coder with that of the basic MPEG-4 Video VM 3 (29). We also investigate the compression efficiency of the mesh geometry and motion vector encoding schemes for the *scalable* and *simulcast* approaches. Alpha planes (shape) of all VOPs and the texture planes of I-VOPs are always encoded by the video VM in all three architectures. In the case of the scalable and simulcast architectures, all VOPs are encoded using the video VM. In the case of the integrated encoder, 2-D mesh object tracking replaces the motion compensation (prediction) tool in

Video Compression/Manipulation: 2D Mesh Modeling

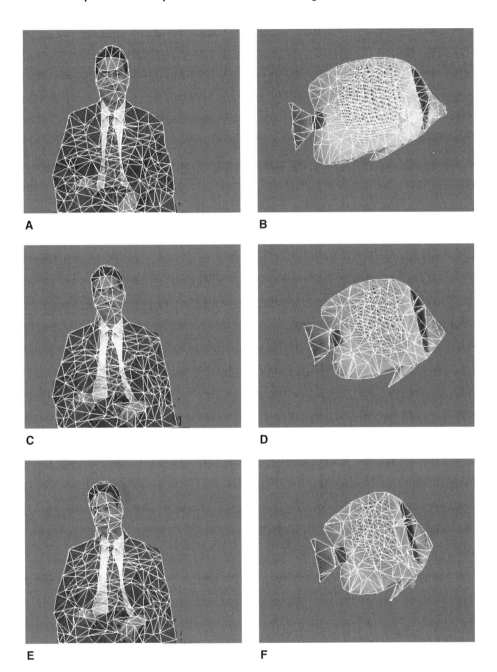

Fig. 14 Illustration of mesh tracking on Bream and Sean video objects. Displayed in (A), (C), and (E) are tracked meshes at time 30, 90, and 180, respectively, overlaid on the corresponding original video object planes of Sean. Displayed in (B), (D), and (F) are tracked meshes at time 60, 100, and 108, respectively, overlaid on the corresponding original video object planes of Bream.

Table 2 Results of Geometry Compression of 2-D Delaunay Meshes

Mesh	No. of nodes	ASCII file	gzip	Proposed algorithm
Akiyo	210	47,568	18,720	2,683
Sean	193	42,584	16,896	2,494
Bream	165	33,960	13,712	2,156
Cyclamen	1000	226,928	86,840	12,298

Note: All numbers are bits.

the video VM using the separate motion–shape–texture mode of the video VM; that is, the intensity residuals are encoded by the object-based texture coding method that is employed in the video VM. Motion compensation is achieved by a forward tracking (from frame $k - 1$ to k) 2-D mesh model and affine warping. The first VOP of a video session is always encoded as an I-VOP. In order to encode the second VOP, a mesh is designed for the first VOP, which is then projected onto the second frame by forward motion vectors. Mesh refinement was not applied in these experiments. Mesh geometry and motion vector coding are lossless based on predictive coding and Huffman coding. Currently, we are using the variable-length coding (VLC) tables for sprite-trajectory and block motion vector coding in the video VM, which are not optimized for mesh geometry and motion vector coding, respectively. The separate motion–shape–texture mode of the MPEG-4 Video Verification Model was used, meaning that motion, shape, and texture data are not grouped together per macroblock, but per VOP. A fixed quantizer $Q = 8$ is used in all experiments. Experiments are performed on several test sequences, and numerical compression results are shown in Tables 4 and Table 5.

The first group of rows of Tables 4 and Table 5 contains results of mesh geometry and mesh motion compression. The middle group of rows contains results of VOP shape and texture compression, and—in the case of the scalable coder—results of block motion vector compression and video (VM) compression. The last group of rows contains the overall results in terms of PSNR and the total number of bits.

The results show that the mesh geometry (node-point coordinates) can be encoded on average with approximately 12.3 bits per mesh node. As can be seen from Tables 4 and 5, mesh motion compression with spatial prediction (method A) performed approximately the same as mesh motion compression with prediction from a block-based motion field (method B). The results show that the mesh motion vectors can be encoded on average

Table 3 Results of Combined Geometry and Motion Vector Compression (Inter) Versus Compression of Sequences of 2D Delaunay Meshes (Intra)

Mesh	Intra	Inter
Akiyo	304,508	56,144
Sean	277,088	66,585
Bream	62,420	30,437

Table 4 Mesh Compression and Video Object Compression Results for Akiyo and Sean, Using the Scalable Approach with Spatial Prediction (Method A), the Scalable Approach with Prediction from Block Motion Vectors (Method B), and the Integrated Approach

	Akiyo			Sean		
	Scalable coding A	Scalable coding B	Integrated coding	Scalable coding A	Scalable coding B	Integrated coding
Mesh geometry bits	2,594	2,594	2,594	2,381	2,381	2,381
Mesh motion bits	49,035	47,484	49,035	59,714	57,826	59,714
Total mesh bits	51,826	50,275	51,826	62,292	60,404	62,292
Video shape bits	164,416	164,416	164,416	176,159	176,159	176,165
Video texture bits	419,149	419,149	526,436	724,719	724,719	889,476
Video block motion bits	115,845	115,845	—	125,007	125,007	—
Total video bits	705,210	705,210	—	1,031,685	1,031,685	—
Total bits (video + mesh)	757,036	755,485	748,478	1,093,977	1,092,089	1,133,733
PSNR Y (dB)	34.8	34.8	34.2	33.5	33.5	33.2
PSNR U (dB)	41.9	41.9	42.5	37.6	37.6	38.8
PSNR V (dB)	41.2	41.2	41.9	37.1	37.1	38.3

Note: several entries are identical, in case identical techniques were used.

Table 5 Mesh Compression and Video Object Compression Results for Bream, Using the Scalable Approach with Spatial Prediction (Method A), the Scalable Approach with Prediction from Block Motion Vectors (Method B) and the Integrated Approach

	Bream		
	Scalable coding A	Scalable coding B	Integrated coding
Mesh geometry bits	2,023	2,023	2,023
Mesh motion bits	27,156	28,863	27,156
Total mesh bits	29,228	30,935	29,228
Video shape bits	61,858	61,858	61,858
Video texture bits	355,683	355,683	444,371
Video block motion bits	48,442	48,442	—
Total video bits	467,491	467,491	—
Total bits (video + mesh)	496,719	498,426	536,965
PSNR Y (dB)	31.4	31.4	31.2
PSNR U (dB)	35.3	35.3	36.3
PSNR V (dB)	40.2	40.2	41.4

Note: Several entries are identical, in case identical techniques were used.

with approximately 2.3 bits per node per VOP for Akiyo, 3.1 bits per node per VOP for Sean, and 6.5 bits per node per VOP for Bream. Note that the results for mesh geometry and motion compression were obtained using VLC tables that were not designed specifically for mesh data.

The results on texture coding show that, apparently, block-based motion compensation provided better temporal VOP prediction than mesh-based motion compensation in these sequences, so that the integrated coder spent more bits on texture coding than the block-based video coder. Comparing the total bit rates of the scalable coders with those of the integrated coder, the results are inconclusive at best, even though in the integrated approach no bits were spent on block-motion coding. In the case of Akiyo, the amount of extra bits spent on texture coding in the integrated approach is smaller than the amount of bits necessary to transmit block motion vectors. Therefore, in this case, the integrated coder performs better than the scalable coders. In the case of Sean and Bream, the amount of bits saved on the block motion vectors is not compensated for by the extra amount of bits spent on texture coding. Note that the integrated coder transmits both alpha planes and mesh boundary geometry; no advantage has been taken of the redundancy between these. The results with scalable coding show that the mesh can be coded as overhead using about 6–7% of the bits spent by the VM on the video object at the intermediate quality setting used in the experiments with Akiyo, Sean, and Bream. Clearly, the percentages will go down significantly as we go to higher-quality video applications, as the number of overhead bits needed to code the mesh data is fixed.

6.4. Compression by VOP Self-Transfiguration

This experiment discusses a special case of mesh-based video coding, in which no prediction error images or alpha planes are transmitted; that is, only the texture plane and the alpha plane of the first VOP is encoded, as well as mesh geometry and motion data. The initial (reference) texture map is simply warped forward from the first VOP to all other VOPs. The meshes corresponding to all VOPs can be reconstructed from the mesh geometry and motion data. These decoded meshes can then be used to warp the first VOP. This coder is also referred to as the *self-transfiguration coder* because it simply animates the first VOP using the motion data obtained by mesh-based tracking of the original video object. This type of coder is useful in certain sequences containing 2-D deformable motion without occlusion effects.

We have compared the compression efficiency of the self-transfiguration coder with that of the MPEG-4 Video VM 5.1 (see Ref. 30), both visually and in terms of rate distortion performance. We have used a subsequence of the Bream fish foreground VO. The VOPs used are 60, 62, 64, ..., 108, 110 (26 VOPs in total, 15 VOPs per second). Results for the VM were obtained in the usual manner by coding the first VOP (number 60) in I-mode and the rest in P-mode, for a number of different quantization stepsize ($Q = Q_I = Q_P$) values. Here, we used the values 5, 10, 15, 20, and 25 for Q. Results for the proposed self-transfiguration coder were obtained by coding the first VOP (number 60) using the VM in I-mode, for a number of different quantization stepsize values. Then, this coded I-VOP was used as the reference texture in the warping stage. Using the 2-D mesh data, the reference I-VOP was warped to all other VOPs. Here, we used the values 7, 12, 17, and 22 for Q.

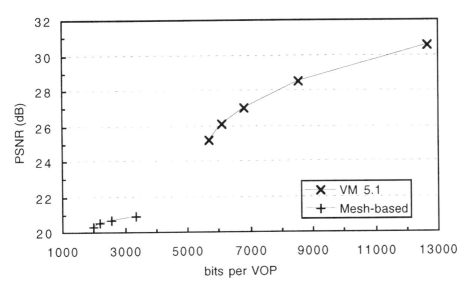

Fig. 15 MPEG-4 Video Verification Model 5 versus mesh-based self-transfiguration coder scheme in terms of rate and distortion.

The rate and distortion results are summarized in Fig. 15. The rate of the VM is defined as the number of bits used to code the I-VOP and P-VOPs, divided by the number of VOPs coded. The rate of the proposed VOP self-transfiguration coder is defined as the sum of the number of bits used to code the first I-VOP and the number of bits used to code the mesh geometry and motion, divided by the number of VOPs coded/animated. Note that those results were not optimized (i.e., we expect the performance to improve, as better VLCs will be used). The PSNR is defined as the average PSNR per VOP of the luminance component.

It can be seen that the VM and the proposed self-transfiguration coder operate in different regimes. The VM operates at higher bit rates and higher PSNRs than the VOP self-transfiguration coder. The latter uses between 2002 and 3355 bits per VOP, whereas the VM uses between 5684 and 28,736 bits per VOP. The PSNR of the result with the VM ranges between 35.18 and 26.06 dB (luminance only). The PSNR of the result with the self-transfiguration coder varies slightly (between 20.90 and 20.26 dB) and is hardly influenced by the Q value used to code the first VOP; it is mainly influenced by geometric distortions related to the accuracy of the mesh tracking. With the self-transfiguration coder, one can easily afford to code the first VOP with a fine quantizer, thereby avoiding visual coding artifacts. Note that the dynamic mesh corresponding to the Bream VOPs mentioned above is coded with 30,437 bits in total, or about 1171 bits per VOP. As can be seen from Fig. 15, this corresponds to 35–60% of the bits used by the proposed self-transfiguration coder.

Despite the differences in the PSNR between the two schemes reported above, we observed that the subjective quality of the results of the proposed self-transfiguration coder is actually better than that of the VM. For instance, the result obtained with the self-transfiguration coder starting from an I-VOP coded with $Q = 7$ is visually very pleasing;

A   B

Fig. 16 Illustration of augmentation of the Bream fish video object with natural and synthetic textures using mesh-based motion tracking; (A) a VOP of the Bream sequence augmented with synthetic texture, showing a piece of text; (B) a VOP of the Bream sequence augmented with natural texture, taken from the Bream sequence itself.

the result obtained with the VM at $Q = 30$ looks heavily distorted—by blocking artifacts in particular. This is because the artifacts caused by the self-transfiguration method are essentially due to spatial shifting of pixels, which are not noticeable, whereas the artifacts caused by the VM are blurring and blocking artifacts that can be highly visible. The PSNR measure seems not very appropriate to evaluate the type of artifact caused by mesh-based animation.

6.5. Augmented Reality Results

Here, we illustrate augmenting a real video object with synthetic and/or natural content via digital postprocessing. In our example, a video of a moving fish is augmented by another image, such that the augmented video is rendered with the natural motion of the original.

The fish is a video object of the MPEG test sequence Bream which is 352×288 pixels, 4:2:0 format. Here, we used every second VOP between VOPs 60 and 110. The motion of the object was represented by a forward tracking content-based triangular mesh. The initial content-based mesh and other meshes tracked on the video object are shown in Fig. 14. The augmentation texture was mapped onto a manually selected region of the reference VOP. Subsequently, the remaining VOPs were obtained by automatic transfiguration from the reference VOP. In the first example, we have used a piece of text as the augmentation texture, as illustrated in Fig. 16a; in the second example, we have used a piece of natural texture from the Bream fish itself, as illustrated in Fig. 16b.

7. CONCLUSIONS

Two- and three-dimensional mesh (wire frame) modeling is a well-known tool in the computer graphics field for representation/animation/rendering of synthetic content. We propose a new 2-D mesh-based representation of motion and shape of video objects, which enables manipulation/animation of natural and synthetic content in a common data structure and their rendering on common software/hardware platforms.

Experimental results show that video object compression based on the 2-D mesh representation is nearly as efficient as the encoder/decoder that is based on the translational-block model and overlapped block motion compensation. Furthermore, the mesh-based encoder/decoder architectures also support several content-based video manipulation and indexing functionalities (e.g., object-tracking, synthetic transfiguration, and augmented reality), which cannot be supported by the block-based encoder/decoder at a comparable accuracy. The potential of mesh-based video compression and manipulation technology has been demonstrated with selected examples.

In addition, mesh-based object modeling and tracking also supports content-based video object indexing and retrieval functionality which is relevant for the upcoming MPEG-7 standardization process. In particular, mesh-based object modeling provides accurate object trajectory information that can be used to retrieve visual objects with specific motion, and vertex-based object shape information which is more efficient than the bitmap representation for shape-based object retrieval. This is subject of further research.

ACKNOWLEDGMENTS

This work is supported in part by a National Science Foundation SIUCRC grant and a New York State Science and Technology Foundation grant to the Center for Electronic Imaging Systems at the University of Rochester.

The authors would like to acknowledge significant contributions by Y. Altunbasak (Hewlett-Packard Labs.), C. Toklu (University of Rochester), and A. Puri (AT&T Research Labs.) to the work described in this chapter. We would also like to thank M.I. Sezan (Sharp Labs America) and A.T. Erdem (Eastman Kodak Company) for their contributions and continuing encouragement during the MPEG-4 standardization process.

Elements of the mesh warping software were provided by A.T. Erdem (Eastman Kodak Company) and C. Toklu (University of Rochester). Delaunay triangulation software was provided by J.R. Shewchuk (Carnegie Mellon University). MPEG-4 Video Verification Model software was provided by M.C. Lee, W. Chen, C. Gu, B. Lin, and S. Winder (Microsoft).

REFERENCES

1. ISO/IEC JTC1/SC29/WG11 N1683, MPEG-4 overview, Bristol, April 1997.
2. L Chiariglione. IEEE Trans Circuits Syst Video Technol CSVT-7(1):5–18, 1997.
3. T Sikora. IEEE Trans Circuits Syst Video Technol CSVT-7(1):19–31, 1997.
4. J Hartman, J Wernecke. The VRML 2.0 Handbook. Reading, MA: Addison-Wesley, 1996.
5. AM Tekalp. Digital Video Processing. Englewood Cliffs, NJ: Prentice-Hall, 1995.
6. V Seferidis, M Ghanbari. Opt Eng 32(7):1464–1474, 1993.
7. MT Orchard, G Sullivan. IEEE Trans Image Process IP-3(5):693–699, 1994.
8. Y Nakaya, H Harashima. IEEE Trans Circuits Syst Video Technol CSVT-4(3):339–356, 1994.
9. J Nieweglowski, TG Campbell, P Haavisto. IEEE Trans Consumer Electron CE-39(3): 141–150, 1993.
10. GJ Sullivan, RL Baker. Proc IEEE ICASSP 4:2713–2716, 1991.
11. Y Wang, O Lee. IEEE Trans Image Process IP-3(5):610–624, 1994.

12. Y Wang, O Lee, A Vetro. IEEE Trans Circuits Syst Video Technol CSVT-6(6):647–659, 1996.
13. C-S Fuh, P Maragos. Affine models for image matching and motion detection. IEEE International Conference on Acoustics, Speech and Signal Processing '91, Toronto, 1991, pp. 2409–2412.
14. Y Wang, O Lee. IEEE Trans Circuits Syst Video Technol CSVT-6(6):636–646, 1996.
15. D Hearn, MP Baker. Computer Graphics. 2nd ed. Prentice Englewood Cliffs, NJ: Hall, 1997.
16. B Lucas, T Kanade. An iterative registration technique with an application to stereo vision. Proceedings DARPA Image Understanding Workshop, 1981, pp. 121–130.
17. Y Altunbasak, AM Tekalp. IEEE Trans Image Process IP-6(9):1255–1269, 1997.
18. JYA Wang, EH Adelson. IEEE Trans Image Process IP-3(5):625–638, 1994.
19. Y Altunbasak, AM Tekalp, G Bozdagi. Two-dimensional object based coding using a content-based mesh and affine motion parameterization. IEEE International Conference on Image Processing, Washington DC, 1995.
20. PJL van Beek, AM Tekalp. Object-based video coding using forward tracking 2-D mesh layers. Visual Communications and Image Processing '97, San Jose, CA, 1997.
21. C Toklu, AT Erdem, MI Sezan, AM Tekalp. Two-dimensional mesh tracking for synthetic transfiguration. IEEE International Conference on Image Processings, Washington, DC, 1995.
22. C Toklu, AM Tekalp, AT Erdem, MI Sezan. 2-D mesh-based tracking of deformable objects with occlusion, IEEE International Conference on Image Processings, Lausanne, 1996.
23. C Toklu, AM Tekalp, AT Erdem. 2-D Triangular mesh-based mosaicking for object tracking in the presence of occlusion. Visual Communications and Image Processing '97, San Jose, CA, 1997.
24. K Wall, PE Danielsson. Computer Graphics, Vision Imaging Process 28:229–227, 1984.
25. Y Altunbasak, AM Tekalp. IEEE Trans on Image Process IP-6(9):1270–1280, 1997.
26. JR Shewchuk. Triangle: Engineering a 2D quality mesh generator and Delaunay triangulator. First Workshop on Applied Computational Geometry. Philadelphia, 1996, pp. 124–133.
27. P J L van Beek, AM Tekalp, A Puri, 2-D Mesh geometry and motion compression for efficient object-based video representation. IEEE International Conference on Image Processing, Santa Barbara, CA, 1997.
28. AV Aho, JE Hopcroft, JD Ullman. Data Structures and Algorithms. Reading, MA: Addison-Wesley, 1983.
29. ISO/IEC JTC1/SC29/WG11 N1277, MPEG-4 video verification model 3.0, July 1996.
30. ISO/IEC JTC1/SC29/WG11 N1469, MPEG-4 video verification model 5.0, Nov. 1996.
31. C Toklu, AT Erdem, MI Sezan, AM Tekalp. Graphical Models Image Process 58(6):553–573, 1996.
32. M de Berg, M van Kreveld, M Overmars, O Schwarzkopf. Computational Geometry—Algorithms and Applications. New York: Springer-Verlaq, 1997.

7
Error Control and Concealment for Video Communication

Qin-Fan Zhu
PictureTel Corporation, Andover, Massachusetts

Yao Wang
Polytechnic University, Brooklyn, New York

1. INTRODUCTION

One inherent problem with any communication system is that information may be altered or lost during transmission due to channel noise. The information-loss problem is manifested even more for video communication because, typically, a very high degree of compression has to be applied before transmission due to the extremely high data rate of raw video signals. Any damage to the compressed bitstream will likely lead to objectionable visual distortion to the reconstructed signal at the decoder. In addition, real-time/interactivity requirements exclude the deployment of some well-known error recovery techniques for certain applications. Finally, issues such as lip synchronization and multipoint configuration further complicate the problem of error recovery.

Transmission errors can be roughly classified into two categories: *random bit errors* and *erasure errors*. Random bit errors are caused by the imperfections of physical channels which results in bit inversion, bit insertion, and bit deletion. Depending on the coding methods and the affected information content, the impact of random bit errors can range from negligible to objectionable. When fixed-length coding is used, a random bit error will only affect one codeword, and the caused damage is generally acceptable. However, if variable-length coding (VLC; for example, Huffman coding) is used, random bit error can desynchronize the coded information such that many of the following bits are undecodable until the next synchronization codeword appears. In some cases, even after synchronization is obtained, decoded information can be still useless, as there is no way to determine which spatial or temporal locations correspond to the decoded information. Erasure errors, on the other hand, can be caused by packet loss in packet networks, burst errors in storage media due to physical defects, or system failure for a short time. Random bit errors in VLC can also cause effective erasure errors because a single bit error can lead to many of the following bits being undecodable, hence useless. The effect of erasure errors (including those due to random bit errors) is much more destructive than random bit errors in general due to the loss or damage of a contiguous segment of bits. Because almost all the state-of-the-art video compression techniques use VLC in one way or another, there is no need to treat random bit errors and erasure errors separately. The generic term "transmission errors" will be used throughout this chapter to refer to both random bit errors and erasure errors.

Various techniques have been developed to combat transmission errors for video communication along two avenues. On one hand, traditional error control and recovery schemes for data communications have been extended for video transmission. These techniques aim at lossless recovery. Examples of such schemes include forward error correction (FEC), or, more generally, error control coding (ECC), and automatic retransmission request (ARQ). On the other hand, signal reconstruction and error concealment techniques have been proposed which try to obtain a close approximation of the original signal or attempt to make the output signal at the decoder least objectionable to the human eyes. Note that unlike data transmission where lossless delivery is required absolutely, the human eye can tolerate a certain degree of distortion in image and video signals. In this chapter, we attempt to summarize and critique what have been proposed in the literature for error control and concealment in video communication.

Different terms have been used to refer to the above problem in video communication, which include *error control, error resilience, error recovery, error concealment, signal/image/video reconstruction,* and so forth. In general, error control and error resilience focus more on reducing the effect of quality degradation due to transmission errors at the encoder side. We will refer to techniques in this category as *forward error concealment* because the video encoder plays the primary role in fulfilling the error concealment task. On the other hand, error concealment and signal reconstruction are often used to refer to those techniques applied at the decoder side. Techniques in this category are referred to as *error concealment by postprocessing*. Finally, error recovery is more related to techniques involving both encoder and decoder. Techniques in this category are referred to as *interactive error concealment*. In what follows, for brevity, we will refer to the entire spectrum of the problems above as error concealment. The chapter is organized as follows. Section 2 formulates the problem and categorizes different approaches. Section 3 describes different methods of error detection. Sections 4–6 present the error concealment techniques in three categories: forward, postprocessing, and interactive. Finally, conclusions are provided in Sec. 7.

2. PROBLEM FORMULATION AND CATEGORIZATION OF APPROACHES

Figure 1 shows a functional block diagram of a real-time video communication system. The input video is compressed by the source encoder to the desired bit rate. The *transport coder* in the figure refers to an ensemble of devices performing channel coding, packetization and/or modulation, and transport-level control using a particular transport protocol. This transport coder is used to convert the bitstream output from the source coder into data units suitable for transmission. On the receiver side, the inverse operations are performed to obtain the reconstructed video signal for display. Note that although we only show a one-way transmission, we use double arrows to emphasize the fact that for some applications, there is a backward channel to convey information from the decoder to the encoder side for system control and error concealment.

Error Control and Concealment

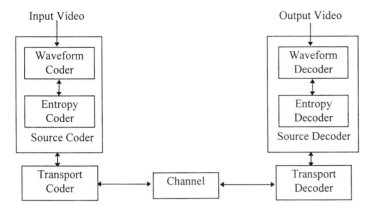

Fig. 1 A real-time video transmission system block diagram.

The *source coder* can be further partitioned into two components: the waveform coder and the entropy coder. The *waveform coder* is a lossy device which reduces the bit rate by representing the original video using some transformed variables and applying quantization. Examples of waveform coding include Discrete Cosine Transform (DCT), wavelet, and vector quantization. The *entropy coder,* on the other hand, is a lossless device which maps the output symbols from the waveform coder into binary codewords according to the statistical distribution of the symbols to be coded. Examples of entropy codings are Huffman coding and arithmetic coding. Although the waveform coder can use any known video coding method, we will mainly focus on the hybrid DCT coding with motion-compensated prediction, as it has been proven to be the most effective for a broad range of applications and is the basis for all current video coding standards (1–3). The transport protocol can vary for different applications. Examples of real-time transport protocols include H.221 in H.320, H.223 in H.324, and H.225 in H.323 (4–9).

The error concealment problem can be formulated loosely as how to find the optimal combination of source coder/decoder and transport coder/decoder so that the signal distortion at the decoder is minimized with a given type of video source signal, bit-rate budget, and channel error characteristics. We assume that the decoder has provisions for error detection and concealment in the presence of transmission errors. The effect of transmission errors is concealed by adding redundancy in the waveform, entropy, or transport coder level. We refer the added redundancy as concealment redundancy. Figure 2 shows a qualitative diagram of reconstructed video quality at the decoder in terms of concealment redundancy and channel error rate. Intuitively as the channel error rate increases, a larger percentage of the total bandwidth should be allocated for increased concealment redundancy so as to achieve the best video quality. Hence, the objective is to design a pair of robust source and transport coders so that the best video quality can be obtained at the decoder, given a total channel bandwidth and channel error distribution.

The above problem is very difficult, if not impossible, to solve due to the many involved variables and the fact that it is difficult to model or describe these variables. First, natural video sources are highly nonstationary in nature and no effective model has been found. The design of a source coder typically requires a good model of the source to improve its performance in terms of both coding efficiency and robustness to transmis-

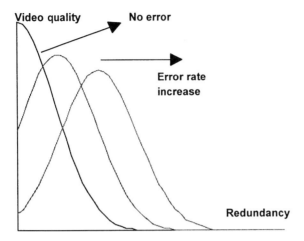

Fig. 2 Video quality versus concealment redundancy.

sion errors. In addition, error characteristics of some video transmission channels are also nonstationary and can change significantly during a service session. For example, an asynchronous transfer mode (ATM) network can become congested with the use of statistical multiplexing for a large number of sources, among other reasons. A mobile videophone may operate at dramatically different error rates, depending on such factors as weather conditions, vehicle moving speeds, and so forth. As mentioned above, other factors such as processing delay, implementation complexity, and application configuration make the problem more difficult to solve.

There have been many error concealment techniques proposed in the literature which attack the transmission error problem from different angles. In most, if not all, cases, some of the variables are fixed first and then a "local" optimal solution is obtained. In this chapter, we categorize these techniques into three groups by whether the encoder or decoder plays the primary role or both are involved in a cooperative way to fulfill the task of error concealment. When the encoder is the primary one involved, we refer these techniques as *forward error concealment*. In these techniques, the source coding algorithm and/or transport control mechanisms are designed to either minimize the effect of transmission errors without any error concealment at the decoder or make the error concealment task at the decoder more effective. Examples of forward error concealment include FEC, joint source, and channel coding, and layered coding. When the decoder is the primary one involved, we refer to these techniques as *error concealment by postprocessing*. In general, these methods attempt to recover the lost information by estimation and interpolation without relying on additional information from the encoder. Spatial and temporal smoothing, interpolation, and filtering fall into this category. Finally, if the encoder and decoder work together to minimize the impact of transmission errors, we refer to these techniques as *interactive error concealment*. Examples in this category include ARQ and selective predictive coding based on feedback from the decoder.

Before delving into the details of various techniques, it is worthwhile mentioning the criteria we use to judge their pros and cons. Obviously, the effectiveness of a technique in terms of image quality is the most important. The required delay is also important for interactive transmission. The third factor is transmission overhead. Finally, the processing complexity is always an issue for any system. Note that the priority of these criteria may change depending on the underlying application. For example, delay is much less important for one-way video transmission such as Internet video streaming and video-on-demand than for two-way and multipoint videoconferencing. In addition, some of the techniques can work for one specific application only, whereas others may be applied to or adapted to suit a broad range of applications. For example, retransmission may work well for point-to-point transmission, but it is difficult to use in multipoint applications. On the other hand, postprocessing error concealment at the decoder can be applied in almost any application.

3. ERROR DETECTION

Before any error concealment technique can be applied at the decoder, it is necessary to first find out whether and where a transmission error has occurred. In this section, we review some of the techniques developed for this purpose. We divide these techniques into two categories: those performed at the transport coder/decoder and those at the video decoder.

One way to perform error detection at the transport coder is by adding header information. For example, in packet-based video transmission, the output of the video encoder is packetized into packets, each of which contains a header and payload field (10). The header contains a sequence number subfield which is consecutive for sequentially transmitted packets. At the transport decoder, the sequence number can be used for packet-loss detection. For example, the multiplex standard H.223 uses such a method for packet-loss detection (6).

Another method for error detection at the transport level is to use FEC (11). In this method, error correction encoding is applied to segments of the output bitstream of the encoder. At the decoder, error correction decoding is employed to detect and possibly correct some bit errors. For example, H.223 uses FEC for both the multiplex packet header and payload to detect errors in the header and payload, respectively (6). In H.261, an 18-bit FEC code is applied to each video frame of 493 bits for error detection and correction (1).

To accomplish error detection at the video decoder, characteristics of natural video signals are usually exploited. In Refs. 12 and 13, the difference of pixel values between two neighboring lines have been used for detecting transmission errors in Pulse Coded Modulation (PCM) and Differential Pulse Coded Modulation (DPCM) coding. When the difference is greater than a threshold, the current image segment is declared to be damaged. In Ref. 14, damage to a single DCT coefficient is detected by examining the difference between the boundary pixels in a block and its four neighbor blocks. At the decoder, four separate difference vectors are formed by taking the differences between the current block and its adjacent blocks over the 1-pixel-thick boundary in four directions, respectively. Then a one-dimensional (1-D) DCT is applied to these difference vectors. Assuming the transition between blocks is smooth, the values of the 1-D DCT vectors should be relatively small in the absence of transmission errors. Hence, if these vectors have a dominate coefficient, then it is declared that one coefficient* is damaged after some statistic test. In addition, the position of the damaged coefficient is also estimated.

---

* This scheme assumes that, at most, one coefficient is damaged. In the event that multiple coefficients are damaged, the algorithm detects and corrects only the coefficient that has the largest error.

Error detection in the frequency domain has also been studied for block-transform coding (15). A synchronization codeword is inserted at the end of each scan line of blocks. When a synchronization codeword is captured at the end of a scan line, the number of blocks decoded is checked against a predetermined number. If a difference is found, then an error is declared and the position of the erroneous block is determined as follows. A weighted mean squared error is calculated between the coefficients of each block in the current line and that in the previous line for an 8×8 block. A larger weight is used for low-frequency coefficients and a smaller weight for high-frequency coefficients, so that the distortion measure correlates more closely to the human visual system. The block with the maximum error is recognized as the erroneous block. This block is split to two blocks or merged with an adjacent block, depending on whether the number of blocks decoded is smaller or larger than the prescribed number. When multiple blocks are damaged, the above detection and splitting/merge procedure repeats until the number of blocks matches the desired one.

As mentioned previously, when variable-length coding (VLC) is used in the source coder, any damage to a single bit can cause desynchronization, resulting in the subsequent bits being undecodable. However, this property can be used as a means to detect transmission errors. Note that in most cases, the VLC being used is not a complete code (i.e., not all the possible codewords are legitimate codewords). Hence, once a video decoder detects a codeword which is not in its decoding table, a transmission error is declared. In addition, the syntax embedded in the bitstream can also be used for error detection. For example, if the decoded quantization step size is 0 or the number of decoded DCT coefficients are more than the maximum number of coefficients (e.g., 64 for an 8×8 DCT transform coder), then a transmission error is detected.

Generally, error detection by adding header information and/or FEC codes at the transport level is more reliable, albeit at the expense of additional channel bandwidth. The benefit of error-detection techniques at the video decoder that rely on the smoothness property of video signals is that it does not add any bits beyond that allocated to the source coder. The use of synchronization codewords and/or incomplete VLC codes offers a compromise: By retaining a small degree of redundancy in the encoding process, it eases the error detection at the decoder. Obviously, these techniques are not mutually exclusive and can be employed jointly in practical systems.

4. FORWARD ERROR CONCEALMENT

In the previous section, we reviewed techniques for detecting transmission errors. From this section onward, we will assume that the locations of errors are known and discuss techniques for concealing the detected errors. In this section, we describe error concealment techniques in which the encoder plays the primary role. There are two kinds of distortion observed at the decoder. The first is the quantization noise introduced by the waveform coder. The second is the distortion due to transmission errors. An optimal pair of source coder and transport coder (including FEC, packetization, and transport protocols) should be designed such that the combined distortion due to both quantization and transmission errors is minimized, given the available bandwidth and channel error characteristics. Typically, the video codec is designed to minimize the quantization error, given the available bandwidth. This practice is guided by the well-known Shannon information theory, which

Error Control and Concealment

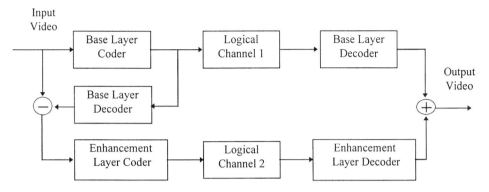

Fig. 3 Layered coding block diagram.

states that one can separately design the source and channel coder to achieve the optimal performance of the overall system. This result was first shown by Shannon for sources and channels that are memoryless and stationary (16) and was later extended to a more general class of sources and channels (17). However, this theorem assumes that the complexity and, hence, processing delay of the source and channel coder can be infinite. In most real-world applications, the above assumptions are not true. First, both the source signals and channel environments can vary rapidly and, hence, are nonstationary. Second, source and channel coders have to be implementable with acceptable complexity and delay. In this situation, joint design of source and channel coder (more generally, transport coder) may achieve better performance.

There are many ways to accomplish forward error concealment. Essentially, they all add a controlled amount of redundancy in either the source coder or the transport coder. In the first case, the redundancy can be added in either the waveform coder or the entropy coder. Some techniques require cooperation between the source and transport coders, whereas others merely leave some redundancy in or add auxiliary information to the coded data that will help error concealment at the decoder. Some techniques require the network to implement different levels of quality of service (QoS) control (e.g., loss and delay) for different substreams, whereas others assume parallel, equal paths. In the following, we review these approaches separately.

4.1. Layered Coding with Transport Prioritization

Until now, the most popular and effective scheme for providing error resilience in a video transport system is layered coding combined with transport prioritization.* In layered video coding, video information is partitioned into more than one group or layer (3,18–24). Figure 3 shows a block diagram of a generic two-layer coding system. The base layer contains the essential information for the video source and can be used to generate an

---

* This scheme assumes that, at most, one coefficient is damaged. In the event that multiple coefficients are damaged, the algorithm detects and corrects only the coefficient that has the largest error.

output video signal with an acceptable quality. With the enhancement layers, a higher-quality video signal can be obtained. To combat channel errors, layered coding must be combined with transport prioritization so that the base layer is delivered with a higher degree of error protection. Different networks may implement transport prioritization using different means. In ATM networks, there is one bit in the ATM cell header which signals its priority. When traffic congestion occurs, a network node can choose to discard the cells having low priority first. Transport prioritization can also be implemented by using different levels of power to transmit the substreams in a wireless transmission environment. This combination of layered coding with unequal power control has also been studied for video transmission in wireless networks (22). In addition, prioritization can be realized with different error control treatments to different layers. For example, retransmission and/or FEC can be applied for the base layer, whereas no or weaker retransmission/FEC may be applied to the enhancement layers. For example, this approach has been taken in the wireless video transport system proposed in Ref. 23.

Layered coding can be implemented in several different fashions depending on the way the video information is partitioned. When the partition is performed in the temporal domain, the base layer contains a bitstream with a lower frame rate and the enhancement layers contain incremental information to obtain an output with higher frame rates. In spatial-domain layered coding, the base layer codes the subsampled version of the original video sequence and the enhancement layers contain additional information for obtaining higher spatial resolution at the decoder. The base layer can also encode the input signal with a coarser quantizer, leaving the fine details to be specified in the enhancement layers. In general, it can be applied to the input samples directly or to the transformed samples. We refer to the first two techniques as *temporal* and *spatial resolution refinement*, respectively, and the third one as *amplitude resolution refinement*. Finally, in transform- or subband-based coders, one can include the low-frequency coefficients or low-frequency band subsignals in the base layer, leaving the high-frequency signal in the enhancement layer. We call this technique *frequency-domain partitioning*. In a video coder using motion-compensated prediction, the coding mode and motion vectors are usually put into the base layer because they are the most important information. Note that the above schemes do not have to be deployed in isolation; they can be used in different combinations. The MPEG-2 video coding standard provides specific syntax for achieving each of the above generic methods. In MPEG-2 terminology, layered coding is referred to as *scalability*, and the above four types of techniques are known as *temporal scalability, spatial scalability, SNR scalability, and data partitioning,* respectively (3).

Although no explicit overhead information is added in layered coding, the graceful degradation of the image quality due to transmission errors is obtained by trading off the compression gain and system complexity. In general, both the encoder and the decoder has to be implemented with the more complicated multilayer structure. In addition, layering will add more coding overhead in the source coder and the transport layer. The introduced coding overhead depends on several factors, including coding method, source spatial and temporal resolution, and bit rate. For example, with the data partition method, a relatively lower overhead will be needed at a higher bit rate than that at a lower bit rate. The four methods presented above have different trade-offs between the increase in robustness to channel noise and the decrease in coding gain. The study in Ref. 24 has found that the three scalability modes in MPEG-2, namely data partitioning, SNR scalability, and spatial scalability, have increasingly better error robustness, in that order, but also increasing coding overhead. To be more precise, data partitioning requires the least bits (requiring

Table 1 Comparison of Different Scalability Modes in MPEG2 Video Coder[a]

Coding mode	Required base layer to total bit rate ratio	The maximum packet loss rate at which the degradation is invisible
One layer (MP@ML)	N/A	10^{-5}
Data partitioning	50%	10^{-4}
SNR scalability	<20%	10^{-3}
Spatial scalability	<20%	$10^{-3}-10^{-2}$

[a] Summarized from experimental results reported in Ref. 24.

only 1% more bits than a single-layer coder at the bit rate of 6 Mbps) to achieve the same image quality when both layers are error free, whereas the spatial scalability has a better reconstructed image when there exist significant losses in the enhancement layer. SNR scalability is in the middle on both scales. Compared to the one-layer coder, the coder performance is improved significantly over the one-layer coder in the presence of channel errors at a relatively small amount of overheads. Table 1 summarizes the required ratio of the base layer to the total bit rate and the highest packet loss rate at which the video quality is still considered visually acceptable. These results are obtained assuming that the base layer is always intact during the transmission.

When designing a layered coder, a factor that needs to be taken into account is whether the information from the enhancement layers will be used for the prediction in the base-layer coding. When it is used, the coding gain in the base layer will be improved. However, when the enhancement information is lost during transmission, it will cause distortion in the base layer in addition to the distortion in the enhancement layer. Hence, in some systems, the base-layer prediction is performed with information from the base layer only, in order to prevent this prediction memory mismatch in the base layer (18).

4.2. Multiple Description Coding

As described in Sec. 4.1, layered coding can offer error resilience when the base layer is transmitted in an essentially error-free channel. However, in certain applications, it is not feasible or cost-effective to guarantee error-free transmission of a certain portion of the transmitted data. In this case, a loss in the base layer can lead to a disastrous effect in decoded visual quality. An alternative approach to combat transmission errors from the source side is to use *multiple-description coding* (MDC). Such a coding scheme is based on the assumption that there are several parallel channels between the source and destination and that each channel may be temporarily down or suffering from long burst errors. Furthermore, the error events of different channels are independent so that the probability that all channels experience loss simultaneously is negligible. These channels could be physically distinct paths between source and destination, such as in a wireless multihop network or in a packet-switched network. Even when only one single physical path exists between the source and destination, the path can be divided into several virtual channels by using time interleaving, frequency division, and so forth.

With MDC, several coded bit streams (referred to as descriptions) of the same source signal are generated and transmitted over separate channels. At the destination, depending on which descriptions are received correctly, different reconstruction schemes (or decod-

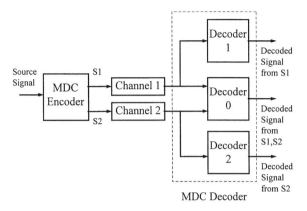

Fig. 4 Block diagram of a multiple-description coder.

ers) will be invoked. The MDC coder and decoder are designed such that the quality of the reconstructed signal is acceptable with any one description and that incremental improvement is achievable with more descriptions. A conceptual-level block diagram for a two description coder is shown in Fig. 4. In this case, there are three decoders at the destination, and only one functions at a time. In order to guarantee an acceptable quality with a single description, each description must carry sufficient information about the original signal. This implies that there will be overlap in the information contained in different descriptions. Obviously, this will reduce the coding efficiency compared to the conventional single-description coder (SDC) that is aimed at minimizing the distortion in the absence of channel loss. This has been shown using a rate-distortion analysis for different types of source (25–27). However, this reduced coding efficiency is in exchange for increased robustness to long burst errors and/or channel failures. With SDC, one would have to spend many error control bits and/or introduce additional latency (in all the bits, or the base layer only in the layered coding case) to correct such channel errors. With MDC, a long burst error or even the loss of an entire description does not have a catastrophic effect, as long as not all the substreams experience failure simultaneously. Thus, one could use fewer error control bits for each substream.

A simple way of obtaining multiple equally important descriptions is by splitting adjacent samples among several channels using an interleaving subsampling lattice and then coding the resulting subimages independently (28,29). If one subimage is lost, it can be recovered satisfactorily based on correlation among adjacent samples in the original image. The problem with this approach is that it requires a large overhead, because the coder cannot make use of the correlation among adjacent samples. Another approach is to use a block-based transform coder and split the coefficients of adjacent blocks among multiple channels. To enable satisfactory recovery of a lost coefficient block, the transform basis should be designed to retain a certain amount of correlation among adjacent coefficient blocks. In Ref. 30, the nonoptimal lapped orthogonal transforms (LOTs) designed in Ref. 31 were evaluated for this purpose. Compared to the spatial subsampling approach, the overhead is reduced, but the reconstruction complexity is higher. Both techniques require some mechanisms of error concealment at the decoder to reconstruct regions where not all descriptions are received. In the following, we review two other approaches which can guarantee a minimally acceptable image quality without relying on error concealment at the decoder.

4.2.1. Multiple-Description Scalar Quantization (MDSQ)

In the approach of Vaishampayan (32), two substreams are obtained by producing two indices for each quantized level. The index assignment is designed so that if both indices are received, the reconstruction accuracy is equivalent to a fine quantizer. On the other hand, if only one index is received, the reconstruction accuracy is essentially that of a coarse quantizer. A simple implementation of this approach is by using two quantizers whose decision regions shift by half of the quantizer interval with respect to each other (known as A2 index assignment (32)). If each quantizer has a bit rate of R, the reconstruction error from two descriptions (i.e., both indices for each quantized samples) is equivalent to that of a single $R + 1$ bit quantizer. On the other hand, if only one description is available, the performance is equivalent to that of a single R bit quantizer. In the absence of channel failure, a total of $2R$ bits are required to match the performance of a single quantizer with $R + 1$ bits; therefore, the loss of coding efficiency is quite significant for large values of R. At lower bit rates, the overhead is smaller. More sophisticated quantizer mappings can be designed to improve the coding efficiency. The MDSQ approach is first analyzed assuming both index streams are coded using fixed-length coding (32). It is later extended to consider entropy coding of the indices (33).

The MDSQ approach described above is developed for memoryless sources. To handle sources with memory, MDSQ can be embedded in a transform coder by coding each transform coefficient using MDSQ (34,35). This approach has been applied to transform-based image and video coders. It was shown that at a packet loss rate as high as 10%, it is possible to achieve reconstructed signal quality that is visually indistinguishable from the original compressed images, with a bandwidth overhead as low as 10–15% (34).

4.2.2. MDC Using Correlation-Inducing Linear Transforms

Another way of introducing correlation between multiple streams is by linear transforms that do not completely decorrelate the resulting coefficients. Ideally, the transform should be such that the transform coefficients can be divided into multiple groups so that the coefficients between different groups are correlated. In this way, if some coefficient groups are lost during transmission, they can be estimated from the received groups. To minimize the loss of coding efficiency, the coefficients within the same group should be uncorrelated. To simplify the design process for a source signal with memory, one can assume the presence of a prewhitening transform so that the correlation-inducing transform operates on uncorrelated samples.

In Refs 36 and 37, Wang et al. and Orchard et al. proposed applying a pairwise correlating transform (PCT) to each pair of uncorrelated variables obtained from the Karhunen-Loeve transform (KLT). The two coefficients resulting from the PCT are split into two streams which are then coded independently. If both streams are received, then an inverse PCT is applied to each pair of transformed coefficients, and the original variables can be recovered exactly (in the absence of quantization errors). If only one stream is received, the missing stream can be estimated based on the correlation between the two streams. In Ref. 36, the PCT uses a 45° rotation matrix, which yields two coefficients having equal variance and therefore requiring the same number of bits. More general classes of PCT using any rotation matrix (i.e., orthogonal) as well as nonorthogonal matrices are considered in Ref. 37. The overhead introduced by this approach can be controlled by the number of coefficients that are paired, the pairing scheme, and the transform parameters (e.g., the rotation angle). This method has been integrated in a JPEG-like coder, in which the PCT is applied to the DCT (similar to KLT in decorrelation capability) coefficients. Only the 45° rotation case has been simulated. It is shown that to guarantee a satisfactory quality from one stream, about 20% overhead is required over the JPEG coder.

Given the relatively large overhead associated with MDC, this approach is appropriate only for channels that have relatively high loss or failure rates. When the channel loss rate is small, the reconstruction performance in the error-free case dominates and the SDC, which is optimized for this scenario, performs best. On the other hand, when the loss rate is very high, the reconstruction quality in the presence of loss is more critical, and the MDC approach will be more suitable. For example, it has been found that under the same total bit rate, the reconstruction quality obtained with the transform coder using PCT exceeds that of the JPEG coder (with even and odd blocks split among two channels) only when the loss rate is larger than 10^{-3} (38). A challenging task is how to design the MDC coder that automatically adapts the amount of added redundancy according to prevailing channel-error characteristics.

4.3. Joint Source and Channel Coding

Given the channel error characteristics, both the waveform coder and entropy coder can be designed to minimize the effect of transmission errors. Spilker noted that when the channel becomes very noisy, a coarse quantizer in the source coding stage outperforms a fine quantizer for PCM-based transmission (39). Kurtenbach and Wintz designed optimal quantizers to minimize the combined mean squared error introduced by both quantization and channel errors, given the input data probability distribution and the channel-error matrix (40). Farvardin and Vaishampayan further extended the design of the optimal quantizer and also proposed a method for performing the code-word assignment to match the channel-error characteristics (41).

Modestino and Daut proposed the use of convolution codes to protect against channel errors for DPCM coding of still images (42). This technique was later extended for DCT still-image coding (43). Three options were proposed to implement combined source and channel coding. In the first option, modulation and error control coding (ECC) are the same for all the bits in every quantized transform coefficient. In the second option, modulation and ECC are the same for all the bits belonging to the same quantized coefficient but can be different for different coefficients. In the third option, modulation and FEC are allowed to vary among different bits of the same coefficient. It was shown that with the first option, for a typical outdoor image, when the channel signal-to-noise ratio (SNR) is smaller than 10 dB, the SNR for the received picture is better with 50% error-correction bits than that without any error-correction bits. The second and the third options can further extend the channel SNR threshold to below 5 dB (43). Another interesting study is to adapt bit allocation for each DCT coefficient based on channel-error characteristics (44). The basic conclusion is that for noisier channels, fewer bits should be allocated to the high-frequency coefficients and more bits should be allocated to the low-frequency coefficients. This conclusion is obtained by assuming the use of fixed-length coding (FLC) for quantized transform coefficients.

4.4. Robust Waveform Coding

In traditional source coder design, the goal is to eliminate both the statistical and visual redundancies of the source signal as much as possible to achieve the best compression gain. This, however, makes the error concealment task at the decoder very difficult. One approach to solve this problem is by intentionally keeping some redundancy in the source coding stage such that better error concealment can be performed at the decoder when transmission errors occur. We refer to techniques in this group as robust waveform coding, which are reviewed in this section.

4.4.1. Adding Auxiliary Information in the Waveform Coder

One simple approach to combat transmission errors is by adding auxiliary information in the waveform coder that can help signal reconstruction in the decoder. Here, we review two such techniques.

An effective technique for error concealment in the decoder is by using motion-compensated temporal prediction (see Sec. 5.1). This requires knowledge of the motion vectors of the lost blocks. One way to help the error concealment task is by sending motion vectors for macroblocks that would not ordinarily use motion-compensated prediction. For example, in MPEG-2, the coder has the option of sending motion vectors for macroblocks in I-frames, so that I-frames can be recovered reliably (3). In the absence of channel errors, these motion vectors are useless. However, when certain macroblocks in an I-frame are damaged, their motion vectors can be estimated from those of the surrounding received macroblocks, and then these macroblocks can be recovered from the corresponding motion-compensated macroblocks in the previous frame.

In Ref. 45, Hemami and Gray proposed adding some auxiliary information in the compressed bitstream so that the decoder can interpolate lost image blocks more accurately. A damaged image block is interpolated at the decoder using a weighted sum of its correctly received neighbor blocks. Determination of the interpolation coefficients is combined with vector quantization in a single step at the encoder, and the resulting quantized weights are transmitted as overhead information to the decoder. For typical JPEG-coded images, the introduced overhead is less than 10%.

4.4.2. Using Partially Decorrelating Transforms

In traditional transform coding, the transform basis is designed to maximize the energy compaction and minimize intercoefficient correlation. To enable the recovery of lost coefficients, one approach is to use suboptimal bases that do not completely decorrelate the transformed coefficients. Hemami investigated the problem of designing robust lapped orthogonal transforms (LOT) (31). A family of LOTs were proposed to achieve different trade-offs between compression gain and reconstruction performance. Based on the channel-error characteristics and available bandwidth, the transform that can achieve the best reconstruction quality can be identified. The reconstruction results using the reconstruction-optimized bases were compared to those obtained by using the DCT–LOT basis (which is optimal in terms of coding efficiency) together with dual transmission of selected blocks. It was found that under a packet loss rate of less than 15%, the dual-transmission method is sufficient. The use of reconstruction-optimized bases is worthwhile only when the packet-loss rate is higher. Note that use of partially decorrelating bases has also been explored for realizing multiple-description coding, as described in Sec. 4.2.

4.4.3. Restricted Prediction Domain

Another way to trade off coding gain and reconstructed picture quality is to restrict the interframe prediction coding within nonoverlapping spatial and temporal subregions. When transmission errors occur, picture-quality degradation will be confined only to the damaged region. Then error concealment can be applied to this region more effectively using the approaches to be described in Sec. 5 and Sec. 6 if a backward channel is available. For example, the independent segment-decoding mode in the H.263 standard only allows spatial and temporal prediction to take place within the same spatial image segment (a Group of Blocks (GOB) or a slice). Here, spatial prediction refers to the prediction of DCT coefficients and motion vectors of one block from adjacent blocks, and temporal prediction is the well-known motion-compensated interframe prediction. To confine the effect of error propagation due to temporal prediction within a temporal subset, input video frames are partitioned into separate groups called threads and each thread is coded without using other threads for prediction (46). This is referred to as *video redundancy coding*. For example, when two threads are used, all the even frames and odd frames are grouped together and temporal prediction is only performed within each group. All threads are started from a sync-frame (e.g., I-frame) and end up in another sync-frame for stopping error propagation. When a transmission error occurs, only one thread will be affected. Between the affected frame and the next sync-frame, a video signal with half of the frame rate is produced.

4.5. Robust Entropy Coding

In the techniques described in Sec. 4.4, redundancies are added during the waveform-coding stage. One can also add redundancy in the entropy-coding stage, in order to help detect bit errors and/or prevent error propagation. We call such techniques robust entropy coding.

4.5.1. Self-Synchronizing Entropy Coding

When variable-length coding (VLC) is used in video coding, a single bit error can lead to the loss of synchronization. First, the decoder may not know a bit error has happened. Furthermore, even when the decoder recognizes that an error has occurred by other means such as the underlying transport protocol, it may not know which bit is in error and, hence, it cannot decode the subsequent bits. One way to prevent this is to designate one codeword as the synchronization codeword in the entropy coder (47–51). A synchronization codeword has the property that the entropy decoder will regain synchronization no matter what happened to the preceding bits once a decoder captures such a codeword. Generally, the resulting entropy coder will be less efficient in terms of bit rate than the ''optimal'' coder without using the synchronization codeword.

Although synchronization can be obtained with a synchronization codeword, the number of decoded symbols may be incorrect. This will typically result in a shift of sequential blocks in a transform coder. To solve this problem, a distinct synchronization codeword can be inserted at a fixed interval either in the pixel domain (50,51) or in the bitstream domain where the number of coded bits is used for measuring the interval (52,53). The synchronization codeword in this case does not carry any information on the encoded video but only plays the role of enabling the decoder to locate where the decoded blocks belong. Several methods have been proposed to minimize the overhead bits introduced by the synchronization codeword (50,51). Although a shorter synchronization codeword introduces less overhead, it also increases the probability that a bit error may generate a fake synchronization codeword. Hence, in practical video coding systems such as H.261 and H.263, relatively long synchronization codewords are used instead (1,2).

For high error-rate environments such as wireless networks, MPEG-4 allows the insertion of an additional synchronization codeword, known as a motion marker, within each coded block between the motion information and the texture information (52,53).

When only the texture information is damaged, the motion information for a block can still be used for better error concealment with techniques to be described in Sec. 5.

4.5.2. Error-Resilient Entropy Coding

In the method described above, the error propagation is limited to the maximum separation between the synchronization codewords. However, in order to reduce the introduced redundancy, these codes have to be used infrequently. Kingsbury et al. have developed error-resilient entropy coding (EREC) methods (54,55). In the method of Ref. 55, variable-length bitstreams from individual blocks are distributed into slots of equal sizes. Initially, the coded data for each image block are placed into their designated slot, either fully or partially. Then, a predefined offset sequence is used to search for empty slots to place any remaining bits of each block. This is done until all the bits are packed into one of the slots. With EREC, the decoder can regain synchronization at the start of each block. It also ensures that the beginning of each block is more immune to error propagation than at the end. Therefore, the error propagation is predominant only in the higher frequencies. The redundancy introduced by using EREC is negligible. In Ref. 55, when EREC is integrated into an H.261-like coder, the reconstruction quality at the bit error rate (BER) of 10^{-4}–10^{-3} is significantly better. Recently, the above EREC method has been used to transcode an MPEG-2 bitstream to make it more error resilient (56). With additional enhancement and error concealment, the video quality at a BER of 10^{-2} was considered acceptable. Kawahara and Adachi also applied the EREC method at the macroblock level together with unequal error protection for H.263 transmission over wireless networks (57). Their simulation results show that the proposed method outperforms the plain FEC both for random bit errors at BER greater than 10^{-3} and for burst errors.

Reversible variable-length code (RVLC) is used in MPEG-4, which can make full use of the available data when a transmission error occurs (52,53). RVLC is designed in such a way that once synchronization is found, the coded bitstream can be decoded backward. With conventional VLC, all data after an erroneous bit are lost until the next synchronization codeword. On the other hand, RVLC can recover data backward from the next synchronization codeword detected, until the first decodable codeword after the erroneous bit. This improved robustness is, however, achieved at a reduced coding efficiency, which is due to the constraint imposed by constructing the RVLC tables.

4.6. Forward Error Control Coding

Forward error correction (FEC) is well known for both error detection and error correction in data communications. However, because FEC has the effect of increasing transmission overhead and therefore reducing usable bandwidth for the payload data, it must be used judiciously in video services which are very demanding in bandwidth but can tolerate a certain degree of loss. FEC has been studied for error recovery in video communication (58–63). In H.261, an 18-bit error-correction code is computed and appended to 493 video bits for error detection and correction for random bit errors in ISDN. For packet video, it is much more difficult to apply error correction because it means several hundred bits have to be recovered due to a packet loss.

In Ref. 58, Lee et al. proposed using Reed-Solomon (RS) codes combined with block interleaving to recover lost ATM cells. As shown in Fig. 5, a RS (32,28,5) code is applied to a block of 28 bytes of data to form a block of 32 bytes. After applying the RS code row by row in the memory up to the 47th row, the payload of 32 ATM cells is

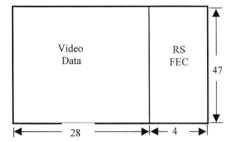

Fig. 5 FEC with interleaving for ATM cell-loss recovery. The numbers in the figure are in bytes.

formed by reading column by column from the memory with the attachment of 1 byte of sequence number. In this way, a detected cell loss at the decoder corresponds to 1 byte erasure in each row of 32 bytes after deinterleaving. Up to 4 lost cells out of 32 cells can be recovered. The Grand-Alliance High-Definition Television (HDTV) broadcast system has adopted a similar technique for combating transmission errors (59). In addition to using the RS code, data randomization and interleaving are also employed for additional protection.

Because a fixed amount of video data has to be accumulated to perform block interleaving as described above, a relatively long delay will be introduced. To reduce the interleaving delay, a diagonal interleaving method is proposed in Ref. 62. At the encoder side, input data are stored horizontally in memory and ATM cells are formed by reading out the data from memory diagonally. In the decoder, the data are stored diagonally in the memory and are read out horizontally. In this way, the delay due to interleaving is halved.

The use of FEC for MPEG-2 in a wireless ATM local area network (LAN) has been studied in Ref. 63. FEC is used at the byte level for random bit error correction and at the ATM cell level for cell-loss recovery. These FEC techniques are applied to both one-layer and two-layer MPEG data. It was shown that the two-layer coder outperforms the one-layer approach significantly, at a fairly small overhead. The authors also compared direct cell-level coding with the cell-level interleaving followed by FEC. It is interesting to note that the authors conclude that the latter introduces a longer delay and a larger overhead for equivalent error recovery performance, and they suggest that direct cell-level correction is preferred.

4.7. Transport-Level Control

Up to this point, we have concentrated on forward error concealment techniques in the source coder and the channel coder. Forward error concealment can also be realized at the transport level. A good example of this is error isolation by structured packetization schemes in packet video. The output of the source coder is assembled into transport packets in such a way that when a packet is lost, the other packets can still be useful because the header and coding mode information is embedded into successive packets (64,65).

A packet often contains data from several blocks. In order to prevent the loss of contiguous blocks because of a single packet or successive packet loss, interleaved packetization can be used, by which successive blocks are put into nonadjacent packets (19,60). In this way, a packet loss will affect blocks in an interleaved order (i.e., a damaged block is surrounded by undamaged blocks), which will ease the error concealment task at the decoder. Note that the use of interleaved packetization in the transport layer requires the source coder to perform block-level prediction only within blocks that are to be packetized sequentially. This will reduce the prediction gain slightly.

Finally, the binary bits in a compressed video bitstream are not equally important. When layered coding is used at the source coder, the transport controller must assign appropriate priority to different layers, which is a form of transport-level control. Even with a nonlayered coder, the picture header and other side information are much more important than the block data. These important bits should be protected so that they can be delivered with much less error rate. One way to realize this is by using dual transmission of important information. In Ref. 66, dual transmission for picture header information and quantization matrix was proposed for MPEG video. In Ref. 67, Civanlar and Cash considered video-on-demand services over an ATM network where the servers and the clients are IP-based and are connected to the network via a fiber distributed data interface (FDDI). They proposed using TCP for transmission of a very small amount of high-priority data before a service session and to use UDP for the remaining low-priority data during the session.

4.8. Summary

Table 2 summarizes the various techniques that have been developed for forward error concealment. All of these techniques achieve error resilience by adding a certain amount of redundancy in the coded bitstreams, in either the source coder or transport coder. Among the techniques that add redundancy in the source coder, some are aimed at guaranteeing a basic level of quality and providing a graceful degradation upon the occurrence of transmission errors (layered coding and multiple-description coding), some help the decoder to perform error concealment upon detection of errors (robust waveform coding), whereas others help to detect bit errors and/or prevent error propagation (robust entropy coding). The transport-level protection (e.g., by using FEC), robust packetization, and so forth must cooperate with the source coder so that more important information bits are given stronger protection and that a single bit error or cell loss does not lead to a disastrous effect. It is also noteworthy that some technique requires close interaction between the source and transport coders (e.g., layered coding with priorities transport, interleaved packetization with restricted prediction domain), whereas others assume different substreams are treated equally (e.g., multiple description coding). Note that these techniques are not mutually exclusive; they can be used together in a complementary way.

5. ERROR CONCEALMENT BY POSTPROCESSING AT THE DECODER

It is well known that images of natural scenes have predominantly low-frequency components (i.e., the image intensities of spatially and temporally adjacent pixels vary smoothly), except in regions with sharp edges. In addition, the human eyes can tolerate more distortion to the high-frequency components than to the low-frequency components. These facts can be used to conceal the artifacts caused by transmission errors at the decoder. In this section,

Table 2 Summary of Forward Error Concealment Techniques

Layered coding with prioritized transport

- Frequency domain partitioning (e.g., MPEG-2 data partitioning)
- Successive amplitude refinement (e.g., MPEG-2 SNR scalability)
- Spatial/temporal resolution refinement (e.g., MPEG-2 spatial/temporal scalability)

Multiple description coding

- Dual scalar quantizer [32]
- Correlation-inducing transforms [36,37]
- Spatial-domain subsampling [28,29]
- Transform-domain subsampling [30]

Robust waveform coding

- Using partially decorrelating transforms [31]
- Adding auxiliary information to help error concealment [74]
- Restricting prediction domain [2]

Robust entropy coding

- Using synchronization codeword to prevent error propagation [48–53]
- Error-resilient entropy coding [54–57]
- Reversible VLC [53]

Source coder design based on channel-error characteristics

- Bit allocation, binary codeword mapping

Transport-level control

- Prioritized transport for layered coding
- Robust packetization [64,65]
- Spatial block interleaving [19]
- Dual transmission of important information

---

we describe several techniques which attempt to perform error concealment at the decoder. Some of these techniques can be used in conjunction with the auxiliary information provided by the source coder to improve the reconstruction quality.

Because of the space limit, we will only review methods that have been developed for video coders using block-based motion-compensation and transform coding (DCT), which is the underlying core technology in all standard video codecs. With such a coder, a frame is divided into macroblocks, which consists of several blocks. There are typically two coding modes at the macroblock level. In the intramode, each block is transformed using block DCT and the DCT coefficients are quantized and entropy coded. In the intermode, a motion vector is found which specifies its corresponding macroblock in a previous frame, and the macroblock is described by this motion vector and the DCT coefficients of the prediction error blocks. By using a self-synchronizing codeword at the end of each row of macroblocks, typically a bit error or cell loss will only cause damage to a single row, so that the upper and lower macroblocks of a damaged block may still be correctly received. If the coded macroblocks are packetized in an interleaved manner, then a damaged macroblock may be surrounded in all four directions by correctly received macroblocks. In addition, if layered coding with frequency-domain partitioning is used, a damaged macroblock may have the coding mode, motion vector, and some low-frequency coefficients correctly received. Finally, the error events between two adjacent frames are usually sufficiently uncorrelated, so that for a given damaged macroblock in the current frame, its corresponding macroblock (as specified by the motion vector) in a previous frame is usually received accurately. All the postprocessing error concealment techniques make use of the correlation between a damaged macroblock and its adjacent macroblocks in the same frame and the previous frame to accomplish error concealment. Some of the techniques only apply to macroblocks coded in the intramode, whereas others, although applicable to intercoded blocks, neglect the temporal information. In the following, we first review techniques that concentrate on recovery of the DCT coefficients or, equivalently, the pixel values. We then present techniques for recovering the coding mode and motion vectors.

5.1. Motion-Compensated Temporal Prediction

One simple way to exploit the temporal correlation in video signals is by replacing a damaged macroblock by the spatially corresponding macroblock in the previous frame. This method, however, can produce adverse visual artifacts in the presence of large motion. Significant improvement can be obtained by replacing the damaged macroblock with the motion-compensated block (i.e., the block specified by the motion vector of the damaged block). This method is very effective when combined with layered coding that includes all the motion information in the base layer (68). Because of its simplicity, this method has been widely used. In fact, the MPEG-2 standard allows the encoder to send the motion vectors for intracoded macroblocks, so that these blocks can be recovered better if they are damaged during transmission (refer to Sec. 4.4.1). It has been found that using motion-compensated error concealment can improve the PSNR of reconstructed frames by 1 dB at a cell loss rate of 10^{-2} for MPEG-2 coded video (24). A problem with this approach is that it requires the knowledge of the motion information, which may not be available in all circumstances. When the motion vectors are also damaged, they need to be estimated from the motion vectors of surrounding macroblocks, and incorrect estimates of motion vectors can lead to large errors in reconstructed images. Another problem with this approach occurs when the original macroblock was coded with the intramode and the coding mode information is damaged. Then concealment with this method can lead to catastrophic results in situations such as a scene change. Recovery of motion vectors and coding modes is discussed in Sec. 5.5.

In Ref. 69, Kieu and Ngan considered the error concealment problem in a layered coder that sends the motion vectors and low-frequency coefficients in the base layer and high-frequency coefficients in the enhancement layer. Instead of simply setting the high-frequency components to zero when the enhancement layer is damaged, the authors showed that using the high-frequency component from the motion-compensated macroblock in the previous frame can improve the reconstructed picture quality. It is assumed that the base layer is delivered without error. When the enhancement layer is damaged, for each damaged macroblock, its motion-compensated macroblock is formed and the DCT is applied to the blocks within the macroblock. The resulting high-frequency DCT coefficients are then merged with the base-layer DCT coefficients of the damaged blocks in the current frame and the inverse DCT is applied to the combined blocks to form an error concealed macroblock.

The above techniques only make use of temporal correlation in the video signal. For more satisfactory reconstruction, spatial correlation should also be exploited. The techniques reviewed below either make use of both spatial and temporal information for error concealment or only make use of the spatial information.

5.2. Maximally Smooth Recovery

This approach makes use of the smoothness property of most video signal through an energy minimization approach. To estimate the intensity value of a damaged pixel, the method requires the difference between this value and the intensity values of its neighboring pixels to be small, so that the resulting estimated image is as smooth as possible. Wang et al. first applied this principle to recover damaged blocks in still images coded using a block-transform-based coder (70). Zhu et al. later extended this method to video coders using motion compensation and transform coding (19). The method formulates the estimation problem as an optimization problem: finding the optimal estimate of the lost coefficients from the available information such that a smoothness measure is maximized (or more precisely, a roughness measure is minimized).

Let $\tilde{\mathbf{a}}_r$ represent the subvector containing the correctly received coefficients in a damaged block, and $\hat{\mathbf{a}}_l$ be the subvector to be estimated for the lost coefficients. Further, let $\mathbf{T}_l$ and $\mathbf{T}_r$ be the submatrices composed of the basis vectors of the DCT transform corresponding to the entries of $\hat{\mathbf{a}}_l$ and $\tilde{\mathbf{a}}_r$, respectively, and let $\mathbf{f}_p$ be the vector consisting of pixels in the motion-compensation block (this vector is set to zero for an intracoded block). Then, the reconstructed prediction error and the original image block can be described by

$$\hat{\mathbf{e}} = \mathbf{T}_r\tilde{\mathbf{a}}_r + \mathbf{T}_l\hat{\mathbf{a}}_l \quad \text{and} \quad \hat{\mathbf{f}} = \hat{\mathbf{e}} + \mathbf{f}_p \tag{1}$$

To determine $\hat{\mathbf{a}}_l$, we require that the reconstructed image block be as smooth as possible. This is accomplished by minimizing a smoothness measure, which is a weighted sum of a spatial smoothness function and a temporal smoothness function, given by

$$\psi(\mathbf{a}_l) = \tfrac{1}{2}[w(\|\mathbf{S}_w\hat{\mathbf{f}} - \mathbf{b}_w\|^2 + \|\mathbf{S}_e\hat{\mathbf{f}} - \mathbf{b}_e\|^2 + \|\mathbf{S}_n\hat{\mathbf{f}} - \mathbf{b}_n\|^2 + \|\mathbf{S}_s\hat{\mathbf{f}} - \mathbf{b}_s\|^2) + (1-w)\hat{\mathbf{e}}^T\hat{\mathbf{e}}]$$
$$= \tfrac{1}{2}[w(\hat{\mathbf{f}}^T S\hat{\mathbf{f}} - 2b^T\hat{\mathbf{f}} + c) + (1-w)\hat{\mathbf{e}}^T\hat{\mathbf{e}}] \tag{2}$$

The first term measures the spatial smoothness and the second term measures the temporal smoothness. The spatial term is essentially a weighted sum of squared pixel wise differences in different directions, the minimization of which forces the pixels to be connected smoothly with each other within the block and with boundary pixels outside the block. The temporal term is the energy of the error vector, the minimization of which enforces a smooth transition between corresponding regions in adjacent frames. The matrices $\mathbf{S}_w$, $\mathbf{S}_e$, $\mathbf{S}_n$, and $\mathbf{S}_s$ depend on the desired amount of smoothing to be imposed between two samples in the directions toward west, east, north, and south, respectively. The vectors $\mathbf{b}_w$, $\mathbf{b}_e$, $\mathbf{b}_n$, and $\mathbf{b}_s$ are composed of the pixels in the one-pixel-wide boundaries outside the block in the four directions above. Figure 6 shows two examples of how the spatial smoothness measure is formed for an 8×8 block. An arrow indicates that a pixel difference should be included between the two pixels. In Fig. 6a, the smoothing constraint is applied only along the block boundary. It has been found that if only the DC coefficient

Error Control and Concealment

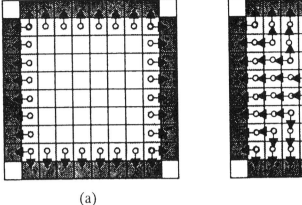
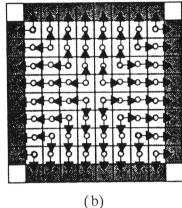

Fig. 6 Illustration of two smoothing constraints. (From Ref. 70, Fig. 1; © 1993 IEEE.)

is lost, applying smoothing along the boundary alone can yield satisfactory results. Figure 6b shows a stronger smoothing constraint which is effective for all loss cases, including the case when only the DC is lost.

The weighting factor w controls the relative contribution of the spatial and temporal smoothing constraints and is chosen according to the possible degrees of spatial and temporal smoothness in the damaged block. This information can be derived from the coding mode (intra/inter, motion compensation or not, etc.) and other available information. For intracoded blocks, a larger w should be used to emphasize the spatial smoothness constraint. On the other hand, for blocks coded in the motion-compensated mode, a smaller w should be used to emphasize the temporal smoothing. When $w \neq 0$, the optimal solution $\hat{\mathbf{a}}_{\text{opt}}$ is given by

$$\hat{\mathbf{a}}_{\text{opt}} = \left(\frac{1-w}{w}\mathbf{I} + \mathbf{T}_l^T \mathbf{S} \mathbf{T}_l\right)^{-1} \mathbf{T}_l^T \left[\mathbf{b} - \mathbf{S}\mathbf{f}_p - \left(\mathbf{S} + \frac{1-w}{w}\mathbf{I}\right)\mathbf{T}_r \tilde{\mathbf{a}}_r\right] \qquad (3)$$

Note that the above solution essentially consists of three linear interpolations, in the spatial, temporal, and frequency domains, from the boundary vector $\mathbf{b}$, the prediction block $\mathbf{f}_p$, and the received coefficient subvector $\tilde{\mathbf{a}}_r$, respectively. It can maintain the information from the received coefficients while enforcing the reconstructed image to be as smooth as possible. When all the coefficients are lost in a damaged block, the solution reduces to spatial and temporal interpolation only. When $w = 0$, which makes use of only the temporal correlation, the optimal solution is such that the damaged block is simply replaced by the prediction block in the previous frame. In this case, the optimal solution reduces to that in Sec. 5.1. On the other hand, with w set to 1, only the spatial correlation is used and the optimal solution is a linear interpolation from the received coefficients and the neighbor pixel data. This can be used for intracoded blocks or still images. Figure 7 shows the PSNR curves for video sequences reconstructed from the above smoothing technique and the copying algorithm, which simply repeats the corresponding block in the previous frame. The figure also contains curves for the case when even–odd block interleaving and prioritized transmissions are used.

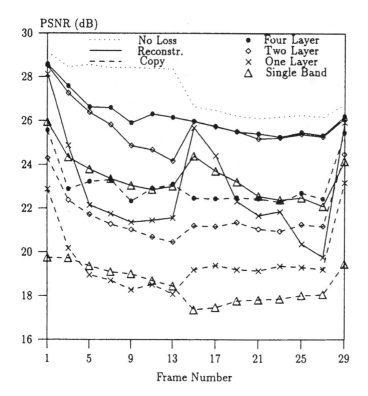

Fig. 7 PSNR comparison for different reconstruction methods and layered coding. (From Ref. 9, Fig. 7; © 1993 IEEE.)

In the above work, the first-order smoothness criterion was employed to reconstruct a lost block. Second-order smoothness criteria have also been investigated (71). A combined quadratic variation and Laplacian operator is used to reduce the blurring across strong edges, which cannot be recovered well with the first-order smoothness measure. The reconstructed images using this second-order measure are visually more pleasing than those obtained with the first-order measure presented here, with sharper edges that are smooth along the edge directions. To further improve the reconstruction quality, an edge-adaptive smoothness measure can be used, so that the variation along the edges is minimized but not across the edges. Several techniques have been developed along this direction (72). This approach requires the detection of edge directions for the damaged blocks. This is a difficult task and a mistake can yield noticeable artifacts in the reconstructed images. The method using the second-order smoothness measure is, in general, more robust and can yield satisfactory images with lower computational cost.

5.3. Projection Onto Convex Sets

Another way to enforce the smoothness of the estimated image is by using the method of projection onto convex sets (POCS). Sun and Kwok proposed using this method to restore a damaged image block in a block-transform coder (73). The convex sets are derived by requiring the recovered block to have a limited bandwidth either isotropically (for a block in a smooth region) or along a particular direction (for a block containing a straight edge). With this method, a combined block is formed by including eight neighboring blocks with the damaged block. First, this combined block is subject to an edge existence test by using the Sobel operator. The block is either classified as a monotone block (i.e., with no discernible edge orientations) or as an edge block. The edge orientation is quantized to one of the eight directions equally spaced in the range of 0° to 180°. Then, two projection operators are applied to the combined block, as shown in Fig. 8. The first projection operator implements a bandlimitedness constraint, which depends on the edge classifier output. If the block is a monotone block, then the block is subject to an isotropic bandlimitedness constraint, accomplished by an isotropic low-pass filter. On the other hand, if the block classifier output is one of the eight edge directions, then a bandpass filter is applied along that direction. The filtering operation is implemented in the Fourier transform domain. The second projection operator implements a range constraint and truncates the output value from the first operator to the range of [0, 255]. For pixels in the edge blocks which are correctly received, their values are maintained. These two projection operations are applied alternatingly until the block does not change any more under further projections. The authors found that 5–10 iterations are usually sufficient when a good initial estimate is available. Note that this technique only makes use of spatial information in the reconstruction process and is therefore applicable to intracoded blocks or still images. For intercoded blocks, one way to make use of the temporal information is by using the motion-compensated block in the previous frame as the initial estimate, and then using the technique presented here to further improve the reconstruction accuracy. Figure 9 shows PSNR curves for a video sequence called ''Basketball'' reconstructed with this scheme, and a method using motion-compensated temporal replacement and the previously described copying algorithm, respectively.

5.4. Spatial- and Frequency-Domain Interpolation

One implication of the smoothness property of the video signal is that a coefficient in a damaged block is likely to be close to the corresponding coefficients (i.e., with the same frequency index) in spatially adjacent blocks. In Ref. 74, Hemami and Meng proposed

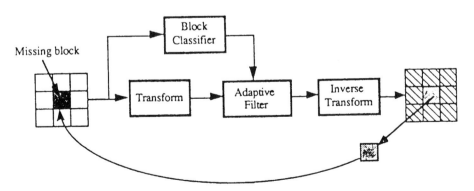

Fig. 8 Adaptive POCS iterative restoration process. (From Ref. 73, Fig. 6; © 1995 IEEE.)

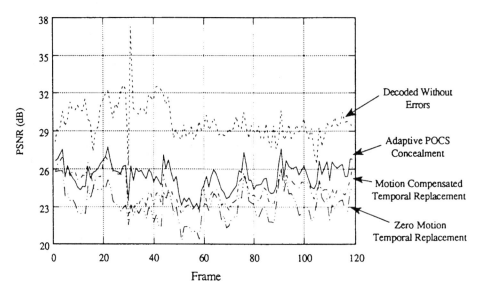

Fig. 9 PSNR comparison on sequence "Basketball." (From Ref. 23, Fig. 10; © 1995 IEEE.)

interpolating each lost coefficient in a damaged block from its corresponding coefficients in its neighbor blocks. The reconstructed coefficient block vector $\hat{C}_Z$ is generated as

$$\hat{C}_Z = \tilde{C}_Z + w_T\overline{C}_T + w_B\overline{C}_B + w_L\overline{C}_L + w_R\overline{C}_R \tag{4}$$

where $\tilde{C}_Z$ is the vector containing coefficients of the current block with the lost coefficients being substituted with zeros, and $\overline{C}_X$ is the vector containing coefficients from the adjacent block X ($X = T$ for top, B for bottom, L for left, and R for right); in which the positions corresponding to correctly received coefficients in the current block are set to zero. Taking the inverse DCT on both sides, the above relation can also be expressed in the spatial domain as

$$\hat{P}_Z = \tilde{P}_Z + w_T\overline{P}_T + w_B\overline{P}_B + w_L\overline{P}_L + w_R\overline{P}_R \tag{5}$$

where each pixel vector is the inverse transform of the corresponding coefficient vector in Eq. (4). If $\tilde{C}_Z = 0$, which corresponds to the case when all the coefficients are lost in the damaged block, the above equation avoids a final inverse transform to obtain the reconstructed block and therefore is computationally more efficient. The weighting coefficients are estimated by minimizing the same smoothing function shown in Fig. 6a. Figure 10 shows the PSNR comparison of three methods for the image Lena with different block-loss rates. Here, it is assumed that a transport protocol is available such that the loss of a block does not affect any other blocks. The curve labeled "mean" is obtained by using the same weights for all the four directions, whereas the curve labeled "optimal" is obtained with the method described in Sec. 4.4.1 (45), which requires about 10% overhead bits.

A problem with the interpolation solution given in Eq. (5) is that, in the case when $\tilde{C}_Z = 0$, a pixel in a damaged block is interpolated from the corresponding pixels in four adjacent blocks, rather than the nearest available pixels. Because the pixels used for interpolation are eight pixels away in four separate directions, the correlation between

Error Control and Concealment 187

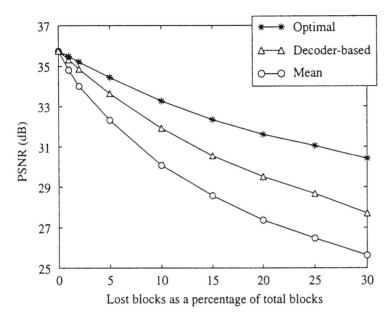

Fig. 10 PSNR comparison for reconstruction techniques applied to 512 × 512 "Lena" with random block loss. (From Ref. 74, Fig. 5; © 1995 IEEE.)

these pixels and the missing pixel is likely to be small and the interpolation may not be accurate. To improve the estimation accuracy, Aign and Fzel proposed interpolating pixel values within a damaged macroblock from its four 1-pixel-wide boundaries (75). Two methods are proposed to interpolate the pixel values. In the first method, a pixel is interpolated from two pixels in its two nearest boundaries, as shown in Fig. 11a. Hence, for an $2N \times 2N$ macroblock, the upper left $N \times N$ block is interpolated from the N pixel in the

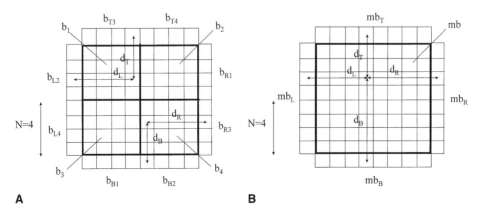

Fig. 11 Spatial interpolation (A) block based; (B) macroblock based. (From Ref. 75, Figs. 1 and 2; © 1995 IEEE.)

upper exterior boundary and N pixels in the left exterior boundary. The interpolation procedure can be described as follows:

$$b_1(i, k) = \frac{d_T b_{L2}(i, N) + d_L b_{T3}(N, k)}{d_L + d_T}$$

$$b_2(i, k) = \frac{d_T b_{R1}(i, 1) + d_R b_{T4}(N, k)}{d_R + d_T}$$

$$b_3(i, k) = \frac{d_B b_{L4}(i, N) + d_L b_{B1}(1, k)}{d_L + d_B}$$

$$b_4(i, k) = \frac{d_B b_{R3}(i, 1) + d_R b_{B2}(1, k)}{d_R + d_B} \tag{6}$$

where $i, k = 1, 2, \ldots, N$.

In the second method, shown Fig. 11b, a pixel in the macroblock is interpolated from the four macroblock boundaries as

$$mb(i, k) = \frac{1}{d_L + d_R + d_T + d_B}$$
$$\times [d_R mb_L(i, 2N) + d_L mb_R(i, 1) + d_B mb_T(2N, k) + d_T mb_B(1, k)]$$

where $i, k = 1, 2, \ldots, 2N$.

As with the POCS method, the above schemes only make use of the spatial smoothness property and are mainly targeted for still images or for intracoded blocks in video. For intercoded blocks, the frequency-domain interpolation described in Eq. (4) and consequently the spatial-domain interpolation formula, Eq. (5), cannot be applied because the DCT coefficients of prediction errors, or prediction errors themselves, of adjacent blocks are not highly correlated. Note that in this case, the spatial-domain pixel vector would correspond to the error vector.

Due to the smoothness properties of natural images, the correlation between high-frequency components of adjacent blocks is small. In Ref. 76, only the DC and the lowest five AC coefficients of a damaged block are estimated from the top and bottom neighboring blocks, whereas the rest of the AC coefficients are forced to be zeros. The DC values are linearly interpolated and the five AC coefficients are synthesized according to the method specified in Ref. 77.

5.5. Recovery of Motion Vectors and Coding Modes

In the techniques described in Secs. 5.1–5.4, we assumed that the coding mode and motion vectors are correctly received. If the coding mode and motion vectors are also damaged, they have to be estimated in order to use these methods for recovering lost coefficients. Based on the same assumption about spatial and temporal smoothness, the coding mode and motion vectors can be similarly interpolated from that of spatially and temporally adjacent blocks.

For the estimation of coding modes, the reconstruction scheme in Ref. 19 simply treats a block with a damaged coding mode as an intracoded block and recovers the block using information from spatially adjacent undamaged blocks only. This is to prevent any catastrophic effect when a wrong coding mode is used for such cases as scene change. Table 3 shows a more sophisticated scheme of estimating the macroblock coding mode from those of its top and bottom neighboring macroblocks for MPEG-2 (76).

Table 3 Estimation of Coding Mode for MPEG-2:
(a) P-Frame; (b) B-Frame

		(a) Top MB	
	MB Type	Forw	Intra
Bottom MB	Forw	Forw	Forw
	Intra	Forw	Intra

		(b) Top MB			
	MB Type	Forw	Back	Inter	Intra
	Forw	Forw	Inter	Inter	Forw
Bottom MB	Back	Inter	Back	Inter	Back
	Inter	Inter	Inter	Inter	Inter
	Intra	Forw	Back	Inter	Intra

Source: Ref. 72 (© 1993 IEEE).

For estimating lost motion vectors, the following methods have been proposed: (a) simply setting the motion vectors to zeros, which works well for video sequences with relatively small motion; (b) using the motion vectors of the corresponding block in the previous frame; (c) using the average of the motion vectors from spatially adjacent blocks; (d) using the median of motion vectors from the spatially adjacent blocks (78). Typically, when a macroblock is damaged, its horizontally adjacent macroblocks are also damaged and, hence, the average or mean is taken over the motion vectors above and below. It has been found that the last method produces the best reconstruction results (78,79). The method in Ref. 80 goes one step further. It selects among essentially the above four methods, depending on which one yields the least boundary-matching error. This error is defined as the sum of the variations along the one-pixel-wide boundary between the recovered macroblock and the one above it, to its left, and below it, respectively. It is assumed that these neighboring macroblocks have been reconstructed previously, and for the damaged macroblock, only the motion vector is missing. In the event that the prediction error for this macroblock is also lost, then for each candidate motion vector, the boundary-matching error is calculated by assuming the prediction error of the damaged macroblock is the same as the top macroblock, the left one, the one below, or zero. The combination of the motion vector and the prediction error that yields the smallest boundary-matching error is the final estimation solution. It was shown that this method yields better visual reconstruction quality than all of the previous four methods.

5.6. Summary

All the error concealment techniques recover the lost information by making use of some a priori knowledge about the image/video signals, primarily the temporal and spatial smoothness properties. The maximally smooth recovery technique enforces the smoothness constraint by minimizing the roughness of the reconstructed signal. In the scheme presented here, which is the simplest version, an isotropic smoothness measure is used everywhere. This is not appropriate near image edges, where smoothness should be enforced along the edges but not across edges. To overcome this problem, one should adapt the smoothness

Table 4 Summary of Postprocessing Error Concealment Techniques

Spatial interpolation [74,75]
    Simple, but can cause over blurring. Require block interleaving at the source end for better quality.
Motion-compensated temporal prediction
    Simple and gives good results if motion vectors of damaged blocks are available. Results are less reliable with estimated motion vectors.
Maximally smooth recovery [19,70]
    Intermediate complexity, generally gives good results, can deal with any loss patterns and exploit spatial/temporal/frequency domain correlation.
POCS [73]
    Complex, can retain edge sharpness, quality depending on robustness of edge detection algorithms.
Fuzzy logic [87]
    Complex, can preserve edge sharpness.
Recovery of motion information [78,79]
    Linear interpolation, median filtering.

(or roughness) measure based on the local image patterns. Several techniques have been developed along this direction (72). An alternative approach is to use alternating projections onto convex sets determined by smoothness constraints along different directions, which is the essence of the POCS method. Depending on the local image pattern, the constraints can be varied from block to block. Compared to the optimization approach, this method is computationally more intensive, as it requires many iterations, although it generally gives more accurate results. The interpolation method can be considered as a special case of the optimization method if only boundary pixels in adjacent blocks are used, and the interpolation coefficients are derived by maximizing the smoothness measure. In fact, the solution of the optimization problem presented in Eq. (3) represents the reconstructed block as a combination of spatial/temporal- and frequency-domain interpolations. The interpolation approaches described in Sec. 5.4, however, only make use of spatial interpolation. The pros and cons of different methods are summarized in Table 4.

The reconstruction methods reviewed here are for transform-based coders using nonoverlapping transforms. Error concealment techniques have also been developed for other coding methods, including subband (81–83), LOT (84,85), and Walsh transform (86). Fuzzy logic has also been used to recover high-frequency components which cannot normally be recovered by the smoothing and interpolation methods presented in this section (87). Finally, besides performing error concealment in the source coder domain as done by the techniques presented in this section, it is also possible to use residual redundancy from the source coder in the channel coder for error concealment. Since the output symbols of any source coder is not completely uncorrelated, this intersymbol correlation can be used to improve the performance of the channel decoder in the presence of transmission errors. In Ref. 88, the authors proposed using a Viterbi decoder in front of the source decoder for this purpose.

6. ENCODER AND DECODER INTERACTIVE ERROR CONCEALMENT

In the previous two sections, we described various techniques for error concealment from either the encoder or the decoder side with little interaction between the two. Conceivably, better overall performance can be achieved if the encoder and decoder can cooperate in the process of error concealment. This cooperation can be realized at either the source coding or transport level. At the source coder, coding parameters can be adapted based on the feedback information from the decoder. At the transport level, the feedback information can be employed to change the percentage of the total bandwidth used for FEC or retransmission. In this section, we first describe several techniques that adapt the source coding strategy based on the feedback information from the decoder. We then present a few schemes that vary transport-level control. Retransmission, when used together with a conventional decoder, leads to decoding delays that may be unacceptable for real-time applications. Two novel schemes that counter this problem are described next. The first approach avoids the decoding delay by remembering the trace of damaged blocks at the decoder. The second scheme sends multiple copies of the lost data in each retransmission trial to reduce the number of retransmissions required. Although this technique can be applied in various video applications, we focus on its application in an Internet video streaming application.

6.1. Selective Encoding for Error Concealment

The error concealment problem would not be such an important issue for most real-time video transmission applications, if the encoder did not use prediction so that a bit error or packet loss would not cause error propagation. If errors only persist for one or two frames, the human eye can hardly perceive the effect because it is too short. However, temporal prediction is an indispensable building block in any video coder, because there is tremendous redundancy between adjacent video frames. Therefore, if the decoder can provide information about the locations of damaged parts to the encoder, the encoder can treat these areas differently so that the effect of error propagation can be either reduced or eliminated. One simple technique along this direction is that whenever the decoder detects an error, it sends a request to the encoder so that the next video frame is coded in the intramode. This way, the error propagation will be stopped in about one round-trip time. However, intracoding typically will reduce the compression gain and hence degrade the video quality under the same bitrate budget.

To reduce the bit rate increase caused by intracoding, only part of the image needs intracoding due to the limited motion vector range (89,90). To further improve the coding efficiency, Wada proposed two schemes to perform selective recovery using error concealment (91). When a packet loss is detected, the decoder sends the identity information of damaged blocks to the encoder. In the same time, error concealment is performed on the damaged blocks. Then, normal decoding continues at the decoder. At the encoder side, two methods are proposed to stop error propagation at the decoder. In the first method (method A), the affected picture area is calculated from the point of damaged blocks up to the currently encoded frame, as shown in Fig. 12. Then, encoding is continued without using the affected area for prediction. Note that encoding without using the affected area does not necessarily mean intracoding. In the second method (method B), shown in Fig. 13, the same error concealment procedure as that performed at the decoder is also carried out for the damaged blocks at the encoder. Then, a local decoding is reexecuted from the point of the concealed blocks up to the currently encoded blocks using the transmitted data stored, so that the encoder's prediction frame buffer matches that at the decoder.

In H.263 (2), one error concealment technique similar to the one described above is called *reference picture selection*. In this method, both the encoder and the decoder

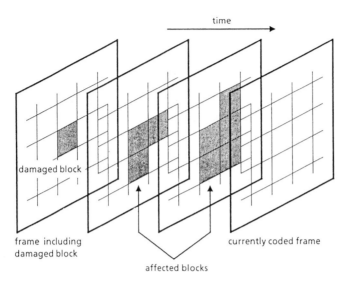

Fig. 12 Method A for selective error concealment. (From Ref. 91; Fig. 1; © 1989 IEEE.)

have multiple prediction frame buffers. Figure 14 shows a block diagram of such an encoder. Besides video data, the encoder and decoder exchange messages about what is correctly received and what is not. From the information, the encoder determines which frame buffers have been damaged at the decoder. Then, the encoder will use an undamaged frame buffer for prediction. The information for the selected prediction frame buffer is also included in the encoded bitstream so that the decoder can use the same frame buffer for prediction.

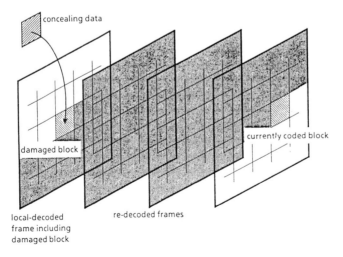

Fig. 13 Method B for selective error concealment. (From Ref. 91, Fig. 2; © 1989 IEEE.)

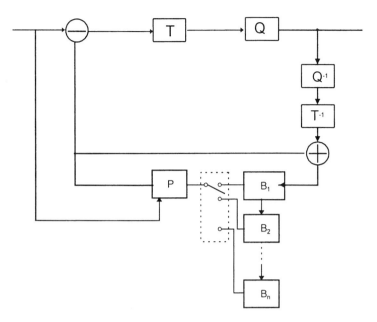

Fig. 14 Reference picture selective coding. T—transform; Q—quantizer; P—motion-compensated predictor; B—frame buffer.

Horne and Reibman (92) proposed sending observed cell-loss statistics back to the encoder, which then adapts its coding parameters to match the prevailing channel conditions. More intrablocks and shorter slices are used when the loss rate is high, for enhanced error resilience, whereas fewer intrablocks and longer slices are invoked when the error rate is low, for improved compression efficiency.

6.2. Adaptive Transport Error Concealment

In the last section, several techniques were described which adapt the source coding strategy at the encoder based on feedback information from the decoder. In this section, we present several schemes that employ the feedback information for adjusting transport-level decisions. First, the transport controller can negotiate with the destination for the retransmission of critical information that is lost. Retransmission has been used very successfully for non-real-time data transmission, but it has been generally considered as unacceptable for real-time video transmission because of the delay incurred. However, this viewpoint has changed slightly in the last few years. It has been realized that even for a coast-to-coast interactive service, one retransmission adds less than 70-ms delay (93). For one-way real-time video applications such as Internet video streaming and broadcast, the delay allowance can be further relaxed to a few seconds so that several retransmissions are possible. Retransmission has also been considered as inappropriate for multipoint videoconferencing because the retransmission requests from a large number of decoders can overwhelm the encoder. However, when a multipoint control unit (MCU) is used in a multipoint conference, the path between the encoder and the MCU is simply point to point. Retransmission can be applied in this path, whereas postprocessing error concealment can be applied at the decoders for the errors between the MCU and the decoders. Another concern about using retransmission is that retransmission may worsen the problem, because it will add more traffic on the network and, thus, further increase the packet-loss rate. However, if retransmission is controlled appropriately, the end-to-end quality can be improved. For example, the encoder can reduce its current output rate so that the sum of the encoder output and the retransmitted data is kept below a given total data rate.

In spite of the above considerations, retransmission has not been used in most video communication systems. This is mainly because most video applications are currently carried over ISDN networks, where the transmission error rate is relatively low. The error propagation problem is circumvented by coding an entire frame in the intramode at the encoder when it is informed of the occurrence of a transmission error by the decoder. Recently, there has been an increasing interest in video communications over very lossy networks such as the Internet and wireless networks, and retransmission is expected to be deployed under these environments. In fact, both H.323 and H.324 standards have defined mechanisms of using retransmission for combating transmission errors (7,9).

Because of its ubiquity, the Internet has been envisioned as the future platform for carrying various digital video services. However, the current Internet is a packet-based network with a best-effort delivery service. There is no end-to-end guaranteed quality of service (QoS). Packets may be discarded due to buffer overflow at intermediate network nodes such as switches or routers, or considered as lost due to excessive long queuing delay. Without any retransmission, experiments show that the packet-loss rate is in the range 2–10%, whereas the round-trip delay is about 50–100 ms on average and can be more than 2 s in coast-to-coast connections (94). With such error and delay characteristics, the achievable QoS is usually poor. Marasli et al. proposed achieving better service quality in terms of delay and loss rate by using retransmission over an unreliable network (95). Instead of trying retransmission indefinitely to recover a lost packet, as done in TCP, the number of retransmission trials is determined by the desired delay. Smith proposed a cyclic UDP protocol, which places the base-layer packets of a layered coder in front of the transmission queue to increase the number of retransmission trials for the base layer (96). Cen et al. and Chen et al. proposed reducing the video output rate at the encoder when the network is congested (97,98). The feedback information about the network condition can be obtained by using the delay and loss-rate statistics at the decoder.

6.3. Retransmission Without Waiting

In order to make use of the retransmitted data, a typical implementation of the decoder will have to wait for the arrival of the requested retransmission data before processing subsequently received data. This will not only freeze the displayed video momentarily but also introduce a certain form of delay. If the decoder chooses to decode faster than its normal speed after the arrival of retransmission, then only a few video frames will be displayed later than its intended display time, and this delay is known as transit delay. On the other hand, the decoder can decode and display all the subsequent frames with a fixed delay, called accumulation delay. In the following, we describe a scheme which uses retransmission for recovering lost information for predictive video coders, but does not introduce the delay normally associated with retransmission (99). A similar technique was developed by Ghanbari, which is motivated by the desire of making use of late cells in an ATM network (100).

With this scheme, when a video data unit is damaged, a retransmission request is sent to the encoder for recovering the damaged data. Instead of waiting for the arrival of the retransmission, the affected video part is concealed by any postprocessing concealment technique described in the last section. Then, normal decoding is continued. Upon the arrival of the retransmitted data, the contained information is added to the corrupted video signal so that the decoder output from this point onward will be restored to the level as if no transmission errors have occurred. In this subsection, we first show how the method works for any generic predictor waveform coder. Then, we describe one low-complexity implementation for video coding when only integer-pixel motion compensation is used.

Consider a generic linear predictive waveform coder with a predictor of order one, which is true for most popular hybrid DPCM/DCT based video codecs using motion compensation. Note that motion-compensated prediction is a linear operator at the picture level. This can be proved as follows: arrange an image as a 1-D vector, then the prediction vector of the current frame can be constructed as the multiplication of a prediction matrix and the previous frame vector. The elements in each row of the prediction matrix are determined by the coding mode and the motion vectors of the corresponding pixel. For example, if a pixel is intercoded without motion compensation, the diagonal element will be 1 and rest of the elements will be 0 in that row. If a pixel is coded with motion-compensated prediction, then the positions and values of the nonzero elements in that row will be determined by the motion vector. If the motion vector is integer, then only one entry has a nonzero value (being 1). On the other hand, if the motion vector is fractional, then several entries are nonzero, with values equal to bilinear interpolation coefficients. Although the following derivation is only done for predictors of order one, it also holds for predictors with order greater than one.

Let $\mathbf{y}_r$ be a vector consisting of prediction errors of all pixels in frame r, and $\mathbf{P}_r$ represent the predictor at time r, which depends on the motion vectors of the current frame. Without transmission errors, the output at sample time r is

$$\mathbf{z}_r = \mathbf{y}_r + \mathbf{P}_r(\mathbf{z}_{r-1}) \tag{7}$$

where $\mathbf{z}_{r-1}$ is the decoded vector at sample time $r - 1$. Let $\hat{\mathbf{y}}_r$ and $\hat{\mathbf{P}}_r$ be the concealment vector and predictor used to replace $\mathbf{y}_r$ and $\mathbf{P}_r$, respectively, to obtain the concealment vector $\hat{\mathbf{z}}_r$ when $\mathbf{y}_r$ and $\mathbf{P}_r$ (i.e., the motion vectors) are damaged due to transmission errors. Then, the concealed output at sample time r is

$$\hat{\mathbf{z}}_r = \hat{\mathbf{y}}_r + \hat{\mathbf{P}}_r(\mathbf{z}_{r-1}) \tag{8}$$

Let d be the delay in sample time between the data damage and retransmission arrival. Then, from sample time $r + 1$ to $r + d$, the error-corrupted outputs at the decoder can be derived as follows:

$$\hat{\mathbf{z}}_{r+1} = \mathbf{y}_{r+1} + \mathbf{P}_{r+1}(\hat{\mathbf{z}}_r)$$

$$\begin{aligned}\hat{\mathbf{z}}_{r+2} &= \mathbf{y}_{r+2} + \mathbf{P}_{r+2}(\hat{\mathbf{z}}_{r+1}) \\ &= \mathbf{y}_{r+2} + \mathbf{P}_{r+2}(\mathbf{y}_{r+1}) + \mathbf{P}_{r+2}(\mathbf{P}_{r+1}(\hat{\mathbf{z}}_r)) \\ &= \mathbf{y}_{r+2} + \mathbf{P}_{r+2}(\mathbf{y}_{r+1}) + \mathbf{P}_{r+1,r+2}(\hat{\mathbf{z}}_r)\end{aligned} \tag{9}$$

$$\vdots$$

$$\hat{\mathbf{z}}_{r+d} = \mathbf{y}_{r+d} + \mathbf{P}_{r+d}(\mathbf{y}_{r+d-1}) + \cdots + \mathbf{P}_{r+1,r+d}(\hat{\mathbf{z}}_r)$$

where the following notation has been used for the multistage linear predictor for brevity:

$$\mathbf{P}_{r+1,r+d}(\mathbf{x}) = \mathbf{P}_{r+d}(\mathbf{P}_{r+d-1}(\mathbf{P}_{r+d-2}(\cdots \mathbf{P}_{r+1}(\mathbf{x})\cdots))) \tag{10}$$

Similarly, the following relation can be derived for the output sample vector $\mathbf{z}_{r+d}$ when there is no transmission error:

$$\mathbf{z}_{r+d} = \mathbf{y}_{r+d} + \mathbf{P}_{r+d}(\mathbf{y}_{r+d-1}) + \cdots + \mathbf{P}_{r+1,r+d}(\mathbf{y}_r) + \mathbf{P}_{r,r+d}(\mathbf{z}_{r-1}) \tag{11}$$

Thus, the correction error vector to be added into $\hat{\mathbf{z}}_{r+d}$ in order to recover $\mathbf{z}_{r+d}$ is

$$\begin{aligned}
\mathbf{c}_{r+d} &= \mathbf{z}_{r+d} - \hat{\mathbf{z}}_{r+d} \\
&= \mathbf{P}_{r,r+d}(\mathbf{z}_{r-1}) + \mathbf{P}_{r+1,r+d}(\mathbf{y}_r) - \mathbf{P}_{r+1,r+d}(\hat{\mathbf{z}}_r) \\
&= \mathbf{P}_{r+1,r+d}(\mathbf{P}_r(\mathbf{z}_{r-1})) + \mathbf{P}_{r+1,r+d}(\mathbf{y}_r) - \mathbf{P}_{r+1,r+d}(\hat{\mathbf{z}}_r) \\
&= \mathbf{P}_{r+1,r+d}(\mathbf{P}_r(\mathbf{z}_{r-1}) + \mathbf{y}_r - \hat{\mathbf{z}}_r) \\
&= \mathbf{P}_{r+1,r+d}(\hat{\mathbf{s}}_r)
\end{aligned} \tag{12}$$

Therefore, after the arrival of the retransmitted information for $\mathbf{y}_r$ and $\mathbf{P}_r$, we can generate the above correction error vector and add it to the reconstructed vector at time $r + d$. From this point on, the decoder output will be identical to that without any transmission error.

In the above derivation, if no prediction is used (i.e., $\mathbf{P}_i = \mathbf{0}$) in encoding any of the sample vectors from sample time $r + 1$ to $r + d$, then there is no need to perform a signal correction at time $r + d$, as the error propagation stops at the sample vector which is coded without prediction. Furthermore, if it is found that no prediction was used to encode the vector at sample time r after receiving the retransmitted information, then the correction vector $\hat{\mathbf{s}}_r$ is simplified to the following form as a result of $\mathbf{P}_r = \mathbf{0}$:

$$\hat{\mathbf{s}}_r = \mathbf{y}_r - \hat{\mathbf{z}}_r \tag{13}$$

The multistage predictor $\mathbf{P}_{r+1,r+d}$ can be obtained by two methods. In the first scheme, the linear predictor operation is accumulated at each sample time from $r + 1$ to $r + d$. Therefore, upon receiving the retransmitted information at sample time $r + d$, the operator $\mathbf{P}_{r+1,r+d}$ is ready to be used to generate the correction error vector $\mathbf{c}_{r+d}$. However, in some cases, this procedure may be quite complicated to implement. In the second scheme, nothing is done during the period of sample times from $r + 1$ to $r + d$ except to save all the linear predictors $\mathbf{P}_i$, $i = r + 1, \ldots, r + d$. After receiving the retransmitted information, the vector $\hat{\mathbf{s}}_r$ is first formed and then all the linear predictors $\mathbf{P}_{r+1}, \ldots, \mathbf{P}_{r+d}$ are sequentially applied to $\hat{\mathbf{s}}_r$ to obtain $\mathbf{c}_{r+d}$.

In the above discussion, we have treated the entire image as one large vector. In practice, one can process one macroblock at a time. This is possible because in most codecs, the macroblock is the basic unit for prediction. When fractional motion compensation is used, it is difficult to trace the motion-compensated predictor on the fly, as one pixel will contribute to the prediction of several pixel positions in the next frame (each with a fractional weight). However, if only integer-pixel motion compensation is used in the codec (e.g., H.261 without the loop filter), then it is possible to form the multistage predictor on the fly. The advantage of tracing the predictor lies in the reduced memory required to store the predictor information and the reduced processing time at frame $r + d$ when the retransmitted packet is received at the decoder. To simplify the description below, all operations are described on a pixel basis instead of on a macroblock basis. For example, $y_i(m, n)$ represents the prediction error signal for pixel (m, n) in frame i.

Tracing the prediction of a pixel is equivalent to summing the motion vectors along its route from the start frame to the end frame. Note that during this process, if a pixel is coded in the intramode, then the error propagation stops. In this case, instead of storing the sum of the motion vectors, a flag G is stored. A frame buffer is created to store the

Error Control and Concealment 197

motion vectors for each pixel from frame $r + 1$ to $r + d$. Each element in the buffer consists of two components, one for the x-direction and one for the y-direction motion vector. In frame r, the buffer is initialized as follows: If pixel (m, n) is undamaged, then $p_x(r, m, n) = G$; otherwise, $p_x(r, m, n) = p_y(r, m, n) = 0$. From frame i, for i from $r + 1$ to $r + d$, the motion vector sums for each pixel are traced as follows:

(a) If pixel (m, n) in frame i is intracoded, then $p_x(i, m, n) = G$.
(b) If the corresponding pixel in the previous frame has a G flag [i.e., $p_x(i - 1, m + v_y(i, k), n + v_x(i, k)) = G$], then $p_x(i, m, n) = G$, where $v_x(i, k)$ and $v_y(i, k)$ are the motion vectors in the horizontal and vertical directions for macroblock k in frame i which contains pixel (m, n).
(c) If otherwise, accumulate the motion vector sums

$$p_x(i, m, n) = p_x(i - 1, m, n) + v_x(i, k)$$
$$p_y(i, m, n) = p_y(i - 1, m, n) + v_y(i, k).$$

Upon the arrival of the retransmission information, the correction error signal for each pixel is generated as follows:

(a) If $p_x(r + d, m, n) = G$, then $c(r + d, m, n) = 0$.
(b) If pixel $(m + p_y(r + d, m, n), n + p_x(r + d, m, n))$ was intracoded in frame r, then $c(r + d, m, n) = y(r, m + p_y(r + d, m, n), n + p_x(r + d, m, n)) - \hat{z}(r, m, n)$.
(c) Otherwise,

$$c(r + d, m, n) = y(r, m + p_y(r + d, m, n), n + p_x(r + d, m, n)) - \hat{z}(r, m, n) + z(r - 1, m + v_y(r, k) + p_y(r + d, m, n), n + v_x(r, k) + p_x(r + d, m, n))$$

In summary, the method described here can achieve lossless recovery except for the time between the information loss and the arrival of the retransmission. During that interval, any postprocessing error concealment techniques described in last section can be applied to the damaged regions to generate a concealed video signal. The beauty of this scheme is that the delay associated with conventional retransmission schemes is eliminated and the video quality can be restored to that without signal damage. The price paid here is the relative high implementation complexity. However, when motion compensation is only applied at the integer-pixel level, then the complexity can be reduced significantly as demonstrated above. Figure 15 shows three PSNR curves for the Salesman sequence with a 20% loss rate at frame 2 and the transmission arrival at frame 5 (100). The top curve is for the case with no transmission error. The middle curve is obtained with the error recovery method described above, and the bottom curve is for the sequence without error concealment.

6.4. Prioritized, Multicopy Retransmission with Application to Internet Video Streaming

Given the stringent delay requirement for real-time video transmission, the residual error rate can be still high after the admissible retransmission trials for applications over very lossy networks, such as the Internet or wireless networks. To reduce the residual error rate, multiple copies of a lost packet can be retransmitted in each single trial (101). With

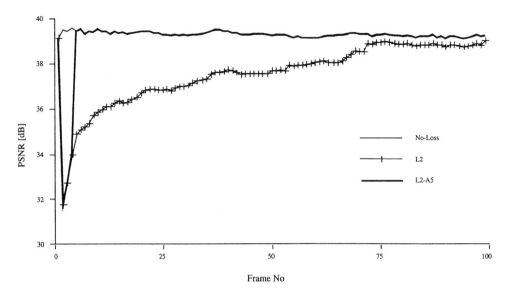

Fig. 15 PSNR comparison of the Salesman sequence with 20% loss at frame 2 and the retransmission arrival at frame 5. (From Ref. 100, Fig. 3; © 1996 IEEE.)

a network loss rate of l, the residual error rate can be reduced to $l^{1+\Sigma_{i=1}^{L} M_i}$, where L is the number of retransmission trials and M_i is the number of copies used for retransmission for the ith trial. For example, if $l = 0.1$, the residual error rate is reduced to 10^{-5} if $L = 2$ and $M_1 = M_2 = 2$. But in order to keep the overall output rate from the encoder under a given budget, the output rate from the source coder has to be reduced to accommodate the retransmission traffic. One way to realize this is to use layered coding at the source coder. When the network loss rate increases, the enhancement layers are partially transmitted or omitted entirely. For a lost packet, the number of retransmission trials and the number of retransmission copies are proportional to the importance of the layer to which the packet belongs. We call this scheme *prioritized multicopy retransmission*. In the remainder of this subsection, we use Internet real-time streaming as an example to explain how to implement this scheme.

Consider an Internet video streaming application using the configuration shown in Fig. 16. The multimedia server sits on the Internet and the client accesses files stored on the server through the dialup modem link via the PSTN (Public Switched Telephone

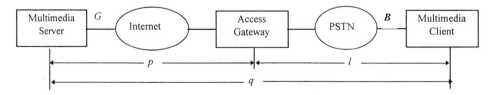

Fig. 16 Internet video streaming with dialup modem.

Network). Instead of downloading the whole file and then playing it back, the file is played out as it is downloaded after a few seconds of initial delay. In this configuration, there are two main factors that affect the video quality at the client side. The first is the packet loss and delay jitter introduced in the Internet. The second is the relatively low channel capacity on the low-speed access link from the client to the access gateway. In general, the loss rate in the PSTN is much lower than that in the Internet.

In this configuration, it is desirable to fully utilize the available bandwidth on the access link. To cope with the high loss rate in the network, layered coding and multicopy retransmission are used. For a lost packet, more than one copy of retransmission can be applied to both increase the probability of successful retransmission and reduce the number of retransmission trials, thus reducing delay. The number of retransmission copies for each layer is dependent on its importance to the reconstructed video quality. To avoid packet discarding at the access gateway, traffic arriving at the gateway cannot be greater than the access link capacity. In addition, the combined data output from the server (which consists of the streaming data and retransmission data) should be smaller than a certain value so as not to jam the Internet.

Let M be the number of layers used in coding the original multimedia data, where $M \geq 2$, and let R_i be the original data rate for layer i. Further, let C_{ij} represent the number of copies for retransmission for layer i in the jth retransmission attempt, and L the maximum retransmission trials allowed. Because of retransmission, the streaming rate, which is defined as the data rate excluding the retransmission data, used for each layer R_i may be different from the original coded data rate R_i. To determine the above parameters, M, L, C_{ij}, and R'_i, note that the combined data output from the server has to satisfy the following relationship:

$$\sum_{i=1}^{M} R'_i + \sum_{i=1}^{M} \left(\sum_{j=1}^{L} q^j C_{ij} \right) R'_i = \min\left(G, \frac{B}{1-p} \right)$$

where q is the end-to-end packet loss rate, G is the maximum data allowed from the server, B is the channel capacity of the access link, and p is the packet-loss rate in the Internet. In the above equation, the left side represents the combined traffic output from the server with the first term accounting for the streaming rate and the second term for the retransmission rate. In general, p is unknown at both the server and the client. However, q can be measured at either the client or the server, and the packet-loss rate in the access link l can also be obtained from the underlying physical layer or link layer. Then, p can be obtained as

$$p = \frac{q-l}{1-l}$$

To obtain the best video quality, the parameters L, R'_i, and C_{ij} have to be chosen jointly. The maximum number of retransmissions, L, is typically determined by the acceptable initial playout delay and the round-trip delay. For the more important base layers, C_{ij} should be bigger than that for the enhancement layers. In addition, as the number of retransmission trials increases and, hence, the remaining retransmission window narrows down, C_{ij} should increase for the base layers and decrease for the enhancement layers to yield more bandwidth for the base layers. For example, assume $M = 3$, $q = 0.1$, $l = 0.001$, $R_1 = 5$ kb/s, $R_2 = 10$ kb/s, $R_3 = 10$ kb/s, $L = 3$, $B = 25$ kb/s, and $G = 30$ kb/s. With C_{ij} chosen as $\{\{2, 3, 5\}, \{2, 1, 1\}, \{1, 1, 0\}\}$, we can achieve the stream rates

Table 5 Summary of Interactive Error Concealment Techniques

Retransmission without waiting [99,100]
   Can achieve lossless recovery without the associated delay. High complexity. Less complex when integer-only motion vectors are used.
Selective encoding [2,91]
   Small overhead. Can be very effective when combined with restricted prediction coding.
Prioritized multicopy retransmission [101]
   Flexible trade-off between delay and reconstruction quality. Can be effective for very lossy channels.

of $R'_1 = 5$ kb/s, $R'_2 = 10$ kb/s, and $R'_3 = 7.4$ kb/s, with the residual error rates for the three layers being 10^{-11}, 10^{-5}, and 10^{-3}, respectively.

6.5. Summary

In this section, we reviewed several techniques in the area of interactive error concealment. For applications which have a backward channel available from the decoder to encoder, this class of methods should give the best performance because the redundancy is added only when an error occurs. This is especially important for channels which have bursty error characteristics. Table 5 summarizes the techniques we presented in this section. Note that these techniques can be used with methods in the other two categories. In fact, the prioritized multicopy retransmission scheme is a combination of retransmission and layered coding.

7. CONCLUSION

In this chapter, we have described various techniques for performing error concealment in real-time video communication. Depending on the channel-error characteristics, system configuration, and requirements, some techniques are more effective than others. The burstiness of transmission errors has a significant impact on the choice of algorithms. For a channel with bursty errors, forward error concealment techniques may not be appropriate because most of the time when there is no error, the overhead introduced by forward error concealment is simply wasted. In this case, retransmission might be more suitable because it only introduces the overhead when needed. The existence of a backward channel from the decoder to the encoder also affects the deployment of some schemes. In applications such as broadcast, where there is no backward channel, none of the interactive error concealment techniques can be applied. The postprocessing techniques can be applied in any circumstances. However, the effectiveness of such techniques is limited by the lack of available information. Also, some of these techniques may be either too complicated for cost-effective implementation or introduce too much processing delay for real-time applications. Aside from the delay and complexity issues, one important criterion for comparing different schemes is the required overhead (in terms of the total bandwidth including bits required by both the source and transport coders) to achieve the same degree of error protection. A fair comparison is, however, difficult to obtain, because these techniques are usually developed to handle different types of transport environments. In real-world applications, typically several techniques are used concurrently and cooperatively to fulfill the error concealment function.

For future research, although more effective error concealment approaches are still called for, we believe that more emphasis should be placed into the system-level design and optimization where the encoding algorithm, transport protocol, and postprocessing method should be designed jointly to minimize the combined distortion due to both compression and transmission. An optimal system should adapt its source coding algorithm and transport control to the network conditions so that the best end-to-end service quality is achieved. In fact, a transport protocol with several levels of error resilience performance has been defined for mobile multimedia communication (102). The system can hop among the several levels adaptively based on the error characteristics of the channel. However, there is very little interaction between the source coder and transport layer in terms of error concealment. An optimal system should allocate the concealment redundancy between the source coder and transport layers adaptively based on the channel environment, so as to optimize the reconstructed video quality for a given decoder error concealment capability. This remains a challenging task for future research and standardization efforts.

ACKNOWLEDGMENT

The authors would like to thank Dr. Amy Reibman, AT&T Labs, Red Bank, NJ, for carefully reading through the manuscript and providing many useful suggestions and critiques.

REFERENCES

1. ITU-T Recommendation H.261, Video Codec for Audiovisual Services at $p \times 64$ kbits, 1993.
2. ITU-T Recommendation H.263, Video Coding for Low Bitrate Communication, 1998.
3. ISO/IEC DIS 13818-2, Information Technology—Generic Coding of Moving Pictures and Associated Audio Information—Part 2: Video, 1994.
4. ITU-T Recommendation H.221, Frame Structure for a 64 to 1920 kbit/s Channel in Audiovisual Teleservices, 1993.
5. ITU-T Recommendation H.320, Narrow-Band Visual Telephone Systems and Terminal Equipment, 1993.
6. ITU-T Recommendation H.223, Multiplexing Protocol for Low Bitrate Multimedia Communication, 1995.
7. ITU-T Recommendation H.324, Terminal for Low Bitrate Multimedia Communication, 1995.
8. ITU-T Recommendation H.225.0, Media Stream Packetization and Synchronization on Non-Guaranteed Quality of Services LANs, 1996.
9. ITU-T Recommendation H.323, Visual Telephone Systems and Equipment for Local Area Networks Which Provide a Non-Guaranteed Quality of Service, 1996.
10. N Ohta. Packet Video: Modeling and Signal Processing. Boston, MA: Artech House, 1994.
11. S Lin, DJ Costello. Error Control Coding: Fundamentals and Applications. Englewood Cliffs, NJ: Prentice-Hall, 1983.
12. KN Ngan, R Stelle. IEEE Trans Commun COM-30:257–265, 1982.
13. KM Rose, A Heiman. IEEE Trans Commun COM-37:373–379, 1989.
14. OR Mitchell, AJ Tabatabai. IEEE Trans Commun COM-29:1754–1762, 1981.
15. W-M Lam, A Reibman. IEEE Trans Image Process 4(s):533–542, 1995.

16. CE Shannon. Bell Syst Tech J 27:379–423, 623–656, 1948.
17. S Vembu, S Verdu, Y Steinberg. IEEE Trans Inform Theory IT-41:44–54, 1995.,
18. M Ghanbari. IEEE J Select Areas Commun 7:801–806, 1989.
19. Q-F Zhu, Y Wang, L Shaw. IEEE Trans CAS for Video Technol 3(3):248–258, 1993.
20. K Ramchandran, A Ortega, KM Uz. IEEE J Select Areas Commun 11:6–23, 1993.
21. Y-Q Zhang, YJ Liu, RL Pickholtz. IEEE Trans Vehic Technol 43:786–796, 1994.
22. M Khansari, M Vetterli. Layered transmission of signals over power-constrained wireless channels. Proc. ICIP-95, Washington, DC, 1995.
23. M Kansari, A Jalali, E Dubois, and P Mermelstein. IEEE Trans CAS Video Technol 6:1–11, Feb. 1996.
24. R Aravind, MR Civanlar, AR Reibman. IEEE Trans CAS Video Technol 6(5):426–435, 1996.
25. JK Wolf, A Wyner, J Ziv. Bell Syst Tech J 59:1417–1426, 1980.
26. L Ozarow. Bell Syst Tech J 59:1921, 1980.
27. AA El-Gamal, TM Cover. IEEE Trans Inform Theory IT-28:851–857, 1982.
28. Y Wang, D Chung. Non-hierarchical signal decomposition and maximally smooth reconstruction for wireless video transmission. In: D Goodman and R Raychaudhuri, eds. Mobile Multimedia Communications. New York: Plenum Press, 1997.
29. AS Tom, CL Yeh, F Chu. Packet video for cell loss protection using deinterleaving and scrambling. Proc. ICASSP'91, Toronto, 1991, pp 2857–2850.
30. D Chung, Y Wang. To appear in IEEE Trans. CAS Video Technol.
31. SS Hemami. IEEE Trans CAS Video Technol 6(2):168–181, 1996.
32. VA Vaishampayan. IEEE Trans Inform Theory IT-39:821–834, 1993.
33. VA Vaishampayan, J Domaszewicz. IEEE Trans Inform Theory IT-40:245–251, 1994.
34. VA Vaishampayan. Application of multiple description codes to image and video transmission over lossy networks. Proceedings 7th International Workshop on Packet Video, Brisbane, 1996.
35. JC Batllo, VA Vaishampayan. IEEE Trans Inform Theory IT-43:703–707, 1997.
36. Y Wang, MT Orchard, AR Reibman. Multiple description image coding for noisy channels by pairing transform Coefficients. Proceedings IEEE 1997 First Workshop on Multimedia Signal Processing, Princeton, NJ, 1997.
37. MT Orchard, Y Wang, AR Reibman, V Vaishampayan. Redundancy rate–distortion analysis of multiple description coding using pairwise correlating transforms. Proceedings IEEE International Conference on Image Process (ICIP-97), Santa Barbara, CA, 1997, vol I, pp 608–611.
38. Y Wang, MT Orchard, AR Reibman, V Vaishampayan. Submitted to IEEE Trans. Image Processing.
39. JJ Spilker Jr. Digital Communications by Satellite. Englewood Cliffs, NJ: Prentice-Hall, 1977.
40. J Kurtenbach, PA Wintz. IEEE Trans Commun Technol CT-17:291–302, 1969.
41. N Farvardin, V Vaishampayan. IEEE Trans Inform Theory IT-38:827–838, 1987.
42. JW Modestino, DG Daut. IEEE Trans Commun. COM-7:1644–1659, 1979.
43. JW Modestino, DG Daut, AL Vickers. IEEE Trans Commun COM-29:1261–1273, 1981.
44. VA Vaishampayan, N Farvardin. IEEE Trans, Commun. COM-38:327–336, 1990.
45. SS Hemami, RM Gray. Image reconstruction using vector quantized linear interpolation, Proc. ICASSP'94, Australia, 1994, pp V-629–V-632.
46. S Wenger. Video redundancy coding in H.263+. Proc. AVSPN, Aberdeen, 1997.
47. JC Maxted, JP Robinson. IEEE Trans Inform Theory IT-31:794–801, 1985.
48. TJ Ferguson, JH Ranowitz. IEEE Trans Inform Theory IT-30:687–693, 1984.
49. PG Neumann. Bell Syst Tech J 50:951–981, 1971.
50. S-M Lei, M-T Sun. IEEE Trans CAS Video Technol 1:147–154, Mar. 1991.
51. W-M Lam, AR Reibman. Self-synchronizing variable-length codecs for image transmission. Proc. ICASSP'92, San Francisco, 1992, pp. III477–III480.

52. R Koenen. Overview of the MPEG-4 Standard, ISO/IEC JTC1/SC29/WG11 N1730, July 1997.
53. T Ebrahimi, MPEG-4 Video Verification Model Version 8.0, ISO/IEC JTC1/SC29/WG11 N1796, July 1997.
54. NT Cheng, NG Kingsbury. IEEE Trans Commun. COM-40(1):140–148, 1992.
55. DW Redmill, NG Kingsbury. IEEE Trans Image Process IP-5(4):565–574, 1996.
56. R Swann, NG Kingsbury. Transcoding of MPEG-II for enhanced resilience to transmission errors. Proc. ICIP-96, Laussanne, 1996, vol 2, pp 813–816.
57. T Kawahara, S Adachi. Video transmission technology with effective error protection and tough synchronization for wireless channels. Proc. ICIP-96, Laussanne, 1996, pp 101–104.
58. SH Lee, PJ Lee, R Ansari. Cell loss detection and recovery in variable rate video. Proceedings 3rd International Workshop on Packet Video, Morriston, NJ, 1990.
59. K Chapplapali, X Lebegue, JS Lim, WH Paik, R Saint Girons, E Petajan, V Sathe, PA Snopko, and J Zdepski. Proc IEEE 83(2):158–174, 1995.
60. T Kinoshita, T Nakahashi, M Maruyama. IEEE Trans CAS Video Technol 3(3):230–237, June 1993.
61. V Parthasarathy, JW Modestino, KS Vastola. IEEE Trans CAS Video Technol 7(2):358–376, 1997.
62. J-Y Cochennec. Method for the correction of cell losses for low bit-rate signals transport with the AAL type 1. ITU-T SG15 Doc. AVC-538, July 1993.
63. E Ayanoglu, R Pancha, AR Reibman, S Talwar. ACM/Baltzer Mobile Networks Applic 1(3): 245–258, 1996.
64. T Turletti, C Huitema. RTP payload format for H.261 video streams. IETF RFC 2032, Oct. 1996.
65. C Zhu. RTP payload format for H.263 video streams. IETF Draft, Mar. 1997.
66. H Sun, J Zdepski. Error concealment strategy for picture-header loss in MPEG compressed video. Proceedings SPIE High-Speed Networking and Multimedia Computing, San Jose, CA, 1994, pp 145–152.
67. MR Civanlar, GL Cash. Image Commun Vol. 8(2), April 1996.
68. M Ghanbari. IEEE Trans CAS Video Technol 3(3):238–247, 1993.
69. LH Kieu, KN Ngan. IEEE Trans Image Processing 3(5):666–677, 1994.
70. Y Wang, QF Zhu, L Shaw. IEEE Trans Commun COM-41(10):1544–1551, 1993.
71. W Zhu, Y Wang. A comparison of smoothness measures for error concealment in transform coding. Proceedings SPIE Visual Communication and Image Processing Taipei, 1995.
72. W Kwok, H Sun. IEEE Trans Consumer Electron CE-39(3):455–460, 1993.
73. H Sun, W Kwok. IEEE Trans Image Process IP-4(4):470–477, 1995.
74. SS Hemami, TH-Y Meng. IEEE Trans Image Process IP-4(7):1023–1027, 1995.
75. S Aign, K Fazel. Temporal and spatial error concealment techniques for hierarchical MPEG-2 video codec. Proceedings Globecom '95, 1995, pp 1778–1783.
76. H Sun, K Challapali, J Zdepski. IEEE Trans Consumer Electron CE-38(3):108–117, 1992.
77. WB Pennebaker, JL Mitchell. JPEG Still Image Data Compression Standard, New York: Van Nostrand Reihold, 1992.
78. P Haskell, D Messerschmitt. Resynchronization of motion compensated video affected by ATM cell loss. Proceedings ICASSP '92, San Francisco, 1992, pp III545–III548.
79. A Narula, JS Lim. Error concealment techniques for an all-digital high-definition television system. SPIE Visual Communication and Image Processing Cambridge, MA, 1993, pp 304–315.
80. W-M Lam, AR Reibman, B Liu. Recovery of lost or erroneously received motion vectors. Proceedings ICASSP '93, Minneapolis, MN, 1993, pp V417–V420.
81. Y Wang, V Ramamoorthy. Signal Proc: Image Commun. 3(2–3):197–229, 1991.
82. SS Hemami, RM Gray. IEEE Trans Image Process IP-6(4):523–539, 1997.

83. G Gonzalez-Rosiles, SD Cabrera, SW Wu. Recovery with lost subband data in overcomplete image coding. Proceedings ICASSP '95, Detroit, MI, 1995.
84. KH Tzou. Post filtering for cell loss concealment in packet video. Proceedings SPIE Visual Communiation and Image Processing Philadelphia, PA, 1989, pp 1620–1627.
85. P Haskell, D Messerchmitt. Reconstructing lost video data in a lapped orthogonal transform based coder. Proceedings ICASSP '88, Albuquerque, NM, 1990, pp 1985–1988.
86. WC Wong, R Steele. Electron Lett, pp. 298–300, 1978.
87. X Lee, Y-Q Zhang, A Leon-Garcia. IEEE Trans Image Process IP-4(3):259–273, 1995.
88. K Sayood, JC Borkenhagen. IEEE Trans Commun COM-39(6):838–846, 1991.
89. H Yasuda, H Kuroda, H Kawanishi, F Kanaya, H Hashimoto. IEEE Trans Commun COM-25:508–516, 1977.
90. N Mukawa, H Kuroda, and T Matuoka. IEEE Trans Commun COM-32:280–287, 1984.
91. W Wada. IEEE J Select Areas Commun 7, 807–814, 1989.
92. C Horne, AR Reibman. Adaptation to cell loss in a 2-layer video codec for ATM networks. Proceedings 1993 Picture Coding Symposium, 1993.
93. G Ramamurthy, D Raychaudhuri. Performance of packet video with combined error recovery and concealment. Proceedings IEEE Infocomm '95, 1995, pp 753–761.
94. D Sanghi, et al. Experimental assessment of end-to-end behavior on internet, Proceedings IEEE Infocomm '93, 1993.
95. R Marasli, PD Amer, PT Conrad. Retransmission-based partially reliable transport service: an analytic model. Proceedings IEEE Infocomm '96, San Francisco, 1996, pp 621–628.
96. BC Smith. Implementation techniques for continuous media systems and applications. PhD dissertation, University of California, Berkeley, 1994.
97. S Cen, C Pu, R Staehli, C Cowan, J Walpole. A distributed real-time MPEG video audio player. Proceedings 5th International Workshop on Network and Operating System Support for Digital Audio and Video, Durham, NH, 1995, pp 151–162.
98. Z Chen, S-M Tan, RH Campbell, Y Li. Real time video and audio in the world wide web. Proceedings Fourth World Wide Web Conference, 1995.
99. Q-F Zhu. U.S. Patent 5,550,847, Aug. 1996.
100. M Ghanbari. IEEE Trans CAS Video Technol, 6(6):669–678, 1996.
101. Q-F Zhu, MR Sridhar, MV Eyuboglu. U.S. Patent 5,768,528, July 1998.
102. ITU-T Recommendation H.223/Annex A–Annex C, Multiplexing protocol for low bitrate mobile multimedia communications, 1998.

8
Multipoint Videoconferencing

I-Ming Pao and Ming-Ting Sun
University of Washington, Seattle, Washington

1. INTRODUCTION

Videoconferencing is a natural extension to voice-only communications. Due to the rapid progress in digital video compression, low-cost video codecs, networking technologies, and international standards, videoconferencing is becoming more and more widely used.

Analog video requires a bandwidth of about 4 MHz and cannot be carried over Public Switched Telephone Network (PSTN) which is designed to carry analog voice of about 3 kHz bandwidth. With the advantage of digital transmission and the advent of networking technology, digital networks such as Integrated Service Digital Network (ISDN), Local Area Networks (LANs), and wireless network are becoming ubiquitous. Straight digitization of the analog video may result in a bit rate as high as 166 Mb/s, which is expensive to transport over current digital networks. The need for compression becomes obvious. With digital video compression techniques developed recently, the digital video can be compressed to below 128 kb/s for transmission over ISDN, or to about 20 kb/s for transmission over PSTN, LANs, and wireless networks. This makes low-cost ubiquitous videoconferencing possible.

Videoconferencing can be over a circuit-switched network such as PSTN and ISDN, or over a packet-switched network such as Internet Protocol (IP)-based network and Asynchronous Transfer Mode (ATM) network. For circuit-switched network, the major limitation is the bandwidth. Currently, over PSTN using the fastest modem, the transmission rate is 56 kb/s. Basic rate narrow-band ISDN (N-ISDN) can only support up to 128 kb/s in two B-channels. These bit rates are too low for transporting high-quality real-time video. Although N-ISDN can support a higher bit rate by using more B-channels, it is more expensive. For the current IP-based packet-switched network, due to the sharing of network resources, Quality of Service (QoS) is difficult to guarantee. For real-time communication like videoconferencing, network congestion, packet delay, delay variation, and packet loss will greatly affect the video quality. ATM network can guarantee QoS; however, it is not as widely available as PSTN, ISDN, or IP-based networks.

The International Telecommunication Union (ITU) has established various videoconferencing terminal standards for different network environments. Some ITU standards are H.320 (1) for N-ISDN, H.310 (2) and H.321 (3) for ATM, H.322 (4) for guaranteed QoS LANs such as ISLAN16-T (5) (formerly known as Isochronous Ethernet), H.323 (6) for nonguaranteed QoS LANs, and H.324 (7) for PSTN. Although some of the standard terminals are available for some time, there are continuing research efforts on videoconferencing over these networks (especially the packet-switched networks). Open research issues include improving video quality, reducing bit rate, transport over networks which do not guarantee QoS, variable-bit-rate video coding for transport over ATM and LAN, scalable video coding for transport over ATM, LAN, and wireless network, error concealment, software-only codec for reducing cost, and so forth.

A multipoint video conference is a video conference that involves three or more terminals. Multipoint videoconferencing is a natural extension of point-to-point videoconferencing. It involves networking, combining, and presenting multiple coded video signals. The same technology can also be used for distance learning, remote collaboration, and video surveillance, which involve multiple sites.

Multipoint videoconferencing is more challenging than the point-to-point videoconferencing because it involves more issues than those open research issues in the point-to-point videoconferencing. A multipoint video conference can either be centralized, where the conference participants are connected to a Multipoint Control Unit (MCU) in a central office, or distributed, where there is no server involved. In centralized multipoint videoconferencing, a video bridge is needed in the MCU to combine the video signals from the conference participants. The video bridge can be implemented using a coded-domain approach or a pixel-domain approach. In a distributed multipoint video conference, user terminals need to handle and decode multiple video bitstreams and, thus, involve more processing than those for point-to-point videoconferencing. Because multiple video signals are involved, efficient network resources utilization is important. Scalable video coding, variable-bit-rate video coding, and statistical multiplexing for saving network resources are interesting topics worthwhile for investigation. At the decoder side, because multiple videos need to be displayed on a monitor, presentation control and human interface is an issue. In distributed multipoint videoconferencing over packet networks, due to the multipoint connection and the lack of common network clock, clock-recovery in the user terminals is more tricky.

This chapter is organized as follows. Section 2 discusses the MCU in the centralized multipoint videoconferencing. The network configurations for supporting multipoint videoconferencing are briefly discussed in Section 3. A distributed videoconferencing system is discussed in Section 4. Some continuous-presence video bridges reported in the literature are reviewed in Section 5. The comparison of coded-domain combining and pixel-domain combining is discussed in Section 6. Finally, a summary is provided in Section 7.

2. MULTIPOINT CONTROL UNIT

Depending on the system configuration, a multipoint video conference can be classified into a centralized or a distributed multipoint conference. A centralized multipoint conference is one in which all participating terminals communicate in a point-to-point fashion with a multipoint control unit (MCU). The terminals transmit their control, audio, video, and data streams to the MCU. The MCU consists of two parts: a mandatory Multipoint Controller (MC) and an optional Multipoint Processor (MP). The MC manages the conference centrally. The MP processes the audio, video, and data streams, and returns the processed streams to each terminal. When the multipoint videoconferencing involves user terminals with multiple video coding standards, the MCU can be used as a transcoding

Multipoint Videoconferencing

gateway to perform the transcoding between different standards, adaptation of bit rates, and enable interoperability between different video conferencing environments (8). A distributed multipoint conference is one in which the participating terminals multicast their audio and video to all other participating terminals without using an MCU (6).

For a multipoint video conference over a Wide Area Network (WAN) such as the ISDN, each user connects to a central office, which results in a centralized multipoint videoconferencing. A centralized multipoint videoconferencing can be of a "switched-presence" type or a "continuous-presence" type. A typical switched-presence MCU permits the selection of a particular video signal from one user for transmission to all users. This can be accomplished through, for example, a signal selected by the conference chairman or a signal selected based on the audio channel activity. Switched-presence MCU generally does not require the processing of video signals to generate a combined signal and is, therefore, relatively simple to implement. Additional details regarding switched-presence videoconferencing can be found in ITU-T Recommendations H.231 (9) and H.243 (10).

The switched-presence MCU is not ideal for multipoint videoconferencing applications because only one participant can be seen at one time, so the reactions of all the participants cannot be seen. A better type of MCU would be those based on the continuous-presence scheme. Continuous-presence videoconferencing generally involves processing the input video signals from the conference participants to generate a combined output video signal. The combined video signal may provide each user with, for example, a partitioned display screen having separate portions corresponding to the video signals of the individual users. Each participant in a continuous-presence conference can then see more than one participants in real time. Therefore, a continuous-presence conference more closely simulates an actual in-person conference than a switched-presence conference. Examples of the switched-presence video conference and the continuous-presence video conference are shown in Figs. 1 and 2, respectively. In Fig. 2, the video combiner is also often referred to as a video bridge.

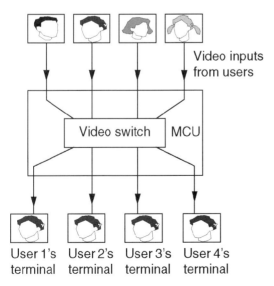

Fig. 1 Switched-presence multipoint videoconferencing.

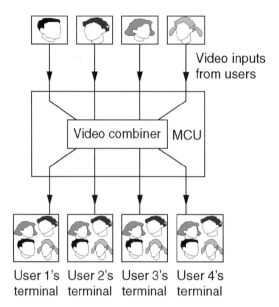

Fig. 2 Continuous-presence multipoint videoconferencing.

There are two possible approaches to implementing a video combiner for continuous-presence multipoint videoconferencing: coded-domain combining and pixel-domain combining. The coded-domain combining is defined as a process that does not decode the compressed video data down to the pixel domain. In the pixel-domain combining, the compressed video is completely decoded to the pixel domain for combining, then the

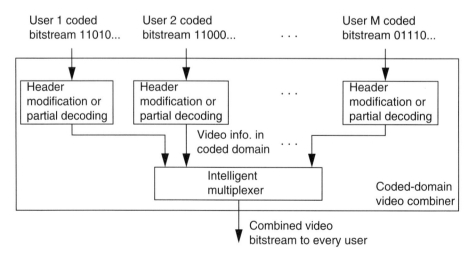

Fig. 3 Coded-domain video combiner.

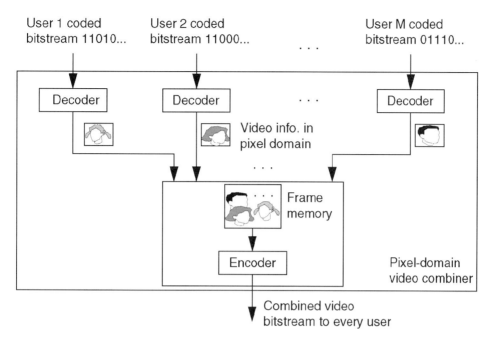

Fig. 4 Pixel-domain video combiner.

combined video is encoded again for transport over the network. The pixel-domain combining approach is more flexible in terms of picture presentation manipulation (e.g., sizing, positioning, chroma keying) and allows interoperability of participants with different types of codecs and different communication bit rates. However, coded-domain combining can offer lower MCU cost than pixel-domain combining. The low implementation cost is due to the fact that no additional encoder and decoder are required, and the combining can be achieved mainly by software. Block diagrams of a coded-domain video bridge and a pixel-domain video bridge are shown in Figs. 3 and 4, respectively. To compare these two approaches, several issues such as flexibility, video quality, processing delay, and implementation cost need to be carefully considered. These will be discussed in more details in later sections.

3. NETWORK CONFIGURATIONS

In a multipoint video conference, video, audio, and user data from all or a selected subset of sites must be made available to each participant to simulate a conference room environment. To achieve this, transporting and combining of all signals to and from all sites have to be done efficiently.

Various network configurations are possible for supporting multipoint videoconferencing (11). Different networks have different constraints. These network constraints affect the requirements and the designs of the video bridge. Some practical network configura-

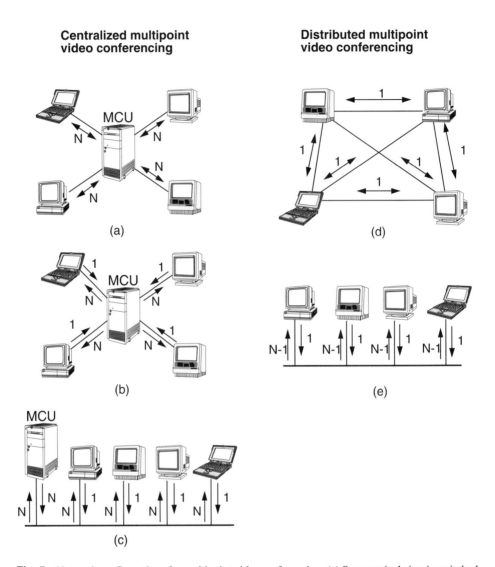

Fig. 5 Network configurations for multipoint videoconferencing. (a) Symmetrical circuit-switched WAN such as ISDN and PSTN (the upstream bandwidth may not be fully utilized); (b) networks which can provide asymmetrical communication links and guaranteed QoS, such as ATM, ADSL, cable; (c) bus-oriented multicast packet-switched LANs; (d) fully connected dedicated networks; (e) bus-oriented multicast packet-switched LANs.

tions for supporting multipoint videoconferencing are shown in Fig. 5. In the figure, the number 1 on the communication channels indicates one unit of channel bandwidth, defined as the bandwidth needed to transport the video signal of a person. The number N indicates that N units of channel bandwidth are needed to transport the combined video signal of multiple persons. In general, to achieve similar video quality, the combined video of multiple persons requires a higher bit rate than that for the video of a single person. Thus,

N is usually greater than 1 in order to achieve good quality with the combined video. The specific value of N depends on the design of the video bridge and the trade-off between the video quality and the bit rates.

Figure 5a shows the configuration for multipoint video conferencing over a symmetrical circuit-switched WAN such as ISDN and PSTN. The network supports constant-bit-rate applications with guaranteed QoS. In this network configuration, each user has dedicated telephone or ISDN lines connected to a central office. In the central office, an MCU with a video bridge combines all the users' video streams. The combined video is sent back to the users to provide multipoint videoconferencing services. From the users' point of view, the network offers a full-duplex symmetrical channel for the upstream and the downstream signals. Even though the upstream signal (which contains the video of only a single person) does not require the bandwidth N, a communication link with the bandwidth N needs to be used in order to be able to support the downstream combined signal (which contains the video of multiple persons). Thus, the upstream channel bandwidth is usually not fully utilized. To fully utilize the upstream communication channel, it is possible to design a scalable system so that the upstream signal contains a higher-resolution video of the user. With the downstream channel, the users can choose to display a higher-resolution video of a specific person or to display the combined video of multiple persons with each individual in a reduced resolution. This is reasonable for practical applications because the user terminal display has a fixed size. The display can naturally show a person in a higher resolution or multiple persons with each individual in a lower resolution.

Figure 5b shows the configuration of a network which can offer an asymmetrical communication channel to support a centralized multipoint conference. Examples include ADSL (12) (Asymmetrical Digital Subscriber Line), ATM, and cable networks. The major difference from the configuration shown in Fig. 5a is that, in addition to the features discussed above, it can provide an asymmetrical communication link for the upstream and the downstream channels. The ATM network can also support variable-bit-rate services. These features offer extra flexibility for supporting multipoint videoconferencing applications.

Figure 5c shows the configuration of a multicast packet-switched LAN with an MCU. It has a bus-oriented configuration. Each person sends the video to the MCU. The MCU combines the video streams and then multicasts to the users. The network supports asymmetrical communication channels but does not guarantee QoS. The available channel bit rate and the end-to-end delay may fluctuate depending on the traffic condition of the network. The network may result in lost packets and relatively long end-to-end delay due to network congestions. It also supports variable-bit-rate applications.

Figures 5d and 5e show network configurations for the distributed multipoint conference. Figure 5d is a dedicated mesh network which requires full connections between the users. For an M-point video conference, it requires $M(M-1)/2$ full-duplex links and is not very resource efficient. It is not practical for a large number of users over a wide area. Figure 5e shows a packet-switched local area network supporting multipoint videoconferencing without using an MCU. In this case, each user multicasts the video to the conference participants. Each user terminal will need to implement the functions performed by the MCU and, thus, may be more expensive compared to the user terminals in the configuration shown in Fig. 5c. Also, similar to the configuration in Fig. 5c, the QoS is an issue compared to the configurations with circuit-switched networks, especially when the multipoint video conference is over a wide area and the traffic is heavy on the network.

As no central bridge is involved, distributed multipoint conference may have several advantages:

1. It offers shorter end-to-end delay, i.e., reduces latency.
2. It takes the least network bandwidth and allows a different bandwidth (video signal resolution) for each user.

On the other hand, the use of a central bridge simplifies the multipoint connections to multiple point-to-point connections. The central bridge provides the following features:

1. By combining video and performing adaptation in the central bridge, it is possible to keep using existing standard video codecs and to achieve the highest level of standards compatibility.
2. Compared with the configuration in Fig. 5d which does not use a central bridge for an M-point conference, the total number of full-duplex links required can be reduced from $M(M - 1)/2$ to M. Because M is normally larger than 3, the number of links can be reduced, and the networking complexity is expected to reduce proportionally. This is especially important when M is large.
3. A central bridge performs a lot of processing which, otherwise, would need to be performed by all the user terminals. By sharing a bridge that takes care of the processing associated with multipoint connections, it is expected that the hardware complexity of the user terminals can be reduced. This is significant because only a central bridge is needed while there are a large number of user terminals.
4. With a centralized bridge, it is easier to maintain and to introduce new features as the technology progresses.

In the next section, we will give an example for the configuration shown in Fig. 5e. In Secs. 5 and 6, we will focus on the configuration shown in Fig. 5a, which can provide ubiquitous multipoint videoconferencing services over a wide area network today.

4. EXAMPLE OF A DISTRIBUTED MULTIPOINT VIDEOCONFERENCING SYSTEM

Distributed multipoint videoconferencing systems do not need an MCU for video combining. The video signals are distributed to all the conference participants through multicast. Decoding, combining, and displaying of multiple video and audio streams are all done together at every end-station. The Multimedia Multiparty Teleconferencing (13) (MMT) system is a Joint Photographic Experts Group (JPEG)-based (14) system designed for distributed multipoint video conferencing over a packet-switched network.

In Ref. 13, user data are multicast from each station to other stations in the conference. The audio and video combining functions are performed in each individual end-station. The synchronization of sender and receiver codecs in the absence of a network reference clock is taken care of in each individual end-station with robustness to packet jitter and loss. Due to the possible network congestion and packet loss, the relatively error-resilient video coding standard JPEG is used. The network has a relatively wide bandwidth to support the JPEG video which has relatively high bit rates.

A feature of the packet-based environment is the capability of routers and switches in the network to multicast a packet or cell to multiple outgoing links. This capability is particularly important because multiple high-bit-rate video streams are involved and the efficient network resource usage is critical. With the distributed configuration, it provides direct paths for delay-sensitive audio and video streams between sources and destinations without having to route the audio and video streams through a centralized MCU.

In the MMT system, the video combining is implemented as follows. Each video window is partitioned into a grid of coding intervals (CIs). The size of a CI is a 16×8-pixel block. The display is partitioned into a grid of CIs. The system requires the video windows of each user to be placed exactly on the grid of display. At each end-station, an "overlay table" (OT) is used to specify the overlay configuration of the video windows. The OT has one entry for each CI on the display screen. Each entry maps one CI on the display screen to a specific CI of a specific video window. This information is derived from a user's definition of the positions and overlay placement of the video windows. To create a composite frame, the composer goes through every entry in OT to read out the specified CI from the specified queue. The JPEG decoder then reads in the composite compressed data stream, decompresses it, and outputs the pixels in a form that is suitable for display. Because JPEG is an intraframe coding scheme and supports a flexible coding format, this video combining scheme is relatively easy to implement.

In a packet network, each station runs off an independent clock. If no provision is made to synchronize these clocks, then because the audio/video data are generated and rendered at a fixed rate based on the local system clocks at the encoder sides, playback buffers at the receivers may eventually underflow or overflow, as the encoder system clocks may be different from the decoder system clocks. The MMT system employs a scheme to solve the clock mismatch and lip synchronization problems. The MMT system does not generate audio packets during silence periods. As a result, the audio stream is not a continuous flow but contains sequences of talk spurts separated by silence. Audio samples are played back freely at the pace of the receiver audio clock. The audio clock mismatch between the sender and receiver is absorbed automatically during the silence intervals. Lip synchronization is achieved using time stamps and an audio master and video slave strategy, in which the video is forced to synchronize to the audio at the beginning of each talk spurt. In this way, the video clock mismatches are also absorbed in silence periods.

In the current video compression standards, JPEG is the most robust encoding scheme in the presence of packet loss because it is intraframe coded. The error caused by packet loss in the current frame will not propagate to the next frame. Due to the intraframe coding and the flexible format, it is also most suitable for video combining in the coded domain. However, because it does not reduce the temporal redundancy, the compression ratio of JPEG is much lower than MPEG and H.261 (15), which are interframe coding standards. As the LAN for MMT has a bandwidth up to 10 M/bits, JPEG can be used in MMT. For transport over PSTN or ISDN, JPEG will not be suitable for compressing the video signals due to the relatively high data rates.

5. OVERVIEW OF SOME VIDEO BRIDGES REPORTED IN THE LITERATURE

For wide area multipoint videoconferencing over a circuit-switched network such as PSTN or ISDN, the centralized configuration is more appropriate, as usually each subscriber only has one or two PSTN or ISDN lines connected to a telephone central office. In the early nineties, several switched-presence MCUs have been developed (16,17). They all used H.261 (15) as the video coding standard. The MCU in Ref. 16 can connect up to eight terminals and can simultaneously handle the multipoint conferences of two groups that are independent of each other. Two MCUs cascaded can connect up to 14 terminals. The MCU in Ref. 17 can connect up to 16 terminals. One hundred twenty-eight terminals can participate in a broadcasting-type conference supported by a number of MCUs.

Several groups have also implemented continuous-presence MCUs (18–21). Both code-domain and pixel-domain video bridges are implemented. These systems are briefly discussed as follows.

5.1. Coded-Domain Video Bridge

Coded-domain video bridges do not decode the incoming bitstream to the pixel domain for combining. Some coded-domain video bridges just modify the syntax of the coded bitstreams and multiplex the modified bitstreams together. Others perform partial decoding, combining, and reencoding on the bitstreams. Currently, digital video signals are usually coded by two major types of video coding standards: intraframe coding (e.g., motion JPEG) and interframe coding (e.g., H.261, MPEG-1 (22)). The kind of video coding standard used has a major impact to the complexity and the flexibility which can be provided by the video bridge. In this subsection, we will discuss the coded-domain video bridges based on these two kinds of video coding standard.

5.1.1. Video Bridge for JPEG-Coded (Intraframe-Coded) Video

Montage (21), as shown in Fig. 6, is a multipoint video bridging architecture which is capable of composing and displaying an arbitrary number of motion JPEG-compressed

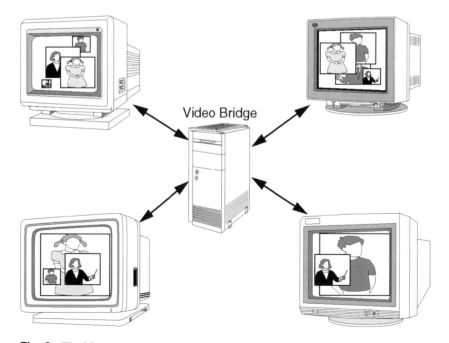

Fig. 6 The Montage system.

video streams with the capability that the video windows can overlap with each other. The system supports end-user customization, where each end point can control and customize the video composition for his/her screen independently of any other participants. The Montage architecture consists of three components: the encoder, the decoder, and the video bridge.

The multiresolution encoder compresses the incoming video stream using an intraframe-coding scheme based on the JPEG algorithm. The filtering and subsampling section in each encoder generates four independent streams of resolutions from full size (640×480) to 1/64 size (80×56) using a macroblock size of 16×8. The packetizer in the encoder packetizes the four resolution streams into packets or cells for transmission. Each macroblock includes an identifier for the decoder to place the macroblock on the receiver's screen correctly.

In the decoder, a special macroblock eliminator is used to determine if a given macroblock is visible or is occluded on the output display. If a macroblock is not visible, it is discarded. A visible macroblock is passed onto the intraframe decoder for decompressing. The Montage system limits the placement of video windows to macroblock boundaries, assuring that a macroblock is either completely visible or completely invisible.

Three possible bridge structures are discussed for the Montage system:

1. No Endpoint Multicasting: The bridge performs virtual circuit routing and multicast control functions. In this structure, each endpoint sets up a point-to-point virtual circuit to the bridge, for both input and output. The bridge receives the multiple-resolution video streams from the encoders and distributes the appropriate resolution streams to the correct receivers. The functions of the bridge are merely redirecting incoming streams to the appropriate receivers and handling some connection management functions.

2. Bandwidth Pruning Bridge: This configuration is aimed at simplifying the decoder and lowering the overall bandwidth requirement at the receivers. This is accomplished by moving some decoder functions (e.g., macroblock eliminator) into the bridge. The bridge now consists of multiple logical bridges, one for each receiver. One major difficulty in this configuration is that the bridge now must understand the size, structure, and details of each participant's output display. This increases both the cost and complexity of the network bridge. When a video stream is moved or occluded differently, the bridge must be notified and it must make the appropriate changes. An advantage of this configuration is that the decoder now becomes less complex.

3. Single-View Bridge: This configuration is similar to the above bandwidth pruning bridge case except that there is only one logical bridge for all participants in a given conference. The single bridge takes the incoming streams, composes them, and multicasts the resultant stream to all participants. This simplifies the bridge structure, but as there is only a single view, end-user customization is lost in this configuration. Usually, the multiple video signals will be displayed in a predefined style.

The intraframe coded video has the advantage that each block is coded independently, which makes the coded-domain combining task much easier than the interframe coded video. Because each macroblock is an independently coded unit, it can be manipulated and decoded without reference to other macroblocks or previous frames. The drawback is that the compression ratio is much lower than the interframe coding. As a result, it is not suitable for transporting over a narrow-band communication network such as N-ISDN to support multipoint videoconferencing over a WAN. The higher bit-rate cost is partly offset by the fact that these data streams can be easily manipulated, composed, and clipped, without the need to decompress them.

5.1.2. Video Bridge for H.261-Coded (Interframe-Coded) Video

H.261 (15) supports two video formats: QCIF (176×144) and CIF (352×288). In each format, the video is constructed based on a number of group of blocks (GOBs). As shown in Fig. 7, the number of GOBs for QCIF is 3, and 12 for CIF. The H.261 video combining can be achieved in the coded domain by combining four QCIF videos into a CIF video (19). The video bridge modifies the syntax and the GOB numbers of the incoming video streams and multiplexes them together into a legal combined CIF bitstream. This is possible because each GOB has a clear code delimiter which can be detected by a preprocessor without doing variable-length decoding. The video bridge processes two QCIF inputs at a time by interleaving their GOBs.

In practice, the frame rates of the QCIF videos may be variable depending on the scene content and the implementation of the video codecs. The video bridge has to align the QCIF frames based on their temporal references (TRs). TRs are numbers carried in the bitstream indicating the transmitted frame numbers. In multiplexing the GOBs to form the CIF frame, for those inputs not having that TR number, GOB headers are generated by the video bridge. These empty GOBs will cause the decoder to repeat the corresponding GOBs in the previous decoded frames.

The flexibility of QCIF combing described above is limited because the combined CIF video sequence can only support four conference participants. If the system needs to support five or more conference participants, a GOB-based combiner (19) has to be used.

Figure 8 illustrates the scenario in which five different video sources are combined via different groups of GOBs. In the figure, Uij denotes the jth GOB unit from the ith user. Through combining at the GOB level, up to 12 users can be simultaneously displayed

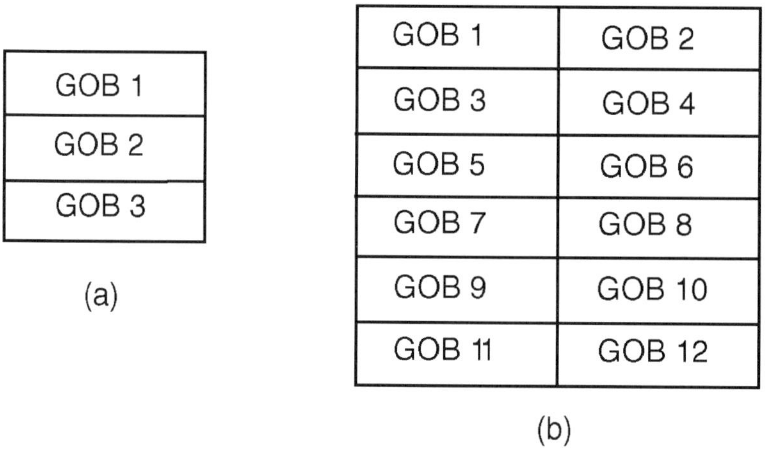

GOB 1
GOB 2
GOB 3

(a)

GOB 1	GOB 2
GOB 3	GOB 4
GOB 5	GOB 6
GOB 7	GOB 8
GOB 9	GOB 10
GOB 11	GOB 12

(b)

Fig. 7 GOB structure of (a) a QCIF frame and (b) a CIF frame.

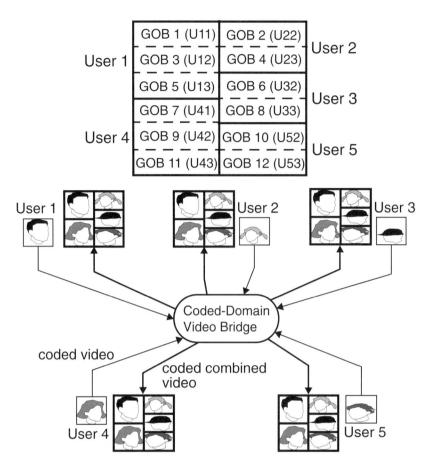

Fig. 8 An example of a GOB combiner combining five users.

with 1 GOB area for each user. In practice, however, the aspect ratio of a single GOB is not suitable for displaying a user's picture. It is found that two GOBs can provide a pleasant looking display of a user. Thus, the system can accommodate up to six users.

When the number of participants is four or less, the coded-domain combiner can take all the GOBs in each input frames without dropping any of them to compose an output CIF picture (e.g., combine four QCIF pictures to form a CIF picture). However, when more than four participants are involved in a videoconferencing session, the display portion of some of the participants will have to be limited to two GOBs instead of three GOBs. For example, when there are five users, the resulting CIF picture may look like the one shown in Fig. 9, with two users having three GOBs and three users having two GOBs. When two GOBs instead of three GOBs are used, the QCIF input frame needs to be modified by dropping one of the three GOBs. When a GOB is dropped from a QCIF frame, it may cause degradation when motion vectors point to areas in the GOB which has been dropped. One way to avoid this problem is to restrict the motion search range in the video encoders to within the two GOBs that are going to be combined. However, this requires the encoders to enforce this restriction during the encoding process. Thus, not

Fig. 9 A sample frame of a video sequence displaying five users.

all the commercially available H.261 codecs can be used for the application. Fortunately, it has been observed from the experimental prototype that in real operations, because the users can see their position in the areas provided, they will naturally position themselves to appear in the middle of the space provided by the two GOBs. If the top GOB in the QCIF frame is dropped, usually that GOB only contains still background above the users' head. Usually, there are no motion vectors pointing into this area. So, even without constraining the motion search range, it usually does not introduce motion-compensations errors (19). Due to the simplicity, the system can be implemented in software using a regular PC. However, the flexibility of the system is also very limited. Both the sizes and the positions of the video windows cannot be changed.

5.2. Pixel-Domain Video Bridge

A pixel-domain video bridge decodes the incoming bitstreams down to the pixel domain before combining the videos. It offers much more flexibility in video combining and manipulation. However, it needs more computation power to decode the bitstream and to encode the combined video. Also, doing pixel-domain manipulation in real time usually requires special hardware. An example of a pixel-domain video bridge is the Personal Presence System (PPS) (20), which is a real-time prototype system developed to demonstrate the flexibility that can be offered using the pixel-domain combining approach.

The PPS takes the flexibility advantage of the pixel-domain combining. Each participant can personalize the view they receive from other participants by having complete custom control over which images they receive, as well as controlling how each image is presented. The size and location of each image can be controlled, overlaid like windows on a PC, cropped so that only the important picture information is displayed, and even lifted from its background and placed in virtual environments. All these operations can be performed down to the pixel level. Conference participants can control the images they receive by making the image that they think is important large while retaining small images of others. They can also change the sizes, positions, and overlaps of the images on their

Fig. 10 The image of four classes.

screens at any time. In Fig. 11 the images of the four classes (Fig. 10) have been removed from their backgrounds and the extracted images have been arranged to make it appear as if all the students were sitting in the same room. To achieve this, the front students' images are enlarged to make it appear that they are closer than the rear classes, and the "class" background is inserted behind the class images to make the virtual meeting environment appear real.

One of the major difficulties of implementing a flexible pixel-domain video combiner such as PPS is that the system needs to handle a large video bandwidth. Because the video

Fig. 11 A combined video frame from the Personal Presence System.

streams have been decoded into the pixel domain, each video will require a high bandwidth. For example, a video with a CCIR 601 resolution will require about 166 Mb/s. In order to support the combining of a large number of video inputs, the video combiner will need to handle a very large bandwidth, making the implementation of the system expensive. Another difficulty is that many operations such as chroma keying operate at the pixel level which are quite computation intensive. The bandwidth and computation requirement will increase as the number of users increases. It is necessary that the architecture be scalable and extensible in order to support a large number of users. To achieve this goal, the PPS uses a special hardware architecture as shown in Fig. 12 that is scalable to handle the large bandwidth and computation required for manipulating multiple video sources when the number of users increase.

The hardware architecture is composed of many small atomic modules called Video Composing Modules (VCMs) (23). These atoms are connected into Video Composing Chains (VCCs). One such VCC is dedicated to each user for the life of the conference. The length of a user's VCC is determined by the number of video streams that a particular user wishes to view simultaneously and this length can change from moment to moment as a user changes their display.

This architecture is extensible in length and number of inputs. Each incoming video stream (B input in Fig. 12) can be manipulated and synchronized in the memory associated with the corresponding module. The combining function is achieved by controlling the multiplexing of either the A input or the B input in the frame memory based on their assigned priorities so that the output of the multiplexer (MUX) contains the combined video. Raster and synchronization information is passed down the chain from left to right, adding less than a microsecond delay per VCM. The leftmost A input in Fig. 12 sets the raster and clock timing for the entire chain and thus for the output. This signal may come from a generator or from the input video signal sent by the customer who would receive the VCC output.

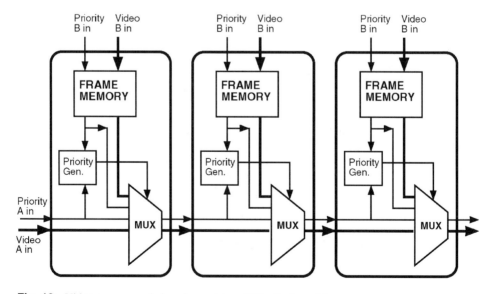

Fig. 12 Video composer chain using a string of identical modules.

Within each VCM, the priority information is passed along with each pixel of the input signal and provides the means for defining the stacking order of the video image streams and also the means for object extraction via combinations of windowing and chroma keying. In other words, the pixel with the highest priority for a given output location takes preference over pixels from other images that cover the same output location. When two video windows overlap, for example, you only see the pixels from the top image at the overlap locations. The priority of the pixels for a B input may vary. In the overlapping window situation, the pixels of the image which is partially covered by another window are given a lower priority than those from the top window. When an object is extracted from a background, the object's pixels are given a high priority and the pixels which make up the background portion of the image are given the lowest priority (these pixels are not displayed; the A input background image is displayed instead of the image's original background).

6. COMPARISON OF CODED-DOMAIN AND PIXEL-DOMAIN VIDEO COMBINING

6.1. Flexibility

The pixel-domain approach can be highly flexible because the output picture can be composed on a pixel-by-pixel basis. For example, the background of the input images can be removed and a different background can be inserted. Also, the position of the conference participants can be moved to an arbitrary position and can be occluded by other objects. The combined video can be easily encoded to different bit rates to fit the available bandwidth of the communication channels or can be encoded to different video coding standards formats for the interworking of terminals supporting different standards. There is less flexibility for the coded-domain approach. Opaque overlapping and translation of video objects can be done in the coded domain (24); however, for complicated manipulations, the operations in coded domain can be cumbersome compared to the pixel-domain approach. Also, if the video rates or formats need to be changed due to the constraints from the communication channels and interworking, it cannot be done as easily as the pixel-domain approach.

Current video coding standards allow the terminals of conference participants to have different capabilities (e.g., terminals can support a frame rate of 30 frames/s, 15 frames/s, or 7.5 frames/s). For the coded-domain combining MCU, when sending a bitstream from terminals with high-frame-rate capabilities to terminals with low-frame-rate capabilities, the MCU has to reduce the frame rate to that it can be handled by the low-frame-rate capability terminals. However, this cannot be easily done if the video is coded by an interframe coding scheme such as H.261. For the pixel-domain combining MCU, this does not present any problem. The MCU decodes the incoming bitstreams into pixel domain and stores them in the individual frame buffers. The frames in these individual frame buffers form a combined frame. The MCU just encodes whatever video frames are available in the combined frame buffers at 30 frames/s for the terminals with the capability of 30 frame/s and encodes the frames at 7.5 frames/s for the terminals with the capability of 7.5 frames/s.

Similarly, the communication links in the multipoint conferencing may be of different rates. For the coded-domain combining, in order to send the combined video to the terminals with lower communication channel bit rates, all the terminals may need to encode the video in low bit rates because it is not easy for the MCU to adapt the rates in the coded domain. In general, for the coded-domain combining, the video quality is more limited by the low-capability terminals and communication channels because, in the MCU, it is more difficult to adjust the frame rates and bit rates in the coded domain.

In practical applications, the MCU may need to support the feature that each participant can choose to display a specific person in higher resolution or to display the combined video of multiple persons with each individual in a reduced resolution. Using pixel-domain combining, this can be easily done, as video frames of individual persons have been stored in the frame buffers. The MCU just needs to encode the videos in proper resolutions. However, using the coded-domain combiner, it is more difficult to change resolutions in the coded domain.

For coded-domain video combining, because the main function of the video bridge is multiplexing the video bitstreams, it usually results in a highly asymmetrical input/output bit-rate (i.e., the combined video has a much higher bit rate than the individual video input). Thus, it is more suitable for network configurations supporting asymmetrical bandwidth. To adapt the bit rate in the coded domain for symmetrical input/output rates is cumbersome. For pixel-domain video combining, the combined video can be easily encoded to any bit rates required by the applications.

6.2. Video Quality

In the pixel-domain combiner, the video bitstreams are decoded then reencoded, this reencoding process causes video-quality degradation. However, it does not mean that the video quality of the pixel-domain combining will necessarily be poorer than the coded-domain combining. In many situations, the pixel-domain combining allows users to encode their videos at higher input bit rates. Consider the situation of a four-point video conference over an ISDN line which can support 128 kb/s. Using the coded-domain combining, in order to keep the combined video to 128 kb/s, each input video can only be encoded at about 32 kb/s. Using the pixel-domain combining, each input video can be encoded at 128 kb/s. In the multipoint conference environment, the motion activities of each participant will vary. Usually when one of the conference participants is speaking, other participants will be listening. In this situation, only the active speaker part will have large motion activities in the combined video frame. For the pixel-domain combiner, it is possible to perform dynamic bit allocation to assign more bits to the portion of video with high motion activities (which usually needs more bits to code). With the higher input bit rates and using the dynamic bit allocation in the reencoding, the video quality of the pixel-domain combining can be much more consistent compared with video combining in the coded domain.

Figure 13 shows the result of a four-point video conference where each user encodes their video in 64 kb/s using H.263 (25). The person at the lower-right corner is the active person. The video bridge combines the video into a 256-kb/s stream. In Fig. 14, each person encodes the video in 256 kb/s. The video bridge decodes the videos, combines the videos, and encodes the combined video at 256 kb/s. It can be seen the quality of the active person is much better with the pixel-domain approach than the coded-domain approach due to the dynamic bit allocation in the encoding process.

Multipoint Videoconferencing 223

Fig. 13 A four-point video conference with each user coded at 64 kb/s and the combined video at 256 kb/s using a coded-domain video bridge.

Fig. 14 A four-point video conference with each user coded at 256 kb/s and the combined video at 256 kb/s using a pixel-domain video bridge.

6.3. End-to-End Delay

The pixel-domain approach needs to decode and reencode the video bitstream. Thus, the delay introduced by the video bridge is the same as the delay of a video encoder–decoder pair. In the GOB-based coded-domain combining, the input videos are buffered before they are multiplexed. In order to prevent the buffer underflow, usually the video bridge will check to make sure there is a complete GOB in the buffer before it starts the multiplexing process. Thus, the worst-case delay is the buffer delay of a worst-case GOB. This delay may also be quite significant.

6.4. Complexity

The coded-domain MCU is usually less complicated to implement. For simplest coded-domain MCU, its main task is just to multiplex the coded bitstreams together and produce a legal bitstream of the combined video. These operations are relatively easy and can be implemented in software using regular PCs. For pixel-domain combining, the MCU needs to perform the decoding and encoding of the video which is much more computationally intensive.

It is possible to reduce the complexity of a decoder–encoder pair. By moving the discrete cosine transform (DCT) and inverse DCT (IDCT) functions around and removing redundant components, it can be shown that the complexity of a decoder–encoder pair can be reduced from the one shown in Fig. 15 to the one shown in Fig. 16, which is only slightly more complicated than a single decoder (26). Similar techniques can be used in the pixel-domain video combiner to reduce the system's complexity.

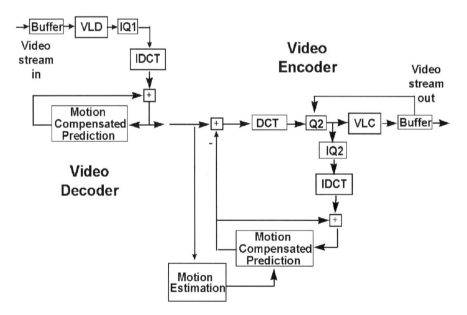

Fig. 15 A decoder–encoder pair.

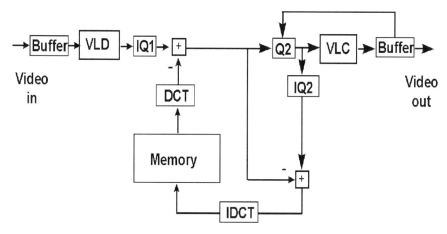

Fig. 16 A decoder–encoder pair after simplification. (From Ref. 26.)

7. SUMMARY

As videoconferencing become more and more popular and video compression techniques keep progressing, it is possible that multipoint videoconferencing applications may become as popular as the telephone in the near future. In this chapter, we gave an overview of multipoint videoconferencing. A multipoint video conference can be a centralized one or a distributed one. Many different network configurations supporting multipoint videoconferencing are discussed. There are two types of multipoint video conferences: switched presence and continuous presence. We focus our discussion on the continuous presence. In the continuous-presence multipoint video conference, we can use the pixel-domain approach or the coded-domain approach. We discussed various aspects of multipoint videoconferencing using these two combining schemes. The pixel-domain approach provides more flexibility but is more computation intensive. The coded-domain approach may be easier to implement, but several issues related to the rate-adaptation may need further investigation.

ACKNOWLEDGMENTS

The authors would like to thank the reviewers for valuable suggestions.

REFERENCES

1. ITU-T Recommendation H.320, Narrow-band visual telephone systems and terminal equipment.
2. ITU-T Recommendation H.310, Broadband audiovisual communication systems and terminals.
3. ITU-T Recommendation H.321, Adaptation of H.320 visual telephone terminals to B-ISDN environments.
4. ITU-T Recommendation H.322, Visual telephone systems and terminal equipment for local area networks which provide a guaranteed quality of service.

5. IEEE Standard 802-9a-1995, IEEE standard for local and metropolitan area networks—supplement to Integrated Services (IS) LAN interface at the Medium Access Control (MAC) and Physical (PHY) Layers: specification of ISLAN16-T, 1995.
6. ITU-T Recommendation H.323, Visual telephone systems and equipment for local area networks which provide a non-guaranteed quality of service.
7. ITU-T Recommendation H.324, Terminal for low bit rate multimedia communication.
8. MH Willebeek-LeMair, DD Kandlur, Z-Y Shae. On multipoint control units for videoconferencing. Proceedings, 19th Conference on Local Computer Networks, Minneapolis, MN, 1994, pp 356–364.
9. ITU-T Recommendation H.231, Multipoint control units for audiovisual systems using digital channels up to 1920 kb/s.
10. ITU-T Recommendation H.243, Procedures for establishing communication between three or more audiovisual terminals using channels up to 2 Mb/s.
11. T-C Chen, W-D Wang. Low-delay center-bridged and distributed combining schemes for multipoint videoconferencing. Visual Communications and Image Processing '94, Chicago, IL, Proceedings of the SPIE 1994, vol 2308, pt 2, pp 850–61.
12. PJ Kyees, RC McConnell, K Sistanizadeh. IEEE Commun Mag 33(4):52–60, 1995.
13. MH Willebeek-LeMair, Z-Y Shae. IEEE Selected Areas Commun 15(6):1101–1114, 1997.
14. GK Wallace. Commun ACM 34(4):30–44, 1991.
15. ITU-T Recommendation H.261, Video codecs for audiovisual services at p $\times$ 64 kb/s.
16. T Arakaki, E Kenmoku, T Ishida. Development of multipoint teleconference system using multipoint control unit (MCU). Pacific Telecommunications Council Fifteenth Annual Conference. Proceedings, Honolulu, 1993, vol 1, pp 132–137.
17. S Oka, Y Misawa. Multipoint teleconference architecture for CCITT standard videoconference terminals. Visual Communications and Image Processing '92, Boston, 1992, pp 1502–1511.
18. SM Lei, TC Chen, MT Sun. IEEE Trans Circuits Syst Video Technol CSVT-4(4):425–437, 1994.
19. MT Sun, A Loui, T-C Chen. IEEE Trans Circuits Syst Video Technol CSVT-7(6):855–863, 1997.
20. DG Boyer, ME Lukacs, M Mills. The personal presence system experimental research prototype. 1996 IEEE International Conference on Communications, 1996, vol 2, pp 1112–1116.
21. R Gaglianello, G Cash. Montage: continuous presence teleconferencing utilizing compressed domain video bridging. 1995 IEEE International Conference on Communications, Seattle, WA, 1995, vol 1, pp 573–581.
22. ISO/IEC 11172, Coding of moving pictures and associated audio for digital storage media at up to about 1,5 Mbit/s.
23. ME Lukacs D/G Boyer. IEEE Commun Mag 33(11):36–43, 1995.
24. SF Chang, DG Messerschmitt. IEEE J Selected Areas Commun 13(1):1–11, 1995.
25. ITU-T Recommendation H.263, Video coding for low bit rate communication.
26. G Keesman, R Hellinghuizen, F Hoeksema, G Heideman. Signal Process: Image Commun 8(6):481–500, 1996.

9
Video Shot Detection and Analysis: Content-Based Approaches

Qi Tian
Kent Ridge Digital Labs, National University of Singapore, Kent Ridge, Singapore

HongJiang Zhang
Hewlett-Packard Company, Palo Alto, California

1. INTRODUCTION: VIDEO MEDIA ANALYSIS AND RECOGNITION

In the last few years, substantial research efforts have been devoted to video data analysis and recognition, including shot detection, shot classification and grouping, story structure parsing, audio and speech analysis, and so forth. This is in response to the need in multimedia applications, including interactive video and digital libraries, where efficient video indexing, browsing, and interaction tools are essential. What we are searching for is a solution to convert video into a digital content-based representation, similar to that used to convert document image into ASCII characters and textual information in character recognition and document image analysis. Broadly speaking, we can refer this research area as digital video media analysis and recognition (DVMAR) (1–7). The goal of this effort is to develop algorithms, tools, and systems that are able to extract and analyze basic elements and structures (both syntactic and semantic) of video, to make access, compression, and interaction of video more effective and content based.

There are three main areas to be explored in DVMAR applications: more flexible access methods, better video compression, and exploration of full flexibility in interacting with digital video. Today, accessing video media is far less flexible than accessing text documents, be it in digital or traditional analog forms. For text documents, structure analysis and recognition have been well developed, and methods of representing information in computer applications have been well studied. For video data, in either analog or digital form of various formats, we are still limited to Video Cassette Recorders (VCR) and similar linear interfaces and there is no easy way to browse a large volume of video documents. This is partially due to the fact that it is only in the last several years that digital video has become affordable as a result of rapid advances in computer hardware/software and communication technologies. What we need is a content-based representation for video data and algorithms that are capable of deriving attributes of this representation from raw video data. Realization of this objective will allow automatic construction of tables of contents, abstracts, and indices of video data, with minimal human intervention, and eventually make video access as easy as accessing text data today. How to model video data and represent video content and how to automatically extract content attributes are the research topics for DVMAR. This chapter reviews and describes the algorithms for shot detection and shot-based video structure analysis, video browsing, and content-based compression, which can be considered the first step in addressing this challenging problem.

Content-based analysis is useful for video compression too. In the last two decades, great advances have been achieved in the developments of video compression algorithms, and several standards have been established, such as MPEG-1/2 and H.26X (8–10). However, all of these standard compression algorithms are based on low-level features that are not semantically or physically related to scene structures of video; thus, they take no advantage of the structure information of video data. With content analysis and content-based representation, video can be decomposed into *objects*, which are more *meaningful* than frame blocks; and compression can be designed based on the objects instead of the frame blocks. This will lead to a much higher compression ratio than MPEG-1/2. For example, in many video programs, there are often many consecutive still frames with no change across them, and there is, therefore, no need at all to code and transmit these frames one by one, as done in today's MPEG or JPEG.

Furthermore, content-based methods are essential to providing high flexibility in interacting with digital video, which has not been addressed by current video compression and transmission standards. Although compared with analog video, we already have much higher flexibility in handling digital video, it is still very limited compared to what we can do with digital text documents. For instance, with content-based approaches, video can be considered as a collection of video objects, instead of frames. Video objects may have different frame rates, resolutions, and aspect ratios at different stages such as production source, compression and transmission, storage, and playback; that is, there will be no such constraints as in today's video formats, in which fixed frame rates and spatial resolutions are used throughout the whole process from source sides to end-user sides. To meet the need of flexible interactivity, high compression ratios, and content-based access, MPEG-4 has adapted the concept of the video object, which will open a new era of digital video applications.

It is clear that achieving the final goal of DVMAR is a great research challenge for many years to come. Although developing algorithms that are able to understand general semantic contents of video is not feasible today, the first step in content analysis can be to *identify structures* of video and to *recognize basic components*. Such structural and low-level semantic information is already useful in representing video content for indexing, compression, and interaction (11). We argue that because shot is a very important, universal structure component of video documents, shot detection, attribute extraction, and analysis are feasible starting points to achieve the ultimate goals of DVMAR.

A shot is a sequence of image frames recorded contiguously and representing a continuous action in time or space. From the point of view of video production, shots are the basic logical component of video. One or more adjoining shots focusing an object or a group of objects of interest forms a scene. Several scenes can then be combined together to form a sequence or segment on the basis of some semantic significance. Although there is no universal and rigid structure of video media, a shot is a physical level, whereas the rest are more logical and are constructed based primarily on semantic meanings. A shot is a good choice as a generic component for video content analysis, which can be considered as words or sentences in text documents. From the algorithmic point of view, it is also feasible to detect shots automatically based on temporal structure analysis of video documents. As illustrated in more detail in the next several sections, shot-based analysis, classification, and grouping form the basis for content-based video indexing, browsing, retrieval, and compression.

In summary, DVMAR is a challenging research area and there are many issues to be addressed that will require many years of research efforts to meet its objectives. We believe that shot analysis is the first step for DVMAR. In Sec. 2, we describe in more detail the tasks of DVMAR by drawing an analogy between text document image analysis and video data analysis, from which we conclude that shots are a generic component for video content analysis. This is followed by brief descriptions of attributes associated with shots, which are useful for shot classification and grouping. In Sec. 3, we describe in detail a number of algorithms for video shot detection. Finally, in Sec. 4, we discuss some applications based on shot detection, ranging from video compression to content-based interaction.

2. SHOT-BASED VIDEO CONTENT ANALYSIS

In this section, we first discuss the tasks of DVMAR by drawing an analogy between text document image analysis and video document analysis. From this discussion, we will see that shots are a basic unit for video content analysis. We, then, further discuss some basic attributes of shots and their applications in video indexing and browsing.

2.1. Analogy Between Text and Video Media Analysis and Recognition

To have a better perspective regarding to the tasks and roles of DVMAR, we can draw an analogy between *document image analysis and recognition* (DAR) and DVMAR. DAR is a well-developed research area with numerous applications in text documentation analysis and recognition, including printed/handwritten character recognition (12). There have been three forms of media: analog, digital, and content-recognized or digital content representation form. In Fig. 1, we show these three forms for both textual document and video media. The analog form is the traditional one that we have been using for many years and is still dominant today, such as newspapers, books, films, and videotapes. The digitized form is becoming an important media form in computer-based applications in the last several years, and it will become even more important in the coming years due to the fast advance in CD-ROM, DVD, and internet application. Today, one can easily scan and digitize paper-based text documents into digital images, and then compress them in various image formats such as TIFF, JPEG, and so forth. This has allowed computers to access text documents for display, transmission, printing, and so forth. However, digitizing document images is only the first step, and the ultimate solution is to allow the computer to access contents of documents in terms of document structures, graphics, symbols, and sentences. This will require converting document images into the digital content representation as shown in Fig. 1. This conversion process is a subject of DAR, and its major tasks include document layout analysis (paragraph, chapter, word, and graphic separation and recognition), and character and symbol recognition (12,13).

It is only when media are stored in the third form that the *content*, not just the bits, of the media becomes computer readable. For instance, text information is no longer stored

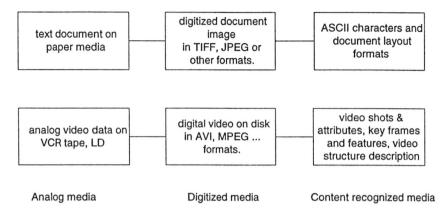

Fig. 1 Three types of text and video representation form—analog, digital, and digital content based.

as pixel maps, but in structured symbols and strings. Such a representation not only supports flexible access and interaction but also allows a much higher compression rate for the text document, as a recognized character only takes 1 byte to store. For example, a typical 8.5×11 in. page of text document, scanned at 300 dpi (dots per inch), will yield an image of 2550×3300 pixels. Assume 8 bits/pixel is used, then it will take about 8 Mbytes of storage. In contrast, a typical page of text document consists of about 1000–1500 words, or 6000–9000 characters; if coded as ASCII, it only takes about 9 Kbytes, plus a much smaller storage size for layout information of the document. This is about a factor of 1000 compression in terms of storage space.

Similarly, digitization of video from videotapes and compression of them into MPEG-1 or MPEG-2 is equivalent to text document scanning. Such digitized video can be read by computers in terms of image frames but not in terms of content, similar to a JPEG image for a text document. To achieve content-based accessibility and a high compression rate for video, digitized video needs to be further analyzed and converted into digital *content* representation form, as shown in Fig. 1.

To better understand tasks of video content analysis, it is very helpful to see how these similar tasks are carried out for its counterpart—text documents. Figure 2 shows a comparison of content analysis for both text document images and video images, where two the main tasks for both media are structure analysis and component recognition. In general, the first task concerns understanding structural compositions of a media, whereas the second concerns recognition of individual components. This is based on the fact that the media usually consists of two parts: components and composition. By no means are they totally independent, but rather related, in practice. For instance, to recognize a character, very often we need contextual information, both spatial and logical.

In structure analysis of text document images, the major tasks includes separating text from graphics and images and identifying the *spatial layout structure* of text such as chapters, paragraphs, sentences, words, and characters. It is possible to further identify various components of text documents such as titles, abstracts, types of documents—memos, faxes, reports and so forth—based on spatial layout structures and character recognition results. On the other hand, structural analysis of video has to handle both

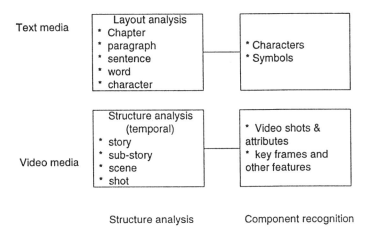

Fig. 2 Tasks of content recognition for text and video media.

temporal and spatial structures, with the former being more fundamental. A number of models have been proposed to model temporal video structures, one of which is the stratification model proposed by Aguierre- Smith and Davenport (14). Figure 2 shows a typical structure of video—story, segment, scene, and shot—in comparison with the text document structure. Obviously, structures of video documents are very closely related to a specific application domain. For instance, the structure of news video is very different from that of feature movies or documentaries. However, shots are always a universal basic structure and physical element, whereas other structure elements or levels may vary from one type of video to another. Therefore, detecting shots from a video frame sequence is the basic task in video structure analysis or parsing.

In component recognition of text document images, the most important components are characters and words. In general, a given language defines a character set, which contains only a limited number of characters. Other components include symbols, tables, logos, and so forth. Many algorithms and technologies for recognition of these components have been developed in the last two decades, which makes it one of the most developed fields in pattern recognition research. However, the video medium is a reproduction of visual reality in time and space, hence there are far more possible objects or components in video which form a virtually unlimited set. In general, to recognize all of them is well beyond the capacity of the current state of the art in computer vision and will remain a highly challenging research area in the near future.

Knowing the scale and amplitude of the problem, the question then is, What can be done in content analysis of video? We argue that video content can be considered in different levels, shot detection, and analysis to be a feasible starting point to achieve our ultimate goals.

2.2. Shot and Shot Attributes

From the comparison between text document image analysis and video document analysis, it is observed that a shot is a generic component for video content analysis, similar to sentences in a text document. There are three basic tasks for shot-based video content analysis—shot detection, shot attributes extraction, and classification or grouping. The first two tasks belong to component recognition, whereas the last one belongs to video structure analysis. This section provides a brief description of shot attributes and their importance in video content analysis before we describe, in detail, a number of shot detection algorithms in Sec. 3.

There are a variety of attributes associated with shots which are useful for shot-boundary detection, shot clustering and classification, and content-based indexing and retrieval. In general, there are two types of shot attributes: generic attributes and application-specific attributes. The former is not specific to particular video applications, whereas the latter is associated with video programs of a particular application domain. For instance, camera motion is a generic shot attribute, whereas anchorperson shots are application specific because it is only meaningful in the news video context. We should also point out that there are many shot attributes, but we only discuss a few typical ones that will be used for video indexing and browsing applications.

The first physical and generic attribute of a shot is its length. Experimental results show that for most of video data, including feature movies, news, cartoon, and so forth the average length of shots ranges from 2 to 6. We have manually labeled shot boundaries for several news videos and a classical movie title; each lasts about 1 h. Figure 3 shows

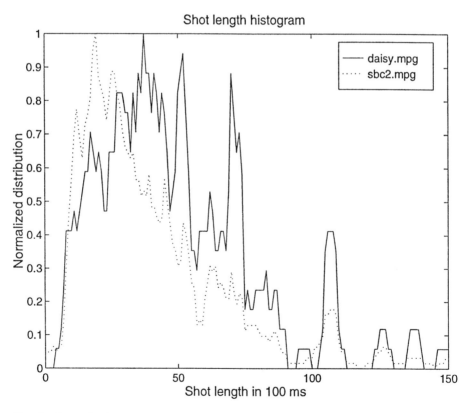

Fig. 3 Normalized shot length statistical distribution for a news video and a classical feature film video, each lasts about 1 h.

Fig. 4 Three frames across a sharp cut.

the normalized shot length distributions. As we can see, that in general news videos, the peak shot length is 2 s; contrast, the classical feature movie has much longer shot lengths. There are several peaks of shot lengths: 4, 5, 7, and 11 s. Experiments in Ref. 15 showed similar results: two movie titles of length 66 and 70 min have 845 and 859 shots, respectively, producing an average shot length of 5 s. This fact may indicate that humans do need a few seconds of visual exposure to comfortably obtain certain perceptions of the outside world.

We can consider shot boundary as another physical and generic attribute of shots. There are several types of transitions between consecutive shots—simple cut and gradual transition—which have to be handled differently in the shot detection process in order to achieve good detection accuracy. In a simple cut, there is an abrupt image frame change, which occurs across one or two boundary frames. Figure 4 shows three consecutive video image frames for a typical simple cut: it can be easily seen that the second frame is very different from the first frame because the camera took two different scenes, resulting in a sharp boundary between these two frames. A typical example of dissolve transitions is shown in Fig. 5; the last frame of the current shot just before the dissolve begins, three frames within the dissolve, and the frame in the following shot immediately after the dissolve. The actual dissolve occurs across about 30 frames, resulting in small changes between every two consecutive frames in the dissolve. There are several types of gradual transitions, including *fades*, *dissolves*, and *wipes*. Due to the rapid development of computer-based nonlinear editing systems for postproduction of video, there are more variations of these types of transition available, as we can see in many music videos. A fade is a gradual change in brightness of images; it can be further classified as fade-in and fade-out processes. The former refers to the process in which a sequence of video image frames gradually appears from all dark images to a regular brightness, and the latter refers a reverse process in which the sequence goes from a regular brightness to all dark. A dissolve is a combination of a fade-out and a fade-in overlapped in time. A wipe is different from the above three types of shot boundaries; in it, a portion of frame shows a previous video sequence and the rest of the frame shows a new video sequence, and this partition pattern is a function of time. Various patterns are possible in wipes: from left to right,

Fig. 5 An example of dissolve sequences.

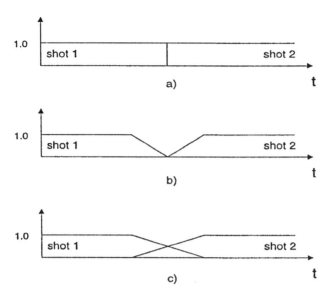

Fig. 6 Three types of gradual transition: (a) a simple cut; (b) a fade-out and fade-in shot boundary; (c) a dissolve shot boundary.

top to bottom, or in a diagonal direction from a corner to another, and so forth. The first three types of shot boundaries are illustrated in Fig. 6.

Apart from the above two physical attributes, there are a number of important content attributes associated with shots: camera motion, objects and their motion, and color and texture of background and objects. Camera motions include panning, tilting, zooming, tracking, booming, dollying, and their combinations. There are many different objects in a shot, such as people, cars, or animals, often with different motion patterns. One derived attribute of a shot is the activity level: a shot can have a high activity level like a car chasing, or a very low activity level, or no activity at all, such as in a shot consisting of still images. We can consider *key frames* another attribute of shots: for each shot, we can choose one or a few frames which represent shot content visually in an abstracted manner. Due to their abstract nature, key frames are a very effective visual representation of shots in the video browsing process. Other shot attributes include text captions, faces and names of main actors, and other important objects with the shot.

We should point out here that some of the above shot attributes can be automatically detected, like shot boundaries, color histograms, and camera motion, whereas many other attributes are difficult to detect or recognize automatically and need to be annotated manually.

3. SHOT DETECTION ALGORITHMS

Detecting shot boundaries is based on the fact that consecutive frames on either side of a boundary generally display a significant change in visual content. Therefore, what is required is some suitable quantitative measure that can capture the quantitative difference between such a pair of frames. Then, if that difference exceeds a given threshold, it may be interpreted as indicating a shot boundary. Hence, establishing suitable metrics is the key issue in automatic shot detection. The optimal metric for video shot detection should be able to differentiate the following three different factors of image change:

- Shot change, abrupt or gradual
- Motion, including those introduced by both camera operation and object motion
- Luminosity changes and noise

In this section, a number of algorithms for shot detection of video data either in original format or in compressed domain representations will be presented.

3.1. Shot Cut Detection Metrics Using Original Format of Digital Video Data

In the three consecutive video frames with a cut occurring between the first and second frames, as shown in Fig. 4, the significant difference in content between the first two frames is readily apparent. If that difference can be expressed by a suitable metric, then a segment boundary can be declared whenever that metric exceeds a given threshold. Therefore, the major difference among a variety of automatic video shot detections is the difference metrics used to quantitatively measure changes between consecutive frames and schemes to apply the metrics.

Difference metrics used in shot detection can be divided into two major types: those based on local pixel feature comparison, such as pixel values and edges, and those based on global features, such as pixel histograms and statistical distributions of pixel-to-pixel change. These types of metrics may be implemented in a variety of ways to accommodate the idiosyncrasies of different video sources and have been successfully used in shot-boundary detection.

3.1.1. Pixel Value Comparison

A simple way to detect a qualitative change between a pair of frames is to compare the spatially corresponding pixels in two consecutive frames to determine how many pixels have changed: this is known as *pairwise pixel comparison*. In the simplest case of monochromatic images, a pixel is judged as changed if the difference between its intensity values in the two frames exceeds a given threshold t. This algorithm simply counts the number of pixels changed from one frame to the next. A shot boundary is declared if more than a given percentage of the total number of pixels (given as a threshold T) have changed.

A potential problem with this metric is its sensitivity to camera movement. For instance, in the case of camera panning, a large number of objects will move in the same direction across successive frames; this means that a large number of pixels will be judged as changed even if the pan entails a shift of only a few pixels.

To make the detection of camera breaks more robust, instead of comparing individual pixels, we can compare corresponding *regions* (blocks) in two successive frames. One such approach applies motion compensation at each block (16); that is, each frame is divided into a small number of nonoverlapping blocks. Block-matching within a given search window is then performed to generate a motion vector and match value, normalized to lie in the interval [0, 1], with 0 representing a perfect match. A common block-matching approach used in video coding is to find for each block in frame j, the best fitting region in image ($j + 1$) in a neighborhood of the corresponding block according to the matching function

$$\text{Diff} = \frac{1}{F_{\max}x} \frac{1}{mn} \sum_{l=-n/2}^{n/2} \sum_{k=-m/2}^{m/2} |F_j(k, l) - F_{j+1}(k + dx, l + dy)| \quad (1)$$

where

$F_j(k, l)$ represents the value at pixel (k, l) of a $(m \times n)$ block in the current frame j

$F_{j+1}(k, l)$ represents the value at pixel (k, l) of the same $(m \times n)$ block in the next frame $(j + 1)$

(dx, dy) is a vector representing the search location and the search space is $dx = \{-p, +p\}$ and $dy = \{-p, +p\}$

$F_{\max}$ is the maximum pixel value of frames and is used to normalize the matching function

The set of (dx, dx) that produces the minimum value of Diff defines the motion vector from the center of block i in the current frame to the next frame. Then, the difference between two frames can be defined as

$$D_M = \sum_{i=1}^{K} c_i R_i \quad (2)$$

where i is the block number in a frame, K is the total number of blocks, R_i is the match value for block i, which equals the minimum Diff in a given searching space, and c_i is a set of predetermined weights for each block. To eliminate the effect of noise, the two highest and two lowest match values are discarded in calculating D_m. A cut is declared if D_m exceeds a given threshold.

3.1.2. Histogram Comparison

An alternative to comparing corresponding pixels or blocks in successive frames is to compare some statistic and global features of the entire image. One such feature is the histogram of the intensity levels. The principle behind this is that the two frames having an unchanging background and objects will show little difference in their respective histograms. The histogram comparison algorithm should be less sensitive to object motion than the pairwise pixel comparison algorithm, because it ignores the spatial changes in a frame.

Let $H_j(i)$ denote the histogram value for frame f, where i is one of the G possible pixel levels: then, the difference between frame f and its successor $(f + 1)$ may be given by the following χ^2-test formula:

$$D_h = \frac{1}{G} \sum_{i=1}^{G} \frac{|H_f(i) - H_{f+1}(i)|^2}{(H_f(i) + H_{f+1}(i))^2} \quad (3)$$

If the overall difference D_h is larger than a given threshold T, a segment boundary is declared. To be more robust to noise, each frame is divided into a number of regions of same size, (e.g., 16 regions); that is, instead of comparing global histogram, histograms for corresponding regions in the two frames are compared and the 8 largest differences are discarded to reduce the effects of object motion and noise (17).

3.1.3. Edge Pixel Comparison

Edges in images provide useful information about the image content, and changes in edge distributions between successive frames are a good indication of content changes. When a cut or a graduation transition occurs, new intensity edges appear far from the location of old edges and, similarly, old edges disappear far from the locations of new edges. Based on this observation, an effective video shot detection algorithm that can detect both cuts and gradual transitions has been developed as described in Ref. 18.

First, we define an edge pixel in the current frame that appears far from (a distance r) an existing edge pixel in the last frame as an *entering* edge pixel, and an edge pixel in the last frame that disappears far from an existing edge pixel in the current frame as an *exiting* edge pixel. Then, by counting the fraction of the entering edge pixels (p_{in}) and that of the exiting edge pixels (p_{out}) over the total number of pixels in a frame, we can detect transitions between two shots; that is, the difference metric between frame f and $f + 1$ can be defined as

$$D_e(f, f + 1) = \max(p_{in}, p_{out}) \tag{4}$$

$D_e(f, f + 1)$ will assume a high value across shot boundaries and generate peaks in the time sequence. Once a peak is detected, it can be further classified as corresponding to a cut or gradual transitions because cuts usually correspond to sharp peaks occurring over two or three frames, whereas gradual transitions usually correspond to low but wide peaks over a larger number of consecutive frames. Experiments have shown that this algorithm is very effective in detecting both sharp cuts and gradual transitions. However, as it requires more computation, this algorithm is slower than others.

3.2. Gradual Transition Detection Using Original Format of Digital Video Data

As described in Sec. 2, cuts are the simplest shot boundary and easy to detect using the difference metrics described above. Figure 7 illustrates a sequence of interframe differences resulting from the histogram comparison. It is easy to select a suitable cutoff threshold value (such as 50) for detecting the 2 cuts represented by the 2 high pulses. However, sophisticated shots boundaries such as dissolve, wipe, fade-in, and fade-out are much more difficult to detect because they involve more gradual changes between consecutive frames than does a sharp cut. Furthermore, changes resulting from camera operations may be of the same order as that from gradual transitions, which further complicate the detection.

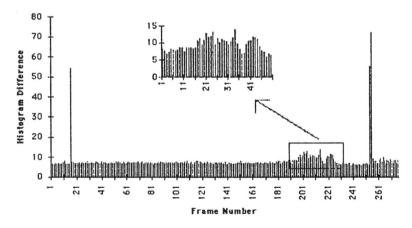

Fig. 7 A sequence of histogram-based interframe differences.

The inset in Fig. 7 is the sequence of interframe differences, resulting from the dissolve sequence shown in Fig. 5, defined by the color histograms. The values are higher than those of their neighbors but significantly lower than the cutoff threshold. This sequence illustrates that gradual transitions will downgrade the power of a simple difference metric and a single threshold for camera break detection algorithms.

The simplest approach to this problem would be to lower the threshold. Unfortunately, this cannot be effectively employed because noise and other sources of changes often introduce the same order of difference between frames as fact of gradual transitions, resulting in "false positives." In this subsection, we discuss three video partition algorithms that are capable of detecting gradual transitions with acceptable accuracy while achieving very high accuracy in detecting sharp cuts.

3.2.1. Twin-Comparison Approach

This algorithm was the first published one that achieved high accuracy in detecting both cuts and gradual transitions (19). As shown in Fig. 5, it is obvious that the first and the last frames across the dissolve are different, even if all consecutive frames are very similar. In other words, the difference metric with the threshold shown in Fig. 8 would still be effective it is was applied directly to the comparison between the first and the last frames. Thus, the problem becomes one of detecting these first and last frames. If they can be determined, then the period of gradual transition can be isolated as a segment unto itself. If we look at the inset of Fig. 7, it can be noted that the difference values between most of the frames during the dissolve (as well as wipes and fades) are higher, although only slightly, than those in the preceding and following segments. What is required is a threshold value that can detect this *sequence* and distinguish it from an ordinary camera shot. Based on this observation, the *twin comparison* algorithm was developed, which introduces two comparisons with two thresholds.

The algorithm uses two thresholds: T_b for sharp cut detection in the same manner as was described in the last subsection, and the second and lower threshold T_S is introduced for gradual transition detection. As illustrated in Fig. 8, whenever the difference value exceeds

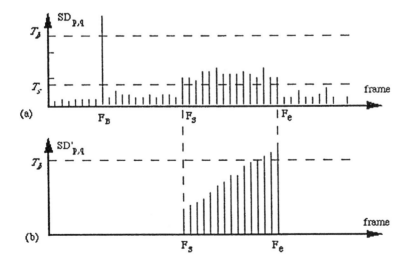

Fig. 8 Twin-comparison approach.

T_b, a camera cut is declared (e.g., F_B in Fig. 8). However, the twin comparison also detects differences that are smaller than T_b but larger than T_S. Any frame which exhibits such a difference value is marked as the potential start (F_S) of a gradual transition. Such a frame is labeled in Fig. 8. This frame is then compared against subsequent frames, which is called an *accumulated comparison*. The end frame (F_E) of the transition is detected when the difference between consecutive frames falls below threshold T_S while the accumulated difference exceeds T_b. Note that the accumulated comparison needs only to be computed when the difference between consecutive frames exceeds T_S. If the consecutive difference value drops below T_S before the accumulated comparison value exceeds T_b, then the potential start point is dropped and the search continues for other gradual transitions.

A potential problem with this algorithm is that camera panning and zooming and large object motion may introduce similar gradual changes as gradual transitions, which will result in "false positives." This problem can be solved by global motion analysis; that is, every potential transition sequence detected will be passed to a motion analysis process to further verify if it is actually a global motion sequence (19). Experiments show that the twin-comparison algorithm is very effective and achieves a very high level of accuracy.

3.2.2. Edge Pixel Comparison

The edge pixel comparison algorithm as defined by Eq. (4) can also be applied in detecting gradual transitions. This is because gradual transitions usually introduce relatively lower but wide peaks of $D_e(f, f')$ values over a number of consecutive frames, different from sharp and narrow peaks for cuts. Fades and dissolves can be distinguished from each other by looking at relative values of local regions. During a fade-in, p_{in} will be much higher than p_{out} because there should be more entering edge pixels and fewer exiting edge pixels while a new shot progressively appears into the frames from back frames. In contrast, during fade-out, p_{out} will be much higher than p_{in} because the current shot progressively disappears into the back frames. A dissolve, on the other hand, consists of an overlapping fade-in and fade-out: During the first half of dissolve, p_{in} will be greater, whereas during the second half, p_{out} will be greater. Wipes can be distinguished from dissolves and fades by looking at the spatial distribution of entering and exiting pixels, because frames of a wipe sequence usually have a portion of the current shot and a portion of the new shot. Therefore, if we take the location into the analysis when calculating the fraction of changed edge pixels, wipes can be detected and distinguished from other types of transitions.

3.2.3. Editing Model Fitting

Hampapur et al, have studied algorithms for detecting different types of gradual transitions by fitting sequences of interframe changes to editing models, one for each given type of gradual transition (20). However, the potential problem with such model-based algorithms is that as more and more different types of editing effects (which still fall in mainly three basic classes: dissolve, wipe, and fade) become available, it is hard to model each one of them. Furthermore, transition sequences may not be following any particular editing model, due to possible noises and/or combination of editing effects. Such problems may exist in other detection algorithms as well, although this particular algorithm may be more prone to this problem.

3.3. Shot Detection Using Compressed Video Representation

As JPEG, MPEG, and H.26X (8) have become industrial standards, more and more video data have been and will continuc to be stored and distributed in one of these compressed formats. It would, therefore, be advantageous for the tools we envisage to operate directly on compressed representations, saving on the computational cost of decompression. More importantly, the compressed-domain representation of video defined by these standards provides features that may be more effective in detecting content changes.

Discrete cosine transform (DCT) coefficients are the basic compressed-domain features encoded in JPEG, MPEG, and H.26X. Another important feature encoded in the latter two standards is motion vectors. These are the two main features to be utilized in algorithms for video shot detection, with the most effective one being the one that combines both features. In this subsection, we discuss in detail three basic types of algorithms for video shot detection using compressed video data.

3.3.1. DCT Coefficient-Based Algorithms (21)

In MPEG, the compression of a video frame begins by dividing a frame into a set of 8×8-pixel blocks (22), as shown in Fig. 9. The pixels in the blocks are then transformed by the forward DCT (FDCT) into 64 DCT coefficients; that is,

$$F(u, v) = \frac{1}{4} C(u)C(v) \left(\sum_{x=0}^{7} \sum_{y=0}^{7} p(x, y) \cos\frac{(2x + 1)u\pi}{16} \cos\frac{(2y + 1)v\pi}{16} \right) \quad (5)$$

where $C(k) = 1/\sqrt{2}$ if $k = 0$ and 1 otherwise. $F(u, v)$ are DCT coefficients and $p(x, y)$ is the value of pixel (x, y) in a block. $F(0, 0)$ is the DC term or DC coefficient of a block, which is the average value of the 64 pixels, and the remaining 63 coefficients are termed the AC coefficients. These DCT coefficients are then quantized and encoded in a zigzag order by placing the low-frequency coefficients before the high-frequency coefficients, as shown in Fig. 6. The coefficients are finally Huffman entropy encoded. The process can then be reversed for decompression.

Because the DCT coefficients are mathematically related to the spatial domain and represent the content of each frame, they can be used to detect the difference between two video frames. Based on this idea, the first DCT comparison metric for shot-detection JPEG video was developed by Arman et al. (21). In this algorithm, a subset of the blocks in each frame (e.g., take out the boundary blocks of a frame and use every other one of the rest) and a subset of the DCT coefficients for each block were used as a vector

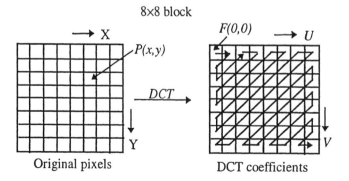

Fig. 9 An 8×8 block-based DCT and zigzag encoding.

representation (V_f) for each frame; that is,

$$V_f = (c_0, c_1, \ldots, c_i, \ldots, c_m) \tag{6a}$$

where c_i is the ith coefficients of the selected subset. The members of V_f are randomly distributed among all AC coefficients. One way to chose the subset is to use every other coefficient. The members of V_f remain the same throughout the video sequence to be segmented.

The difference metric between frames is then defined by content correlation in terms of a normalized inner product:

$$D_{\text{DCTC}} = 1 - \frac{|V_f \cdot V_{f+\varphi}|}{|V_f||V_{f+\varphi}|} \tag{6b}$$

where φ is the number of frames between the two frames being compared.

It has been observed that for detecting shots boundaries, DC components of DCTs of video frames provide sufficient information (23). Based on the definition of DCT, this is in equivalent to a low-resolution version of frames, averaged over 8×8 nonoverlap blocks. Applying this idea makes the calculation of Eq. (5) much faster while maintaining a similar detection accuracy.

Using DC sequences extracted from JPEG or MPEG data also makes it easy to apply histogram comparison; that is, each block is treated as a pixel with its DCT value as the pixel value. Then, histograms of DCT–DC coefficients of frames are calculated and compared using metrics, Eq. (3). This algorithm has been proven to be very effective, achieving both high detection accuracy and speed in detecting sharp cuts (24,25).

The DCT-based metrics can be directly applied to JPEG video, where every frame is intracoded. However, as shown in Fig. 10, in MPEG, there are three types of frames used: intraframes (I), predicted frames, (P) and bi-directional predicated and interpolated frames. (B) Only the DCT coefficients of I-frames are transformed directly from original images, whereas for P- and B-frames DCT coefficients are, in general, residual errors from motion-compensated prediction. This means that DCT-based metrics can only be applied in comparing I-frames in MPEG video. Because only a small portion of frames in MPEG are I-frames, this significantly reduces the amount of processing which goes

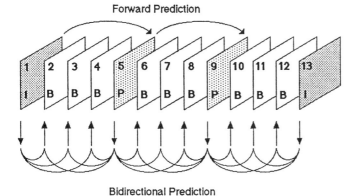

Fig. 10 Frame types in MPEG video.

into computing differences. On the other hand, the loss of temporal resolution between I-frames will introduce a large fraction of false positives in video shot detection which have to be handled with subsequent processing.

3.3.2. Motion-Vector-Based Algorithms

Apart from pixel value, motion resulting from either moving objects, camera operations, or both represents another important visual content in video data. In general, the motion vectors should show continuity between frames within a camera shot and show discontinuity between frames across two shots. Thus, a continuity metric for a field of motion vectors should serve as an alternative criterion for detecting segment boundaries.

The motion vectors between video frames are, in general, obtained by block-matching between consecutive frames, which is an expensive process. However, if the video data are compressed using either MPEG-1 or MPEG-2 standards, motion vectors can be obtained from the bitsteams of the compressed images. In the MPEG data stream, as shown in Fig. 10, there is one set of motion vectors associated with each P-frame representing the prediction from the last or next I-frame; and there may be two sets of motion vectors associated with each B-frame, forward and backward; that is, each B-frame is predicted and interpolated from its preceding and succeeding I/P-frames by motion compensation. If there is a significant change (discontinuity) in content between two frames, such as two B-frames, a B-frame and an I/P-frame, or a P-frame and an I-frame, there will be many blocks in the frame whose residual error from motion compensation is too high to tolerate. For those blocks, MPEG will not apply the motion-compensation prediction but instead intracode the pixels using DCT. Subsequently, there will be either no or only a few motion vectors associated with those blocks in the B- or P-frames. Therefore, if there is a shot boundary falling in between two frames, there will be a smaller number of intercoded blocks but a larger number of intracoded blocks, due to the discontinuity of the content between the frames. Based on this observation, we can detect a shot boundary by counting the number of intercoded blocks in P- and B-frames (24); that is, if

$$\frac{N_{\text{inter}}}{N_{\text{intra}}} < T_b \qquad (7)$$

is lower than a given threshold, then a shot boundary is declared between the two frames.

3.3.3. Hybrid Algorithms

Combining the DCT-based and motion-based metrics into a hybrid algorithm will improve the detection accuracy as well as processing speed in shot detection of MPEG-compressed video; that is, DCT–DC histograms of every two consecutive I-frames are first compared to generate a difference sequence (24). Given the large temporal distance between two consecutive I-frames, it is assumed that if we set the threshold relatively low, all shot boundaries, including gradual transitions, will be detected by looking at the points where there is a high difference value. Of course, this first pass will also generate some false positives. Then, B- and P-frames between two consecutive I-frames, which have been detected as potentially containing shot boundaries, will be examined in a second pass using a motion-based metric, Eq. (7); that is, the second pass is only applied to the neighborhood of the potential boundary frames. In this way, both high processing speed and detection accuracy are achieved at the same time. The only potential problem with this hybrid algorithm is that it may detect false motion sequences as transition sequences, just like the twin-comparison algorithm, which requires a motion-analysis-based filtering process, as will be discussed later.

In summary, experimental evaluations have shown that compression-domain feature-based algorithms perform with at least the same order of accuracy as those using video data in the original format, although the detection of sharp cuts are more reliable than that of gradual transitions. On the other hand, compression-feature-based algorithms achieve much high processing speed, which make software-only real-time video shot detection possible.

3.4. Algorithm Design and Performance Considerations

Finally, in this subsection, we briefly discuss performance issues of shot-detection algorithms. When we evaluate performances of a shot-detection algorithm, there are two measures: accuracy and complexity. We will focus our discussion on detection accuracy.

There are two quantitative measures for shot-detection accuracy: detection rate S_d, and false detection rate S_f. S_d is defined as the percentage ratio of the number of correctly detected shots to the number of total shots in a video sequence; S_f is defined as the percentage ratio of the number of falsely detected shots to the number of total shots detected. Obviously, an ideal shot-detection algorithm should have $S_d = 100\%$ and $S_f = 0\%$. However, in practice, we can only expect such ideal performance in some very simple video sequences that consist of simple and sharp cuts with no or little camera or object motion. In Refs. 26 and 27, a number of shot-detection algorithms have been tested and compared with both S_d and S_f measures.

From the algorithmic point of view, the main factor which determines the detection accuracy of an algorithm is the threshold values used in comparing frames, assuming that an effective content representation of frames and a comparison metric have been chosen. As we discussed earlier, even in the same video sequence, there is a large variation in frame-to-frame differences regardless of which representation schemes and difference measures are used. In general, if we set the threshold to capture shot boundaries that produce low frame-to-frame difference so as to increase the value of S_d, then, most likely, this will also increase the false detection rate S_f. Therefore, a major issue in designing a detection algorithm based on a given content representation and a set of difference metrics is to find a set of thresholds that will achieve the optimal trade-off between the required values of S_d and S_f. How to set this trade-off depends, to a large extent, on particular applications. Because ideal accuracy is unlikely to be achieved, manual verification and correction may be needed in many applications to refine the automatic shot-detection results. It is easy to understand that it is more difficult for human operators to find missing shot boundaries because one needs to look through every frame of a video sequence to identify all missing shots. On the other hand, in most of cases, one needs only to view the detected boundary points to identify false ones. Therefore, the manual identification of missing boundary points is often more costly than the manual correction of false alarms. Human efforts in verification and correction of false alarms could be reduced even further if we introduce a confidence measure in shot-detection algorithms; that is, if a detector produces a confidence measure value lower than a given threshold at a frame of a video sequence, it will mark a rejection flag to indicate that this boundary point needs human verification. In this way, human assistance will be directed only to these points where it is needed. In many applications, this approach may be acceptable and is definitely better than making a wrong decision. The similar approach has been widely used in document image analysis and character recognition.

Now, let us give a brief account of shot-detection algorithms described above and try to draw some general guideline for algorithm design. In general, for simple camera cuts, most of the algorithms can provide a fairly good performance. For gradual transition detection, multiframe-based algorithms such as the twin comparison produce better results. In general, there are four main components in the shot-detection algorithms: image content features, number of video frames involved, difference measures, and decision methods. These four components are closely related, especially the first three, and this division is only for convience of discussion here.

As presented earlier, image content features used in shot detection can be divided into pixel features such as corresponding pixel values and histograms; derived features such as edges and DCT coefficients, and motion features, among others. In general, the simple pairwise pixel difference between two frames is not an effective feature for content comparison between frames due to the fact that such differences are very sensitive to camera or object motion. Histograms of frames are global feature and less sensitive to camera or object motion, compared with pairwise pixel difference. An improvement of this histogram method is to use a set of local histograms of each frame instead of global histograms. The advantage of using such block histograms is that it takes the spatial distribution of pixel values into account. Edges are a good indication of image content, and experiments have shown that it is an effective feature for achieving good performance in shot detection. However, it is computationally expensive. The combination of two or more content features often makes a shot-detection algorithm more robust. A good example is the integration of DCT coefficients and motion vectors in the shot detection of video compressed with MPEG. However, as in other pattern-recognition problems, selection of an optimal feature or feature set depends, to a large extent, on applications and performance requirements. Experience play an important role in making such selections.

The number of frames used in shot detection is another major factor for detection accuracy. Many proposed algorithms compare content features of only two consecutive frames in deciding whether there is a shot boundary across the two image frames. However, two-frame-based algorithms are not very reliable except in detecting the simple camera cut. This is because there are many factors that contribute to image content variations between two consecutive frames, such as motions introduced by camera operation, object movement, flashing light, and so on. Furthermore, two-frame-based algorithms will fail to detect gradual transitions, as there is, in general, no significant change over two frames during a transition. Therefore, multiple-frame comparison may be needed for robust shot-detection algorithms, especially for detecting gradual transitions. In addition, it is necessary to take the motion of both camera and object into account. In most algorithms, shot detection and camera/object motion estimation are treated separately. Shot detection is usually performed first, followed by detecting camera/object motion within a detected shot. To achieve robust shot detection, it is necessary to integrate these two processes together, as motions may introduce content changes over a sequence of frames that may confuse the detector. A good solution to this problem is to apply motion analysis to a frame sequence and use motion information to detect shot boundaries. A good example of this approach has been presented in Ref. 19. However, detecting camera and/or object motion reliably is computationally expensive and difficult to be very robust.

Difference measures defined with a given image content representation and decision schemes based on these difference measures is another important factor on detection accuracy. Most detection algorithms use simple first-order difference metrics based on a given image feature to measure difference between two or more frames. Also, detection decisions are often based on simple thresholding or a finite-state machine where a sequence of decision processes is performed to make the decision where multiple thresholds may be involved. For these types of algorithm, it is crucial to choose appropriate thresholds. However, it is often impossible to determine a threshold or a set of thresholds to achieve high detection accuracy, and the determination is often only based on experiments over a variety of video data. To avoid problems of simple thresholding and to design a robust shot-detection algorithm, more work is needed in developing more elaborate classification algorithms, such as supervised training-based classification methods or neural network-based methods. Also, multiple content features, instead of single one, can be combined in shot detection. As a rule of thumb, the feature or difference metric that produces a strong bi-modal distribution in frame-to-frame difference, one representing nontransition differences and the other representing transitions, will, in general, produce better shot-detection results. This is because a strong bi-modal distribution usually makes it possible to choose an optimal threshold.

An important task in the development of robust shot-detection algorithms is comprehensive evaluations and benchmarking. For this purpose, we need to build a large test video database consisting of video sequences of all types, including feature movies, documentary, news, and cartoons, with complete data of shot boundaries of all possible types, camera operation and object motion, and so forth, annotated by human inspection. Only with the availability of such test databases can we perform comprehensive benchmarking of all different shot-detection algorithms.

4. APPLICATIONS OF SHOT CONTENT ANALYSIS AND REPRESENTATION

Today, accessing video is less flexible than accessing text. To read a book, we have a variety of means to quickly browse it, including table of contents, key words, or author indexes. For video media, whether analog or digital, all we have is a VCR-like interface—play, pause, fast forward, and rewind. There are many reasons for the lack of flexible access methods for video. The lack of a well-defined structure in video data and tools to automatically derive such structure are main obstacles. Digital video will improve this situation by allowing nonlinear and random access. However, it is far less than sufficient to only store and transmit video in digital format such as MPEG or various AVI formats. Only when we have a set of power tools that are able to extract video structure and build indexes of video streams will more efficient video browsing methods become possible and will video eventually be accessed as easy as today's text document.

In this section, we discuss applications of shot-based video content representation, including content-based video browsing, news video parsing and indexing, and object-based video compression. We will show that content features extracted in shot-based content analyses, such as shots and their attributes and temporal structure, can be utilized in more effective, content-based video indexing, browsing, and compression.

4.1. Video Browsing—A Structure-Tree-Based Approach

There are two major issues in video browsing—video structure analysis and user-interface design. In the first task, we need to define a model to represent video structure and then to develop algorithms to extract the structure information. In user-interface design, we need to decide on how to present the extracted video structure and how users can interact with the structured video data. Shot detection is the first step in video structure analysis, decomposing a video sequence into basic temporal units. This should be followed by classification and grouping of shots based on shot attributes presented in Sec. 2.2 and other application-specific requirements. For instance, shots can be grouped into dialogs or actions according to a set of shot attributes and temporal constraints and further grouped into stores (28). With such structure information, a video title can be organized into a structured hierarchy essentially similar to a structured book with chapters, sections, and paragraphs. With this structure in place, new browsing tools can be developed that enables viewers to navigate through video quickly to locate what he or she may want to see. Viewing video will no longer be the passive viewing process to which we are limited today.

In feature movies or documentary video, in general, there are at least three structure levels: chapters, segments, and shots, where shots can be further abstracted into key frames. Based on the statistics, the average shot length is about 3–5 s, or 120 frames/pshot. Considering a video title of about 1 h in length, it will have about 110,000 frames and will consist of about 900 shots. Manually searching for any particular shot from these 900 will be a tedious job. However, this difficulty will be largely reduced if we organize the video into a structured hierarchy. Figure 11 shows how a video title can be decomposed into a hierarchy tree, with the approximate numbers of chapters, segments, shots, and frames present at each level (28). This structural tree provides an effective way to perform content-based access of video, which can be built based on shot detection and classification.

In the SWIM (Show What I Mean) video retrieval and browsing system (28,29), a hierarchical video browser has been implemented based on this method. Figure 12 shows several levels of a documentary video of about 10 min. Only three levels of the hierarchy are presented in this example to illustrate the power of this browsing tool. When one clicks a pictorial icon at a higher level, images at the lower level will be presented in several groups. This process can be iterated until the finest detail is reached or when the user has found the segment in which he/she is interested. This type of nonlinear interface provides a way for users to browse video more flexibly and efficiently.

This hierarchical browser can be further enhanced if we can construct it based not only on structural but also semantic information; that is, if we can group shots into seg-

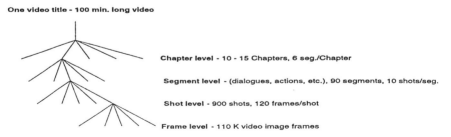

Fig. 11 A hierarchy structure organization for a video title; this organization is very similar to a book organized into a structural form. As we can see, there is a two-order-of-magnitude reduction from the number of frames to the number of shots (100 K to 1 K) and one order reduction from shot to segment, and from segment to chapter.

Video Shot Detection/Analysis

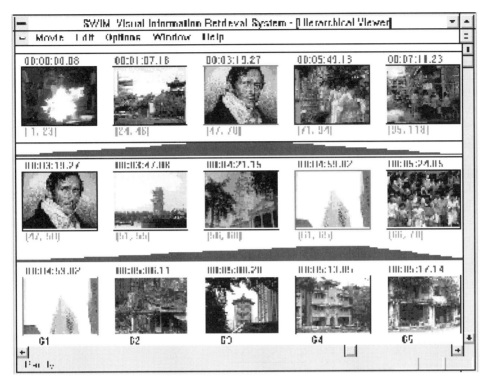

Fig. 12 SWIM model: key-frame-based, hierarchical views.

ments, and segments into chapters, we will be able to construct a chapter–segment–shot structure hierarchy as illustrated in Fig. 11. For feature movies, shots can be grouped into dialogues where two or more persons have conversions. An average dialogue lasts about 20–40 s, or 5–10 shots. Similarly, shots can also be grouped into actions where the actions' length varies considerably compared with dialogues. There are also other ways or criteria to group shots. Based on these dialogues, actions, and other possible types of shot groups, the segment level of a more semantic hierarchy as illustrated in Fig. 11 can be formed. The segments can be further grouped into stories or chapters for certain type of video programs. This is most likely done manually, based on semantic understanding.

In Ref. 15, an algorithm is presented to automatically group shots into stories. Shots are detected and labeled based on time-constrained clustering, which is based on visual similarity such as color, texture, and image correlation. After clustering, a label is designated to each cluster of shots, called a meta-shot. A story is defined as a collection of contiguous meta-shots such that the associated label sequence is the minimum sequence that contains all the identical labels. For instance, for a video sequence with 11 shots (A, B, A, B, A, C, D, F, E, D, F), we consider that there are two story units, one is A, B, A, B, A, and another is C, D, F, E, D, F. A story can have several dialogues and actions segments. How to construct stories from segments is, to a large extent, application dependent. Although tools for automatic extraction of story structure are useful, we believe the process can only be done semiautomatically because complete semantic understanding of video content is still beyond the capability of current video-parsing algorithms

4.2. News Video Parsing, Indexing, and Browsing

Although it is not feasible to automatically parse the content of video in general, it is possible if we constrain it to a particular domain. A good example is the news video parsing, indexing, and browsing system presented in Refs. 11 and 30. The parsing and indexing in this system consists of shot detection, classification, and still-frame detection modules. The system segments news videos into individual news items, which can be stored in and retrieved from a video database. Such a system is very useful in a news information and archival system for broadcasting companies and information providers, because it can help to reduce the labor-intensive activity of manually extracting news stories from news programs. The system provides a fairly accurate automatic process for the segmentation of news items from the news program and the facility to browse and retrieve such news items from a database.

The basis of the news video parsing and indexing system is the temporal syntax of news video programs, which has a relative simple and straightforward structure. Usually, a news program consists of a sequence of news items (possibly interleaved with commercials), each usually led and concluded by an anchorperson shot. In addition, the news program has certain regular features, such as weather, sports, and business reports. Based on this temporal structure information, we can classify all shots into anchorpersons, news, commercial breaks, weather forecasts, financial indexes shots, and starting and ending sequences, and build a structural index of a given news program.

The automatic news parsing process consists of two steps: temporal segmentation of a video sequence into shots, and shot classification based on anchorperson shot model and the temporal structure model. In general, news shots do not have any fixed temporal or spatial structures. However, anchorperson shots can be distinguished easily by their spatial structures; that is, we can model anchor frames by positions and movements of anchorpersons in the frame. Usually, an anchorperson shot consists of one or more anchorpersons and a pictorial news icon or a static background, which can be easily detected by temporal and frame model matching. Identifying an anchorperson shot involves testing if every frame in the shot satisfies a frame model, which, in turn, means testing each frame against a set of region models. A shot of financial indexes or weather maps tend to have still frames displayed across the entire shot. These shots usually contain low activity with minimal temporal changes between frames, which can be detected using their temporal features. The special temporal and spatial structures of sequences that lead into and conclude the commercial break can be used to detect the commercial break sequences.

The automatic news video parsing and indexing system presented in Fig. 13 is a fully functional system designed based the algorithms presented above. The input to the system is a MPEG, compressed video data stream, which is automatically segmented into shots directly without decompression using the hybrid shot-detection algorithm presented in Sec. 3. After the news video has been partitioned into individual shots, the system classifies shots into anchorperson, news, and other special shots including the weather report and financial reports. The parsing and classification processes can be run at a real-time speed on a low-end workstation because there is no need to decompress the MPEG stream in the parsing process.

The prototype system consists of the following modules:

- Movie player—a MPEG player which supports common video playback functions and random access to MPEG video files

Video Shot Detection/Analysis

Fig. 13 News video structure-based video browsing. News video has been parsed into news items; the top part is a video player with various play modes, the lower part shows first video frames of each news stories (30).

- News parser—the parsing module which automatically segment a news video to shots and identify all the news items using the domain-knowledge-based classification algorithms presented above
- News browser—the viewer which shows anchorperson shots and the associated news shots and allows playback of the news items
- News logging—a logging function which enable the segmented news items to be further classified semantically according to news categories and stored into a video database. The category consists of local news, international news, sports, business, and commercials.
- News retrieval—a retrieval module which enables specific news items to be retrieved from the video database by key word, category, and visual content

Table 1 Test Result of the News Parsing Algorithms

	Items	Detected	Falsely Detected
		News Programs	
sbc2.mpg	20	20	1
sbc4.mpg	19	18	3
		Finance/Weather	
sbc2.mpg	2	2	0
sbc4.mpg	1	1	0

search, that is, a user can choose a news category to find a list of news and select a desired news clip for playback.

Table 1 lists some test result of the news video-parsing algorithms implemented in the system. The test video data consist of two half-hour broadcast news programs by Singapore Broadcasting Corporation (SBC) in MPEG-1 format. It is shown that the system can achieve very high parsing and classification accuracies.

4.3. Content-Based Video Compression—Video-Object-Based Approach

Content-based representation of video also provides a new approach for video compression. When a video is decomposed into more meaningful components or objects with more homogeneous attributes, compression could be optimized based on the objects. This will support both a higher compression ratio and a higher content accessibility compared with current MPEG-1/2 standards. In this section, we discuss a new content-based video compression approach. In this new scheme, a shot or a group of consecutive shots are considered as an object for video compression, and video is no longer considered as a sequence of frames with a fixed frame rate. Based on the attributes of video shots such as activity levels, appropriate frame rates can be assigned for video compression and play. In other words, in this new scheme, frame rates for video production, compression, transmission, and playback are decoupled.

4.3.1. Fixed Frame Rate Versus Variable Frame Rate

In current analog video systems, fixed frame rates (25 frames/s for PAL and 30 frames/s for NTSC) are used in production, transmission and playing back. These frame rates were chosen based on several factors, including bandwidth limitation and reproduction of motion pictures. It is clear that in many situations, using a fixed frame rate is a waste of valuable resources. For instance, we often see many shots consisting of still frames in news and documentary programs, and it is obvious that there is no need to send these still frames at 30 frames/s because all the frames are exactly the same. For digital video, the frame rate can be changed or adjusted easily, as frame rate is not related to any particular device mechanism, as in analog video/film devices. The question is we when can use a lower frame rate and still maintaining appropriate playback video quality. The advantages of the use of lower frame rates are savings in transmission bandwidth, storage space, and so forth.

Video object is an important concept in MPEG-4 standard, in which video is no longer represented only as a sequence of video frames at a fixed frame rate, as in MPEG1/2. Instead, video consists of many video objects, which will be coded at their native resolution; then, the object can be scaled and presented in various spatial resolutions and frame rates for a particular need. Following the same methodology, in our new scheme, there will be three frame rates used in the object-based video compression: F_{source}, the frame rate of the video source, F_{trans}, the frame rate for transmission/compression, and $F_{playback}$, for playing back or presentation. The three frame rates can be different. F_{trans} is determined based on various factors such as activity levels of shots and communication channel capacity. A key idea is that, based on an understanding of video content such as video activity level, an optimal F_{trans} can be determined such that only essential information is transmitted from video source, perhaps at a much lower frame rate than that for playback.

4.3.2. Content-Based Variable-Frame-Rate Video Compression

In our proposed content-based scheme, the shot-detection process is first performed for a given video, and, at the same time, attributes of shots are extracted. Based on these attributes such as activity level of shots, appropriate frame rates can then be determined for each video object according to the activity level.

The reason we chose activity levels of shots in determining the frame rate for video compression is simple: When there is no motion or little motion in a given shot, then there is no need to sample the shot with a high frame rate. In general, we consider that video activity tends to be more homogeneous within a shot. In other words, the activity level should be more homogeneous within a shot than across different shots. Based on activity levels, shots can be classified into the following four categories: *still-frame shots,* where no change is taking place within the shot; *near-still-frame shots,* where very little change in the frames is presented; *moderate-activity shots,* where a moderate amount of video activities are presented either due to object movement in images or camera movement or zooming; and *high-activity shots,* where a large amount of scene changes is taking place, such as car chasing, high-speed ball kicking, and so forth. These four levels can be measured by motion statistics calculated from shots.

Based on the activity levels of video objects, consecutive shots with similar activity levels can be grouped together, and appropriate resampling can be performed to convert the source video into a lower frame rate F_{trans}. Then, these video objects can be compressed with the assigned frame rate. For instance, if the video object consists of still video frames only, then a very low frame rate can be used for resampling and subsequent compression (e.g., 1 frame/s). Furthermore, we also can take into account communication channel capacity in choosing F_{trans}. Once F_{trans} is determined and video objects have been resampled, they can be compressed using MPEG-1/2. As long as both transmission and receiving ends agree on a set of video object protocols, which can be established in an negotiation phase, in the receiving end the video object with a transmission frame rate F_{trans} can be decoded and restored into a video stream with an appropriate frame rate of $F_{playback}$. Figure 14 illustrates the processes of such schemes. This video-object-based video compression can be conceived as a front-end processor or postprocessor of MPEG-1/2 compression. This architecture also follows MPEG-4 architecture very well.

There are many advantages in this video-object-based video compression scheme. First, this frame-rate decoupling is an important step in taking advantage of the flexibility of digital video. In digital video, the frame rate can be easily manipulated to achieve a

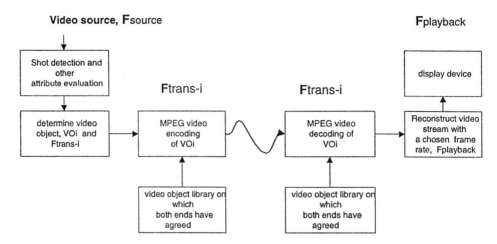

Fig. 14 Variable frame rate video for video-object-based video compression—sender and receiver end, where three video frame rates are F_{source}, F_{trans}, and $F_{playback}$.

higher compression rate. Second, because video is represented as a collection of video objects instead of a sequence of frames, it is much easier to interact. This is valuable for many applications where access and interaction with video are essential. Finally, because video objects can be manipulated individually, it is possible that we can adjust F_{trans} based on communication capacity. For instance, for a slow communication channel, we can use a small F_{trans} to save bandwidth with lower video quality; in this way, we can achieve universal accessibility.

5. CONCLUSION

As digital video is progressively penetrating into all multimedia applications, the need for power tools to manage and access the vast amount of video data becomes increasingly urgent. Fundamentally, what we need are new technologies for video content analysis and representation to facilitate organization, storage, retrieval, and compression of mass collections of video data in an efficient and user-friendly way.

In this chapter, we have discussed a number of techniques for video shot detection, and their applications in content-based video browsing and compression. Shot detection and classification have been aimed at recovering structure and low-level information of video sequences, although these techniques are basic and very useful in facilitating intelligent video browsing and content-based video compression.

Shot detection is the first step toward video content analysis and content-based retrieval. After detection, attributes of shots can be analyzed, based on which shots can be further grouped and classified for a variety of applications such as browsing, compression, and video structure analysis. Some of the applications have been described briefly in this chapter, including video browsing for feature movies, news video parsing and indexing, and object-based video compression.

We believe that in the process of video structure extraction, low-level processing such as shot detection and classification can be performed automatically or semiautomatically. However, from a low level to a middle level, and further to a high level, most of processing can still only be done semiautomatically or manually. This is because semantic understanding is needed to carry out these tasks, making it unlikely for computers to carry out these tasks in the foreseeable future. Therefore, the research of the video structure extraction should aim at developing tools for facilitating shot classification and forming of higher-level video structure, instead of automating the whole extraction process. Also, it should be pointed out that shot classification and the clustering process should be studied for specific types of video, as different types of video have very different structures.

By observing the progress in document image analysis and recognition in the last four decades, we will see that there are many more challenges ahead to achieve the goal of DVMAR—to extract the content of video and build digital content representation of video. Successful efforts in extracting semantic content of video are still very limited and there is a long way to go before we can achieve our goal of automatic video content understanding and content-based retrieval. Many of the bottlenecks and challenges in video content analysis are the classic ones in pattern recognition and computer vision. The new MPEG-7 effort is in response to this need and tries to set necessary standards regarding digital representation of multimedia. Although it is too early to see how successful the MPEG-7 effort would be, it is very clear that digital content representation and extraction have become very important topics for digital video applications.

REFERENCES

1. VN Gudivada, VV Raghavan. IEEE Computer 18–22, Sept. 1995, vol 28, pp 18–22.
2. RW Picard, AP Pentland, IEEE Trans PAMI Pattern Analysis and Machine Intelligence-18(8), 1996, pp 769–770.
3. R Jain, A Pentland, D Petkovic, eds. Workshop Report: NSF-ARPA Workshop on Visual Information Management Systems, Cambridge, MA 1995.
4. P Aigrain, HJ Zhang, D Petkovic. Int J Multimedia Tools Applic 3(3), 1996, pp 179–202.
5. HJ Zhang, JH Wu, D Zhong, SW Smoliar. Pattern Recogn 1997, vol 30, No. 4, pp 643–658.
6. E Wold, T Blum, D Keislar, J Wheaton. IEEE Multimedia, Fall 1996.
7. HJ Zhang, Q Tian. ACM Comput Surv 27(4):643–644, 1995.
8. MPEG1-7 web site: http://drogo.cselt.stet.it/mpeg/.
9. MPEG-7: Context and Objectives, ISO/ IEC JTC1/SC29/WG11/N1425, Nov. 1996, Macao, http://drogo.cselt.stet.it/mpeg/mpeg-7.htm.
10. C Reader. MPEG4: Coding for content, interactivity and universal accessibility. Op Eng 35(1): 104–108, 1996.
11. HJ Zhang, et al. Multimedia Syst 2:256–266, 1995.
12. L Ggorman, R Kasturi, eds. Document Image Analysis. IEEE Computer Society, 1995, Los Alamitos, California.
13. Proceedings First, Second, Third International Conferences on Document Analysis and Recognition, 1991, 1993, 1995.
14. TG Aguierre-Smith, G Davenport. The stratification system: A design environment for random access video. Proceedings 3rd International Workshop on Network and Operating System Support for Digital Audio and Video, La Jolla, CA, 1992, pp 250–261.
15. MM Yeung, BL Yeo. Video content characterization and compaction for digital library applications. Proceedings of SPIE, Storage and Retrieval of Image and Video Databases, 1997, vol 3022, pp 45–58.

16. B Shahraray. Scene change detection and content-based sampling of video sequences, SPIE Proceedings Digital Video Compression: Algorithm and Technologies. SPIE—The International Society for Optical Engineering Bellingham, Washington, USA. San Jose, 1995, vol 2419, pp 2–13.
17. A Nagasaka, Y Tanaka. Automatic video indexing and full-search for video appearances, in E Knuth, IM Wegener, eds. Visual Database Systems. Amsterdam: Elsevier Science Publishers, 1992, pp 113–127.
18. R Zabih, K Mai, J Miller. A robust method for detecting cuts and dissolves in video sequences. Proceedings ACM Multimedia '95, San Francisco, 1995.
19. HJ Zhang, A Kankanhalli, SW Smoliar. Multimedia Syst 1(1):10–28, 1993.
20. A Hampapur, R Jain, TE Weymouth. Multimedia Tools Applic 1(1):9–46, 1995.
21. F Arman, A Hsu, MY Chiu. Feature management for large video databases. SPIE—The International Society for Optical Engineering Bellingham, Washington, USA. Proceedings SPIE Conference on Storage and Retrieval for Image and Video Databases I, 1993, vol 1908, pp 2–12.
22. B Furht, SW Smoliar HJ Zhang. Image and Video Processing in Multimedia Systems. Boston: Kluwer Academic Publishers, 1995.
23. B-L Yeo, B Liu. A unified approach to temporal segmentation of motion JPEG and MPEG compressed video. Proceedings IEEE International Conference on Multimedia Computing and Networking, Washington DC, 1995, pp 81–88.
24. HJ Zhang, CY Low, Y Gong, SW Smoliar. Video parsing using compressed data. Proceedings SPIE '94 Image and Video Processing II, San Jose, CA, 1994, pp 142–149.
25. IS Sethi, N Patel. A statistical approach to scene change. Proceedings SPIE Conference on Storage and Retrieval for Video Databases III, San Jose, CA, 1995.
26. JS Boreczky, LA Rowe. Comparison of video shot boundary detection techniques. Proceedings SPIE Conference on Storage and Retrieval for Video Databases III, San Jose, CA, 1996
27. G Ahanger, TDC Little. Journal of Visual Communication and Image Representation, Special Issue on Digital Libraries, Vol. 7, No. 1, March 1966, pp. 28–43.
28. HJ Zhang, CY Low, SW Smoliar, JH Wu. Video parsing, retrieval and browsing: An integrated and content-based solution. Proceedings ACM Multimedia 95, San Francisco, 1995, pp 15–24.
29. HJ Zhang. SWIM: A prototype environment for image/video retrieval. Proceedings Second Asian Computer Vision, Singapore, 1995; web site: http://www.iss.nus.sg/RND/MS/Projects/vc/project1.html.
30. CY Low, Q Tian, HJ Zhang. An automatic news video parsing, indexing and browsing system. Proceedings ACM Multimedia '96, Boston, 1996.

10
MPEG-4: An Object-Based Standard for Multimedia Coding

Atul Puri
AT&T Labs–Research, Red Bank, New Jersey

Alexandros Eleftheriadis
Columbia University, New York, New York

1. INTRODUCTION

The increasing diversity of applications requires communications not only in the form of speech and data but also with synthetic and natural images and video. However, multimedia is expensive in the sense of its bandwidth requirement, with video being highly bandwidth intensive. Efficient compression of video is therefore critical to making any multimedia application feasible. The success of multimedia terminals, products, or services (1) depends on many factors; of particular significance is interworking, which is facilitated by standardization.

Toward this end, considerable progress has been made, in particular, in the ISO Moving Picture Experts Group (MPEG), which has completed the first (MPEG-1) and second (MPEG-2) of its work items. The ISO MPEG-1 video standard (2) was primarily optimized for coding of noninterlaced video at bit rates of 1.2–1.5 Mbits/s and the ISO MPEG-2 video standard (3,4) was primarily optimized for coding of interlaced video at bit rates of 4–9 Mbits/s. The MPEG-1 standard was intended for interactive video on digital storage media (e.g., compact disk) type of application, whereas the MPEG-2 standard was intended for digital TV (e.g., broadcast, cable, satellite) applications. Both the MPEG-1 and MPEG-2 video standards employ a motion-compensated discrete cosine transform (DCT) coding framework. Further, the MPEG-2 video standard also supported scalability in picture quality and spatial and temporal resolutions.

In addition the ISO standards, the ITU-T (formerly CCITT) has also developed video and audio coding standards. The recently completed ITU-T H.263 standard (4) is optimized for video coding at low bit rates of 10–24 kbits/s. It is based on the earlier ITU-T H.261 video standard (5) which was optimized at 64 kbits/s (although it allows a range of 64 kbit/s to 2 Mbits/s). In a general sense, the H.263 standard (6) uses the motion-compensated DCT coding framework which is also common to the H.261, MPEG-1, and MPEG-2 standards. It employs, however, a number of features beyond those in H.261 but similar to features in MPEG-1, including half-pixel motion compensation as well as several additional modes, such as unrestricted motion vector, advanced 8×8 block prediction, PB-frames, and syntax-based arithmetic coding. These modes are options that are negotiated between a decoder and an encoder. However, the ITU-T has continued work on further embellishing H.263 by adding many more features and optional modes; the resulting standard, currently in development, is referred to as H.263+ (7).

The current ongoing MPEG standard (MPEG-4) (6,8–10) was started in 1993 with intended completion by late 1998. Its original focus was modified in July 1994 from that of coding videophone scenes with high efficiency at very low bit rates, to flexible coding of generic scenes facilitating a number of important functionalities not supported by other standards. Among the functionalities (8) that were considered important for MPEG-4 were content-based coding, universal accessibility (which includes robustness to errors), and good coding efficiency. Further, MPEG-4 video is being optimized for bit rates ranging from about 10 kbits/s to around 1.5 Mbits/s and is expected to be applicable to even higher bit rates. It is worthwhile pointing out that the MPEG standards (2,4) are essentially decoding standards and, thus, only specify the bitstream representation and the semantics of the decoding process; in other words, the encoding algorithm is not standardized. Because MPEG-4 is designed to truly be a multimedia standard, the ongoing development of the specification (11–13) goes much beyond that of previous MPEG standards and addresses not only audio coding (13), video coding (12), and multiplexing of coded data (11) but also coding of text/graphics and synthetic images (12) as well as representation of audio–visual scene descriptions and interactivity (11).

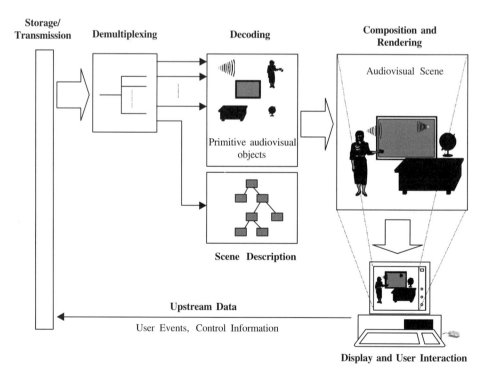

Fig. 1 A high-level view of an MPEG-4 terminal.

Figure 1 shows a high level view of an MPEG-4 terminal (14–16). We use the term "terminal" in a generic sense, including both stand-alone hardware as well as software running on general-purpose computers. A set of individually coded audio–visual objects (natural or synthetic) are obtained multiplexed from a storage or transmission medium. They are accompanied with scene description information that describes how these objects should be combined in space and time in order to form the scene intended by the content creator. The scene description is thus used during composition and rendering, which results in individual frames or audio samples being presented to the user. In addition, the user may have the option to interact with the content, either locally or with the source, using an upstream channel (if available).

The rest of the chapter is organized as follows. In Sec. 2, we present an overview of the ITU-T video and systems standards with emphasis on mobile applications. In Sec. 3, we review the applications, requirements, tests, and the organization of the MPEG-4 standard. Next, Sec. 4 discusses MPEG-4 visual tools with emphasis on error resilience tools. In Sec. 5, the MPEG-4 audio standard is briefly discussed. In Sec. 6, MPEG-4 systems are discussed in detail. In Sec. 7, we discuss the issue of profiles, and in Sec. 8, the plans for verification tests are presented. Section 9 discusses the work in progress for version 2 of MPEG-4 video, as well as the plans for the MPEG-7, the next MPEG standard. Section 10 summarizes the key points presented.

2. RELATED ITU-T STANDARDS

The ITU-T H.263 (6) standard, since it is derived from ITU-T H.261, is based on the framework of block-motion-compensated DCT coding. Both the ITU-T H.261 (5) and the H.263 (6) standards, like the ISO MPEG-1 (2) and the MPEG-2 (3,4) standards, specify bitstream syntax and decoding semantics. However, unlike the MPEG-1 and MPEG-2 standards, these are only video coding specifications and thus do not specify audio coding or systems multiplexing, each of which can be chosen from a related family of ITU-T standards to develop applications requiring full systems for audio–visual coding. Also, unlike the MPEG standards, the ITU-T standards are primarily intended for conversational applications (at low bit rates with low delay) and do not include coding techniques that facilitate interactivity with stored data.

The H.324 (17) standard is the multimedia communication standard for low-bit-rate circuit-switched networks, including ordinary analog telephone lines which builds on industry's experience with H.320, the ITU-T standard for ISDN videoconferencing. H.324's architecture consists of a multiplexer which combines different media streams into a single stream (H.223) (18), a control protocol for capability negotiation (H.245), and audio and video decoding (G.723.1 and H.263). Because H.263 and H.223 are key components of ITU-T H.324 multimedia communications, we discuss them next.

2.1. H.263 (and H.263+)

The H.263 standard (6) specifies decoding with the assumption of block-motion-compensated DCT structure for encoding, similar to H.261. There are, however, some significant differences in the H.263 decoding process as compared to the H.261 decoding process, which allow encoding to be performed with higher coding efficiency. For developing the H.263 standard, an encoding specification called the Test Model (TMN) was used for optimizations. TMNs progressed through various iterative refinements; the final test model was referred to as TMN5. The H.263 standard is significantly optimized for coding at low bit rates (of a few tens of kbits/s) while maintaining good subjective picture quality at higher bit rates as well. We now discuss the encoding structure of TMN5, which can provide efficient encoding compliant to the H.263 decoding standard.

2.1.1. TMN5 Encoding

Figure 2 shows the block diagram of the simplified encoder allowing video coding as per TMN5. Video coding is performed by partitioning each picture into macroblocks, where a macroblock consists of 16×16 luminance (Y) block (composed of four 8×8 blocks) and the corresponding 8×8 chrominance blocks of Cb and Cr. Each macroblock can be coded as intra (original signal) or as inter (prediction error signal). Spatial redundancy is exploited by DCT coding. Temporal redundancy is exploited by motion compensation, which is used to determine the prediction error signal. Block DCT coefficients are quantized by using a quant_scale parameter resulting in transmission of quant_index for every nonzero coefficient.

Besides quant_index for nonzero coefficients, the macroblock motion vector and a number of coding control flags and parameters are included in the bitstream generated by the video multiplex coder. In fact, the general coding structure of TMN5 described thus far applies not only to the H.263 standard but also to all MPEG video standards. In addition, TMN5 coding also includes motion estimation and compensation with half-pixel accuracy and bi-directionally coded macroblocks. Both of these features are also present in MPEG video standards, which contain additional tools and features as well. Additional features of TMN5 coding are 8×8 overlapped block-motion compensation, unrestricted motion

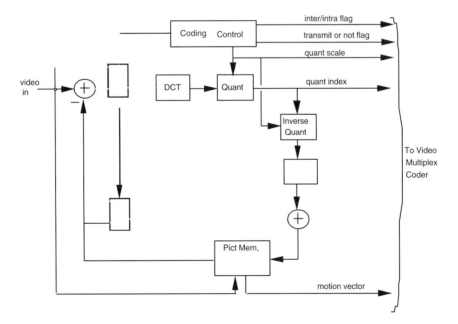

Fig. 2 TMN5 encoder.

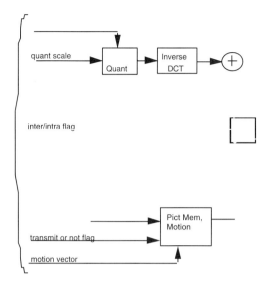

Fig. 3 H.263 decoder.

vector range at picture boundary, and arithmetic coding; these features are mainly useful for low-bit-rate applications and were not included in MPEG-1 and MPEG-2 standards.

2.1.2. H.263 Decoding

The H.263 decoder decodes the self-contained bitstream generated by TMN5 or a similar encoder resulting in reconstructed video. (see Fig. 3.)

The video decoding algorithm of H.263 is based on H.261 with refinements/modifications to support enhanced coding efficiency. Four negotiable options are supported to allow improved performance. One difference with respect to H.261 is that instead of full-pixel motion compensation and a loop filter, H.263 supports half-pixel motion compensation (as discussed during TMN5 encoding), providing improved prediction. Another difference is in the group-of-blocks (GOB) structure, the header for which is now optional. The four negotiable options of H.263 mentioned while discussing TMN5 encoding are as follows:

- Unrestricted motion vector mode. This mode allows motion vectors to point outside a picture, with edge pixels used for prediction of nonexisting pixels.
- Syntax-based arithmetic coding mode. This mode allows use of arithmetic coding instead of variable length (Huffman) coding.
- Advanced prediction mode. This mode allows use of overlapped block-motion compensation (OBMC) with four 8×8 block motion vectors instead of a single 16×16 macroblock motion vector.
- PB-frames mode. In this mode, a P-picture and a B-picture are coded together as a single PB-picture unit.

2.1.3. H.263+ Features and Modes

As mentioned earlier, the H.263+ (7) standard further adds to H.263 (6) a number of features and negotiable additional modes which are listed as follows:

- Scalability—spatial, temporal, and signal-to-noise ratio (SNR) scalability
- Custom source formats
- Advanced intra coding (AIC): a mode which improves the compression efficiency for intra macroblock encoding by using spatial prediction of DCT coefficient values
- Deblocking filter (DF): a mode which reduces the amount of block artifacts in the final image by filtering across block boundaries using an adaptive filter;
- Slice structure (SS): a mode which allows a functional grouping of a number of macroblocks in the picture, enabling improved error resilience, improved transport over packet networks, and reduced delay
- Reference picture selection (RPS): a mode which improves error resilience by allowing a temporally previous reference picture to be selected which is not the most recent encoded picture that can be syntactically referenced
- Reference picture resampling (RPR): a mode which allows a resampling of a temporally previous reference picture prior to its use as a reference for encoding, enabling global motion compensation, predictive dynamic resolution conversion, predictive picture area alteration and registration, and special-effect warping
- Reduced-resolution update (RRU): a mode which allows an encoder to maintain a high frame rate during heavy motion by encoding a low-resolution update to a higher-resolution picture while maintaining high resolution in stationary areas
- Independent segment decoding (ISD): a mode which enhances error resilience by ensuring that corrupted data from some region of the picture cannot cause propagation of error into other regions
- Alternate inter VLC (AIV): a mode which reduces the number of bits needed for encoding predictively coded blocks when there are many large coefficients in the block
- Modified quantization (MQ): a mode which improves the bit-rate control by changing the method for controlling the quantizer step size on a macroblock basis, reduces the prevalence of chrominance artifacts by reducing the step size for chrominance quantization, increases the range of representable coefficient values for use with small quantizer step sizes, and increases error detection performance and reduces decoding complexity by prohibiting certain unreasonable coefficient representations
- Supplemental enhancement information, including chroma-key to provide transparency information for implicitly representing shapes of arbitrary regions in pictures

2.2. H.223 Multiplexer

As mentioned earlier, ITU-T H.324 (17) is a toolkit standard consisting of component standards, such as V.34 modem, H.223 multiplexer, H.245 control protocol, G.723.1 audio decoder, and H.263 (or H.261) video decoder. We have already discussed H.263; we now introduce H.223, the multiplexer used to mix audio, video, data, and control channels together for transmission on a V.34 modem.

The ITU-T H.223 (18) multiplexer combines features from time-division multiplexers (TDM) and packet multiplexers and new ideas. However, it has less delay than TDM and packet multiplexers and also has less overhead. It is byte oriented for ease of implementation, can match different data rates using stuffing, and uses a marker for resynchronization recovery. Further, each protocol data unit (PDU) can carry a mix of different data streams in different proportions, hence allowing fully dynamic allocation of bandwidth to the different channels. In addition to the multiplex layer, H.223 also provides a set of three adaptation layers (AL1–AL3). AL1 is primarily intended for variable-rate framed information such as control (e.g., H.245 channel). AL2 is primarily intended for audio (G.723.1), whereas AL3 is intended for video, such as H.263 or H.261. The ITU-T H.223 Annex A multiplexer has been designed for error-prone channels and therefore features robust packet synchronization and a constant packet length. On the protection sublayer, framing, error detection, and forward error correction tools using convolutional coding are included. Further, ITU-T H.324 Annex C specifies features of multimedia terminals operating in mobile radio environments, in terms of differences with normal terminals.

3. MPEG-4 OVERVIEW

MPEG-4 was originally intended for very high compression coding of audio–visual information at very low bit rates of 64 kbits/s or lower. When MPEG-4 video was started, it was anticipated that with continuing advances in advanced (nonblock-based) coding schemes; for example, in region-based and model-based coding, a scheme capable of achieving very high compression, mature for standardization, would emerge. By mid-1994, two things became clear. First, video coding schemes that were likely to be mature within the time frame of MPEG-4 were likely to offer only a moderate increase in compression (say, by factor of 1.5 or so) over existing methods as compared to the original goal of MPEG-4. Second, a new class of multimedia applications were emerging that required increasing levels of functionality compared to that provided by any other video standard at bit rates in the range 10–1024 kbits/s. This led to broadening of the original scope of MPEG-4 to a larger range of bit rates and important new functionalities (8). Basically, three important trends were identified:

- The trend toward interactive computer applications
- The trend toward wireless communications
- The trend toward integration of audio–visual data into a number of applications

The focus and scope of MPEG-4 was redefined as the intersection of the traditionally separate industries of telecommunications, computer, and TV/film where audio–visual applications exist. It was felt that the existing standard or emerging audio–visual standards were not adequately addressing the expectations and requirements of these industries in combination, leading to incompatible solutions for similar applications. Hence, MPEG-4 was aimed at addressing these new expectations and requirements by providing audio–visual coding solutions to allow interactivity, universal accessibility, and sufficient compression. The mission and the focus statement of MPEG-4 explaining the trends leading up to MPEG-4 and what can be expected in the future are documented in the MPEG-4 Proposal Package Description (PPD) document (8). Figure 4 shows the application areas of interest to MPEG-4 arising at the intersection of the aforementioned industries.

To make the discussion a bit more concrete, we now provide a few examples of applications or application classes (15) to which MPEG-4 standard is aimed:

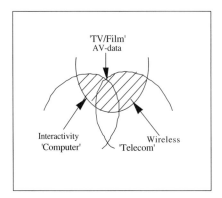

Fig. 4 Applications areas addressed by MPEG-4 (shaded region).

- Internet and Intranet video
- Wireless video
- Video databases
- Interactive home shopping
- Video e-mail, home movies
- Virtual-reality games, simulation, and training

With this revised understanding of the goals of MPEG-4, the work was subsequently reorganized and partitioned into the following subgroups:

- *Requirements*—develops requirements, application scenarios, and meaningful clustering of coding tool combinations for interoperability (profiles)
- *Tests*—develops methods for subjective and objective assesment and conducts tests
- *Video*—develops coded representation of moving pictures of natural origin
- *Synthetic and Natural Hybrid Coding (SNHC)*—develops coded representation of synthetic audio, graphics, and animations
- *Audio*—develops coded representation of audio of natural origin
- *Systems*—develops techniques for scene description, interactivity, as well as multiplexing/demultiplexing and presentation of moving images, audio, graphics, and data
- *Digital Media Integration Framework (DMIF)*—develops interfaces between digital storage media, networks, servers, and clients for delivery bitstreams in networked environments
- *Implementation Studies*—evaluates realizability of coding tools and techniques.

Although each of the subgroups has been given its own charter (indicated by their name), they work toward the common goal in a synchronized manner via shared technical documents and joint meetings. The typical process in MPEG development activity starts out with partial collection of requirements, definition of the PPD, and a call for technical proposals for evaluation. For instance, in video and audio subgroups, the submitted proposals undergo evaluation via subjective testing, objective analysis, and study of implementation aspects; this is often referred to as the *competitive* phase. At the end of the competitive phase, the top few proposals are selected and the collaborative phase begins and consists

Table 1 MPEG-4 Version 1 Workplan

Working draft	Committee draft	Final committee draft	Draft international standard	International standard
Nov. 1996	Nov. 1997	July 1998	Nov. 1998	Jan. 1999

of the development of reference coding description, which in case of MPEG-4 is called the verification model (VM) and is employed for evaluating the performance of competing tools via core experiments (CE) and for subsequent optimization of tools. A CE, when successful, replaces part of the VM (if a similar tool exists) or, in other cases, simply extends the VM. Thus, VMs undergo an iterative refinement; for instance, VMs in MPEG-4 video have undergone eight major revisions and the latest one is called VM8. However, the MPEG-4 standard only specifies bitstream and decoding semantics, and before becoming an International Standard, it undergoes a sequence of iterative Working Drafts, followed by a Committee Draft, a Final Committee Draft, and the Draft International Standard. At a recent MPEG-4 meeting, the more mature portions of the ongoing work were labeled as MPEG-4 Version 1, with the remaining portions to be released in a year as Version 2. In Table 1, we provide the schedule of MPEG-4 Version 1.

The MPEG-4 standard (ISO/IEC 14496) (16) is planned to consist of the following basic parts. Other parts may be added when the need is identified.

- ISO/IEC 14496-1: Systems
- ISO/IEC 14496-2: Visual (Natural and Synthetic Video)
- ISO/IEC 14496-3: Audio (Natural and Synthetic Audio)
- ISO/IEC 14496-4: Conformance
- ISO/IEC 14496-5: Software
- ISO/IEC 14496-6: DMIF

3.1. MPEG-4 Functionalities and Requirements

Now that we have some idea of the type of applications MPEG-4 is aimed at, we clarify the three basic functionality classes (3,8) that the MPEG-4 standard is addressing:

- *Content-Based Interactivity* allows the ability to interact with important objects in a scene. Currently, such interaction is, typically, only possible for synthetic objects; extending such an interaction to natural and hybrid synthetic/natural objects is important to enable new audio–visual applications.
- *Universal Accessibility* means the ability to access audio–visual data over a diverse range of storage and transmission media. Due to an increasing trend toward mobile communications, it is important that access be available to applications via wireless networks. This acceptable performance is needed over error-prone environments and at low bit rates.
- *Improved Compression* is needed to allow an increase in efficiency in transmission or decrease in the amount of storage required. For low-bit-rate applications, high compression is very important for enabling new applications.

Although we have looked at general classes of functionalities being addressed by MPEG-4, it is desirable to look at specific functionalities that MPEG-4 Version 1 expects

Table 2 Functionalities Expected to Be Supported by MPEG-4 Version 1

Content-Based Interactivity

Hybrid Natural and Synthetic Data Coding: The ability to code and manipulate natural and synthetic objects in a scene including decoder controllable methods of compositing of synthetic data with ordinary video and audio, allowing for interactivity.

Improved Temporal Random Access: The ability to efficiently access randomly in a limited time and with fine-resolution parts (frames or objects) within an audio–visual sequence. This also includes the requirement for conventional random access.

Content-Based Manipulation and Bitstream Editing: The ability to provide manipulation of contents and editing of audio–visual bitstreams without the requirement for transcoding.

Universal Access

Robustness in Error-Prone Environments: The capability to allow robust access to applications over a variety of wireless and wired networks and storage media. Sufficient robustness is required especially for low-bit-rate applications under severe error conditions.

Content-Based Scalability: The ability to achieve scalability with fine granularity in spatial, temporal, or amplitude resolution, quality, or complexity. Content-based scaling of audio–visual information requires these scalabilities.

Compression

Improved Coding Efficiency: The ability to provide subjectively better audio–visual quality at bit rates compared to existing or emerging video coding standards.

---

to offer; in Table 2, we show a list of six such functionalities (3,8,19) clustered into three functionality classes.

Besides the new functionalities, MPEG-4 is also supporting basic functionalities such as synchronization of audio and video, auxiliary data-stream capability, multipoint capability, low-delay mode, coding of a variety of audio types, interoperability with other audio–visual systems, support for interactivity, ability to efficiently operate in the 9.6–1024-kbit/s range, ability to operate in different media environments, and the ability to operate in a low-complexity mode.

The process of collection of requirements for MPEG-4, although it was started in late 1993, is continuing (14) at present, in parallel with other work items. This is because development of an MPEG standard is intricate, tedious, thorough, and, thus, time intensive (about 3–5 years per standard with overlap between standards). To keep up with marketplace needs for practical timely standards and to follow evolving trends, the requirements collection process is kept flexible. The major restructuring of the MPEG-4 effort in July 1994 to expand its scope was a response to the evolving trends in the marketplace. Evaluating requirements for MPEG-4 is an ongoing complex exercise that uses both a top-down (common requirements of related applications) and bottom-up approach (related functions provided by a tool that may be needed in various applications).

The collected requirements are clustered and translated into a set of individual requirements for MPEG-4 Video, Audio, SNHC, and System groups as general directions for developing coding methods/tools. In the advanced stages of development, clustering of coding tools takes place to define meaningful profiles (see Sec. 7) that could satisfy application clusters with similar requirements.

3.2. Tests and Evaluation

We now provide details of the testing and evaluation that took place in the competitive phase to determine the potential of the proposed technologies for MPEG-4. We also discuss how the outcome of tests and evaluation was used to initiate the collaborative phase.

3.2.1. Video Tests

The competitive phase was initiated with an open call for proposals in November 1994 (and subsequently revised (19)), inviting technical proposals for the first testing and evaluation (20) which took place in October 1995. A proposal package description (PPD) was developed describing the focus of MPEG-4, the functionalities being addressed, general applications MPEG-4 was aimed at the expected work plan, the planned phases of testing, information on verification models (VM) development, and the time schedule for MPEG-4. The MPEG-4 PPD, although started in November 1994, underwent successive refinements until July 1995. In parallel to the PPD development (8), a document describing the MPEG-4 Test/Evaluation Procedures was started in March 1995 and was iteratively refined until July 1995 (20).

The proposers were asked to submit either complete proposals for formal subjective testing or simpler tool proposals. As not all functionalities were tested in the first evaluation, tool proposals were invited for the other functionalities. In a few cases, proposers also used tool submissions as an opportunity to identify and separately submit the most promising components of their coding proposals. Because tools were not formally tested, they were evaluated by a panel of experts and this process was referred to as *evaluation by experts*.

The framework of the first evaluation (20) involved standardizing test material to be used in the first evaluation. Toward that end, video scenes are classified from relatively simple to more complex by categorizing them into three classes: Class A, Class B, and Class C. Two other classes of scenes, Class D and Class E, were defined; Class D contained stereoscopic video scenes and Class E contained hybrids of natural and synthetic scenes.

Because MPEG-4 is addressing many types of functionalities and different classes of scenes, it was found necessary to use three types of test methods. The first test method was Single Stimulus (SS) and involved rating the quality of coded scene on an 11-point scale from 0 to 10. The second test method was Double Stimulus Impairment Scale (DSIS) and involved presenting to assessors a reference scene (coded by a known standard) and after a 2-s gap, a scene coded by a candidate algorithm, with impairment of candidate algorithm compared to reference using a five-level impairment scale. The third test method was Double Stimulus Continuous Quality Scale (DSCQS) and involved presenting two sequences with a gap of 2 s. One of the two sequences was coded by the reference and the other was coded by the candidate algorithm, and blind tests performed. In the DSCQS method, a graphical continuous quality scale was used and was later mapped to a discrete representation on a scale of 0 to 100.

Table 3 summarizes the list of formal subjective tests (20), including an explanation of each test and the type of method employed.

To facilitate comparison of candidate proposals for MPEG-4 to the existing standards, the latter were used as anchors in the subjective tests. In tests involving Class A

Table 3 List of MPEG-4 First Evaluation Formal Tests and Their Explanations

Compression

Class A sequences at 10, 24, *and* 48 *kbits/s*: Coding to achieve the highest compression efficiency. Input video resolution is CCIR-601, and although any spatial and temporal resolution can be used for coding, the display format is CIF on a windowed display. The test method employed is SS.

Class B sequences at 24, 48, *and* 112 *kbits/s*: Coding to achieve the highest compression efficiency. Input video resolution is CCIR-601, and although any combination of spatial and temporal resolutions can be used for coding, the display format is CIF on a windowed display. The test method employed is SS.

Class C sequences at 320, 512, *and* 1024 *kbits/s*: Coding to achieve the highest compression efficiency. Input video resolution is CCIR-601, and although any combination of spatial and temporal resolution can be used for coding, the display format is CCIR-601 on a full display. The test method employed is DSCQS.

Error Robustness

Error resilience at 24 *kbits/s for Class A*, 48 *kbits/s for Class B, and* 512 *kbits/s for Class C*: Test with high random bit error rate (BER) of 10^{-3}, multiple burst errors with three bursts of errors with 50% BER within a burst, and a combination of high random bit errors and multiple burst errors. The display format for Class A and Class B sequences is CIF on a windowed display and for Class C sequences is CCIR-601 on full display. The test method employed for Class A and Class B is SS and that for Class C is DSCQS.

Error recovery at 24 *kbits/s for Class A*, 48 *kbits/s for Class B, and* 512 *kbits/s for Class C*: Test with long burst errors of 50% BER within a burst and a burst length of 1–2 s. Display format for Class A and Class B is CIF on a windowed display and Class C is CCIR-601 on full display. The test method employed for Class A and Class B is SS and that for Class C is DSCQS.

Scalability

Object scalability at 48 *kbits/s for Class A*, 320 *kbits/s for Class E, and* 1024 *kbits/s for Class B/C sequences*: Coding to permit dropping of specified objects resulting in remaining scene at lower than total bit rate; each object and the remaining scene is evaluated separately by experts. The display format for Class A is CIF on a windowed display and for Class B/C and Class E is CCIR-601 on a full display. The test method employed for Class A is SS; for Class B/C, it is DSCQS; and for Class E, it is DSIS.

Spatial scalability at 48 *kbits/s for Class A and* 1024 *kbits/s for Class B/C/E sequences*: Coding of a scene as two spatial layers with each layer using half of the total bit rate; however, full flexibility in choice of spatial resolution of objects in each layer is allowed. The display format for Class A is CIF on a windowed display and that for Class B/C/E is CCIR-601 on a full display. The test method employed for Class A is SS, and that for Class B/C/E is DSCQS.

Temporal scalability at 48 *kbits/s for Class A, and* 1024 *kbits/s for Class B/C/E sequences*: Coding of a scene as two temporal layers with each layer using half of the total bit rate; however, full flexibility in choice of temporal resolution of objects in each layer is allowed. The display format for Class A is CIF on a windowed display and that for Class B/C/E is CCIR-601 on a full display and Class B sequences, the H.263 standard (with TMN5-based coding) was used for coding the anchors; likewise, for Class C sequences the MPEG-1 standard was used as the anchor. To reduce the number of variables that could influence the outcome, the down-sampling and up-sampling filters were specified for conversion from input formats to lower-resolution formats used in coding. To facilitate scalability of arbitrary shaped objects (object scalability, spatial scalability, and temporal scalability), standardized segmentation masks were generated and used by all.

The proposers were required to submit D1 tapes of the coded results, detailed description of their proposal, coded bitstreams, and an executable version of decoder software. Although proposers were encouraged to participate in the entire set of tests listed in Table 3, they were allowed to participate in individual tests. About 34 proposers registered for the formal subjective tests. Also, about 40 tools submissions were received for evaluation by experts, but not formally tested. The results of the individual tests and a thorough analysis of the trends were made available (21) during the November 1995 MPEG meeting.

The results (22) of tests in various categories revealed that the anchors performed quite well, usually among the top three or four proposals. In several cases, the statistical difference between top performing proposals was insignificant. In a few specific cases, the new proposals outperformed the anchor or performed similarly but provided additional functionalities. It was also found that because spatial and temporal resolutions were not prefixed, there was some difficulty in comparing subjective and objective [signal-to-noise ratio (SNR)] results, due to differences in the choices made by each proposer. It seemed that subjective viewers had preferred higher spatial quality at the expense of temporal resolution. In addition to the results from subjective tests, the tools evaluation experts presented their results; about 16 tools or so were judged to be promising for further study.

Soon after the analysis of the results of the first subjective testing and evaluation in November 1995, the collaborative phase began by collecting tools for the purpose of defining a VM and core experiments. A second test and evaluation of proposals, scheduled for mid-1996, was divided into two parts: an extension of the first test which was held in January 1996 in the form of evaluation by experts, and a formal second test which was scheduled for July 1997. A few new proposers participated in the January 1996 evaluation and promising new tools were proposed.

3.2.2. Audio Tests

The MPEG-4 Audio also conducted subjective testing, similar to video. Three classes of audio test sequences, Class A, Class B, and Class C were identified:

- Class A: Single-source sequences consisting of a clean recording of a solo instrument
- Class B: Single source with background sequences consisting of a person speaking with background noise
- Class C: Complex sequences consisting of an orchestral recording

All sequences were originally sampled at 48 kHz with 16 bits/sample and were monophonic in nature. For generating reference formats, filters were specified to down-sample them to 24, 16, and 8 kHz. A number of bit rates such as 2, 6, 16, 24, 40, and 64 kbits/s were selected for testing of audio/speech. The first three bit rates are obviously only suitable for speech material. The audio test procedures used were as defined in ITU-R Recommendation 814.

The proposals submitted for testing included variants of MPEG-2 Advanced Audio Coding (AAC), variations of MPEG-1 audio coding, and new coding schemes. For specific bit rates, some candidates outperformed the reference coding schemes, although for all combinations tested, no single scheme was the clear winner.

After subjective testing, the collaborative work started and an initial MPEG-4 Audio VM was developed. MPEG-4 Audio development underwent a core experiments process similar to that of MPEG-4 Video development process.

3.2.3. SNHC Tests

The SNHC group started its work much later than the video group. Its focus was primarily on coding for storage and communication of 2-D and 3-D scenes involving synthetic images, sounds, and animated geometry and its integration into scenes that contain coded natural images/video and sound. Further, it was sought that the coded representation should also facilitate various forms of interactions.

In its Call for Proposals (23) and PPD (22), it sought proposals allowing efficient coding and interactivity in the following areas:

- Compression and simplification of synthetic data representations—synthetic and natural texture, panoramic views, mapping geometry, mapping photometry, animation, and deformation
- Parameterized animated models—encoding of parameterized models and encoding of parameter streams
- New primitive operations for compositing of natural and hybrid objects
- Scalability—extraction of subsets of data for time-critical use and time-critical rendering
- Real-time interactivity with hybrid environments
- Modeling of timing and synchronization
- Synthetic Audio

In the competitive phase, for the purpose of standardized evaluation, a database of a test data set was established. The actual evaluation of proposals by a group of experts took place in September 1996. The evaluation criteria was based on the functionality addressed, such as coding efficiency, quality of decoded model, real-time interactivity, anticipated performance in the future, and implementation cost. Similar to the tests/evaluations in video, each proposer was required to submit the technical description detailing scope advantages, details and statistics, coded bitstreams and an executable software decoder, a D1 tape showing results, and simplification or modification of test data.

At the time, due to participation by a small number of organizations, only a limited number of topics were covered. After the evaluation, the collaborative phase was begun by harmonizing the selected proposals and tools for definition of the first version of SNHC VM and a number of core experiments. The SNHC VM included visual and the audio tools addressing one or more aspects of synthetic data from the perspective of compression, scalability, interactivity, and other functionalities. The SNHC effort is thus expected to contribute to tools and algorithms in Parts 2 and 3 of the MPEG-4 standard. Further, at that time, it was expected that MPEG-4 systems would provide the framework needed for composing decoded natural and synthetic objects in the same scene (see Sec. 6 for the extensive scene description capabilities of MPEG-4 systems).

3.3. Video Development

In the period from November 1995 through January 1996, the process of the definition of core experiments was initiated. A total of 36 core experiments were defined prior to the January 1996 MPEG meeting and another 5 experiments were added at the meeting, bringing the total number of experiments (24) to 41. These experiments were classified into a number of topics and four ad hoc groups were formed to coordinate the core experiment process; each ad hoc group was assigned one or more topics as follows:

- Coding Efficiency—prediction, frame texture coding, quantization and rate control.
- Shape and Object Texture Coding—binary and grey scale shape coding, object texture coding
- Robust Coding—error resilience and error concealment
- Multifunctional Coding—bandwidth and complexity scalability, object manipulation, pre and post processing

The result of work of ad hoc group (25) on defining the VM resulted in the first MPEG-4 Video VM (VM1) and was released on 24 January 1996. It supported the following features:

- Coding of arbitrary shaped objects using Video object planes (VOPs)
- Coding of binary and gray-scale shape of arbitrary shaped objects
- Padding of pixels to fill the region outside of the object to full blocks for motion compensation and DCT
- Macroblock-based motion-texture (motion-compensated DCT) coding derived from H.263
- A mode allowing separation of motion and texture data for increased error resilience

The ad hoc groups undertook the responsibility of producing a detailed description of core experiments, finding organizations to independently verify results, and finalize the experimental conditions. The purpose of the initial series of core experiments was to either offer alternative tools allowing higher coding performance or extra needed functionality compared to VM1. Based on results of core experiments and/or to satisfy additional needed functionality not included in VM1, during the March 1996 MPEG meeting a number of additional features were added to VM1, and thus VM2 (23) was released on 29 March 1996. The additional features were as follows:

- Bidirectional VOPs derived from combination of H.263 PB-frames mode and MPEG-1/2 B-pictures
- DC coefficients prediction for intra macroblocks as per MPEG-1/2
- Extended motion vector range
- Quantization visibility matrices as per MPEG-1/2

Since then, there have been six more iterations on the video VM and the process of iterative development and refinement of video VMs via core experiments has continued. At the July 1997 meeting, a number of mature tools from VM7 (26) have been accepted for MPEG-4 Version 1, which is expected to become the Committee Draft in November 1997. The remaining tools of VM7 with tools added at that meeting will be considered for MPEG-4 Version 2. In Sec. 4, we describe the basic coding methods formed by tools accepted for Part 2 of the MPEG-4 Version 1 standard.

3.4. Audio Development

The MPEG-4 Audio coding effort occurred in parallel with the MPEG-2 AAC (formerly, Non-Backward Compatible—NBC) coding effort. The MPEG-2 standard originally had an audio coding mode called the backward compatible (BC) mode which, as the name suggests, was backward compatible with MPEG-1 Audio coding. However, at a late stage in MPEG-2 it was discovered that the BC audio coding was rather inefficient compared to noncompatible solutions and thus work on the NBC mode was begun and overlapped with the MPEG-4 schedule. The NBC mode was renamed AAC and became a new part of MPEG-2, achieving International Standard status in April 1997 (although it had reached a mature status in mid-1996).

Toward the very low-bit-rate end, a valid question to ask is why not use the existing ITU-T coders? As an answer to this question, the ITU-T speech coders currently operate at 6.3/5.3 kbits/s (G.723), 8 kbits/s (G.729), 16 kbits/s (G.728), 32 kbits/s (G.721), and 48/56/64 kbits/s (G.722). In comparison, MPEG-4 Speech coding is being designed to operate at bit rates between 2 and 24 kbits/s for the 8-kHz mode and 14–24 kbits/s for the 16-kHz mode, whereas ITU-T coders do not operate at bit rates as low as 2 kbits/s for the 8-kHz mode, or 14–24 kbits/s for the 16-kHz mode. Furthermore, MPEG-4 Speech coders are being designed for bit-rate scalability, complexity scalability, and multi-bit-rate operation from 2 to 24 kbits/s. The coding quality of the coder is comparable to that of the ITU coder at corresponding bit rates. MPEG-4 is standardizing a speech coder which can operate down to 2 kbits/s. This will be the lowest-bit-rate international standard. ITU standards do not support this low bit rate. The quality at 2 kbits/s is "communication quality" and could be used for usual conversation, performing better than a FS1016 4.8-kbits/s coder.

Therefore, the MPEG-4 Audio VMs have targeted bit rates from 2 to 64 kbits/s; a number of coding schemes are used to cover portions of this range. Besides coding efficiency, content-based coding of audio objects and scalability are being investigated. There have been a total of four iterations of audio VM, from VM1 to VM4; the last VM was released in July 1997. In fact, the more mature tools of Audio VM3 have been accepted for the audio part (13) of the MPEG-4 Version 1 standard.

In Sec. 5, we briefly discuss the basic coding techniques accepted for Part 3 of the MPEG-4 Version 1 standard.

3.5. SNHC Development

There have been a total of four iterations of SNHC VM, from VM1 to VM4; the last VM was released in July 1997. In fact, the more mature tools of SNHC VM3 have been accepted for the visual part of the MPEG-4 Version 1 standard; the remaining tools have been left in VM4 for consideration for next version of MPEG-4.

In Sec. 4, we describe the SNHC tools expected to be included in the visual part of the MPEG-4 Version 1 standard. In Sec. 5, we discuss SNHC tools expected to be included in the audio part of the MPEG-4 Version 1 standard.

3.6. Systems Development

The Systems layer in MPEG has been traditionally responsible for integrating media components into a single system, providing multiplexing and synchronization services for audio and video streams.

In MPEG-2 (38), for example, these are the primary functionalities and were designed for two types of transport facility. The first, Program Stream, is intended for reliable media such as storage devices, and can only carry a single program (combinations of synchronized audio and video streams). It is a direct extension of MPEG-1 Systems (6). The second, Transport Stream, is intended for potentially unreliable media and can carry multiple programs. This is shown in Fig. 5. The object-based nature of MPEG-4 necessitates a much more complex Systems layer because, in addition to still addressing multiplexing and synchronization, it must also provide for ways to combine simple audio or visual objects into meaningful scenes.

The Systems specification has a long history of evolutionary development (3,27–30), starting from the very early stages of MPEG-4 in 1994. Initially, the MPEG-4 project was investigated within the Applications and Operating Environments Group (AOE). The focus of the project was to examine how one could substantially change the paradigm of creation and delivery of audio–visual content and break away from the limitations of frame- or pixel-based content. New terminal architectures were investigated, favoring programmable architectures that could potentially provide a very high degree of flexibility for application and content developers. It was foreseen that different components of such a terminal could be made to work together by using a special language, termed MPEG-4 Syntactic Description Language (MSDL). This language would describe the syntax of a bitstream and allow different "tools" to be combined together in various ways to form "algorithms," which would perform particular coding tasks. It is interesting to note that these concepts were articulated before Java became widely known. A Syntactic Description Language, extending C++ and Java, was introduced in November 1995 and underwent several revisions (27,31–33).

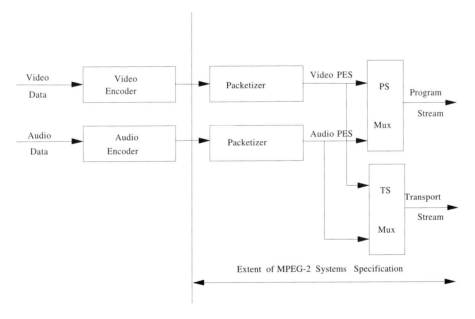

Fig. 5 MPEG-2 Systems.

Considering the object-based nature of MPEG-4, a key requirement from the Systems part is the capability to combine individual audio–visual objects in scenes. In late 1995, this was accomplished by using Java (34). Issues of performance and compliance soon arose. Clearly, it is essential for a content creator to be assured that the content generated will be shown in an identical (or nearly so) way regardless of the terminal used, if both such terminals comply to the standard. A three-step approach was adopted, involving three different flexibility levels, as shown in Fig. 6. In Level 0, no programmability was allowed. In Level 1, facilities were provided to combine different tools into algorithms, whereas in Level 2, even individual tools were considered as targets for programmable behavior. The group was also renamed MPEG-4 System and Description Languages, separating system and syntactic description (27). After further examination, in late 1996 it was decided that any meaningful operation of Level 1 would require the complexity of implementing a Level 2 system, and hence this intermediate level was eliminated.

The group subsequently focused on a parametric (bitstream oriented, nonprogrammable) solution for describing how objects should be combined together. Using Fig. 6, that would be a Level 0 design. The group also reverted to the use of the traditional term "Systems," reflecting the varied components that it addresses. A programmable approach is still being considered and is discussed in more detail in Sec. 6.2.4.

In addition to the overall architectural issues, the issue of multiplexing in MPEG-4 also underwent several stages of evolution. The H.223 Annex A multiplexer was used as a basis, including error protection tools (interleaving and ARQ). A key requirement (14), however, for MPEG-4 was the need to be transport independent. As a result, services that belong to a transport layer were subsequently removed from the set of specified tools, so that efficient implementation of MPEG-4 systems could be performed in a very broad range of environment (broadcast, ATM, IP, and wireless).

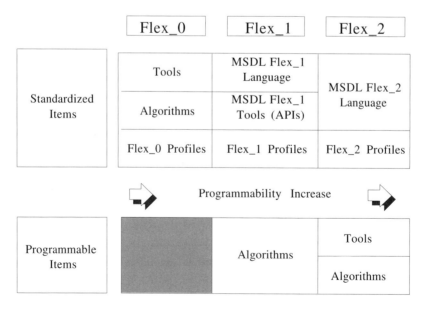

Fig. 6 Evolution of MPEG-4 Systems Architecture (1996).

3.7. DMIF Development

At a recent MPEG meeting, the significance of DMIF activity has been recognized and DMIF has been given the status of a new group (35). Previously, DMIF was an ad hoc group operating under the systems group. The charter of DMIF is to develop standards for interfaces between Digital Storage Media (DSM), networks, servers, and clients for the purpose of managing DSM resources and controlling the delivery of bitstreams and associated data. The ongoing work of this group is expected to result in Part 6 of the MPEG-4 standard.

4. MPEG-4 VISUAL

The ongoing work on MPEG-4 Visual standard specification (12) consists of tools and methods from two major areas—coding of (natural) video and coding of synthetic video (visual part of the SNHC work). We address both these areas; in Secs. 4.1 through 4.4, we discuss tools and techniques relevant to natural video coding, and in Secs. 4.5 through 4.8, we discuss tools and techniques relevant to synthetic video coding.

4.1. MPEG-4 Video Coding Basics

In this and the next subsection, we describe the coding methods and tools of MPEG-4 Video; the encoding description is borrowed from Video VM7 (26), the decoding description follows (12). An input video sequence can be defined as a sequence of related snapshots or pictures, separated in time. In MPEG-4, each picture is considered as consisting of temporal instances of objects that undergo a variety of changes such as translations, rotations, scaling, brightness, color variations, and so forth, Moreover, new objects enter a scene and/or existing objects depart, leading to the presence of temporal instances of certain objects only in certain pictures. Sometimes, scene change occurs, and thus the entire scene may either get reorganized or replaced by a new scene. Many of MPEG-4 functionalities require access not only to an entire sequence of pictures but also to an entire object, and further, not only to individual pictures but also to temporal instances of these objects within a picture. A temporal instance of a video object can be thought of as a snapshot of arbitrary shaped object that occurs within a picture, such that, like a picture, it is intended to be an access unit, and, unlike a picture, it is expected to have a semantic meaning.

The concept of video objects (VOs) and their temporal instances, video object planes (VOPs), is central to MPEG-4 Video. A VOP can be fully described by texture variations (a set of luminance and chrominance values) and (explicit or implicit) shape representation. In natural scenes, VOPs are obtained by semiautomatic or automatic segmentation, and the resulting shape information can be represented as a *binary shape* mask. On the other hand, for hybrid (of natural and synthetic) scenes generated by blue screen composition, shape information is represented by an 8-bit component, referred to as the *gray-scale shape*. In Fig. 7, we show the decomposition of a picture into a number of separate VOPs. The scene consists of two objects (head and shoulders view of a human, and a logo) and the background. The objects are segmented by semiautomatic or automatic means and are referred to as VOP1 and VOP2, where as the background without these objects is referred to as VOP0. Each picture in the sequence is segmented into VOPs in this manner. Thus,

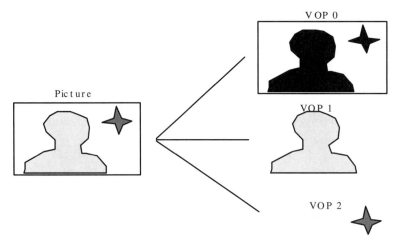

Fig. 7 Semantic segmentation of a picture into VOPs.

a segmented sequence contains a set of VOP0s, a set of VOP1s and a set of VOP2s; in other words, in our example, a segmented sequence consists of VO0, VO1, and VO2.

Each VO is encoded separately and multiplexed to form a bitstream that users can access and manipulate (cut, paste, etc.). The encoder sends, together with VOs, information about scene composition to indicate where and when VOPs of a VO are to be displayed. However, this information is optional and may be ignored at the decoder which may use user specified information about composition. As scene description is provided by the Systems layer (see Sec. 6), this functionality is expected to be removed from the video bitstream definition.

In Fig. 8, we show a high-level logical structure of a VO-based coder. Its main components are a VO Segmenter/Formatter, VO Encoders, Systems Multiplexer, VO De-

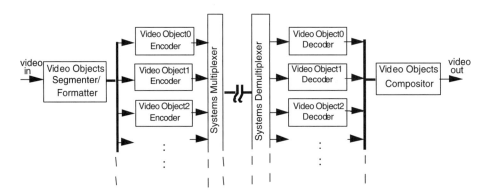

Fig. 8 Logical structure of video-object-based codec of MPEG-4 Video.

MPEG-4: Standard for Multimedia Coding

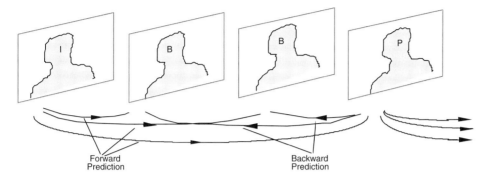

Fig. 9 An example prediction structure when using I-, P-, and B-VOPs.

coders, and a VO Compositor. The VO Segmenter segments the input scene into VOs for encoding by VO Encoders. The coded data of various VOs is multiplexed for storage or transmission; they are then demultiplexed and decoded by VO decoders and offered to the compositor, which composes and renders the decoded scene.

To consider how coding takes place in a video object encoder, consider a sequence of VOPs. Now, extending the concept of intra (I) pictures, predictive (P), and bidirectionally predictive (B) pictures of MPEG-1/2 to VOPs, I-VOP, P-VOP and B-VOP result. If two consecutive B-VOPs are used between a pair of reference VOPs (I- or a P-VOP), the resulting coding structure is as shown in Fig. 9.

In Fig. 10, we show the internal structure of the VM-based encoder which codes a number of VOs of a scene. Its main components are Motion Coder, Texture Coder, and

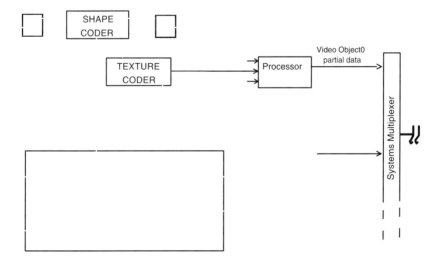

Fig. 10 Detailed structure of video objects encoder.

Shape Coder. The Motion Coder uses macroblock- and block-motion estimation and compensation, similar to H.263 and MPEG-1/2 but modified to work with arbitrary shapes. The Texture Coder uses block DCT coding based on H.263 and MPEG-1/2 but much better optimized; further, it is also adapted to work with arbitrary shapes. An entirely new component is the Shape Coder. The partial data of VOs (such as VOPs) is buffered and sent to the Systems Multiplexer.

From a top-down perspective, the organization of coded MPEG-4 Video data can be described by the following class hierarchy:

- Video Session: A video session represents the highest level in the class hierarchy and simply consists of an ordered collection of video objects. This class has only been a place holder for video VM and core experiments work and, because the composition of objects is now handled by systems, it is not needed.
- Video Object: A video object (two dimensions + time) represents a complete scene or a portion of a scene with a semantic meaning.
- Video Object Layer (VOL): A video object layer (two dimensions + time) represents various instantiations of an video object. For instance, different VOLs may correspond to different layers, such as in the case of scalability.
- Group of Video Object Planes (GOV): A group of video object planes consists of optional entities and is essentially access units for editing, tune-in, or synchronization.
- Video Object Plane (VOP): A video object plane represents a snapshot in time of a video object. A simple example may be an entire frame or a portion of a frame. Different coding methods from MPEG-1/2 such as intra (I) coding, predictive (P) coding and bidirectionally predictive (B) coding can now be applied to VOPs.

The class hierarchy used for representation of coded bitstream described above is shown by the tree structure of Fig. 11.

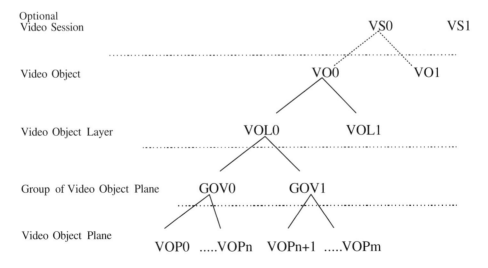

Fig. 11 Class hierarchy for structuring coded video data.

4.2. Video Coding Details

4.2.1. Binary Shape Coder

Compared to other standards, the ability to represent arbitrary shapes is an important capability of the MPEG-4 video standard. In general, shape representation can be either implicit (based on chroma-key and texture coding) or explicit (boundary coding separate from texture coding). Implicit shape representation, although it offers less encoding flexibility, can result in quite usable shapes while being relatively simple and computationally inexpensive. Although explicit shape representation can offer flexible encoding and somewhat better quality shapes, it is more complex and computationally expensive. Regardless of the implications, the explicit shape representation was chosen in MPEG-4 Video; we now briefly describe the essence of this method (12,26) without its many details.

For each VO given as a sequence of VOPs of arbitrary shapes, the corresponding sequence of binary alpha planes is assumed to be known (generated via segmentation or via chroma-key). For the binary alpha plane, a rectangular bounding box enclosing the shape to coded is formed such that its horizontal and vertical dimensions are multiples of 16 pixels (macroblock size). For efficient coding, it is important to minimize the number of macroblocks contained in the bounding box. The pixels on the boundaries or inside the object are assigned a value of 255 and are considered opaque, whereas the pixels outside the object but inside the bounding box are considered transparent and are assigned a value of 0. If a 16×16 block structure is overlaid on the bounding box, three types of binary alpha blocks exist: completely transparent, completely opaque, and partially transparent (or partially opaque). Figure 12 shows an example of an arbitrary shape VOP with a bounding box and the overlaid 16×16 block structure: the opaque area is shown shaded, whereas the transparent area is shown unshaded.

Coding of each 16×16 binary alpha block representing shape can be performed either lossy or losslessly. The degree of lossiness of coding the shape of a video object is controlled by a threshold which can take values of 0, 16, 32, ..., 256. The higher the value of this threshold, the more lossy the shape representation; a zero value implies lossless shape coding. Within the global bound of specified lossiness, local control, if needed, can be exerted by selecting a maximum subsampling factor on a 16×16 binary alpha that results in just acceptable distortion. The estimation of this factor is iterative and consists of using the same subsampling factor in both dimensions and determining the acceptability of the resulting shape quality. To be specific, a 4:1 down-sampled binary

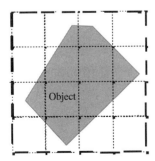

Fig. 12 A VOP in a bounding box.

alpha block is used first and if the shape errors are higher than acceptable, a 2:1 downsampled binary alpha block is used next; again, if it is found unacceptable, an unsubsampled binary alpha block is used.

Further, each binary alpha block can be coded in intra mode or in inter mode, similar to coding of texture macroblocks. In the intra mode, no explicit prediction is performed. In inter mode, shape information is differenced with respect to the prediction obtained using a motion vector, the resulting binary shape prediction error may or may not be coded. The motion vector of a binary alpha block is estimated at the encoder by first finding a suitable initial candidate from among the motion vectors of three previously decoded surrounding texture macroblocks as well as the three previously decoded surrounding shape binary alpha blocks. Next, the initial candidate is either accepted as the shape motion vector or is used as the starting basis for a new motion vector search, depending on whether the resulting prediction errors of the initial motion vectors are below a threshold. The motion vector is coded differentially and included in the bitstream. Following this procedure, a binary alpha block is assigned a mode from among the following choices:

1. Zero differential motion vector and no inter shape update
2. Nonzero differential motion vector and no inter shape update
3. Transparent
4. Opaque
5. Intra shape
6. Zero differential motion vector and inter shape update
7. Nonzero differential motion vector and inter shape update

Depending on the coding mode and whether it is an I-, P-, or B-VOP, a variable-length code word is assigned, identifying the coding type of the binary alpha block. The entropy coding of shape data is performed by using the context information determined on a pixel basis to drive an adaptive arithmetic coder. The pixels of binary alpha blocks are raster scanned; a binary alpha block may, however, be transposed. Next, a context number is determined and is used to index a probability table, and further, the indexed probability is used to drive the arithmetic coder. To determine the context, different templates of surrounding pixels are used for intra and inter coded binary alpha blocks, as shown in Fig. 13.

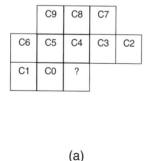
(a)

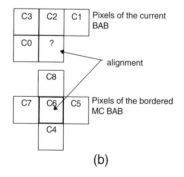
(b)

Fig. 13 Pixel templates used for (a) intra and (b) inter context determination of a binary alpha block (BAB). Pixel to be coded is marked with ?.

The decoding of binary alpha blocks follows the inverse sequence of operations with the exception of encoder specific information such as motion estimation, subsampling factor determination, mode decision, and so forth which are readily extracted from the coded bitstream.

4.2.2. Motion Coder

The Motion Coder (12,26) consists of a motion estimator, motion compensator, previous/next VOPs store, and motion vector (MV) predictor and coder. In case of P-VOPs, the motion estimator computes motion vectors using the current VOP and the temporally previous reconstructed VOP available from the previous reconstructed VOPs store. In the case of B-VOPs, the motion estimator computes motion vectors using the current VOP and the temporally previous reconstructed VOP from the previous reconstructed VOP store, as well as the current VOP and temporally next VOP from the next reconstructed VOP store. The motion compensator uses these motion vectors to compute motion-compensated prediction signals using the temporally previous reconstructed version of the same VOP (reference VOP). The MV predictor and coder generate prediction for the MV to be coded. We now discuss the details of padding needed for motion compensation of arbitrary shaped VOPs, as well as the various modes of motion compensation allowed.

In the reference VOP, based on its shape information, two types of macroblocks require padding: those that lie on the boundary and (depending on encoding choice, some or all of) those that lie outside of the VOP. Macroblocks that lie on the VOP boundary are padded by first replicating the boundary pixels in the horizontal direction, followed by replicating the boundary pixels in the vertical direction, making sure that if a pixel can be assigned a value by both horizontal and vertical padding, it is assigned an average value. Next, the macroblocks that lie outside of the VOP are padded by extending the boundary macroblock pixels, up, down, left, and right, and averaging wherever a pixel is assigned a value from more than one directions. The processing for the previous step can be reduced by only padding macroblocks that are outside of the VOP but right next to the boundary pixels.

The basic motion estimation and compensation is performed on a 16×16 luminance block of a macroblock. The motion vector is specified to half-pixel accuracy. The motion estimation is performed by a full search to an integer-pixel accuracy vector and using it as the initial estimate, a half-pixel search is performed around it. The luminance block-motion vector is scaled by a factor of 2 for each component and rounded for use on 8×8 chrominance blocks.

MPEG-4 video, like H.263, supports an unrestricted range for motion estimation and compensation. Basically, motion vectors are allowed to point out of the VOP bounding box, by extending the reference VOP bounding box in all four directions. Further, a larger range of motion vectors is supported for motion vector coding in MPEG-4 as compared to H.263.

Often a single motion vector for a 16×16 luminance block does not reduce the prediction errors sufficiently, or when dealing with boundary macroblocks, motion vectors can be sent for individual 8×8 blocks. Further, the 8×8 block-motion vectors are used to generate an overlapped block-motion-compensated prediction. Both the 8×8 block-motion compensation and overlapped-motion-compensated prediction are referred to as advanced prediction in H.263 and are adapted in MPEG-4 to work with arbitrary shaped VOPs.

An intra versus inter coding decision is performed to determine if motion vector(s) need to be sent for the macroblock being coded. Further, a decision is also performed to determine if 16×16 or 8×8 block-motion vectors will be sent for the macroblock being coded. All motion vectors are coded differentially using the median of the neighboring three decoded macroblock- (or block in case of 8×8 coding) motion vectors as the prediction.

As mentioned earlier, a B-VOP is a VOP which is coded bidirectionally. For example, macroblocks in a B-VOP can be predicted using the forward, the backward, or both the forward and backward motion vectors. This has similarities to MPEG-1/2 in which B-pictures can use such motion vectors. However, MPEG-4 Video also supports an H.263-based mode for motion compensation, referred to as the direct mode. In the direct mode, the motion vector for a macroblock in a B-VOP is obtained by scaling of the P-VOP motion vector, and further correcting it by a small (delta) motion vector. The actual motion-compensation mode to be used for a macroblock is decided by taking into account the motion-compensated prediction errors produced by various choices and the coding overhead of any additional motion vectors. All motion vectors (except delta) are coded differentially with respect to motion vectors of the same type.

MPEG-4 also supports efficient coding of interlaced video. It combines the macroblock-based frame/field motion compensation of MPEG-2 with the normal motion compensation of MPEG-4, resulting in overall improved motion compensation. Furthermore, it allows motion compensation of arbitrary shaped VOPs of interlaced video, whereas MPEG-2 only supports rectangular pictures of interlaced video.

4.2.3. Video Texture Coder

The Texture Coder (12,26) codes the luminance and chrominance variations of blocks forming macroblocks within a VOP. Two types of macroblocks exist: those that lie inside the VOP and those that lie on the boundary of the VOP. The blocks that lie inside the VOP are coded using DCT coding similar to that used in H.263 but optimized in MPEG-4. The blocks that lie on the VOP boundary are first padded and then coded similar to the block that lies inside the VOP. The remaining blocks are transparent (they lie inside the bounding box but outside of the coded VOP shape) and are not coded at all.

The Texture Coder uses block DCT coding and codes blocks of size 8×8 similar to H.263 and MPEG-1/2, with the difference that because VOP shapes can be arbitrary, the blocks on the VOP boundary require padding prior to texture coding. The general operations in the texture encoder are DCT on original or prediction error blocks of size 8×8, quantization of 8×8 block DCT coefficients, scanning of quantized coefficients, and variable-length coding of quantized coefficients. For inter (prediction error block) coding, the texture coding details are similar to that of H.263 and MPEG-1/2. However, for intra coding of texture data, a number of improvements are included. We now discuss the quantization for intra and inter macroblocks, followed by coefficient prediction, scanning, and entropy of intra macroblocks, and, finally, the entropy coding of inter blocks.

Typically, the DC coefficients of DCT blocks belonging to an intra macroblock, are scaled by a constant scaling factor of 8. However, in MPEG-4 video, a nonlinear scaler (36) as per Table 4 is used to provide a higher coding efficiency while keeping the blockiness artifacts under the visibility threshold. The characteristics of nonlinear scaling are different between the luminance and chrominance blocks and further depend on the quantizer used for the block.

Table 4 Nonlinear Scaler for DC Coefficients of DCT Blocks

Component	dc_scaler for Quantizer (Qp) Range			
	1–4	5–8	9–24	25–31
Luminance	8	2Qp	Qp + 8	2Qp − 16
Chrominance	8		(Qp + 13)/2	Qp − 6

MPEG-4 Video supports two techniques of quantization, one referred to as the H.263 quantization method (with a dead zone for intra and inter), and the other, the MPEG quantization method (no dead zone for intra but uses dead zone for inter and intra quantization matrices). Further, the quantization matrices are downloadable as in MPEG-1/2, but with the difference that it is possible to update matrices partially.

Unlike H.263, the quantized intra DC coefficients are predicted (36) with respect to three previous decoded DC coefficients (e.g., quantized DC coefficients of blocks A, B, and C when predicting a quantized DC value for block X in Fig. 14). Although MPEG-1/2 also allows the prediction of DC coefficients, the gradient-based prediction of MPEG-4 is more effective. In computing the prediction for block X, if the absolute value of a horizontal gradient ($|QDC_A - QDC_B|$) is less than the absolute value of a vertical gradient ($|QDC_B - QDC_C|$), then the QDC value of block C is the prediction, otherwise, the QDC value of block A is used as a prediction. This process is repeated independently for every block of an intra macroblock using horizontally and vertically adjacent blocks. Further, the procedure is identical for luminance and chrominance blocks.

Not only are the DC coefficients of intra blocks predicted and coded differentially, so are some of the AC coefficients (36). In particular, on a block basis, either the first row or the first column of AC coefficients of DCT blocks of each intra macroblock are predicted. The direction (horizontal or vertical) used for prediction of the DC coefficient of a block is also used for predicting the corresponding first column or row of AC coefficients. The prediction direction can differ from block to block within each intra macroblock. Further, the AC coefficient prediction can be disabled for a macroblock when it does not work well. Figure 15 shows the prediction of quantized AC coefficients belonging to the first column or the first row of block X from the corresponding quantized AC coefficients of block A or C.

The predicted DC and AC coefficients (as well as the unpredicted AC coefficients) of DCT blocks of intra macroblocks are scanned by one of the three scans (36); alternate-

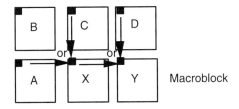

Fig. 14 Prediction of DC coefficients of blocks in an intra macroblock.

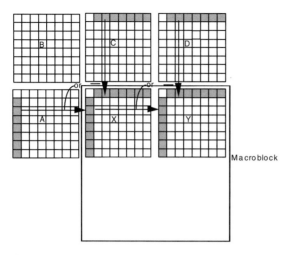

Fig. 15 Prediction of AC coefficients of blocks in an intra macroblock.

horizontal, alternate-vertical (MPEG-2 interlace scan), and the zigzag scan (normal scan used in H.263 and MPEG-1). The actual scan used depends on the coefficient predictions used. For instance, if the AC coefficient prediction is disabled for a intra macroblock, all blocks in that macroblock are zigzag scanned. If AC coefficient prediction is enabled and DC coefficient prediction was selected from the horizontally adjacent block, alternate-vertical scan is used. Likewise, if AC coefficient prediction is enabled and DC coefficient prediction was selected from the vertically adjacent block, an alternate-horizontal scan is used. Figure 16 shows the three scans used.

A three-dimensional variable-length code is used to code the scanned DCT events of intra blocks. An event is a combination of three items (last, run, level). The ''last'' indicates if a coefficient is the last nonzero coefficient of a block or not, the ''run'' indicates the number of zero coefficients preceding the current nonzero coefficient, and ''level'' indicates the amplitude of the quantized coefficient.

The DCT coefficients of inter blocks, unlike DCT coefficients of intra blocks, do not undergo any prediction or adaptive scanning; in fact, they use the fixed zigzag scan of Fig. 16c. The scanned coefficients of inter blocks are also coded by a three-dimensional variable-length code table with similar structure as the intra variable-length code table but with code entries optimized for inter statistics.

0	1	2	3	10	11	12	13
4	5	8	9	17	16	15	14
6	7	19	18	26	27	28	29
20	21	24	25	30	31	32	33
22	23	34	35	42	43	44	45
36	37	40	41	46	47	48	49
38	39	50	51	56	57	58	59
52	53	54	55	60	61	62	63

(a)

0	4	6	20	22	36	38	52
1	5	7	21	23	37	39	53
2	8	19	24	34	40	50	54
3	9	18	25	35	41	51	55
10	17	26	30	42	46	56	60
11	16	27	31	43	47	57	61
12	15	28	32	44	48	58	62
13	14	29	33	45	49	59	63

(b)

0	1	5	6	14	15	27	28
2	4	7	13	16	26	29	42
3	8	12	17	25	30	41	43
9	11	18	24	31	40	44	53
10	19	23	32	39	45	52	54
20	22	33	38	46	51	55	60
21	34	37	47	50	56	59	61
35	36	48	49	57	58	62	63

(c)

Fig. 16 Scans for intra blocks: (a) alternate-horizontal; (b) alternate-vertical; (c) zigzag.

Finally, as mentioned earlier, MPEG-4 also supports efficient coding of interlaced video. It combines the macroblock-based frame/field DCT coding of MPEG-2 with the improved DC coefficient coding, quantization, scanning, and variable-length coding of normal MPEG-4 Video coding resulting in improved coding efficiency. Furthermore, it allows DCT coding of arbitrary shaped VOPs of interlaced video, whereas MPEG-2 only supports rectangular pictures of interlaced video.

4.2.4. Sprite Coding

In computer games, a sprite refers to a synthetic object that undergoes some form of transformation (including animation). Also, in connection with a highly efficient representation of natural video, the term "mosaic" or "world image" is used in the literature to describe a large image built by integration of many frames of a sequence spatially and/or many frames of a sequence temporally. In MPEG-4 terminology, such an image is referred to as a static sprite. Static sprites can improve the overall coding efficiency; for example, by coding the background only once and warping it to generate the rendition required at a specific time instance.

A static sprite (12,26) is usually built off-line and can be used to represent synthetic or natural objects. It is quite suitable for natural objects that undergo rigid motion and where a wallpaper like rendering is sufficient. One of the main components in coding using natural sprites is generation of the sprite itself. For generating a static sprite, the entire video object is assumed to be available. For each VOP in the VO, the global motion is estimated according to a transformation model (say, perspective transformation) using a VOP which is then registered with the sprite by warping the VOP to the sprite coordinate system. Finally, the warped VOP is blended with the sprite which is used for the estimation of the motion of the subsequent VOP.

A number of choices regarding the transformation models exist such as stationary, translation, magnification–rotation–translation, affine, and perspective transformation. Each transformation can be defined as either a set of coefficients or the motion trajectories of some reference points; the former is convenient for performing the transformations, whereas the later is used for encoding the transformations. If four reference points are used, perspective transformation can be employed for warping and is defined by

$$x' = \frac{(ax + by + c)}{(gx + hy + 1)}$$

$$y' = \frac{(dx + ey + f)}{(gx + hy + 1)}$$

where $\{a, b, c, d, e, f, g, h, l\}$ are the coefficients of the transformation and (x, y) is one of the reference points of interest in the current VOP which corresponds to point (x', y') in the sprite, expressed in the sprite coordinate system.

Once the sprite is available, global motion between the current VOP and the sprite is estimated, using the perspective transform, for example. The reconstructed VOPs are generated from the sprite by directly warping the quantized sprite using specified motion parameters. Residual error between the original VOP and the warped sprite is not sent.

4.3. Scalable Video Coding

Scalability of video is the property that allows a video decoder to decode portions of the coded bitstream to generate decoded video of quality commensurate with the amount of data decoded. In other words, scalability allows a simple video decoder to decode and produce basic quality video, whereas an enhanced decoder may decode and produce enhanced quality video, all from the same coded video bitstream. This is possible because scalable video encoding ensures that input video data is coded as two or more layers, an independently coded base layer, and one or more enhancement layers coded dependently, thus producing scalable video bitstreams. The first enhancement layer is coded with respect to the base layer, the second enhancement layer with respect to the first enhancement layer, and so forth.

Scalable coding offers a means of scaling the decoder complexity if processor and/or memory resources are limited and often time varying. Further, scalability also allows graceful degradation of quality when the bandwidth resources are also limited and continually changing. It also allows increased resilience from errors under noisy channel conditions.

MPEG-4 offers a generalized scalability (12,26,36) framework supporting both the temporal and the spatial scalabilities (the primary type of scalabilities). Temporally scalable encoding offers decoders a means to increase temporal resolution of decoded video using decoded-enhancement-layer VOPs in conjunction with decoded-base-layer VOPs. Spatial scalability encoding, on the other hand, offers decoders a means to decode and display either the base layer or the enhancement layer output; typically, because the base layer uses one-quarter resolution of the enhancement layer, the enhancement layer output provides the better quality, albeit requiring increased decoding complexity. The MPEG-4 generalized scalability framework employs modified B-VOPs that only exist in an enhancement layer to achieve both temporal and spatial scalability; the modified enhancement layer B-VOPs use the same syntax as normal B-VOPs but with modified semantics that allow them to utilize a number of interlayer prediction structures needed for efficient scalable coding.

Figure 17 shows an example of the prediction structure used in temporally scalable coding. The base layer is shown to have one-half of the total temporal resolution to be

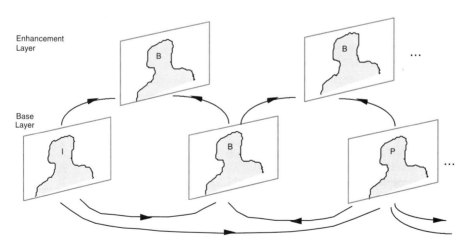

Fig. 17 Temporal scalability.

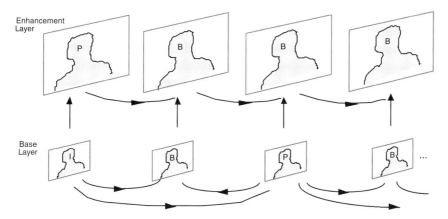

Fig. 18 Spatial scalability.

coded; the remaining one-half is carried by the enhancement layer. The base layer is coded independently, as in normal video coding, whereas the enhancement layer uses B-VOPs that use both an immediate temporally previous decoded-base-layer VOP as well as an immediate temporally following decoded-base-layer VOP for prediction.

Next, Fig. 18 shows an example of the prediction structure used in spatially scalable coding. The base layer is shown to have one-quarter resolution of the enhancement layer. The base layer is coded independently as in normal video coding, whereas the enhancement layer mainly uses B-VOPs that use both an immediate previous decoded enhancement layer VOP as well as a coincident decoded-base-layer VOP for prediction.

Some flexibility is allowed in the choice of spatial and temporal resolutions for base and enhancement layers as well as the prediction structures allowed for the enhancement layer in order to cope with a variety of conditions in which scalable coding may be needed. Further, both spatial and temporal scalability with rectangular VOPs and temporal scalability of arbitrary shape VOPs is expected to be supported in MPEG-4 Version 1. Figures 17 and 18 are applicable not only to rectangular VOP scalability but also to arbitrary shape VOP scalability (in this case, only the shaded region depicting the head and shoulder view is used for predictions in scalable coding, and the rectangle represents the bounding box).

4.4. Robust Video Coding

Truly robust video coding requires a diversity of strategies. MPEG-4 video offers a number of tools which an encoder operating in the error resilient mode (26) can employ. MPEG-4 Video also offers other tools (not specific to error resilience) that can be used to provide robust video coding. We now discuss the various available tools and how they can be used by themselves or in conjunction to provide robust video coding.

4.4.1. Object Priorities

The object-based organization of MPEG-4 Video potentially makes it easier to achieve a higher degree of error robustness due to the possibility of prioritizing each semantic object based on its relevance. Further, because MPEG-4 Systems offers scene description and composition flexibilities (see Sec. 6), it can ensure that the reconstructed scenes are meaningful even if low-priority objects are only partially available or become unavailable (say, due to data loss or corruption). Currently, MPEG-4 systems offers a way of providing priorities to each stream; if these are found insufficient, priorities may also be assigned to VOs and VOLs in the video bitstream (the location and semantics have not yet been resolved). Further, VOP types lend themselves to a form of automatic prioritization because B-VOPs are noncausal and do not contribute to error propagation and thus can be assigned a lower priority and perhaps even be discarded in case of severe errors.

Although, in principle, coding of scenes as arbitrary shaped objects can be advantageous, the downside can be shape overhead, increase in decoding complexity, and the sensitivity of shape coding (which can be interframe and context dependent) to errors.

4.4.2. Resynchronization

Video object planes already offer a means of resynchronization to prevent accumulation of errors. Further, it is possible for an encoder to offer increased error resilience by placing resynchronization (resync) markers in the bitstream with approximately constant spacing. The resync marker is a unique 17-bit code that normally cannot be emulated by any valid combination of VLC code words that may precede it. The VM7 error-resilient encoding recommends the spacings of Table 5, in bits as a function of the coding bit rate.

In fact, to enable recovery from errors, a video packet header is used which, in addition to resync marker, contains macroblock number, quantizer scale, and extension header (optional, when present, it includes VOP time and coding type information), the timing information ensures that the decoder can determine the VOP to which the video packet header belongs.

4.4.3. Data Partitioning

Data partitioning allows a mechanism to provide increased error resilience by separating the normal motion and texture data of all macroblocks in a video packet and sending all of the motion data followed by a motion marker, followed by all of the texture data. The motion marker is a unique 17-bit code that cannot be emulated by any valid combination of variable-length code words VLC that may precede it.

The motion data per macroblock is arranged to contain coded/noncoded information, followed by combined macroblock type and coded block pattern information, followed by motion vector(s); the motion data of the next macroblock follows that of the previous macroblock until the motion data for all macroblocks in the video packet can be sent. The texture data per macroblock is arranged in two parts. The first part contains coded block information of luminance blocks in a macroblock followed by optional differential quantizer information. This is repeated for all macrolocks in the video packet. The second part

Table 5 Recommended Spacings for Resync Markers

Bit rate (kbits/s)	Spacing (bits)
≤24	480
25–48	736

contains coded DCT coefficients of a macroblock, followed by that of the next macroblock until DCT coefficients for all macroblocks in the video packet can be sent.

4.4.4. Reversible VLCs

The reversible VLCs offer a mechanism for a decoder to recover additional texture data in the presence of errors since the special design of reversible VLCs enables decoding of code words in both the forward (normal) and the reverse directions. The encoder indicates whether DCT coefficient coding will use the reversible VLCs or normal VLCs (depending on the coding efficiency versus error resilience trade-offs needed) by signaling this information as part of the bitstream. It is possible to invoke the error-resilience mode independent of whether reversible VLCs are used.

The process of additional texture recovery in a corrupted bitstream starts by first detecting the error and searching forward in the bitstream to locate the next resync marker. Once the next resync marker is located, from that point, due to the use of reversible VLCs for texture coding, the texture data can be decoded in the reverse direction until an error is detected. Further, when errors are detected in texture data, the decoder can use correctly decoded motion vector information to perform motion compensation and conceal these errors.

4.4.5. Intra Update

Intra coding of macroblocks can provide refresh from coding errors but is expensive when used very frequently due to higher coding cost when video data are coded in the intra mode. In MPEG-4, considerable effort has been placed in improving the efficiency of intra coding and the resulting scheme offers higher efficiency than H.263 or MPEG-1-based intra coding. Thus, encoders requiring higher error resilience can choose to code an increased number of macroblocks in the intra coding mode than with previous standards, providing an improved refresh from coding errors.

4.4.6. Scalability for Robustness

Scalable coding can offer a means of graceful degradation in quality when packet errors due to noisy conditions or packet losses due to congestion on the network are likely. Because scalable coding involves independent coding of the base layer, the base-layer data can be assigned a higher priority and be better protected. Because the enhancement layers only offer improvement in spatial or temporal resolution, the enhancement-layer data can be assigned a lower priority.

In discussing scalability and the performance trade-offs/benefits it offers, it is important to distinguish between spatial scalability of video, which usually incurs some performance degradation and increase in complexity, and temporal scalability of video, which does not involve any degradation of performance or increase in complexity. Further, even for spatial scalability, the loss depends on many factors such as the choice of resolutions in the base and the enhancement layer(s), down-sampling filter, prediction configuration, complexity of scenes, and so forth and can be minimized with some care. Thus, in particularly noisy conditions, selective use of scalability can increase robustness significantly with very little degradation in performance or increase in complexity.

4.4.7. Correction and Concealment

Due to the channel-specific nature of the degree and type of error correction needed, MPEG-4 will not recommend a specific error-correction method but will leave it up to the chosen data transport layer to implement the needed technique. Furthermore, error concealment strategies (although encouraged) are not standardized by MPEG-4; perhaps the work done on this topic in MPEG-2 can be useful.

4.5. Facial Animation Coding

The facial animation parameters (FAPs) and the facial definition parameters (FDPs) (12) are sets of parameters designed to allow animation of faces, reproducing expressions, emotions, and speech pronunciation, as well as definition of facial shape and texture. The same set of FAPs when applied to different facial models result in reasonably similar expressions and speech pronunciation without the need to initialize or calibrate the model. The FDPs, on the other hand, allow the definition of a precise facial shape and texture in the setup phase. If the FDPs are used in the setup phase, it is also possible to precisely produce the movements of particular facial features. Using phoneme-to-FAP conversion, it is possible to control facial models, accepting FAPs via text-to-speech (TTS) systems; this conversion is not standardized. Because it is assumed that every decoder has a default face model with default parameters, the setup stage is not necessary to create face animation but for customizing the face at the decoder.

The FAP set contains two high-level parameters: *visemes* and *expressions*. A viseme is a visual correlate to a phoneme. The viseme parameter allows viseme rendering (without having to express them in terms of other parameters) and enhances the result of other parameters, ensuring the correct rendering of visemes. Only static visemes which are clearly distinguished are included in the standard set; examples of such visemes are shown in Table 6. The expression parameter similarly allows definition of high-level facial expressions. The facial expression parameter values are defined by textual descriptions such as joy, sadness, anger, fear, disgust, and surprise.

All the parameters involving translational movement are expressed in terms of the facial animation parameter units (FAPUs). These units are defined in order to allow interpretation of the FAPs on any facial model in a consistent way, producing reasonable results in terms of expression and speech pronunciation. The measurement units are shown in Fig. 19 and are defined as follows.

IRISD0	Iris diameter (by definition, it is equal to the distance between the upper and lower eyelids) in neutral face; IRISD = IRISD0/1024
ES0	Eye separation; ES = ES0/1024
ENS0	Eye–nose separation; ENS = ENS0/1024
MNS0	Mouth–nose separation; MNS = MNS0/1024
MW0	Mouth width; MW = MW0/1024
AU	Angle unit = 10^{-5} rad

Table 6 Viseme Number, its Related Phoneme Set, and Examples

1	2	3	4	5	6	7	8	9	10	11	12	13	14
(p,b,m)	(f,v)	(T,D)	(t,d)	(k,g)	(tS,dZ,S)	(s,z)	(n,l)	(r)	(A:)	(e)	(I)	(Q)	(U)
put	far	think	tip	call	chair	sir	lot	red	car	bed	tip	top	book
bed	voice	that	doll	gas	join	zeal	not						
mill					she								

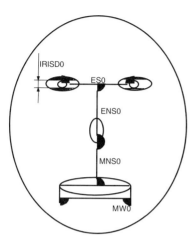

Fig. 19 Facial animation parameter units.

The FDPs are used to customize the proprietary face model of the decoder to a particular face or to download a face model along with the information of how to animate it. The FDPs are normally transmitted once per session, followed by a stream of compressed FAPs. However, if the decoder does not receive the FDPs, the use of FAPUs ensures that it can still interpret the FAP stream. This ensures minimal operation in broadcast or teleconferencing applications.

The FDP set is specified using the FDP node (in MPEG-4 Systems, see Sec. 6), which defines the face model to be used at the receiver. Two options are supported:

- Calibration information is downloaded so that the proprietary face of the receiver can be configured using facial feature points and, optionally, a 3-D mesh or texture.
- A face model is downloaded with the animation definition of the facial animation parameters. This face model replaces the proprietary face model in the receiver.

4.6. Object Mesh Coding

For general natural or synthetic visual objects, mesh-based representation (12) can be useful for enabling a number of functions such as temporal rate conversion, content manipulation, animation, augmentation (overlay), transfiguration (merging or replacing natural video with synthetic), and others. MPEG-4 includes a tool for triangular mesh-based representation of general-purpose objects.

A visual object of interest, when it first appears (as a 2-D VOP) in the scene, is tassellated into triangular patches resulting in a 2-D triangular mesh. The vertices of the triangular patches forming the mesh are referred to as the *node points*. The node points of the initial mesh are then tracked as the VOP moves within the scene. The 2-D motion of a video object can thus be compactly represented by the motion vectors of the node points in the mesh. Motion compensation can then be achieved by texture mapping the patches from VOP to VOP according to affine transforms. Coding of video texture or still

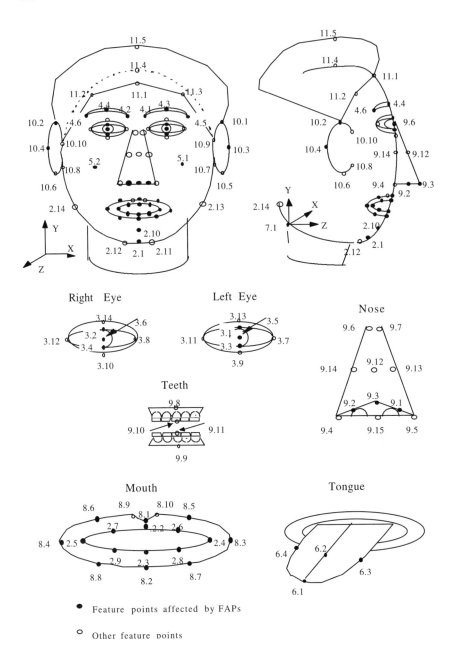

Fig. 20 Facial definition feature set.

texture (to be discussed next) of object is performed by the normal texture coding tools of MPEG-4. Thus, efficient storage and transmission of the mesh representation of a moving object (dynamic mesh) requires compression of its geometry and motion.

The initial 2-D triangular mesh is either a uniform mesh or a Delaunay mesh; the mesh triangular topology (links between node points) is not coded; only the 2-D node-point coordinates $\vec{p}_n = (x_n, y_n)$ are coded. A uniform mesh can be completely specified using five parameters such as the number of nodes horizontally and the number of nodes vertically, the horizontal and the vertical dimensions of each quadrangle consisting of two triangles, and the type of splitting applied on each quadrangle to obtain triangles. For a Delaunay mesh, the node-point coordinates are coded by first coding the boundary node points and then the interior node points of the mesh. To encode the interior node positions, the nodes are traversed one by one using a *nearest-neighbor* strategy. A linear ordering of the node points is computed such that each node is visited only once. When a node is visited, its position is differentially coded with respect to the position of the previous coded node used as the predictor. By sending the total number of node points and the number of boundary node points, the decoder knows how many node points will follow and how many of those are boundary nodes; thus, it is able to reconstruct the polygonal boundary and the locations of all nodes. The mesh-based representation of an object and the traversal of nodes for mesh geometry coding is illustrated in Fig. 21 by an example.

First, the total number of nodes and the number of boundary nodes is encoded. The top-left node $\vec{p}_0$ is coded without prediction. Then, the next clockwise boundary node $\vec{p}_1$ is found and the difference between $\vec{p}_0$ and $\vec{p}_1$ is encoded; then, all other boundary nodes are encoded in a similar fashion. Then, the not previously encoded interior node that is nearest to the last boundary node is found and the difference between these is encoded; this process is repeated until all the interior nodes are covered. The mesh geometry is only encoded when a new mesh needs to be initialized with respect to a particular VOP of the corresponding visual object; it consists of the initial positions of the mesh nodes.

The mesh motion is encoded (37) at subsequent time instants to describe the motion of the corresponding video object; it consists of a motion vector for each mesh node such that the motion vector points from a node point of the previous mesh in the sequence to a node point of the current mesh.

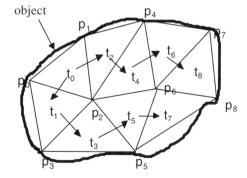

Fig. 21 Two-dimensional mesh representation of an object and coding of mesh geometry.

The mesh bitstream syntax consists of the two parts: mesh geometry and mesh motion. The node coordinates and node motion vectors are specified to one-half pixel accuracy.

4.7. Still Texture Coding

The discrete wavelet transform (DWT) (12,26) is used to code still-image data employed for texture mapping. Besides coding efficiency, an important requirement for coding texture map data is that it should be coded in a manner facilitating continuous scalability, thus allowing many resolution/qualities to be derived from the same coded bitstream. Whereas DCT-based coding is able to provide comparable coding efficiency as well as a few scalability layers, DWT-based coding offers flexibility in organization and the number of scalability layers.

The principle of DWT encoding is shown in Fig. 22.
The basic modules of a zero-tree wavelet-based coding scheme are as follows:

1. Decomposition of the texture using the discrete wavelet transform
2. Quantization of the wavelet coefficients
3. Coding of the lowest-frequency subband using a predictive scheme
4. Zero-tree scanning of the higher-order subband wavelet coefficients
5. Entropy coding of the scanned quantized wavelet coefficients and the significance map

A 2-D separable wavelet decomposition is applied to the still texture to be coded. The wavelet decomposition is performed using a Daubechies (9,3) tap biorthogonal filter which has been shown to provide good compression performance. The filter coefficients are as follows

Low pass
$= [0.03314563036812, \quad -0.06629126073624, \quad -0.17677669529665,$
$0.41984465132952, \quad 0.99436891104360, \quad 0.41984465132952,$
$-0.17677669529665, \quad -0.06629126073624, \quad 0.03314563036812]$

High pass $= [-0.35355339059327, 0.70710678118655, -0.35355339059327]$

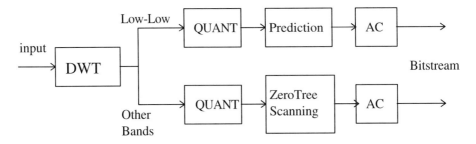

Fig. 22 Block diagram of the DWT encoder of still-image texture.

A group delay is applied to each filter to avoid the phase shift on both the image domain and the wavelet domain.

The wavelet coefficients of the lowest band are coded independently of the other bands. These coefficients are quantized using a uniform midrise quantizer. After quantization of the lowest subband coefficients, an implicit prediction (same as that used for DC prediction in intra DCT coding) is applied to compute the prediction error which is then encoded using an adaptive arithmetic coder that uses min-max coding.

The wavelet coefficients of the higher bands are first quantized by multilevel quantization which provides the flexibility needed to trade off the number of levels, the type of scalability (spatial or SNR), the complexity and the coding efficiency. Different quantization step sizes (one for luminance and one for chrominance) can be specified for each level of scalability. All the quantizers of the higher bands are uniform mid-rise quantizers with a dead zone twice the quantization step size. The quantization step sizes are specified by the encoder in the bitstream. In order to achieve the fine-granularity SNR scalability, a bi-level quantization scheme is used for all the multiple quantizers. This quantizer is also a uniform mid-rise quantizer with a dead zone twice the quantization step size. The coefficients that lie outside the dead zone (in the current and previous pass) are quantized with a 1-bit accuracy. The number of quantizers is equal to the maximum number of bitplanes in the wavelet transform representation. In this bi-level case, instead of the quantization step sizes, the maximum number of the bitplanes is specified in the bitstream.

After quantization, each wavelet coefficient is either zero or nonzero. The coefficients of all bands (except the lowest) are scanned by zero-tree scanning. Zero-tree scanning is based on the observation that a strong correlation exists in the amplitudes of the wavelet coefficients across scales, and on the idea of partial ordering of the coefficients. The coefficient at the coarse scale is called the parent, and all coefficients at the same spatial location, and of similar orientation, at the next finer scale are that parent's children. Since the lowest-frequency subband is coded separately, the wavelet trees start from the adjacent higher bands. In order to achieve a wide range of scalability levels efficiently as needed by the application, a multiscale zero-tree coding scheme is employed. The zero-tree symbols and the quantized values are coded using an adaptive arithmetic encoder which uses a three-symbol alphabet.

The process of scalable decoding the various spatial/SNR layers from a single DWT-coded bitstream is shown in Fig. 23.

4.8. View Dependent Texture Coding

View-dependent texture coding (12) is designed for incrementally streaming the texture data when the viewpoint is moving. This technique is intended for streaming texture across the network allowing low-bandwidth communication of remote virtual environments when very low delay transmission is possible. The streaming of texture data is done using a back-channel from the decoder to the encoder. The back-channel is used to indicate to the encoder the current viewing conditions and thus it becomes possible to send from the encoder to the decoder only the data that is really needed.

The two main stages in transmission are the initialization and the streaming stages. The basic operations that are performed by the coder are as follows. At the encoder, the image to be coded is wavelet transformed; its useful coefficients are selected, coded, and transmitted. At the decoder, the received coefficients are decoded and inverse transformed.

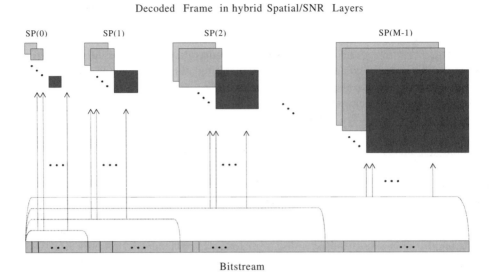

Fig. 23 Scalable decoding of still-object texture.

The resulting image is mapped on to a mesh grid using orthographic or perspective projection.

In orthographic projection, at the decoder side, the knowledge of the 3-D regular mesh grid is needed on which texture is mapped. The texture-mapping operation is defined using the coordinate vertices of the mesh grid. In perspective projection, the same principle is used. In order to take into account perspective distortion, the resolution of texture image is increased as the distance of the viewer and the cell center is decreased. Feedback is sent to the encoder regarding the viewpoint, and only the necessary texture data are sent by the encoder.

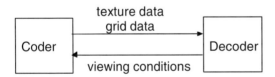

Fig. 24 View-dependent texture system.

5. MPEG-4 AUDIO

The benefits of the MPEG-4 Speech coder can be exploited in a number of applications. As an example, an MPEG-4-based Internet-phone system offers robustness against packet loss or change in transmission bit rates. When applied in an audio–visual system, the coding quality is improved by assigning the audio and the visual bit rates adaptively based on the audio–visual content. As another example, the MPEG-4 Speech coder could also be used in radio communication systems where it can offer higher error robustness by changing the bit allocation between speech coding and error correction, depending on the error conditions. These features have not been realized by any other standard. Furthermore, the low bit rate is useful for ''Party talk.'' Although up to 10 persons can have conversation simultaneously over the Internet, one terminal only has to receive a bit rate of only 18 kbits/s, to hear the conversation of nine other persons.

In addition to speech coding, MPEG-4 also offers multichannel audio coding based on optimized MPEG-2 AAC coding. MPEG-4 also offers solutions for medium-bit-rate audio coding. Furthermore, it supports the concept of audio objects. Just as video scenes are made from visual objects, audio scenes may be usefully described as the spatio-temporal combination of audio objects. An ''audio object'' is a single audio stream coded using one of the MPEG-4 coding tools, like CELP or Structured Audio. Audio objects are related to each other by mixing, effects processing, switching, and delaying them and may be spatialized to a particular 3-D location. The effects processing is described abstractly in terms of a signal-processing language (the same language used for Structured Audio), so content providers may design their own empirically, and include them in the bitstream.

5.1. Natural Audio

As mentioned earlier, Natural Audio coding (13,38) in MPEG-4 consists of the following:

- The lowest bit-rate range between 2 and 6 kbits/s is covered by Parametric Coding (mostly for speech coding).
- The medium bit rates between 6 and 24 kbits/s use Code Excited Linear Predictive (CELP) with two sampling rates, 8 and 16 kHz, for a broader range of audio signals.
- For higher bit rates starting at about 16 kbits/s, frequency-domain coding techniques are applied—e.g., optimized version of MPEG-2 Advanced Audio Coding (AAC). The audio signals in this region typically have bandwidths starting at 8 kHz.

Figure 25 provides a composite picture of the applications of MPEG-4 Audio and Speech coding, the signal bandwidth and the type of coders used.

5.1.1. Parametric Coder

The parametric coder core provides two sets of tools. The HVXC coding tools (Harmonic Vector eXcitation Coding) allow coding of speech signals at 2 kbits/s, whereas the Individual Line coding tools allow coding of nonspeech signals like music at bit rates of 4 kbits/s and higher. Both sets of tools allow an independent change of speed and pitch during the decoding and can be combined to handle a wider range of signals and bit rates.

We now list the encoder and the decoder tools in HVXC and Individual Line coding (13); however, only the decoder tools are normative.

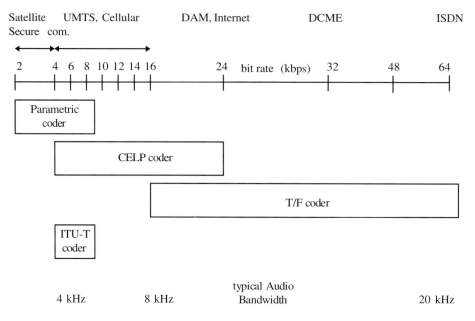

Fig. 25 Natural audio coding in MPEG-4.

HVXC Encoder Tool	HVXC Decoder Tools
1. Normalization	1. LSP Decoder
2. Pitch Estimation	2. Harmonic Decoder
3. Harmonic Magnitude Extraction	3. Time Domain Decoder
4. Perceptual Weighting	4. Speed Control
5. Harmonic Encoding	5. Harmonic Synthesizer
6. V/UV Decision	6. Time Domain Synthesizer
7. Time Domain Coder	7. Postprocessing
8. Short Frame Mode	8. Short Frame Mode

Individual Line Encoder Tools	Individual Line Decoder Tools
1. Individual Line Parameter Extraction	1. Individual Line Bitstream Decoding
2. Individual Line Bitstream Encoding	2. Individual Line Synthesizer

5.1.2. CELP Coder

The CELP coder is designed for speech coding at two different sampling frequencies, namely 8 kHz and 16 kHz. The speech coders using a 8-kHz sampling rate are referred to as narrow-band coders, whereas those using a 16-kHz sampling rate are wide-band coders. The CELP coder includes tools offering a variety of functions, including bit-rate control, bit-rate scalability, speed control, complexity scalability, and speech enhancement. Using the narrow-band and the wide-band CELP coders, it is possible to span a wide range of bit rates (4–24 kbps). Real-time bit-rate control in small steps can be provided.

A common structure of tools have been defined for both the narrow-band and wide-band coders; many tools and processes have been designed to be commonly usable for both narrow-band and wide-band speech coders.

We now list the encoder and the decoder tools in CELP coding (13); however, only the decoder tools are normative.

CELP Encoder Tools	CELP Decoder Tools
1. Preprocessing	1. Bitstream Demultiplexer
2. LPC Analysis	2. LPC Decoder
3. LPC Quantizer	3. Excitation Generator
4. Perceptual Weighting	4. LPC Synthesis Filter
5. Pitch Estimation	5. LPC Postprocessing
6. Analysis Filter	6. Weighting Module
7. Weighting Module	
8. Bitstream Multiplexer	

5.1.3. Time/Frequency Coder

The high-end audio coding in MPEG-4 (13) is based on MPEG-2 AAC coding. MPEG-2 AAC is a state-of-the-art audio compression algorithm that provides compression superior to that provided by older algorithms. AAC is a transform coder and uses a filterbank with a fine-frequency resolution that enables superior signal compression. AAC also uses a number of new tools such as temporal noise shaping, backward adaptive linear prediction, joint stereo coding techniques, and Huffman coding of quantized components, each of which provide additional audio compression capability. Furthermore, AAC supports a wide range of sampling rates and bit rates, from 1 to 48 audio channels, up to 15 low-frequency enhancement channels, multilanguage capability, and up to 15 embedded data streams. MPEG-2 AAC provides a five-channel audio coding capability, while being a factor of 2 better in coding efficiency relative to MPEG-2 BC; because AAC has no such backward compatibility requirement and it incorporates the recent advances, in MPEG formal listening tests for five-channel audio signals it provided slightly better audio quality at 320 kbits/s than MPEG-2 BC can provide at 640 kbits/s.

5.2. Text to Speech

Text-to-speech (TTS) conversion system synthesizes speech as its output when a text is accessed as its input. In other words, when the text is accessed, the TTS changes the text into a string of phonetic symbols and the corresponding basic synthetic units are retrieved from the preprared database. Then the TTS concatenates the synthetic units to synthesize the output speech with the rule-generated prosody. Some application areas for MPEG-4 TTS are as follows:

- Artificial Story Teller (or Story Teller on Demand).
- Synthesized speech output synchronized with facial animation (FA)
- Speech synthesizer for avatars in various virtual-reality (VR) applications
- Voice newspaper
- Dubbing tools for animated pictures
- Voice Internet
- Transportation timetables

The MPEG-4 TTS (13) can not only synthesize speech according to the input speech with a rule-generated prosody but also execute several other functions. They are as follows:

- Speech synthesis with the original prosody from the original speech
- Synchronized speech synthesis with facial animation (FA) tools
- Synchronized dubbing with moving pictures not by recorded sound but by text and some lip-shape information
- Trick mode functions such as stop, resume, forward, backward without breaking the prosody even in the applications with facial animation (FA)/motion pictures (MP)
- Change of the replaying speed, tone, volume, speaker's sex, and age

MPEG-4 TTS can be used for many languages because it adopts the concept of the language code such as the country code for an international call. Presently, only 25 countries (i.e., the current ISO members) have their own code numbers to identify that their own language has to be synthesized, except the International Phonetic Alphabet (IPA) code assigned as 0. However, 8 bits have been assigned for the language code to ensure that all countries can be assigned language codes when asked in the future. IPA could still be used to transmit all languages.

For MPEG-4 TTS, only the interface bitstream profiles are the subject of standardization. Because there are already many different types of TTS and each country has several or a few tens of different TTSs synthesizing its own language, it is impossible to standardize all the things related to TTS. However, it is believed that almost all TTSs can be modified to accept the MPEG-4 TTS interface very quickly by a TTS expert because of the rather simple structure of the MPEG-4 TTS interface bitstream profiles.

5.3. Structured Audio

Structured audio formats use ultra-low bit-rate algorithmic sound models to code and transmit sound. MPEG-4 standardizes an algorithmic sound language and several related tools for the structured coding of audio objects. Using these tools, algorithms which represent the exact specification of a sound scene are created by the content designer, transmitted over a channel, and executed to produce sound at the terminal. Structured audio techniques in MPEG-4 (13) allow the transmission of synthetic music and sound effects at bit rates from 0.01 to 10 kbits/s, and the concise description of parametric sound postproduction for mixing multiple streams and adding effects processing to audio scenes.

MPEG-4 does not standardize a synthesis method, but a signal-processing language for describing synthesis methods. SAOL, pronounced "sail," stands for "Structured Audio Orchestra Language" and is the signal-processing language enabling music synthesis and affects postproduction in MPEG-4. It falls into the music-synthesis category of "Music V" languages; that is, its fundamental processing model is based on the interaction of oscillators running at various rates. However, SAOL has added many new capabilities to the Music V language model which allow for more powerful and flexible synthesis description. Using this language, any current or future synthesis method may be described by a content provider and included in the bitstream. This language is entirely normative and standardized, so that every piece of synthetic music will sound exactly the same on every compliant MPEG-4 decoder, which is a substantial improvement over the great variety in Musical Instrument Digital Interface (MIDI)-based synthesis systems.

The techniques required for automatically producing a Structured Audio bitstream from an arbitrary sound are beyond today's state of the art, although they are an active research topic. These techniques are often called "automatic source separation" or "automatic transcription''; there are many references within the audio-processing literature on the capabilities of today's methods. In the meantime, content authors will use special content creation tools to directly create Structured Audio bitstreams. This is not a fundamental obstacle to the use of MPEG-4 Structured Audio, because these tools are very similar to the ones that content authors use already; all that is required is to make them capable of producing MPEG-4 output bitstreams.

There is no fixed complexity which is adequate for decoding every conceivable Structured Audio bitstream. Simple synthesis methods are of very low complexity, and complex synthesis methods require more computing power and memory. As the description of the synthesis methods is under the control of the content provider, the content provider is responsible for understanding the complexity needs of his bitstreams. Past versions of structured audio systems with a similar capability have been optimized to provide multitimbral, highly polyphonic music and postproduction effects in real time on a 150-MHz Pentium computer or simple Digital Signal Processing (DSP) chip. One ''level'' of capability in the Synthetic Object Audio profile of MPEG-4 Audio may provide for simpler and/or less normative synthesis methods in RAM- or computing-limited terminals.

6. MPEG-4 SYSTEMS

The Systems (11) part of MPEG-4 perhaps represent the most radical departure from previous MPEG specifications. Indeed, the object-based nature of MPEG-4 necessitates a totally new approach on what the Systems layer is required to provide. Issues of synchronization and multiplexing are, of course, still very essential. Note, however, that an MPEG-4 scene may be composed of several objects, and hence synchronization between a large number of streams is required. In addition, the spatio-temporal positioning of such objects forming a scene (or scene description) is a key component. Finally, issues of interactivity are also quite new in MPEG as well.

6.1. System Decoder Model

A key problem in designing an audiovisual communication system is ensuring that time is properly represented and reconstructed by the terminal. This serves two purposes: It ensures (1) that ''events'' occur at designated times as indicated by the content creator and (2) that the sender can properly control the behavior of the receiver. The latter is essentially providing an open-loop flow control mechanism, a requirement for a specification that covers broadcast channels (without an upstream, or feedback, channel). Assuming a finite set of buffer resources at the receiver, by proper clock recovery and time stamping of events, the source can always ensure that these resources are not exhausted.

Clock recovery is typically performed using clock references. The receiving system has a local system clock which is controlled by a Phase Locked Loop (PLL), driven by the differences in received clock references and the local clock references at the time of their arrival. This way the receiver's clock speed is increased or decreased, matching that of the sender. In addition, coded units are associated with decoding time stamps, indicating the time instance in which a unit is removed from the receiving system's decoding buffer. The combination of clock references and time stamps is sufficient for full control of the receiver.

In order to properly address specification issues without making unnecessary assumptions about system implementation, MPEG-4 Systems defines a System Decoder

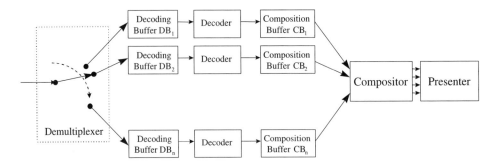

Fig. 26 Systems Decoder model.

Model. This represents an idealized unit in which operations can be unambiguously controlled and characterized. The MPEG-4 System Decoder Model exposes resources available at the receiving terminal and defines how they can be controlled by the sender or content creator. The model is shown in Fig. 26. It is composed of a set of decoders (for the various audio or visual object types), provided with two types of buffers: decoding and composition.

The decoding buffers have the same functionality as in previous MPEG specifications and are controlled by clock references and decoding time stamps. In MPEG-2, each program had its own clock. Proper synchronization was ensured by using the same clock for coding and transmitting the audio and video components. In MPEG-4, each individual object is assumed to have its own clock, or Object Time Base (OTB). Of course, several objects may share the same clock. In addition, coded units of individual objects [Access Units (AUs) corresponding to an instance of a video object or a set of audio samples] are associated with Decoding Time Stamps (DTSs). Note that the decoding operation at DTS is considered (in this ideal model) to be instantaneous.

The composition buffers which are present at the decoder outputs form a second set of buffers. MPEG-4 defines an additional time stamp, the Composition Time Stamp (CTS), which defines the time at which data are taken from the composition buffer for (instantaneous) composition and presentation. As MPEG-4 does not normatively specify composition or rendering, the composition buffers act as the last interface between the MPEG-4 specification and the presentation device(s). Composition time stamps play essentially the same role as presentation time stamps did in MPEG-1 and MPEG-2.

In order to coordinate the various objects, a single System Time Base is assumed to be present at the receiving system. All object time bases are subsequently mapped into the system time base, so that a single notion of time exists in the terminal. For clock recovery purposes, a single stream must be designated as the master. The specification does not mandate the stream that has this role, but a plausible candidate is the one which contains the scene description. Note also that, in contrast with MPEG-2, the resolution of both the STB and the OCRs is not mandated by the specification. In fact, the size of the OCR fields for individual access units is fully configurable, as we discuss later.

In other designs (e.g., IETF's RTP), the assumption of a globally known clock can be made (provided by other network services such as NTP). There is work underway to provide a unified methodology so that mapping of MPEG-4 timing architecture can be seamlessly performed in such an environment as well. This is facilitated by the flexible multiplexing methodology adopted in MPEG-4 and is discussed later.

6.2. Scene Description

Scene Description refers to the specification of the spatio-temporal positioning and behavior of individual objects. It is a totally new component in the MPEG specifications and allows the easy creation of compelling audiovisual content. Scene description involves an architectural component (i.e., the proper way to conceptually organize audiovisual information) and a syntactic component which is the mapping of such architecture into a bitstream. Note that the scene description is transmitted in a separate stream from the individual media objects. This allows one to change the scene description without operating on any of the constituent objects themselves.

6.2.1. Architecture

The architecture of MPEG-4's Scene Description is based on VRML (39,46). Scenes are described as hierarchies of nodes forming a tree. Leaves of the tree correspond to media objects (audio or visual, natural or synthetic), whereas intermediate nodes perform operations on their underlying nodes (grouping, transformation, etc.). Nodes also expose attributes through which their behavior can be controlled. This hierarchical design is shown in Fig. 27.

There are, however, several key differences between the domains that VRML and MPEG-4 address, which necessitate the adoption of slightly different approaches. MPEG-4 is concerned with the description of highly dynamic scenes, where temporal evolution

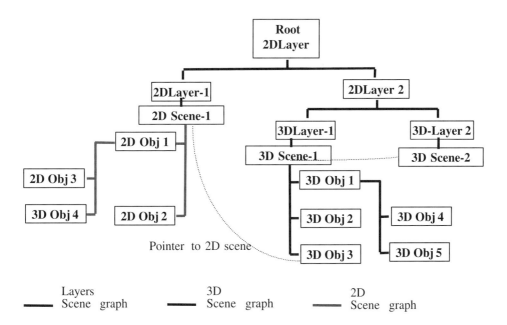

Fig. 27 Scene Description.

is more dominant than navigation. VRML, on the other hand, allows the definition of a static 3-D world (static in the sense that all the objects in that world are predefined and cannot be changed) in which a user is allowed to navigate. As a result, scene descriptions in MPEG-4 actually have their own time base and can be updated at any time. Updates can take the form of a node's replacement, elimination, insertion, or attribute value modification. The presence of a time base and decoding time stamps ensures application of such updates at the correct time instances.

In addition, MPEG-4 also needs to address pure 2-D composition of objects. The complexity and cost of 3-D graphics versus the possibility of developing low-cost or low-power systems makes it desirable to partition the space of graphics capabilities. In Fig. 19, we see an example where both 2-D and 3-D components coexist.

As a result of the close relationship of the MPEG-4 Scene Description with VRML, the types of Scene Description capabilities provided by MPEG-4 are essentially those provided by VRML nodes. It is currently being examined if a number of profiles (including one corresponding to VRML) can be obtained by appropriately defining subsets of MPEG-4 BIFS.

6.2.2. Binary Format for Scenes

The mapping of the scene description into a parametric form suitable for low-overhead transmission has resulted in a scene description format called Binary Format for Scenes (BIFS). This representation format associates each node with a node type. Nodes are then represented by their node type and a set of attribute value specifications. To avoid the specification of all attributes of a node, default values are used, and attributes are individually addressable within a node. This way one can, for example, only specify a nondefault value for attribute X of a node of type Y. To minimize the number of bits spent to code node types, a context-based approach is used where node types are recast as *node data types* which only need to describe nodes that are valid children of a particular node. Because the latter sets of nodes are typically small, the number of bits required to identify its node with a node data type is small.

Furthermore, nodes can optionally be reused. In this case, they are associated with an identifier which can appear in place of a node of the same type. In Fig. 27, for example, the node "3D Obj 3" contains the child node "2D Scene-1," which is used elsewhere in the scene as well. As mentioned earlier, scene descriptions can be updated: nodes can be inserted, deleted, replaced, or their attributes can be modified.

In addition to the VRML set of nodes, MPEG-4 defines its own set of nodes, particularly to handle media objects, including natural audio and video, still images, face animation, text-to-speech synthesizer, streaming text, 2-D composition operators, layout control, and so forth.

6.2.3. Interactivity

Interactivity in MPEG-4 can take two forms: client based and server based. In the client-based case, user operations affect the local scene description. The VRML model of events and routes is used within the BIFS format, thus providing direct support for a large variety of circumstances. In addition, and noting that user events can be transformed to scene updates, one can also have a form of interactivity that does not necessitate normative support by the specification: user events can form a secondary source of scene updates.

Server-based interaction requires the presence of an upstream channel. The details of such a mode of interactivity are still under investigation, as they are slightly complicated by the needs for network independence.

6.2.4. Adaptive Audio–Visual Session (MPEG-Java)

An additional flexible mechanism for describing scenes is also being developed, called the Adaptive Audio–Visual Session (AAVS). This is based on the use of the Java language for constructing scenes, but not for composition, rendering, or decoding. By taking the programmable aspect of the system outside of the main data flow of decoders and the rendering engine (which have to operate extremely fast), overall performance can be kept high. Note that the AAVS approach is being designed as an extension of BIFS. In that sense, a BIFS scene can be a subset of an AAVS scene (but not vice versa).

The benefits of the AAVS approach are in three major categories. First, because of its flexible nature, the scene description is capable of adapting to terminal capabilities, hence providing a form of graceful degradation. Second, when scene description operations can be expressed concisely in a flexible way (e.g., a spline trajectory), using a flexible approach may result in increased compression. Finally, by allowing programmability, user interaction can extend beyond the modes defined in a parametric scene description and become as rich as the content creator wishes. Incidentally, AAVS is now known as MPEG-J, short for MPEG-Java.

Use of programmability in an audio–visual terminal certainly opens up a very broad spectrum of opportunities. If the integration with the requirements of the real-time world of decoders is successful, then this will be a major step forward in information representation.

6.3. Associating Scene Description with Elementary Streams

6.3.1. Object Descriptors

Individual object data as well as scene description information is carried in separate Elementary Streams (ES). As a result, BIFS media nodes need a mechanism to associate themselves with the ESs that carry their data (coded natural video object data, etc.). A direct mechanism would necessitate the inclusion of transport-related information into the scene description. As we mentioned earlier, an important requirement in MPEG-4 is transport independence. As a result, an indirect way was adopted, using Object Descriptors (ODs).

Each media node is associated with an object identifier, which, in turn, uniquely identifies an OD. Within an OD, there is information on how many ESs are associated with this particular object (may be more than one for scalable video/audio coding, or multichannel audio coding) and information describing each of those streams. The latter information includes the type of the stream, as well as how to locate it within the particular networking environment used. This approach simplifies remultiplexing (e.g., going through a wired–wireless interface), as there is only one entity that may need to be modified.

6.3.2. Stream Map Table

The Object Descriptor allows unique reference of an elementary stream by an identifier; this identifier may be assigned by an application layer when the content is created. The transport channel in which this stream is carried may only be assigned at a later time by a transport entity; it is identified by a channel association tag associated to an elementary stream id by a stream map table. In interactive applications, the receiving terminal may select the desired elementary streams, send a request, and receive the stream map table in return. In broadcast and storage applications, the complete stream map table must be included in the application's signaling channel.

6.4. Multiplexing

The key underlying concept in the design of the MPEG-4 multiplexer is network independence. MPEG-4 content may be delivered across a wide variety of channels, from very low-bit-rate wireless, to high-speed asynchronous transfer mode (ATM) and broadcast systems, to DVDs. A critical design question was what should be the tools included in the specification for mandatory implementation. Clearly, the broad spectrum of channels could not allow a single solution to be used. At the same time, inclusion of a large number of different tools and configurations would make implementations extremely complex and—through excessive fragmentation through profiles—make interoperability extremely hard to achieve in practice. Consequently, the decision was made that MPEG-4 would not provide specific transport-layer features but would, instead, make sure that it could be easily mapped to existing such layers. This is accomplished by allowing several components of the multiplexer to be configurable, thus allowing designers to achieve the desired trade-off between functionality and efficiency. In addition, it is assumed that Quality of Service (QoS) guarantees may be made available by the underlying transport service if so desired.

The overall multiplexing architecture of MPEG-4 is shown in Fig. 28. At the topmost level, we have the Access Unit Layer (AL) which provides the basic conveyor of timing and framing information. It is at this level that time stamps and clock references are provided. The AL header, however, is very flexible: The presence of OCR/DTS is optional and their resolution (number of bits) is configurable. In addition, the header contains information about framing (start or end of a coded unit, random access indicator), as well as sequence numbering. The latter is particularly useful in error-prone wireless or broadcast environments, where preemptive retransmission of critical information (e.g., scene description) may be performed. Using a sequence number, a receiver that has already accurately received the information can ignore the duplicate. In order to be able to "bootstrap" the demultiplexing process, the AL header configuration information is carried in a channel that has a predefined configuration.

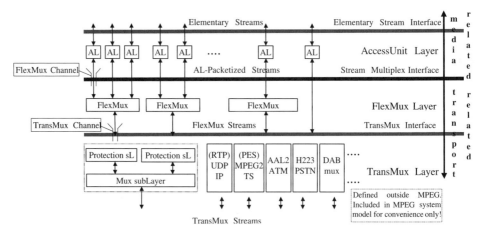

Fig. 28 MPEG-4 Multiplexing Architecture.

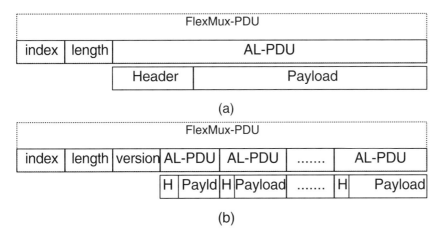

Fig. 29 FlexMux modes: (a) simple (single-object Protocol Data Unit (PDU)) (b) MuxCode (multi-object PDU).

Immediately below the AL, we have the optional Flexible Multiplexer or "FlexMux" layer. This is a very simple design, intended for systems that may not provide native multiplexing services. An example is the data channel available in GSM cellular telephones. Its use, however, is entirely optional and does not affect the operation of the rest of the system. As shown in the right side of Fig. 28, there can be a "null" connection directly from the AL to the lower layers. The FlexMux provides two modes of operation: a "simple" and a "muxcode" mode, as shown in Fig. 29.

In the simple mode, data from a single object (or scene description) is present in the FlexMux payload (appropriately encapsulated as an Access Unit Layer Protocol Data Unit (AL PDU)). In the MuxCode mode, data from multiple objects are placed in predefined positions within the FlexMux payload. The "index" field of the FlexMux header indicates which of the modes is used, depending on its value.

Finally, at the lowest level, we have the Transport Multiplexing or "TransMux" layer, which is not specified by MPEG-4. This can be any current or future transport-layer facility, including MPEG-2 Transport Stream, RTP, H.223, and so forth. For a mobile system, this layer must provide its own protection sublayer to ensure the desired QoS characteristics.

6.5. File Format

For storage-based content delivery (e.g., CD-ROM or DVD), MPEG-4 also defines a file format. This is a simple layer that substitutes the TransMux layer and where random access information is provided in the form of directories. This allows an application or user to rapidly move back and forth in an MPEG-4 file, having immediate access to Access Units. In order to easily identify MPEG-4 files as such, the string "MPEG-4" is used as a magic number (the first 6 bytes in the file), and the use of the '.mp4' extension has been adopted.

The details of file format are given in Ref. 11. Multiplexed elementary streams are encapsulated in a Stored File Format Layer. An MPEG-4 file is expected to contain a

Header followed by Segment data that, in turn, contain a series of zero or more FlexMux PDUs. The segments form a doubly linked list for convenient navigation within the file. Each segment header consists of a list of directory entries which provide pointers to access units of elementary streams that are contained within its segment data. Note that at the discretion of author, random access may be provided to just a subset of access units contained in the segment; thus, not all units need to have directory entries. MPEG-4 is currently investigating expansion of its file format activities beyond this simple structure, and it is expected that further developments will be made on this design.

6.6. Syntactic Description

Syntactic description in past MPEG specifications was performed using ad hoc techniques (pseudo-C). Although this provided a concise way of representing simple bitstreams, it could not fully describe the sophisticated constructs required in source coding (e.g., Huffman codes) without explanatory text, tables, and so forth. Furthermore, the description was not amenable to software tool manipulation. Indeed, in the 50-year history of media representation and compression, the lack of software tools is particularly striking. This has made the task of both codec and application developers much more difficult.

Use of source coding, with its bit-oriented nature, directly conflicts with the byte-oriented structure of modern microprocessors and makes the task of handling coded audio–visual information more difficult. A simple example is fast decoding of variable-length codes; every programmer wishing to use information using entropy coding must hand-code the tables so that optimized execution can be achieved. General-purpose programming languages such as C++ and Java do not provide native facilities for coping with such data. Even though other facilities already exist for representing syntax (e.g., ASN.1—ISO International Standards 8824 and 8825), they cannot cope with the intricate complexities of source coding operations (variable-length coding, etc.).

MPEG-4 Systems has adopted an object-based syntactic description language for the definition of its bitstream syntax [Flavor—Formal Language for Audio-Visual Object Representation (27,31–33)]. It is designed as an extension of C++ and Java in which the type system is extended to incorporate bitstream representation semantics (hence, forming a syntactic description language). This allows the description, in a single place, of both the in-memory representation of data as well as their bitstream-level (compressed) representation. Also, Flavor is a declarative language and does not include methods or functions. By building on languages widely used in multimedia application development, one can facilitate integration with an application's structure.

Figure 30 shows a simple example of a Flavor representation. Note the presence of bitstream representation information right after the type within the class declaration. The map declaration is the mechanism used in Flavor to introduce constant- or variable-length code tables (1-to-n mappings); in this case, binary code words (denoted using the ''0b'' construct) are mapped to values of type unsigned char. Flavor also has a full complement of object-oriented features pertaining to bitstream representation (e.g., ''bitstream polymorphism'') as well as flow control instructions (if, for, do–while, etc.). The latter are placed within the declaration part of a class, because they control the serialization of the class' variables into a bitstream.

A translator has been developed that automatically generates standard C++ and Java code from the Flavor source code (27), so that direct access to, and generation of, compressed information by application developers can be achieved with essentially zero

```
map SampleVLC(unsigned char) {
    0b0,  2,
    0b10, 5,
    0b11, 7
}
class   HelloBits {
    int(8) size;
    int(size) value1;
    unsigned char(   SampleVLC) value2;
}
```

Fig. 30 A simple example of Syntactic Description.

programming. In this way, a significant part of the work in developing a multimedia application (including encoders, decoders, content creation, and editing suites, indexing, and search engines) is eliminated.

7. PROFILES AND LEVELS

The concept of profiles and levels of MPEG-2 Video (3,4) is extended to MPEG-4 such that profiles and levels provide a means of defining subsets of the syntax and semantics of MPEG-4 Video, Audio, and Systems specifications and thereby the decoder capabilities required to decode a particular bitstream. A profile is a defined subset of the entire bitstream syntax that is defined by this specification. A level is a defined set of constraints imposed on parameters in the bitstream. Conformance tests will be carried out against defined profiles at defined levels. The purpose of defining conformance points in the form of profiles and levels is to facilitate bitstream interchange among different applications. Implementers of MPEG-4 are encouraged to produce decoders and bitstreams which correspond to those defined conformance regions. The discretely defined profiles and levels are the means of bitstream interchange between different applications of MPEG-4.

In MPEG-4, it is likely that profiles will be defined for video objects, audio objects, and systems. In addition, it is likely that profiles may also be defined for audio composition and video composition. The work on profiles is ongoing and is thus likely to evolve; we present here the current structure of profiles envisaged and their key requirements (14,41,42).

7.1. Video Profiles

Three video profiles are currently being considered; their requirements are briefly discussed next.

7.1.1. Video Simple Object Profile

The video simple object profile is designed to enable applications such as real-time communication and surveillance. A few of the important requirements are as follows:

- Frame-based visual objects but may also support arbitrary shaped visual objects optimized for head and shoulder views
- Robustness to random errors, burst errors, and ability to detect errors and conceal errors
- Two modes, a simple mode (delay of 250 ms) and a low-delay mode (delay of 150 ms)
- Support a low-complexity mode for real-time decoding of QCIF video at 15 fps and 24 kbits/s.

7.1.2. Video Random Access Object Profile

The video random access object profile is designed to enable the content-based storage and retrieval application. Some important requirements are as follows:

- Support spatio-temporal arbitrary shaped objects with exact shape representation as well as scalable shape representation
- Support random access to any spatio-temporal object and to particular instances of the object
- Support spatial, SNR quality scalability independently for each object

7.1.3. Video Main Object Profile

The video main object profile is designed to enable the broadcast application. A few of the important requirements are as follows:

- Support spatio-temporal arbitrary shaped visual objects and all types of natural and 2-D synthetic visual content of progressive or interlaced formats
- Support efficient methods for random access within a limited time and with fine resolution
- Support object-based scalability such as spatial, SNR, and temporal scalability
- Robustness to random errors, burst errors, and ability to partition data, detect errors, and conceal errors

7.2. Audio Profiles

Three audio profiles are currently being considered; their requirements are briefly discussed next.

7.2.1. Audio Speech Object Profile

The audio speech object profile is designed to enable applications such as real-time communication and surveillance. A few of the important requirements are as follows:

- Support audio objects formed by blocks of composite speech, optimized for speech coding
- Support for two layers of quality scalability and complexity scalability from minimum to maximum
- Ability to withstand random errors, burst errors, detect and conceal errors

7.2.2. Audio Low-Delay Object Profile

Some important requirements are as follows:

- Support speech, audio, music, signaling tones, and computer-generated audio
- Speech quality equivalent to G.722, subjective audio quality equivalent to or better than MPEG-1 and 2

- Speech with bit rates less than to be decided value, music with bit rates less than to be decided value

7.2.3. Audio Main Object Profile

The audio main profile is designed to enable the broadcast application. A few of the important requirements are as follows:

- Support audio objects that can be tied to presentation viewpoint or any visual or data object
- Support all types of natural, synthetic, and hybrid audio content; Monoaural, stereo, and multichannel content supported
- Support real-time decoding with the lowest complexity possible
- Two modes, a real-time encoding mode for low-delay and an off-line mode for optimum coding efficiency

7.3. Systems Profiles

Three systems profiles are currently being considered; their requirements are briefly discussed next.

7.3.1. Systems Simple Profile

The systems simple profile is designed to enable applications such as real-time communication and surveillance. The main requirements for multiplexing are as follows:

- The multiplex shall support four objects (level 1) and 16 objects (level 2) in the same bitstream.
- The multiplex shall support means to dynamically change number of objects in the bitstream.
- The multiplex shall support the extraction of objects from the same bitstream without requiring additional capabilities.

7.3.2. Systems Interactive Profile

The systems interactive profile is designed to enable content-based storage and retrieval application. The main requirements for multiplexing are as follows:

- The multiplex shall support 1024 objects in a bitstream.
- The multiplex shall support access to data associated with the temporal segment without decoding the audio–visual data.
- The multiplex shall support access to data associated with the spatio-temporal segment.

7.3.3. Systems Main Profile

The systems main profile is designed to enable the broadcast application. The main requirements for multiplexing are as follows:

- The multiplex shall support 1024 objects in a bitstream.
- The multiplex shall support means to dynamically change number of objects in the bitstream.

- The multiplex shall support the extraction of objects from the same bitstream without requiring additional capabilities.
- The multiplex shall support the mixing of objects from local and remote sources.

8. VERIFICATION TESTS

MPEG-4 is planning to hold verification tests to confirm the performance of its various combination of tools that will form profiles. The first of the series of such tests (43) will take place in March 1998 and is aimed at verifying error resilience tools.

8.1. Error Resilience Verification Tests

Version 1 of the MPEG-4 video codec will be implemented in software before the October 1997 MPEG meeting. It will be combined with simulations of the Universal Access Layer, a component of the MPEG-4 System Layer, and a TransMux. The particular TransMux to be implemented depends on the application and corresponding network. For wireless applications, the TransMux will be ITU's mobile multiplexing standard, H.223/Mobile. The overall encoding system to be simulated for the demonstration of MPEG-4 over a wireless network consists of an application layer consisting of an MPEG-4 error-resilient video encoder and simulated audio data, an AL consisting of adaptation layers for audio and video data, an H.223/Mobile layer consisting of adaptation layers, and a multiplexer. The multiplexed data are to be stored on a PC hard drive and pass through physical-layer simulation to the decoding system that performs the inverse operations such as H.223/Mobile demultiplexing and adaptation layers for audio and video, AL consisting of adaptation layer for audio and video data, and back to the application layer which consists of simulated audio data and MPEG-4 error-resilient video decoder.

In this system simulation, the audio data will be generated by a random number generator and will be used as a placeholder for multiplexing purposes only. The video data will be encoded and multiplexed and stored on a hard drive off-line (i.e., non-real-time). However, the error simulator and MPEG-4 decoder will be developed to run in real time on a PC. It should be noted that at this MPEG meeting a real-time video decoder with error simulator was demonstrated by the error-resilience ad hoc group.

A similar system configuration and simulation needs to be developed for other error-prone networks that are to be investigated. In particular, the video group would also like to test its error-resilience capabilities over a packet-based network, such as the Internet. Therefore, a similar simulation needs to be developed for this test.

For the March 1998 tests, the test conditions of Tables 7 and 8 are being considered.

In order to simulate these various networks, a physical-layer simulator needs to be developed. Furthermore, any error control strategy utilized by the network must be taken

Table 7 Wireless Network

	PCS	IMT2000 (1)	IMT2000 (2)
Data rate (total)	32 kbits/s	128 kbits/s	384 kbits/s
Error conditions	Random and burst	Random and burst	Random and burst

Table 8 Internet Simulation

	PSTN modem (1)	PSTN modem (2)	ISDN modem
Data rate (total)	28.8 kbits/s	56 kbits/s	64 kbits/s
Error conditions	Random, burst, and packet	Random, burst, and packet	Packet

into account when developing this simulation. A number of other open issues also remain that need to be resolved.

9. BEYOND CURRENT MPEG-4 WORK

As mentioned earlier, the MPEG-4 effort as described in the chapter thus far has recently been labeled to be MPEG-4 Version 1. Within MPEG-4, a number of other tools are currently being investigated and expect to be mature a little later than expected; these tools are expected to be released in a following release and together with MPEG-4 Version 1, to be called MPEG-4 Version 2. Besides MPEG-4, MPEG committee has recently accepted a new work item called "Multimedia content description interface" that is likely to be the basis of MPEG-7, the next MPEG standard.

9.1. MPEG-4 Version 2

MPEG-4 Version 2 is expected to include either tools that provide new functionalities not supported in MPEG-4 Version 1 or tools that even provide the same functionalities but a lot more efficiently, with much higher quality or with lower implementation cost. Also, Version 2 is expected to maintain backward compatibility with Version 1. However, there are many unresolved issues. The algorithmic work for MPEG-4 Version 2 is expected to occur as a continuation of currently ongoing effort using verification models. For example, here is a partial list of tools expected to be considered for the visual standard.

- Multifunctional coding tools—spatial scalability of arbitrary shaped objects, automatic segmentation
- Shape coding tools—gray-scale shape, interlaced shape, scalable shape
- Coding efficiency tools—shape adaptive DCT, matching persuit, global motion compensation, dynamic sprites, $\frac{1}{4}$ pel motion compensation
- SNHC tools—body animation coding, generic 2-D/3-D mesh coding with texture mapping

The schedule of MPEG-4 Version 2 (35) involves a 1-year delay with respect to the schedule for MPEG-4 Version 1 [i.e., (the first) working draft of MPEG-4 is expected to be released by November 1997 and the committee draft by November 1998].

9.2. MPEG-7: Multimedia Content Description Interface

MPEG-7 is the next standard of MPEG family of standards, with the primary goal to extend the limited capabilities of proprietary solutions for identifying content. MPEG-7

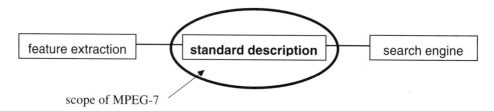

Fig. 31 Scope of MPEG-7.

will specify (44) a standard set of descriptors that can be used to describe various types of multimedia information. This description shall be associated with the content itself, to allow fast and efficient searching for material of a user's interest. Audio–visual material that has MPEG-7 data associated with it can be indexed and searched for; this "material" may include still pictures, graphics, 3-D models, audio, speech, video, and information about how these elements are combined in a multimedia presentation ("scenarios," composition information). Special cases of these general data types may include facial expressions and personal characteristics. Figure 31 shows the current understanding of the scope of MPEG-7.

The words *description* and *feature* represent a rich concept that can be related to several levels of abstraction. Descriptions can vary according to the types of data (e.g., color, musical harmony, textual name, odor, etc.). Descriptions can also vary according to the application (e.g., species, age, number of percussion instruments, information accuracy, people with criminal record, etc.).

MPEG-7 will concentrate on standardizing a *representation* that can be used for categorization. However, development of audio–visual content recognition tools will be a task for industries selling MPEG-7 enabled products. MPEG might build some *coding tools,* just as it did with MPEG-1, MPEG-2, and MPEG-4. The preliminary workplan of MPEG-7 is shown in Table 9.

10. SUMMARY

In this chapter, we have introduced the MPEG-4 standard currently in progress. The necessary background information leading up to the MPEG-4 standard and the multiple facets of this standard were discussed, including the directions for the future. In particular, we have presented the following:

Table 9 MPEG-7 Workplan

Call for proposals	Working draft	Committee draft	Final Draft international standard	International standard
Nov. 1998	Dec 1999	October 2000	July 2001	Sept 2001

- A brief review of the low-bit-rate ITU-T H.263 standard which is the starting basis for the MPEG-4 standard
- Basics of MPEG-4, including discussion of its focus, applications, functionalities, and requirements
- The MPEG-4 Video development process including test conditions, first evaluation, results of evaluation, formulation of core experiments, and definition of the Verification Model
- A brief introduction to the MPEG-4 Audio, SNHC, and Systems development process
- Details of coding tools and methods that are expected to form the MPEG-4 Visual standard
- Introduction to coding methods and tools that are expected to form the MPEG-4 Audio standard
- Discussion of tools and techniques employed by MPEG-4 Systems standard
- The ongoing work on MPEG-4 profiles
- A brief introduction to MPEG-4 Version 2, and the scope of MPEG-7 (the next MPEG standard)

The success of MPEG-4 will eventually depend on many factors, such as the market needs, competing standards, software versus hardware paradigms, complexity versus functionality trade-offs, timing, profiles, and so forth. Technically, MPEG-4 appears to have a significant potential due to the integration of natural and synthetic worlds, computers and communication applications, and the functionalities and flexibilities it offers. Initially, perhaps only the very basic functionalities will be useful. As the demand for sophisticated multimedia grows, the advanced functionalities may be useful. Up-to-date information on progress on various topics in MPEG-4 can be found by visiting MPEG-related web sites (45–49).

REFERENCES

1. R Cox, B Haskell, Y LeCun, B Shahraray, and L Rabiner. On the Applications of Multimedia Processing to Communications. AT&T internal technical memo to be published in IEEE Transactions on Circuits and Systems for Video Technology (CSVT).
2. ISO/IEC 11172 International Standard (MPEG-1), Information Technology—Coding of Moving Pictures and Associated Audio for Digital Storage Media at up to About 1.5 Mbit/s, 1993.
3. BG Haskell, A Puri, AN Netravali. Digital Video: An Introduction to MPEG-2. New York: Chapman & Hall, 1997.
4. ISO/IEC 13818 International Standard (MPEG-2), Information Technology—Generic Coding of Moving Pictures and Associated Audio (also ITU-T Rec. H.262), 1995.
5. ITU-T Recommendation H.261, Video Codec for Audio Visual Services at p × 64 kbit/s, 1990.
6. ITU-T. Draft ITU-T Recommendation H.263: Video Coding for Low Bitrate Communication. December 1995.
7. ITU-T H.263+ Video Group. Draft 12 of ITU-T Recommendation H.263+, May 1997.
8. AOE Group. MPEG-4 Proposal Package Description (PPD)—Rev. 3. ISO/IEC JTC1/SC29/WG11 N0998, Tokyo, July 1995.
9. A Puri. Status and Direction of the MPEG-4 Standard. International Symposium on Multimedia and Video Coding, New York, 1995.
10. A Puri. MPEG-4: A Flexible and Extensible Multimedia Coding Standard in Progress, invited paper in Advances in Multimedia, IEEE Press, 1997.

11. MPEG-4 Systems Group. MPEG-4 Systems Working Draft Version 5.0. ISO/IEC JTC1/SC29/WG11 N1825, Stockholm, July 1997.
12. MPEG-4 Video and SNHC Groups. MPEG-4 Visual Working Draft Version 4.0. ISO/IEC JTC1/SC29/WG11 N1797, Stockholm, July 1997.
13. MPEG-4 Audio Group. MPEG-4 Audio Working Draft Version 4.0. ISO/IEC JTC1/SC29/WG11 N1745, Stockholm, July 1997.
14. Requirements Group. MPEG-4 Requirements version 4. ISO/IEC JTC1/SC29/WG11 N1727, Stockholm, July 1997.
15. Requirements Group. MPEG-4 Applications Document. ISO/IEC JTC1/SC29/WG11 N1729, Stockholm, July 1997.
16. Requirements Group. MPEG-4 Overview. ISO/IEC JTC1/SC29/WG11 N1730, July 1997.
17. D Lindbergh, The H.324 Multimedia Communication Standard. IEEE Commun Mag 34(12): 46–51, 1996.
18. ITU-T. ITU-T Recommendation H.223: Multiplexing Protocol for Low Bitrate Multimedia Communication, 1995.
19. L Chiariglione, MPEG-4 Call for Proposals. ISO/IEC JTC1/SC29/WG11 N0997, Tokyo, July 1995.
20. AOE Group. MPEG-4 Testing and Evaluation Procedures Document. ISO/IEC JTC1/SC29/WG11 N0999, Tokyo, July 1995.
21. H Peterson. Report of the Ad Hoc group on MPEG-4 Video Testing Logistics. ISO/IEC JTC1/SC29/WG11 Doc. MPEG95/0532, November 1995.
22. MPEG-4 Integration Group. MPEG-4 SNHC Proposal Package Description. ISO/IEC JTC1/SC29/WG11 N1199, Florence, March 1996.
23. MPEG-4 Integration Group. MPEG-4 SNHC Call For Proposals. ISO/IEC JTC1/SC29/WG11 N1195, Florence, March 1996.
24. A Puri. Report of Ad hoc Group on Coordination of Future Core Experiments in MPEG-4 Video. ISO/IEC JTC1/SC29/WG11 MPEG 96/0669, Munich, January 1996.
25. T Ebrahimi. Report of Ad hoc Group on Definition of VMs for Content Based Video Representation. ISO/IEC JTC1/SC29/WG11 MPEG 96/0642, Munich, January 1996.
26. MPEG-4 Video Group. MPEG-4 Video Verification Model Version 7.0. ISO/IEC JTC1/SC29/WG11 N1642, Bristol, April 1997.
27. O Avaro, P Chou, A Eleftheriadis, C Herpel, C Reader. The MPEG-4 System and Description Languages. Signal Process: Image Commun (in press).
28. Signal Process: Image Commun, Special Issue on MPEG-4, Part 1: Invited Papers 10(1–3), 1997.
29. Signal Process: Image Commun, Special Issue on MPEG-4, Part 2: Submitted Papers. 10(4), 1997.
30. IEEE Trans. Circuits Syst Video Technol, Special Issue on MPEG-4, CSVT-7(1), 1997.
31. Y Fang, A Eleftheriadis. A syntactic framework for bitstream-level representation of audio-visual objects. Proceedings, 3rd IEEE International Conference on Image Processing (ICIP-96), Lausanne, 1996.
32. A Eleftheriadis. The MPEG-4 System Description Language: From practice to theory. Proceedings, 1997 IEEE International Conference on Circuits and Systems (ISCAS-97), Hong Kong, 1997.
33. A Eleftheriadis. Flavor: A language for media representation. Proceedings, ACM Multimedia '97 Conference, November 1997.
34. J Gosling, B Joy, G Steele. The Java Language Specification. Reading, MA: Addison-Wesley, 1996.
35. L Chiariglione, Resolutions of 40th WG11 meeting. ISO/IEC JTC1/SC29/WG11 N1716, Stockholm, July 1997.
36. A Puri, RL Schmidt, BG Haskell. Improvements in DCT based video coding, Proc. SPIE Visual Communications and Image Processing, San Jose, 1997.

37. PJL van Beek, AM Tekalp, A Puri. 2-D Mesh geometry and motion compression for efficient object based video compression. IEEE Int. Conf. on Image Processing, 1997.
38. BG Haskell, A Puri, J Osterman. Happenings in ISO MPEG: An Introduction to MPEG-4. Data Compression Conference, Snow Bird, 1997.
39. ISO/IEC 14472 Draft International Standard: Virtual Reality Modeling Language, 1997.
40. AL Ames, DR Nadeau, JL Moreland. The VRML Sourcebook. New York: Wiley, 1996.
41. MPEG-4 Requirements Ad hoc Group. Draft of MPEG-4 Requirements. ISO/IEC JTC1/SC29/WG11 N1238, Florence, March 1996.
42. Requirements Group. MPEG-4 Profile Requirements version 4. ISO/IEC JTC1/SC29/WG11 N1728, Stockholm, July 1997.
43. Ad Hoc Group on Error Resilience Core Experiments. Plan for March 1998 Error Resilience Verification Test. ISO/IEC JTC1/SC29/WG11 N1829, Stockholm, July 1997.
44. Requirements Group. MPEG-7 Context and Objectives version 4. ISO/IEC JTC1/SC29/WG11 N1733, Stockholm, July 1997.
45. ISO/IEC JTC1/SC29/WG11 (MPEG) Web Site: http://drogo.cselt.it/mpeg
46. MPEG-4 Systems Web Site: http://garuda.imag.fr/MPEG4
47. Flavor Web Site: http://www.ee.columbia.edu/flavor
48. MPEG-4 Video Web Site: http://wwwam.hhi.de/mpeg-video
49. MPEG-4 SNHC Web Site: http://www.es.com/mpeg4-snhc

11

Video Coding Standards for Multimedia Communication: H.261, H.263, and Beyond

Tsuhan Chen
Carnegie Mellon University, Pittsburgh, Pennsylvania

1. INTRODUCTION

Standards are essential for communication. Without a common language that both the transmitter and the receiver understand, communication is impossible. For multimedia communication that involves transmission of video data, standards play an even more important role. Not only does a video coding standard have to specify a common language, formally known as the *bitstream syntax*, the language also has to be *efficient*. Efficiency has two aspects. One is that the standard has to support a good compression algorithm that brings down the bandwidth requirement for transmitting the video data. The other is that the standard has to allow efficient implementation of the encoder and the decoder, (i.e., the complexity of the compression algorithm has to be as low as possible).

In this chapter, we will discuss technologies used in video coding standards. I will focus on the standards developed by the International Telecommunication Union–Telecommunication Standardization Sector (ITU-T), formerly called the Consultative Committee of the International Telephone and Telegraph (CCITT). These include H.261, H.263, and a recent effort, informally known as H.263+, to provide a new version of H.263, (i.e., H.263 Version 2) in 1998. These video codec standards form important components of the ITU-T H-Series Recommendations that standardize audio–visual terminals in a variety of network environments.

This chapter is outlined as follows. In Sec. 2, we will explain the roles of standards for video coding and provide an overview of standard organizations and video coding standards. In Sec. 3, we will present in detail the techniques used in a historically very important video coding standard, H.261. In Sec. 4, H.263, a video coding standard that has a framework similar to that of H.261 but with superior coding efficiency, will be discussed. Section 5 covers recent activities in H.263+ that resulted in a new version of H.263 with several enhancements. We will conclude this chapter with some remarks in Sec. 6.

2. FUNDAMENTALS OF STANDARDS

Although standards may not be crucial for multimedia storage applications, they are very important for multimedia communication. Suppose some multimedia content needs to be transmitted from a source to a destination. The success of the communication is mainly determined by whether the source and the destination understand the same *language*. Adoption of standards by equipment manufacturers and service providers results in a higher volume and, hence, lowers the cost. In addition, it offers consumers more freedom of choice among manufacturers, and is therefore highly welcomed by consumers.

For transmission of video content, standards play an even more crucial role. Not only does the source and the destination need to speak the same language (i.e., *bitstream syntax*), the language also has to be efficient for compression of video content. This is due to the relatively large amount of bits required to transmit uncompressed video data.

There are two major types of standards. Industrial or commercial standards are mainly defined by mutual agreement among a number of companies. Sometimes, these standards can become very popular in the market and hence become the de facto standards and widely accepted by other companies. The other type is called voluntary standards that are defined by volunteers in open committees. The agreement of these standards have to be based on a consensus of all committee members. These standards are usually driven by market needs. At the same time, they need to stay ahead of the development of technologies. It would be very difficult for the product vendors to agree on a common ground if each one has already developed products based on its own proprietary techniques.

The standards we will discuss in this chapter belong to the second type. For multimedia communication, there are two major standard organizations: the International Telecommunication Union–Telecommunication Standardization Sector (ITU-T) and the International Organization for Standardization (ISO). Recent video coding standards defined by these two organizations are summarized in Table 1. These standards differ mainly in the operating bit rates due to the applications for which they are originally designed, although all standards can essentially be used for all applications in a wide range of bit rates. In terms of coding algorithms, all standards in Table 1 follow a similar framework and, as

Table 1 Video Coding Standards

Standards organization	Video coding standard	Typical range of bit rates	Typical applications
ITU-T	H.261	$p \times 64$ kbits/s, $p = 1, \ldots, 30$	ISDN video phone
ISO	IS 11172-2 MPEG-1 Video	1.2 Mbits/s	CD-ROM
ISO	IS 13818-2 MPEG-2 Video[a]	4–80 Mbits/s	SDTV, HDTV
ITU-T	H.263	≤ 64 kbits/s	PSTN video phone
ISO	CD 14496-2 MPEG-4 Video	24–1024 kbits/s	–
ITU-T	H.263 Version 2	< 64 kbits/s	PSTN video phone
ITU-T	H.263L	< 64 kbits/s	–

[a] ITU-T also actively participated in the development of MPEG-2 Video. In fact, ITU-T H.262 refers to the same standard and uses the same text as IS 13818-2.

we will explain later, differ only in the ranges of parameters and some specific coding modes. In this chapter, we will focus on the standards developed by ITU-T: H.261, H.263, and H.263 Version 2.

There are two approaches to understanding a video coding standard. One approach is to focus on the bitstream syntax and try to understand what each layer of the syntax represents and what each bit in the bitstream indicates. This approach is very important for manufacturers who need to build equipment that is compliant to the standard. The other approach is to focus on coding algorithms that can be used to generate standard-compliant bitstreams and try to understand what each component does and why some algorithms are better than others. Although, strictly speaking, a standard does not specify any encoding algorithms, the latter approach provides a better understanding of video coding techniques as a whole, not just the standard bitstream syntax. Therefore, we will take this approach in this chapter and will describe certain bitstream syntax only when necessary.

3. H.261

H.261 is a video coding standard defined by the ITU-T Study Group XV (SG15)* for video telephony and videoconferencing applications (1). It emphasizes low bit rates and the low coding delay. It was originated in 1984 and intended to be used for audiovisual services at bit rates around $m \times 384$ kbits/s, where m is between 1 and 5. In 1988, the focus shifted and it was decided to aim at bit rates around $p \times 64$ kbits/s, where p is from 1 to 30. Therefore, H.261 also has an informal name called $p \times 64$ (states as p times 64). H.261 was approved in December 1990. The coding algorithm used in H.261 is basically a hybrid of *motion compensation* to remove temporal redundancy and *transform coding* to reduce spatial redundancy. Such a framework forms the basis of all video coding standards that were developed later. Therefore, H.261 has a very significant influence on many other existing and evolving video coding standards.

3.1. Source Picture Formats and Positions of Samples

Digital video is composed of a sequence of pictures, or frames, that occur at a certain rate. For H.261, the frame rate is specified to be 30,000/1001 (approximately 29.97) pictures per second. Each picture is composed of a number of samples. These samples are often referred to as pixels (picture elements), or simply pels. For a video coding standard, it is important to understand the picture sizes to which the standard applies as well as the position of the samples. H.261 is designed to deal with two picture formats:

---

* Note that after a recent reorganization within ITU-T in early 1997, SG16 is now the group for video coding standards. In this chapter, however, we will mostly use SG15 to refer to this study group. This should not cause much confusion as most of the development of standards mentioned here actually happened before the reorganization.

Table 2 Picture Formats Supported by H.261 and H.263

	Sub-QCIF	QCIF	CIF	4CIF	16CIF
No. of pixels per line	128	176	352	704	1408
No. of lines	96	144	288	576	1152
Uncompressed bit rate	4.4Mbs	9.1 Mb/s	37 Mb/s	146 Mb/s	584 Mb/s

the common intermediate format (CIF) and the quarter CIF (QCIF).* Please refer to Table 2, which summarizes a variety of picture formats. Using the terminology of the computer industry, the CIF is close to the CGA format commonly used in computer displays. At such a resolution, the picture quality is not very high. It is close to the quality of a typical video cassette recorder and is much less than the quality of the broadcast television. This is because H.261 is designed for video telephony and videoconferencing, in which typical source material is composed of scenes of talking persons, so-called head and shoulder sequences, rather than general TV programs that contain a lot of motion and scene changes.

In H.261, each sample contains a luminance component, called Y, and two chrominance components, called C_B and C_R. The values of these components are defined as in Ref. 2. In particular, black is represented by $Y = 16$, white is represented by $Y = 235$, and the range of C_B and C_R is between 16 and 240, with 128 representing zero color difference (i.e., gray). A picture format, as shown in Table 2, defines the size of the image, hence the resolution of the Y pels. The chrominance pels, however, typically have a lower resolution than the luminance pels, in order to take advantage of the fact that human eyes are less sensitive to chrominance than to luminance. In H.261, the C_B and C_R pels are specified to have half the resolution, both horizontally and vertically, of that of the Y pels. This is commonly referred to as the 4:2:0 format. Each C_B or C_R pel lies in the center of four neighboring Y pels, as shown in Fig. 1. Note that block edges, to be defined in the next section, lie in between rows or columns of Y pels.

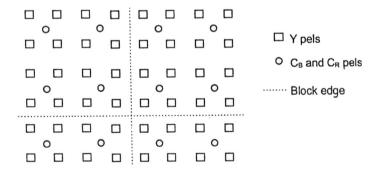

Fig. 1 Positions of samples for H.261.

---

* In the still-image mode as defined in Annex D of H.261, four times the currently transmitted video format is used. For example, if the video format is CIF, the corresponding still-image format is 4CIF.

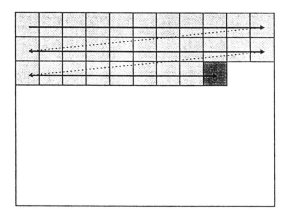

Fig. 2 Illustration of block-based coding.

3.2. Blocks, Macroblocks, and Group of Blocks

Typically, we do not code an entire picture all at once. Instead, it is divided into blocks that are processed one by one, both by the encoder and the decoder, in a scan order as shown in Fig. 2. This approach is often referred to as *block-based coding*.

In H.261, a block is defined as a group of 8×8 pels. Because of the down-sampling in the chrominance components as mentioned earlier, one block of C_B pels and one block of C_R pels correspond to four blocks of Y pels. The collection of these six blocks is called a macroblock (MB), as shown in Fig. 3, with the order of blocks marked as 1 to 6. A MB is treated as one unit in the coding process. A number of MBs are grouped together and called a group of blocks (GOB). For H.261, a GOB contains 33 MBs, as shown in Fig. 4. The resulting GOB structures for a picture, in the CIF case and the QCIF case, are shown in Fig. 5.

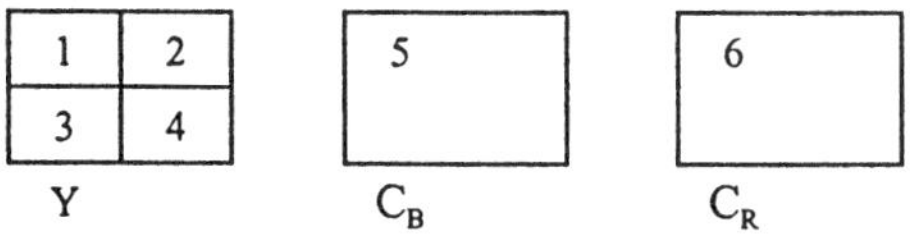

1	2
3	4

Y

5

C_B

6

C_R

Fig. 3 A macroblock.

1	2	3	4	5	6	7	8	9	10	11
12	13	14	15	16	17	18	19	20	21	22
23	24	25	26	27	28	29	30	31	32	33

Fig. 4 A group of blocks.

GOB 1	GOB 2
GOB 3	GOB 4
GOB 5	GOB 6
GOB 7	GOB 8
GOB 9	GOB 10
GOB 11	GOB 12

CIF

GOB 1
GOB 3
GOB 5

QCIF

Fig. 5 GOB structures.

3.3. The Compression Algorithm

Compression of video data typically is based on two principles: the reduction of spatial redundancy and the reduction of temporal redundancy. H.261 uses the discrete cosine transform to remove spatial redundancy, and motion compensation to remove temporal redundancy. We now discuss these techniques in detail.

3.3.1. Transform Coding

Transform coding has been widely used to remove redundancy between data samples. In transform coding, a set of data samples are first linearly transformed into a set of *transform coefficients*. These coefficients are then quantized and entropy coded. A proper linear transform can decorrelate the input samples, hence removing the redundancy. Another way to look at this is that a properly chosen transform can concentrate the energy of input samples into a small number of transform coefficients, so that resulting coefficients are easier to encode than the original samples.

The most commonly used transform for video coding is the discrete cosine transform (DCT) (3,4). Both in terms of objective coding gain and subjective quality, DCT performs very well for typical image data. The DCT operation can be expressed in terms of matrix multiplication:

$$\mathbf{Y} = \mathbf{C}^T \mathbf{X} \mathbf{C}$$

where $\mathbf{X}$ represents the original image block and $\mathbf{Y}$ represents the resulting DCT coefficients. The elements of $\mathbf{C}$, for an 8×8 image block, are defined as

$$C_{mn} = k_n \cos\left(\frac{(2m + 1)n\pi}{16}\right)$$

where

$$k_n = \begin{cases} \dfrac{1}{(2\sqrt{2})} & \text{when } n = 0 \\ \dfrac{1}{2} & \text{otherwise} \end{cases}$$

After the transform, the DCT coefficients in $\mathbf{Y}$ are quantized. Quantization implies loss of information and is the primary source of the compression. The quantization step size

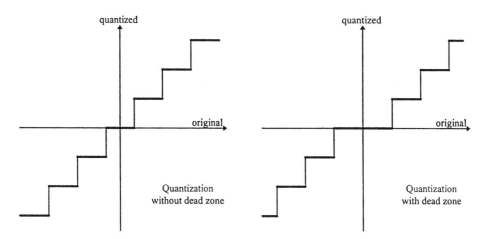

Fig. 6 Quantization with and without the "dead zone."

depends on the available bit rate and can also depend on the coding modes. Except for the intra DC coefficients that are uniformly quantized with a step size of 8, the "dead zone" is used to quantize all other coefficients in order to remove noise around zero. The input–output relations for the two cases are shown in Fig. 6.

The quantized 8×8 DCT coefficients are then converted into a one-dimensional (1-D) array for entropy coding. Figure 7 shows the scan order used in H.261 for this conversion. Most of the energy concentrates on the low-frequency coefficients, and the high-frequency coefficients are usually very small and are quantized to zero before the scanning process. Therefore, the scan order in Fig. 7 can create long runs of zero coefficients, which is important for efficient entropy coding, as we will discuss in the next paragraph.

The resulting 1-D array is then decomposed into segments, with each segment containing one or more (or none) zeros followed by a nonzero coefficient. Let an *event*

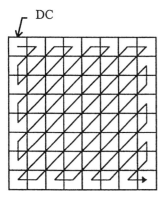

Fig. 7 Scan order of the DCT coefficients.

Table 3 Part of the VLC Table

Run	Level	Code
0	1	1s If first coefficient in block
0	1	11s Not first coefficient in block
0	2	0100 s
0	3	0010 1s
0	4	0000 110s
0	5	0010 0110 s
0	6	0010 0001 s
0	7	0000 0010 10s
0	8	0000 0001 1101 s
0	9	0000 0001 1000 s
0	10	0000 0001 0011 s
0	11	0000 0001 0000 s
0	12	0000 0000 1101 0s
0	13	0000 0000 1100 1s
0	14	0000 0000 1100 0s
0	15	0000 0000 1011 1s
1	1	011s
1	2	0001 10s
1	3	0010 0101 s
1	4	0000 0011 00s
1	5	0000 0001 1011 s
1	6	0000 0000 1011 0s
1	7	0000 0000 1010 1s
2	1	0101 s
2	2	0000 100s
2	3	0000 0010 11s
2	4	0000 0001 0100 s
2	5	0000 0000 1010 0s
3	1	0011 1s
3	2	0010 0100 s
3	3	0000 0001 1100 s
3	4	0000 0000 1001 1s
...	...	...

represent the pair of (run, level), where "run" represents the number of zeros and "level" represents the magnitude of the nonzero coefficient. This coding process is sometimes called "run-length coding." Then, a Huffman coding table is built to represent each event by a specific code word (i.e., a sequence of bits). Events that occur more often are represented by shorter code words, and less frequent events are represented by longer code words. So, the table is often called a variable-length-coding (VLC) table. In H.261, this table is often referred to as a two-dimensional (2-D) VLC table because of its 2-D nature (i.e., each event representing a pair of (run, level)). Some entries of VLC table used in H.261 are shown in Table 3. In this table, the last bit "s" of each code word denotes the sign of the level, "0" for positive and "1" for negative. It can be seen that more likely events (i.e., short runs and low levels) are represented with short code words, and vice versa. After the last nonzero DCT coefficient is sent, the EOB symbol, represented by 10, is sent to indicate the end of block.

At the decoder, all of the above steps are reversed one by one. Note that all the steps can be exactly reversed except for the quantization step, which is where the loss of information arises.

3.3.2. Motion Compensation

The transform coding described in the previous subsection removes spatial redundancy within each frame of picture. It is therefore referred to as *intra* coding. However, for video material, *inter* coding is also very useful. Typical video material contains a large amount of redundancy along the temporal axis. Video frames that are close in time usually have a large amount of similarity. Therefore, transmitting the difference between frames is more efficient than transmitting the original frames. This is similar to the concept of differential coding and predictive coding. The previous frame is used as an estimate of the current frame, and the residual, the difference between the estimate and the true value, is coded. When the estimate is good, it is more efficient to code the residual than to code the original frame.

Consider the fact that typical video material is composed of moving objects. Therefore, it is possible to improve the prediction result by first estimating the motion of each region in the scene. More specifically, the encoder can estimate the motion (i.e., displacement) of each block between the previous frame and the current frame. This is often achieved by matching each block (actually, macroblock) in the current frame with the previous frame to find the best matching area.* This area is then offset accordingly to form the estimate of the corresponding block in the current frame. Now, the residue has much less energy and, therefore, is much easier to code. This process is called motion compensation (MC), or, more precisely, motion-compensated prediction (5,6). This is illustrated in Fig. 8. The residue is then coded using the same process as that of intra coding.

Frames that are coded without any reference to previously coded frames are called intra frames, or simply I-frames. Frames that are coded using a previous frame as a reference for prediction are called predicted frames, or simply P-frames. However, note that a P-frame may also contain intra coded blocks. The reason is as follows. For a certain block, it may be impossible to find a good enough matching area in the reference frame to be used as prediction. In this case, direct intra coding of such a block is more efficient. This situation happens often when there is occlusion in the scene, or when the motion is very heavy.

Motion compensation saves the bits for coding the DCT coefficients. However, it does imply that extra bits are required to carry information about the motion vectors. Efficient coding of the motion vector is therefore also an important part of H.261. Because motion vectors of neighboring blocks tend to be similar, differential coding of the horizontal and vertical components of motion vectors is used; that is, instead of coding motion

---

* Note, however, the standard does not specify how motion estimation should be done.

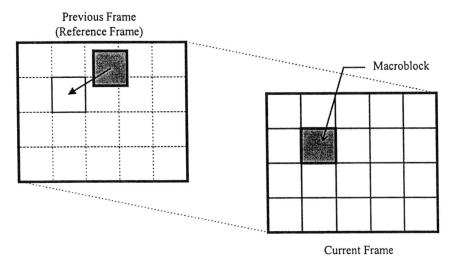

Fig. 8 Motion compensation.

vectors directly, the previous motion vector is used as a prediction for the current motion vector, and the difference, in both the horizontal component and the vertical component, is then coded using the VLC table, part of which is shown in Table 4. Note two things in this table. First, short code words are used to represent a small difference, because these are more likely events. Second, note that one code word can represent up to two possible values for the motion vector difference. Because the range of either the horizontal compo-

Table 4 Part of the VLC Table for Coding of Motion Vectors

MVD	Code	MVD	Code
...	...	1	010
−7 & 25	0000 0111	2 & −30	0010
−6 & 26	0000 1001	3 & −29	0001 0
−5 & 27	0000 1011	4 & −28	0000 110
−4 & 28	0000 111	5 & −27	0000 1010
−3 & 29	0001 1	6 & −26	0000 1000
−2 & 30	0011	7 & −25	0000 0110
−1	011	...	...
0	1		

Video Coding Standards for Multimedia

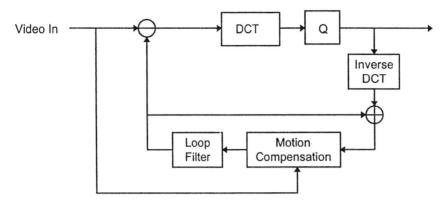

Fig. 9 Block diagram of a video encoder.

nent and the vertical component of motion vectors is between −15 and +15, only one will yield a motion vector with the allowable range.

3.3.3. Summary

The coding algorithm used in H.261 can be summarized into block diagrams in Figs. 9 and 10. At the encoder, the input picture is compared with the previously decoded frame with motion compensation. The difference signal is DCT transformed and quantized, and then entropy coded and transmitted. At the decoder, the decoded DCT coefficients are inverse DCT transformed and then added to the previously decoded picture with motion compensation.

Because the prediction of the current frame is composed of blocks at various locations in the reference frame, the prediction itself (simply called the predicted frame) may contain coding noise and blocking artifacts. These artifacts may cause a higher prediction error. It is possible to reduce the prediction error by passing the predicted frame through a low-pass filter before it is used as the prediction for the current frame. This filter is referred to as a loop filter, because it operates inside the motion-compensation loop.

3.4. Reference Model

As in all video coding standards, H.261 specifies only the bitstream syntax and how a decoder should interpret the bitstream to decode the image. Therefore, it specifies only

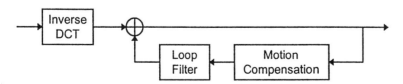

Fig. 10 Block diagram of a video decoder.

the design of the decoder, not how the encoding should be done. For example, an encoder can simply decide to use only zero motion vectors and let the transform coding take all the burden of coding the residual. This may not be an efficient encoding algorithm, but it does generate a standard-compliant bitstream.

Therefore, to illustrate the effectiveness of a video coding standard, an example encoder is often provided by the group that defines the standard. For H.261, such an example encoder is called a reference model (RM), and the latest version is RM 8 (7). It specifies details about motion estimation, quantization, decisions for inter/intra coding, and MC/no MC, buffering, and the rate control.

3.5. ITU-T H-Series Related to H.261

H.261 has been included in several ITU-T H-series terminal standards for various network environments. One example is H.320, which is mainly designed for narrow-band integrated service digital network (ISDN) terminals (8). H.320 defines the systems and terminal equipment that use H.261 for video coding, H.221 for frame multiplexing, H.242 for signaling protocol (9), and G.711, G.722, and G.728 for audio coding. Sometimes, H.320 is also used to refer to this set of standards. H.261 can also be used in other terminal standards, including H.321, H.322, H.323, and H.324.

4. H.263

H.263 (10) was defined by ITU-T SG15, the same group that defined H.261. The activities of H.263 started around November 1993, and the standard was adopted in March 1996. The main goal of this endeavor was to design a video coding standard suitable for applications with bit rates below 64 kbits/s (the so-called very low-bit-rate applications). For example, when sending video data over the public service telephone network (PSTN) and the mobile network, the video bit rates typically range from 10 to 24 kbits/s. During the development of H.263, it was identified that the near-term goal would be to enhance H.261 using the same general framework, and the long-term goal would be to design a video coding standard that may be fundamentally different from H.261 in order to achieve further improvement in coding efficiency. As the standardization activities move along, the near-term effort became H.263 and H.263 Version 2 (to be discussed in the next section), and the long-term effort is now referred to as H.263L.

In this section, we will discuss H.263. In essence, H.263 combines the features of H.261 together with MPEG and is optimized for very low bit rates. In terms of the signal-to-noise ratio (SNR), H.263 can provide 3–4-dB gain over H.261 at bit rates below 64 kbits/s. In fact, H.263 provides superior coding efficiency to that of H.261 at all bit rates. When compared with MPEG-1, H.263 can give a 30% bit-rate saving.

4.1. H.263 Versus H.261

Because H.263 was built on top of H.261, the main structures of the two standards are essentially the same. Therefore, we will focus only on the differences between the two standards. The major differences are as follows:

1. H.236 supports more picture formats and uses a different GOB structure.

2. H.263 uses half-pel motion compensation but does not use loop filtering, as in H.261.
3. H.263 uses 3-D VLC for coding of DCT coefficients.
4. In addition to the basic coding algorithm, four options that are negotiable between the encoder and the decoder provide improved performance.
5. H.263 allows the quantization step size to change at each MB with less overhead.

4.2. Picture Formats, Sample Positions, and the GOB Structure

In addition to CIF and QCIF as supported by H.261, H.263 also supports sub-QCIF, 4CIF, and 16CIF. Resolutions of these picture formats can been found in Table 2. Chrominance subsampling and the relative positions of chrominance pels are the same as those defined in H.261. However, H.263 uses different GOB structures. These are shown in Fig. 11 for various formats. Unlike H.261, a GOB in H.263 always contains at least one full row of MBs.

4.3. Half-Pel Prediction and Motion Vector Coding

A major difference between H.261 and H.263 is the half-pel prediction in the motion compensation. This concept is also used in MPEG. Although the motion vectors in H.261 can have only integer values, H.263 allows the precision of motion vectors to be at a half of a pel. For example, it is possible to have a motion vector with values (4.5, −2.5). When a motion vector has noninteger values, bilinear interpolation is used to find the corresponding pel values for prediction.

The coding of motion vectors in H.263 is more sophisticated than that in H.261. The motion vectors of three neighboring MBs (the left, the above, and the above-right,

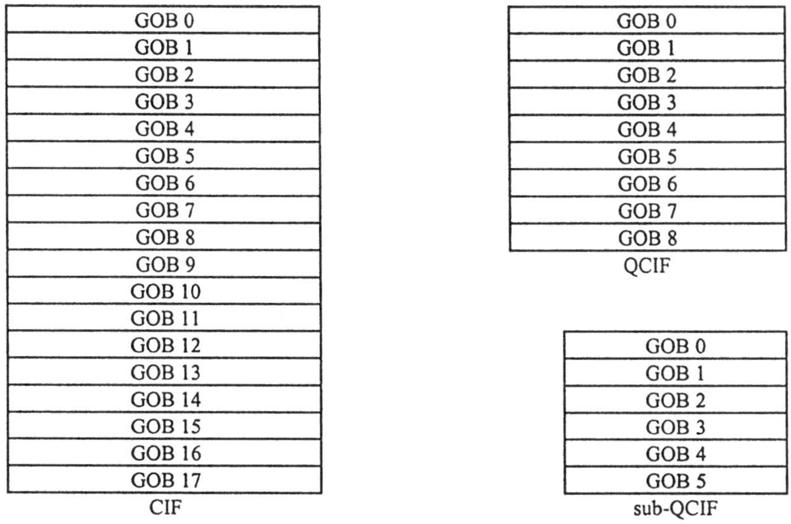

GOB 0
GOB 1
GOB 2
GOB 3
GOB 4
GOB 5
GOB 6
GOB 7
GOB 8
GOB 9
GOB 10
GOB 11
GOB 12
GOB 13
GOB 14
GOB 15
GOB 16
GOB 17

CIF

GOB 0
GOB 1
GOB 2
GOB 3
GOB 4
GOB 5
GOB 6
GOB 7
GOB 8

QCIF

GOB 0
GOB 1
GOB 2
GOB 3
GOB 4
GOB 5

sub-QCIF

Fig. 11 GOB structures for H.263.

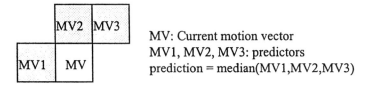

Fig. 12 Prediction of motion vectors.

as shown in Fig. 12) are used as predictors. The median of the three predictors is used as the prediction for the motion vector of the current block, and the prediction error is coded and transmitted. However, around a picture boundary or GOB boundary, special cases are needed. When only one neighboring MB is outside the picture boundary or GOB boundary, a zero motion vector is used to replace the motion vector of that MB as the predictor. When two neighboring MBs are outside, the motion vector of the only neighboring MB that is inside is used as the prediction. These are shown in Fig. 13.

4.4. Run-Length Coding of DCT Coefficients

H.263 improves the run-length coding used in H.261 by giving an extra term "last" to indicate whether the current coefficient is the last nonzero coefficient of the block. Therefore, a set of (run, level, last) represents an event and is mapped to a code word in the VLC table, hence the name 3-D VLC. With this scheme, the EOB (end of block) code used in H.261 is no longer needed. Tables 5 and 6 show some entries of the VLC table.

4.5. Negotiable Options

H.263 specifies four options that are negotiable between the encoder and the decoder. At the beginning of each communication session, the decoder signals the encoder which of these options the decoder has the capability to decode. If the encoder also supports some of these options, it may enable those options. However, the encoder does not have to enable all the options that are supported by both the encoder and decoder. The four options in H.263 are the unrestricted motion vector mode, the syntax-based arithmetic coding mode, the advanced prediction mode, and the PB-frame mode.

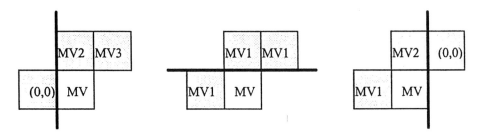

Fig. 13 Motion vector prediction at picture/GOB boundaries.

Table 5 Partial VLC Table for DCT Coefficients

Last	Run	Level	Code
0	0	1	10s
0	0	2	1111 s
0	0	3	0101 01s
0	0	4	0010 111s
0	0	5	0001 1111 s
0	0	6	0001 0010 1s
0	0	7	0001 0010 0s
0	0	8	0000 1000 01s
0	0	9	0000 1000 00s
0	0	10	0000 0000 111s
0	0	11	0000 0000 110s
0	0	12	0000 0100 000s
...	...	...	...

4.5.1. Unrestricted Motion Vector Mode

This is the first one of the four negotiable options defined in H.263. In this option, motion vectors are allowed to point outside of the picture boundary. In this case, edge pels are repeated to extend to the pels outside so that prediction can be done. Significant coding gain can be achieved with unrestricted motion vectors if there is movement around picture edges, especially for smaller picture formats like QCIF and sub-QCIF. In addition, this mode allows a wider range of motion vectors than H.261. Large motion vectors can be very effective when the motion in the scene is caused by heavy motion (e.g., motion due to camera movement).

4.5.2. Syntax-Based Arithmetic Coding

In this option, arithmetic coding (11) is used, instead of VLC tables, for entropy coding. Under the same coding condition, using arithmetic coding will result in a bitstream different

Table 6 Partial VLC Table for DCT Coefficients

Last	Run	Level	Code
...	...	...	...
1	0	1	0111 s
1	0	2	0000 1100 1s
1	0	3	0000 0000 101s
1	1	1	0011 11s
1	1	2	0000 0000 100s
1	2	1	0011 10s
1	3	1	0011 01s
1	4	1	0011 00s
1	5	1	0010 011s
1	6	1	0010 010s
1	7	1	0010 001s
1	8	1	0010 000s

from the bitstream generated by using a VLC table, but the reconstructed frames and the SNR will be the same. Experiments show that the average bit-rate saving is about 3–4% for inter frames and about 10% for intra blocks and frames.

4.5.3. Advanced Prediction Mode

In the advanced prediction mode, overlapped block-motion compensation (OBMC) (12) is used to code the luminance of P-pictures, which typically results in less blocking artifacts. This mode also allows the encoder to assign four independent motion vectors to each MB; that is, each block in a MB can have an independent motion vector. In general, using four motion vectors gives better prediction, because one motion vector is used to represent the movement of an 8×8 block instead of a 16×16 MB. Of course, this implies more motion vectors and, hence, requires more bits to code the motion vectors. Therefore, the encoder has to decide when to use four motion vectors and when to use only one. Finally, in the advanced prediction mode, motion vectors are allowed to cross picture boundaries, as is the case in the unrestricted motion vector mode.

When four vectors are used, the prediction of motion vectors has to be redefined. In particular, the locations of the three "neighboring" blocks of which the motion vectors are to be used as predictors now depend on the position of the current block in the MB. These are shown in Fig. 14. It is interesting to note how these predictors are chosen. Consider the situation depicted in the upper left of Fig. 14. When the motion vector corresponds to the upper left block in a MB, note that the third predictor (MV3) is not even connected to the current block. What happens if we were to use the motion vector of a closer block, say the one marked with MV* in Fig. 14? In that case, MV* would very likely be the same as MV2 (because they belong to the same MB) and the median of the three predictors would be equal to MV2. Therefore, the advantage of using three predictors would be lost.

4.5.4. PB-Frame Mode

In the PB-frame mode, a PB-frame consists of two pictures coded as one unit, as shown in Fig. 15. The first picture, called the P-picture, is a picture predicted from the last decoded picture. The last decoded picture can be either an I-picture, a P-picture, or the P-picture

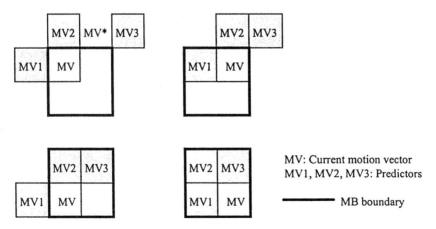

Fig. 14 Redefinition of motion vector prediction.

Video Coding Standards for Multimedia

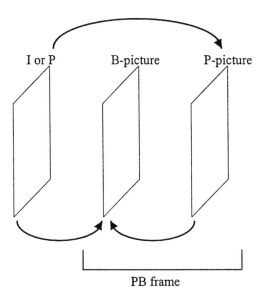

Fig. 15 The PB-frame mode.

of a PB-frame. The second picture, called the B-picture (B for bidirectional), is a picture predicted from both the last decoded picture and the P-picture that is currently being decoded. As opposed to the B-frames used in MPEG, PB frames do not need separate bidirectional vectors. Instead, forward vectors for the P-picture is scaled and added to a small delta-vector, to obtain vectors for the B-picture. This results in less bit-rate overhead for the B-picture. For relatively simple sequences at low bit rates, the picture rate can be doubled with this mode with a minimal increase in the bit rate. However, for sequences with heavy motion, PB-frames do not work as well as B-pictures. Also, note that the use of the PB-frame mode increases the end-to-end delay, so it may not be suitable for two-way interactive communication.

4.6. Test Model Near-Term

Similar to H.261, there are documents drafted by ITU-T SG15 that describe example encoders (i.e., the test models). For H.263, these are called TMN, where N indicates that H.263 is a near-term effort in improving H.261. The latest version is TMN6 (13). TMN6 specifies the details of the advanced motion prediction, the overlapped block-motion compensation, the choice between the 8×8 mode and 16×16 mode for motion vectors, the syntax-based arithmetic coding, and the use of the PB frame mode.

4.7. ITU-T H Series Related to H.263

As for H.261, H.263 can be used in several terminal standards for different network environments. One example is H.324 (14), which defines audio–visual terminals for the traditional public service telephone network (PSTN). In H.324, a telephone terminal uses H.263 as the video codec, H.223 as the multiplexing protocol (15), H.245 as the control

Table 7 Workplan for H.263 Version 2

July '96	Evaluate proposals; begin draft text
Nov. '96	Final proposal evaluations; complete draft written
Feb. '97	Final evaluations completed; finalized text written
Mar. '97	Determination at SG16 meeting
Jan.–Feb. '98	Decision at SG16 meeting

protocol (16), G.723 for speech coding at 5.3–6.3 kbits/s, and V.34 for the modem interface. H.324 is sometimes used to refer to the whole set of standards. H.263 can also be used in other terminal standards, such as H.323, which is designed for local area networks (LAN) without guaranteed quality of service (QoS).

5. H.263 VERSION 2

After the standardization of H.263 was finished, the continued interest in very low-bit-rate video coding made it clear that further enhancements to H.263 were possible, in addition to the four optional modes. ITU-T SG16* therefore established an effort, informally known as H.263+, to meet the need for standardization of such enhancements of H.263. The result is a new version of H.263, H.263 Version 2 (17), which is expected to be *decided* (approved by the Study Group) in January–February 1998. Similar to H.263 Version 1 (the version dated March 1996), H.263 Version 2 is supposed to provide a near-term standardization for the applications of real-time telecommunication and related nonconversational services. These enhancements are either improved quality of functionalities provided by H.263 Version 1, or additional capabilities to broaden the range of applications. For example, the enhancements in H.263 Version 2 include improvement of perceptual compression efficiency, reduction in the video delay, and greater error resilience.

Because H.263 Version 2 was a near-term solution to the standardization of enhancements to H.263, it considered only well-developed proposed enhancements that fit into the framework of H.263 (i.e., motion compensation and DCT-based transform coding). The H.263+ workplan is outlined in Table 7. On the other hand, H.263L is a parallel activity that is intended to be a long-term effort. It considers more radical algorithms that do not necessarily fit in the H.263 framework.

5.1. Development of H.263 Version 2

During the development of H.263 Version 2, proposed techniques are grouped into key technical areas (KTAs). Altogether, there were about 22 KTAs being identified. In November 1996, after consideration of the contributions and after some consolidation of KTAs, 12 KTAs were chosen. These are summarized in a draft text that passed the determination

---

* Toward the end of the H.263+ activities, in early 1997, a reorganization within ITU-T moved the activities in video coding standards from SG15 into SG16.

process in March 1997. Several adopted KTAs result in extra negotiable options. Combined with the original four options in H.263, this makes a total of 16 negotiable coding options in H.263 Version 2, which can be used together or separately. Some KTAs result in extended source formats and a forward error-correction method, and some are adopted as the supplemental enhancement information that may be included in the bitstream to indicate extra functionalities. These new features will be outlined in the next few sections. In addition, a new test model (TMN8) has been prepared by the group for testing, simulation, and comparisons.

5.2. Source Formats

One feature of H.263 Version 2 is that it extends the possible source formats specified in H.263. These extensions include the following:

1. *Higher Picture Clock Frequency* (PCF): This allows picture clock rates higher than 30 frames per second. This feature helps to support additional camera and display technologies.
2. *Custom Picture Formats:* It is possible for the encoder and the decoder to negotiate custom picture formats, not limited by a number of fixed formats anymore. The number of lines can be from 4 to 1152 as long as it is divisible by 4, and the number of pels per line can be from 4 to 2048 as long as it is divisible by 4.
3. *Custom Pixel Aspect Ratios* (PAR): This allows the use of additional pixel aspect ratios other than those used in CIF (11:12) and SIF (10:11), and the square (1:1) aspect ratio. All custom PAR are shown in Table 8.

5.3. New Coding Modes for Coding Efficiency

Among the new negotiable coding options specified by H.263 Version 2, five are intended to improve the coding efficiency:

1. *Advanced Intra Coding Mode:* This is an optional mode for intra coding. In this mode, intra blocks are coded using a predictive method. A block is predicted

Table 8 Custom Pixel Aspect Ratios

Pixel aspect ratio	Pixel width:pixel height
Square	1:1
CIF	12:11
525-Type for 4:3 picture	10:11
CIF for 16:9 picture	16:11
525-Type for 16:9 picture	40:33
Extended PAR	$m:n$, m and n are relatively prime

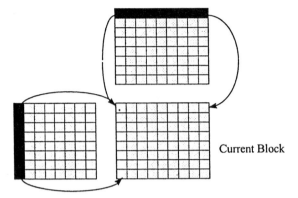

Fig. 16 Advanced intra coding mode.

from the block to the left or the block above, as shown in Fig. 16. For isolated intra blocks for which no prediction can be found, the prediction is simply turned off.
2. *Alternate Inter VLC Mode:* This mode provides the ability to apply a VLC table originally designed for intra coding to inter coding where there are often many large coefficients by simply using a different interpretation of the level and the run.
3. *Modified Quantization Mode:* This mode improves the flexibility of controlling the quantizer step size. It also reduces of the quantizer step size for chrominance quantization, in order to reduce the chrominance artifacts. An extension of the range of values of DCT coefficient is also provided. In addition, by prohibiting certain unreasonable coefficient representations, this mode increases error-detection performance and reduces decoding complexity.
4. *Deblocking Filter Mode:* In this mode, an adaptive filter is applied across the 8 × 8 block edge boundaries of decoded I- and P-pictures to reduce blocking artifacts. The filter affects the picture that is used for the prediction of subsequent pictures and thus lies within the motion prediction loop, similar to the loop filtering in H.261.
5. *Improved PB-Frame Mode:* This mode deals with the problem that the PB-frame mode in H.263 cannot represent large motion very well. It provides a mode with more robust performance under complex-motion conditions. Instead of constraining a forward motion vector and a backward motion vector to come from a single motion vector as in H.263, the improved PB-frame mode allows them to be totally independent, as in the B-frames of MPEG.

5.4. Enhancements for Error Robustness

The following optional modes are especially designed to address the needs of mobile video and other unreliable transport environments:

1. *Slice Structured Mode:* In this mode, a "slice" structure replaces the GOB structure. Slices have more flexible shapes and may appear in any order within

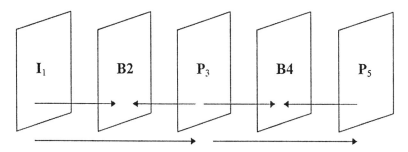

Fig. 17 Temporal scalability.

the bitstream for a picture. Each slice has a specified width. The use of slices allows a flexible partitioning of the picture, in contrast with the fixed partitioning and fixed transmission order required by the GOB structure. This can provide enhanced error resilience and minimize the video delay.

2. *Reference Picture Selection Mode:* In this mode, the reference picture does not have to be the most recently encoded picture. Instead, any temporally previous picture can be referenced. This mode can provide better error resilience in unreliable channels such as mobile and packet networks, because the codec can avoid using an erroneous picture for future reference.

3. *Independent Segment Decoding Mode:* This mode improves error resilience by ensuring that any error in a certain region of the picture does not propagate to other regions.

5.5. Enhancements Related to Scalabilities

The temporal, SNR, and spatial scalability modes support layered-bitstream scalability in three forms, similar to MPEG-2. Bidirectionally predicted frames, the same as those used in MPEG, are used for temporal scalability by adding enhancement frames between other coded frames. This is shown in Fig. 17. A similar syntactical structure is used to provide an enhancement layer of video data to support spatial scalability by adding enhancement information for construction of a higher-resolution picture, as shown in Fig. 18. Finally,

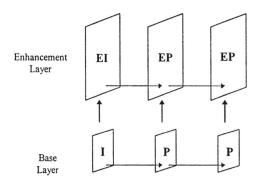

Fig. 18 Spatial scalability.

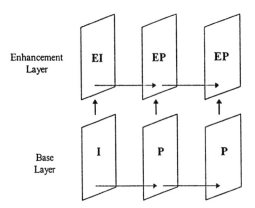

Fig. 19 SNR scalability.

SNR scalability is provided by adding enhancement information for reconstruction of a higher-fidelity picture with the same picture resolution, as in Fig. 19. Furthermore, different scalabilities can be combined together in a very flexible way. Figure 20 gives an example.

5.6. Other Enhancement Modes

There are two other enhance modes described in H.263 Version 2:

1. *Reference Picture Resampling Mode:* This allows a prior coded picture to be resampled, or warped, before it is used as a reference picture. The warping is defined by four motion vectors that specify the amounts of offset of the four

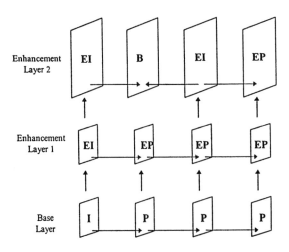

Fig. 20 Multilayer scalability.

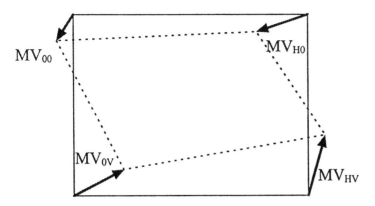

Fig. 21 Reference picture resampling.

corners of the reference picture, as shown in Fig. 21. This mode allows an encoder to smoothly switch between different encoded picture sizes, shapes, and resolutions. It also supports a form of global motion compensation and special-effect image warping.

2. *Reduced-Resolution Update Mode:* This mode allows the encoding of inter frame difference information at a lower spatial resolution than the reference frame. It gives the encoder the flexibility to maintain an adequate frame rate by encoding foreground information at a reduced spatial resolution while holding onto a higher-resolution representation of more stationary areas of a scene.

5.7. Supplemental Enhancement Information

One important feature of H.263 Version 2 is the usage of supplemental information, which may be included in the bitstream to signal enhanced display capabilities or to provide tagging information for external usage. For example, it can be used to signal a full-picture or partial-picture freeze, or freeze-release request with or without resizing. It can be used to label a snapshot, the start and end of a video segment, and the start and end of a progressively refined video. The supplemental information may be present in the bitstream even though the decoder may not be capable of providing the enhanced capability to use it, or even to properly interpret it. In other words, unless a requirement to provide the requested capability has been negotiated by external means in advance, the decoder can simply discard anything in the supplemental information. Another use of the supplemental enhancement information is to specify the chroma-key for representing transparent and semitransparent pixels (18). We will now explain this in more detail.

The Chroma-Keying Information Flag (CKIF) in the supplemental information indicates that the chroma-keying technique is used to represent transparent and semitransparent pixels in the decoded picture. When being presented on the display, transparent pixels are not displayed. Instead, a background picture which is externally controlled is revealed. Semitransparent pixels are rendered by blending the pixel value in the current picture with the corresponding value in the background picture. Typically, an 8-bit number α is used to indicate the transparency: $\alpha = 255$ indicates that the pixel is opaque and $\alpha = 0$ indicates that the pixel is transparent. Between 0 and 255, the displayed color is a weighted sum of the original pixel color and the background pixel color.

When CKIF is enabled, 1 byte is used to indicate the keying color value for each component (Y, C_B, or C_R) which is used for chroma-keying. After the pixels are decoded, the α value is calculated as follows. First, the distance d between the pixel color and the key color value is calculated. The α value is then computed as follows:

```
if          (d < T₁)   then α = 0

else if     (d > T₂)   then α = 255
```

$$\text{else} \quad \alpha = \frac{255(d - T_1)}{T_2 - T_1}$$

where T_1 and T_2 are the two thresholds that can be set by the encoder.

5.8. Levels of Preferred Mode Support

It is interesting to note that although H.263 Version 2 provides a variety of optional modes that are all useful sometimes, not all manufacturers would want to implement all the options. Therefore, H.263 Version 2 contains an appendix that specifies three levels of preferred modes to be supported. Each level contains a number of options to be supported by an equipment manufacturer. This appendix is not a normative part of the standard. It is only used to provide manufacturers some guidelines to determine which modes are more likely to be widely adopted across a full spectrum of terminals and networks.

Three levels of preferred modes are described in H.263 Version 2, and each level supports the optional modes specified in lower levels. The first level is composed of the advanced intra coding mode, the deblocking filtering mode, full-frame freeze as defined in the supplementary information, and the modified quantization mode. The second level supports, in addition to modes supported in Level 1, the unrestricted motion vectors mode, the slice structured mode, and the reference picture resampling mode. In addition to these modes, Level 3 supports the advanced prediction mode, the improved PB-frame mode, the independent segment decoding mode, and the alternative inter VLC mode.

5.9. Further Work

When the proposals for H.263 Version 2 were evaluated and some were adopted, it became apparent that many new proposals fit into the H.263 syntactical framework but would not be ready for determination in March 1997. SG16 therefore considered another round of H.263 extensions, informally called H.263++, that would create a third generation of H.263 syntax. Four key technical areas were identified in which the group has an interest in pursuing further investigation toward possible later standardization. These KTAs are variable transform type, adaptive arithmetic coding, error-resilient VLC tables, and deringing filtering.

6. CONCLUSION

By explaining the technical details of a number of important video coding standards defined by ITU-T, I hope I have provided the readers some insight to the significance of

Table 9 Sources of Further Information

http://www.itu.ch	ITU-T
ftp://standard.pictel.com	General standards
ftp://standard.pictel.com/video-site	ITU-T Video Coding Experts Group (Q.15/16)
ftp://standard.pictel.com/lbc-site	ITU-T CSN Experts Group (Q.11/16)
ftp://standard.pictel.com/avc-site	ITU-T APC Experts Group (Q.12, 13, 14/16)

international standards for multimedia communication. Pointers to more up-to-date information about the video coding standards described in this chapter can be found in Table 9. When this chapter was prepared, activities in H.263++ and H.263L were ongoing. It is recommended that the readers check the resources in Table 9 for more recent updates of H.263++ and H.263L.

ACKNOWLEDGMENT

I would like to thank the anonymous reviewer for very insightful suggestions that helped significantly improve the quality of this chapter.

REFERENCES

1. ITU-T Recommendation H.261. Video Codec for Audiovisual Services at p × 64 kbit/s. Geneva, 1990, revised at Helsinki, March 1993.
2. ITU-R Recommendation BT.601-4. Encoding Parameters of Digital; Television for Studios.
3. N Ahmed, T Natarajan, KR Rao. Discrete cosine transform. IEEE Trans. on Computers C-23:90–93, 1974.
4. KR Rao, P Yip. Discrete Cosine Transform. New York: Academic Press, 1990.
5. AN Netravali, JD Robbins. Motion-compensated television coding: Part I. Bell Syst Tech 58(3):631–670, 1979.
6. AN Netravali, BG Haskell. Digital Pictures. 2nd ed. New York: Plenum Press, 1995.
7. Description of Reference Model 8 (RM8). CCITT Study Group XV, Specialist Group on Coding for Visual Telephony. Doc. No. 525, June 1989.
8. ITU-T Recommendation H.320. Narrow-Band ISDN Visual Telephone Systems and Terminal Equipment. March 1996.
9. ITU-T Recommendation H.242. System for Establishing Communication Between Audiovisual Terminals Using Digital Channels up to 2Mbit/s, 1993.
10. ITU-T Recommendation H.263: ''Video coding for low bitrate communication,'' March 1996.
11. IH Witten, RM Neal, JG Cleary. Arithmetic coding for data-compression. Commun ACM 30(6):520–540, 1987.
12. MT Orchard, GJ Sullivan. Overlapped Block Motion Compensation—An Estimation-Theoretic Approach. IEEE Trans Image Process 3(5):693–699, 1994.
13. ITU-T SG15, H.236+ Ad Hoc Group. Video Test Model, TMN6. LBC-96-141, April 1996.
14. ITU-T Recommendation H.324. Terminal for Low Bitrate Multimedia Communication. March 1996.

15. ITU-T Recommendation H.223. Multiplexing Protocol for Low Bitrate Multimedia Communication, 1995.
16. ITU-T Recommendation H.245. Control Protocol for Multimedia Communication, 1995.
17. ITU-T SG16, Gary Sullivan, ed. Draft Text of Recommendation H.263 Version 2 (''H.263+'') for Decision, September 1997.
18. T Chen, CT Swain, BG Haskell. Coding of sub-regions for content-based scalable video. IEEE Trans Circuits Syst Video Technol CSVT-7(1):256–260, 1997.

12
AM–FM Image Modeling and Gabor Analysis

Joseph P. Havlicek
University of Oklahoma, Norman, Oklahoma

Alan Bovik
University of Texas at Austin, Austin, Texas

Dapang Chen
National Instruments, Austin, Texas

1. INTRODUCTION

In his classic 1946 paper, Gabor proposed that arbitrary time signals could be synthesized and analyzed in terms of an appropriate set of *elementary signals* conjointly localized in time and frequency (1). He was motivated by applications in communications engineering. Specifically, he was interested in the optimal transmission of information. He observed that the human auditory system perceives the energy distribution of acoustical signals simultaneously in both time and frequency. By contrast, classical time-domain and Fourier analysis techniques each characterize the distribution of signal energy in only one of these two domains—in time *or* in frequency.

    It is well known that the amount of information that a signal can transmit is proportional to the product of the bandwidth with the temporal duration. One of the innovative things that Gabor did was to look critically at the joint distribution of a signal's energy in the *time–frequency plane,* which he called the *information diagram.* He reasoned that the information-bearing capacity of a signal is given by its area in the information diagram and, borrowing techniques from the theory of quantum mechanics, deduced that "there are certain 'elementary signals' which occupy the smallest possible area." These elementary signals are the translates of a Gaussian envelope frequency modulated by complex exponentials. Today, they are known as *Gabor functions.* Gabor proved that these signals realize the uncertainty principle lower bound on simultaneous localization in time and frequency and that they are unique in this respect. He partitioned the information diagram into unit area cells, each occupied by a single Gabor function, and observed that the amplitudes of the real and imaginary components of this Gabor function were each capable of carrying one *quantum of information.* It is from this point of view that Gabor explored the synthesis of communications signals as countable sums of Gaussian elementary signals weighted by complex-valued coefficients.

The fact that Gabor functions are optimally localized in both time and frequency makes them attractive tools for time–frequency analysis of general nonstationary signals. Classical Fourier techniques, which represent a signal in terms of stationary sinusoids, are in many respects unsuitable for this purpose. Furthermore, the Gabor transform offers improved joint resolution over both the windowed Fourier transform and the spectrogram. The Gabor expansion has also been generalized to provide nonuniform tiling of the time–frequency space (2). In fact, with the advent of wavelet theory, the Gabor function has been recast as a biorthogonal wavelet, although, strictly speaking, it is not admissible as an analyzing wavelet because its Fourier spectrum does not fall to zero at the frequency origin (3,4). A slight modification to remedy this defect results in the *Morlet wavelet,* which is the kernel of the cycle-octave transform (5).

Over the last two decades, Gabor functions have assumed a special role in image processing and machine vision. This is, in part, a consequence of their optimal conjoint localization. There are deeper biological motivations as well, however. The *only* known vision systems that work well *in general* are biological vision systems. Hence, researchers have relentlessly sought to devise artificial vision systems that emulate biological systems to the highest degree possible. A compelling body of physiological and psychophysical evidence suggests that Gabor-like linear spatial filtering plays a crucial role in the function of mammalian biological vision systems, particularly with regard to the perception of texture. It was Campbell and Robson who first determined that certain aspects of human visual function can be explained by the existence of independent linear channels sensitive to narrow ranges of spatial frequencies (6). This multiple-channels model was corroborated by Graham and Nachmias (7). Blakemore and Campbell established that the channels are orientation selective as well as frequency selective (8). Based on psychophysical experiments involving one-dimensional (1-D) textures, Richards and Polit argued that the texture discrimination capability of the human visual system can be explained by the existence of only four distinct (magnitude) spatial frequency channels (9). Based on similar experiments, Caelli and Bevan suggested that the maximum number of orientation channels is 18 (10). Numerous ensuing studies characterized cortical cell responses as linear spatial frequency filters (11–18). Although there are certainly nonlinear aspects to the function of complex cells, and even simple cells, the validity of linear models for the spatial filtering stages of both simple and complex cells has been reasonably established (19–24).

In 1980, Marčelja realized that the cortical cell spatial receptive fields being measured experimentally were essentially 2-D versions of Gabor's Gaussian elementary signals (25). Research by numerous investigators supported the Gabor function receptive field model, and it became widely accepted (24,26–31). Agreement on the validity of the Gabor receptive field model is not unanimous, and the search for better models continues (32,33). What is clear, however, is that the spatial receptive fields of mammalian visual cortex cells can, to a very close approximation, be modeled by 2-D Gabor functions. An extensive and rigorous study by Jones et al. measuring the receptive fields of simple cortical cells in the striate cortex of cat, both in the 2-D spatial and spatial frequency domains, found the deviations from the Gabor model to be ''. . .devoid of spatial structure and statistically indistinguishable from random noise'' (24,30,31).

There is a fundamental difference between Fourier frequency analysis and the analysis of an image by multiband Gabor filtering. In Fourier analysis, an image $s: \mathbb{R}^n \to \mathbb{C}$ is represented by a set of inner products $\langle s, e^{j\Omega^T \mathbf{x}} \rangle$ between the image and Fourier transform kernel functions $e^{j\Omega^T \mathbf{x}}$. Each kernel function is constrained to have constant amplitude and linear phase (i.e., constant frequency) over the entire image domain $\mathbb{R}^n$. The Fourier kernel functions are infinitely localized in frequency but posses no spatial localization whatsoever. In this sense, we refer to Fourier analysis as *stationary* frequency analysis. In contrast, Gabor filtering quantifies the local similarity between an image and Gabor functions that admit simultaneous localization in both space and spatial frequency. Thus, the Gabor-like spatial filtering performed by biological vision systems analyzes retinal images not in terms of stationary frequencies but in terms of *spatially localized* frequencies. In this regard, Gabor spectral analysis is inherently *nonstationary*.

The use of Gabor analysis has produced dramatic advances both in the understanding of image texture and in the development of machine vision texture processing algorithms. Indeed, surface textures encountered in nature are often decidedly nonstationary. Images arising from natural physical, chemical, biological, and erosive processes typically contain nonstationary textured regions or quasi-repetitive structures. Examples include crystals, a zebra's stripes, wind patterns in sand, and wood grains (4). Even in cases where only stationary surface textures are imaged, the perspective distortion that typically occurs when 3-D surfaces are projected onto the the 2-D retina or focal plane generally gives rise to nonstationary, quasi-regular image textures.

Despite the fact that texture is clearly a fundamental property of both physical surfaces and of their projections in images, a precise, quantitative definition of texture has never been formulated. Nevertheless, the characterization, analysis, and representation of projected surface texture have long been recognized as fundamental problems (34–36). Furthermore, texture-processing algorithms have been devised and applied to a wide variety of artificial vision problems with great success. For example, texture processing has been used to segment images into objects or regions of homogeneous texture, to classify or recognize surface materials from their projections in images, and to infer 3-D surface shape from image texture.

Texture is manifest in patterned surface markings—in macroscopic surface topology variations. This practically self-evident observation has been rigorously validated and refined experimentally (34,37–39). The human visual system processes texture information with remarkable efficacy and efficiency, both at the preattentive and at higher-level stages. Clearly, both pattern density and orientation are significant at a fundamental level in human visual perception. This fact implies that there is an intimate relationship between texture and local spatial frequency. An answer to the question of precisely how texture should be mathematically modeled in machine vision systems has remained elusive, however. What *is* known is that the electrochemical responses of visual cortical cells *do* transmit contrast and spatial-phase information (11). This strongly suggests an interpretation of texture, and indeed of general image information as well, in terms of smoothly varying *modulations* occurring in frequency- and orientation-selective channels.

In this chapter, we review recently developed techniques which use nonstationary, nonlinear *AM–FM models* to analyze and represent images in terms of spatially localized modulations. AM–FM models are introduced in Sec. 2, where we also examine some of the history of how modulations have been used successfully in texture processing. In Sec. 3, we develop the multidimensional Gabor expansion, which is perhaps the best understood and most readily computable nontrivial multicomponent AM–FM image representation. We also present the closely related finite multidimensional Gabor transform. The computation of more general AM–FM image representations is addressed in Sec. 4, where Gabor analysis is once again seen to play a central role. We present a computational paradigm for estimating dominant image modulations and two approaches for computing full AM–FM image representations. Image processing and machine vision applications of AM–FM modeling and Gabor analysis are discussed in Sec. 5, along with several detailed examples. Conclusions appear in Sec. 6.

2. FUNDAMENTALS OF AM-FM MODELING

Computed amplitude, phase, and frequency modulations are generally useful in solving a variety of vision problems. Biological vision systems are able to detect the presence of and make use of smoothly varying narrow-band concentrations of spatially local frequencies, which we interpret as *frequency modulations*. Spatial variations in the energies of such concentrated frequency bundles may be interpreted as *amplitude modulations*.

Many successful modern texture-processing techniques have implicitly utilized computed modulations. For example, Turner analyzed textured images with a bank of 16 Gabor filters arranged in a polar tesselation with four circularly symmetric filters on each of four oriented rays (40). Witkin (41) and Kass and Witkin (42) used demodulated filterbank channel responses to compute shape and orientation from texture. Using features defined on the responses of isotropic, annularly shaped frequency filters and Gaussian orientation filters, Coggins and Jain obtained a correct classification rate of 98% for 200, 64×64 subimages of the Brodatz textures (43). It is possible to establish a clear correspondence between their discrimination features and spatially localized amplitude and frequency modulations. Rao and Schunck (44) and Rao and Jain (45) used $\nabla^2 g$ filters to estimate the dominant local orientation and local coherency at each point in an image and used these to perform texture segmentation. The $\nabla^2 g$ filter responses admit a direct interpretation as computed frequency modulations. Malik and Perona computed texture gradients from Difference of Gaussians (DoG) channel filter responses (34,46), and, in fact, these gradients are proportional to computed frequency modulations. They identified texture boundaries as local peaks in the texture gradient magnitudes. Schachter used smoothly varying amplitude *and* frequency modulations *explicitly* to synthesize realistic looking image textures (47).

In 1986, Bovik et al. proposed an interpretation of image texture as a *carrier* of region information, which could be explicitly *demodulated* (48–50). Their approach was to filter the image with a multiband bank of Gabor filters and subsequently assign a given pixel to a texture-segmented region according to which channel filter produced the greatest magnitude response at the pixel. They also observed that visually discriminable textures often arise when two adjacent regions contain textures with identical spatial frequency- and amplitude-modulation characteristics, but have a spatial-phase displacement relative to one another. Using the frequency-modulated *derivative of Gaussian* and $\nabla^2 g$ filters in concert with their Gabor channel filters, they obtained the Laplacians of the channel filter response phases. Significant zero crossings of these Laplacians correspond to significant phase discontinuities in the texture under analysis, and they used this fact to refine the texture segmentations obtained directly from the channel amplitudes by also segmenting textures with identical amplitude and frequency characteristics but differing phase characteristics. A detailed analysis of the type of Gabor channel filters used in Ref. 50 was subsequently carried out by Bovik (51), who observed that a host of factors including occlusions, surface discontinuities, deformations and defects in surface topology, surface reflectance, shadows, specularities, and noise can all give rise to phase perturbations corresponding to amplitude and phase modulations that are not locally smooth everywhere.

The design of machine vision algorithms to estimate image phase modulations is especially difficult as a consequence of the *phase-wrapping* problem: The value of the phase at any point is ambiguous by an additive factor of $2\pi k$, $k \in \mathbb{Z}$. In a machine vision system, this problem may be conveniently circumvented by *representing* spatial phase in terms of estimates of the local spatial *frequencies*. The explicit joint estimation of nonstationary amplitude and frequency modulations from an image was first demonstrated in 1992 by Bovik (52) and Bovik et al. (4), who used an iterative constrained relaxation algorithm to estimate the spatially dominant amplitude and frequency modulations at each image point.

At about the same time, joint AM–FM models were being used with great success in the study of nonlinearities and nonstationarities in human speech (53–58). The common analytical framework was quickly recognized and exploited for nonstationary AM–FM image modeling and analysis (59–64). For an n-dimensional complex-valued nonstationary multipartite image $s: \mathbb{R}^n \to \mathbb{C}$, consider the K-component AM–FM model

$$s(\mathbf{x}) = \sum_{i=1}^{K} s_i(\mathbf{x}) = \sum_{i=1}^{K} a_i(\mathbf{x}) \exp[j\varphi_i(\mathbf{x})] \tag{1}$$

where $a_i: \mathbb{R}^n \to [0, \infty)$ and $\varphi_i: \mathbb{R}^n \to \mathbb{R}$. A real-valued image $t: \mathbb{R}^n \to \mathbb{R}$ may be analyzed against the model (1) by generating the unique complex extension $s(\mathbf{x}) = t(\mathbf{x}) + j\mathcal{H}[t(\mathbf{x})]$, where $\mathcal{H}$ is the directional multidimensional Hilbert transform (64–67). If the local frequency spectrum of each component $s_i(\mathbf{x}) = a_i(\mathbf{x}) \exp[j\varphi_i(\mathbf{x})]$ is highly concentrated, then the signal $s(\mathbf{x})$ is called *locally coherent*. The local coherency of $s(\mathbf{x})$ may be quantified rigorously in terms of certain Sobolev norms of each $a_i(\mathbf{x})$ and each $\varphi_i(\mathbf{x})$ (4,60,68–70).

For locally coherent images, the functions $a_i(\mathbf{x})$ and $\nabla \varphi_i(\mathbf{x})$ do not vary too wildly: Each component is approximately sinusoidal over *sufficiently small* neighborhoods and may, therefore, be characterized as being *locally narrow band*. On a global scale, however, the individual components and the signal itself are generally *wide band*. Signals characterized by a large time–bandwidth product have traditionally been described as *sophisticated* (71,72). We use the term *sophisticated signal* in a slightly more specialized sense here to describe nonstationary signals that are locally coherent, globally wide band, and, in general, multicomponent. AM–FM models are most useful for characterizing and representing such signals. The multicomponent AM–FM model (1) facilitates the analysis of sophisticated signals *in terms of* modulations. The functions $a_i(\mathbf{x})$ and $\nabla \varphi_i(\mathbf{x})$ are called the amplitude- and frequency-*modulation functions* of $s_i(\mathbf{x})$.

In the next two sections, we examine several approaches for actually computing representations of the form (1) for multidimensional sophisticated signals. This problem is extremely difficult in general. For any real-valued image $t(\mathbf{x})$, there are uncountably infinitely many pairs of functions $a(\mathbf{x})$ and $\varphi(\mathbf{x})$ for which $t(\mathbf{x}) = \text{Re}\{a(\mathbf{x}) \exp[j\varphi(\mathbf{x})]\}$. Thus, *any* real-valued signal can be modeled as the real part of a single-component AM–FM function in infinitely many different ways. For a complex-valued image, the single-component representation $a(\mathbf{x}) \exp[j\varphi(\mathbf{x})]$ is unique. However, the multicomponent decomposition indicated in Eq. (1) is decidedly ambiguous. Therefore, wherever possible, it is of interest to model a sophisticated signal as a sum of AM–FM components, each of which is locally coherent. At least one such multicomponent interpretation exists for any discrete image, as the Discrete Fourier Transform (DFT) *is* a trivial multicomponent AM–FM representation. However, in many cases, there may also exist smooth multicomponent AM–FM representations which capture the essential image structure using only a few components.

3. THE GABOR TRANSFORM

Gabor's elementary signals took the form $f_{x_0, \Omega_0}(x) = g(x - x_0)e^{j\Omega_0 x}$, where the window function $g(x)$ was Gaussian (1). He proposed representing an arbitrary signal $s(x)$ in the countable expansion

$$s(x) = \sum_{m,n \in \mathbb{Z}} a_{m,n} f_{mx_0, n\Omega_0}(x) \tag{2}$$

where the functions $f_{mx_0, n\Omega_0}(x)$ were chosen to tile the time–frequency plane uniformly. Note that each elementary signal is, in fact, a single-component AM–FM function admitting Gaussian amplitude modulation and constant frequency modulation. From an analysis point of view, the difficulty in computing the coefficients required to express a given signal in a Gabor expansion stems from the fact that the elementary signals themselves are not mutually orthogonal. Gabor proposed calculating the coefficients via a successive approximation scheme, which he described as "a rather inconvenient process" (1). A practical analytical technique for determining the expansion coefficients in terms of a biorthogonal set of *auxiliary functions* was devised by Bastiaans and reported in 1980 (73). It has been recently established that these coefficients are not numerically stable for all L^2 signals (74,75), a phenomenon that is most readily explained in terms of the fact that the Zak transform of Gabor's Gaussian window function admits a zero.

However, in proposing his biorthogonal approach for computing the coefficients, Bastiaans observed that the technique can be applied equally well to express a signal $s(x)$ in an analogous expansion

$$s(x) = \sum_{m,n \in \mathbb{Z}} C_{m,n} h_{m,n}(x) \tag{3}$$

wherein the Gaussian window function of Gabor's original expansion is replaced by an arbitrary unity L^2-norm window function $h(x)$ and $h_{m,n}(x) = h(x - mT)e^{jn\Omega x}$ for some fixed temporal and spectral sampling intervals T and Ω (73,76,77). In modern usage, the term *Gabor expansion* refers to any such expansion of a signal into a countable weighted sum of translated and frequency-modulated unity L^2-norm window functions. The jointly amplitude–frequency-modulated signals $h_{m,n}(x)$ in Eq. (3) tile the time–frequency plane uniformly.

Using Bastiaans' technique, the expansion coefficients $C_{m,n}$ in Eq. (3) are computed as inner products between the signal $s(x)$ and frequency-modulated translates of an auxiliary function $\gamma(x)$. Specifically,

$$C_{m,n} = \langle s, \gamma_{m,n} \rangle = \int_{\mathbb{R}} s(x) \gamma_{m,n}^*(x) \, dx \tag{4}$$

where $\gamma_{m,n}(x) = \gamma(x - mT)e^{jn\Omega x}$. Equation (4) is called the *Gabor transform* of $s(x)$ with respect to $h(x)$. For a given window function $h(x)$, the auxiliary function $\gamma(x)$ is found by solving (73,75,78,79)

$$\frac{2\pi}{T\Omega} \int_{\mathbb{R}} h(x) \gamma^* \left(x - \frac{2\pi m}{\Omega} \right) \exp\left(-j \frac{2\pi n x}{T} \right) dx = \delta(m)\delta(n) \tag{5}$$

where $\delta(\cdot)$ is the Kroneker delta. We call Eq. (5) the 1-D *biorthogonality constraint equation*. Its solution $\gamma(x)$ is unique if $T\Omega = 2\pi$.

For $T\Omega < 2\pi$, the Gabor expansion is called *oversampled*. In the oversampled case, the solution of Eq. (5) is not unique, and additional constraints can be placed on the auxiliary function $\gamma(x)$. For example, Qian et al. developed constraints which require $\gamma(x)$ to be similar to the window function $h(x)$ in a certain sense (79–81). The expansion (3) then takes on a quasi-orthogonal character and the coefficients $C_{m,n}$ may be interpreted as roughly indicating the degree of correlation between the signal and $h_{m,n}(x)$. For window functions $h(x)$ having Zak transforms that admit no zeros, there also exist attractive alternative methods for computing the Gabor coefficients [Eq. (4)] (75).

3.1. Multidimensional Gabor Transform

A formulation for the coefficients in a 2-D Gabor expansion analogous to Eq. (3) was developed by Porat and Zeevi (2,82,83), and Gabor's uncertainty relation was extended into two dimensions by MacKay (84) and by Daugman (85). As in the 1-D case, the lower bound on conjoint uncertainty is realized uniquely by frequency-modulated Gaussians known as *2-D Gabor functions*. For separable window functions, the general Gabor expansion may be developed in n dimensions as follows. Let the window function $h: \mathbb{R}^n \to \mathbb{C}$ be defined by $h(\mathbf{x}) = \prod_{i=1}^{n} h_i(x_i)$, where $\mathbf{x} = [x_1 \ x_2 \ldots x_n]^T$ and where each h_i has unit L^2-norm. Define the spatial sampling interval in the direction x_i to be T_i, and let $\mathbf{T} = \text{diag}(T_1, T_2, \ldots, T_n)$. Similarly, define the spectral sampling interval in the direction x_i to be Ω_i, and let $\mathbf{\Omega} = \text{diag}(\Omega_1, \Omega_2, \ldots, \Omega_n)$. Let $\mathbf{m} = [\mathbf{p}^T | \mathbf{q}^T]^T$, where $\mathbf{p}, \mathbf{q} \in \mathbb{Z}^n$. Then the elementary function $h_\mathbf{m}(\mathbf{x})$ may be defined according to

$$h_\mathbf{m}(\mathbf{x}) = h(\mathbf{x} - \mathbf{T}\mathbf{p}) \exp(j\mathbf{q}^T \mathbf{\Omega} \mathbf{x}) \tag{6}$$

An arbitrary signal $s: \mathbb{R}^n \to \mathbb{C}$ can then be written in the multidimensional Gabor expansion

$$s(\mathbf{x}) = \sum_{\mathbf{m} \in \mathbb{Z}^{2n}} C_\mathbf{m} h_\mathbf{m}(\mathbf{x}) \tag{7}$$

The expansion coefficients $C_\mathbf{m}$ are given by

$$C_\mathbf{m} = \langle s, \gamma_\mathbf{m} \rangle = \int_{\mathbb{R}^n} s(\mathbf{x}) \gamma_\mathbf{m}^*(\mathbf{x}) \, d\mathbf{x} \tag{8}$$

where $\gamma_\mathbf{m}(\mathbf{x}) = \gamma(\mathbf{x} - \mathbf{T}\mathbf{p}) \exp(j\mathbf{q}^T \mathbf{\Omega} \mathbf{x})$ and where the auxiliary function $\gamma(\mathbf{x}) = \prod_{i=1}^{n} \gamma_i(x_i)$ is separable. Equation (8) is called the *multidimensional Gabor transform* of $s(\mathbf{x})$ with respect to $h(\mathbf{x})$. The auxiliary function $\gamma(\mathbf{x})$ may be found by solving the multidimensional biorthogonality constraint equation

$$\frac{(2\pi)^n}{\prod_{i=1}^{n} T_i \Omega_i} \int_{\mathbb{R}^n} h(\mathbf{u}) \gamma^*(\mathbf{u} - 2\pi \mathbf{\Omega}^{-1} \mathbf{p}) \exp(-j(2\pi)^n \mathbf{q}^T \mathbf{T}^{-1} \mathbf{u}) \, d\mathbf{u} \tag{9}$$
$$= \delta(\mathbf{p})\delta(\mathbf{q})$$

However, as both $h(\mathbf{x})$ and $\gamma(\mathbf{x})$ are separable, the solution of Eq. (9) is equivalent to independently solving the n 1-D constraints

$$\frac{2\pi}{T_i \Omega_i} \int_{\mathbb{R}} h_i(x) \gamma_i^* \left(x - \frac{2\pi m}{\Omega_i}\right) \exp\left(-j \frac{2\pi n x}{T_i}\right) dx = \delta(m)\delta(n), \quad 1 \leq i \leq n \tag{10}$$

In the multidimensional case, it is possible for the expansion (7) to be oversampled in several of the n dimensions and not in others. If $T_i \Omega_i = 2\pi$, then the expansion is critically sampled in dimension i and a unique solution exists for γ_i. Alternatively, if $T_i \Omega_i < 2\pi$, then the solution of Eq. (10) is not unique and it is generally possible to impose additional desirable constraints on γ_i (78–81).

3.2. Finite Quasi-Orthogonal Discrete Gabor Transform

The multidimensional Gabor transform Eq. (8) developed in Sec. 3.1 may be used to compute Gabor expansions for continuous-domain signals $s(\mathbf{x})$. However, in image processing, video, and multimedia telecommunications applications, it is often desirable to compute Gabor expansions for finitely supported or periodic discrete-domain signals $\tilde{s}(\mathbf{k})$, where $\tilde{s}: \mathbb{Z}^n \rightarrow \mathbb{C}$ and $\mathbf{k} = [k_1 \ k_2 \ldots k_n]^T$. Such expansions based on a novel discrete Poisson formula developed by Wexler and Raz have been investigated in one dimension with great success (78,79,81). Zak transform methods for evaluating the coefficients in finite discrete 1-D Gabor expansions have also been investigated (75). In this section, we develop the multidimensional finite discrete Gabor expansion using separable discrete biorthogonality constraints.

Let $\tilde{s}(\mathbf{k})$ be periodic in k_i with period L_i, and let $\mathbf{L} = [L_1 \ L_2 \ldots L_n]^T$ [for the case of a finitely supported signal, assume that $\tilde{s}(\mathbf{k})$ has been periodically extended]. Also define the periodicity matrix $\mathcal{L} = \mathrm{diag}(L_1, L_2, \ldots, L_n)$. We will express $\tilde{s}(\mathbf{k})$ in a discrete Gabor expansion having $Q_i \in \mathbb{N}$ spectral samples and $P_i \in \mathbb{N}$ spatial samples in dimension i. The spatial sampling interval in dimension i is $\tilde{T}_i \in \mathbb{N}$ and the spectral sampling interval in dimension i is $\tilde{\Omega}_i \in \mathbb{N}$, where $P_i \tilde{T}_i = Q_i \tilde{\Omega}_i = L_i$. Let $\tilde{\boldsymbol{\Omega}} = \mathrm{diag}(\tilde{\Omega}_1, \tilde{\Omega}_2, \ldots, \tilde{\Omega}_n)$, $\tilde{\mathbf{T}} = \mathrm{diag}(\tilde{T}_1, \tilde{T}_2, \ldots, \tilde{T}_n)$, $\mathbf{P} = [P_1 \ P_2 \ldots P_n]^T$, $\mathbf{Q} = [Q_1 \ Q_2 \ldots Q_n]^T$, and $\mathbf{N} = [\mathbf{P}^T | \mathbf{Q}^T]^T$.

Let $\tilde{h}(\mathbf{k}) = \prod_{i=1}^{n} \tilde{h}_i(k_i)$ be a discrete separable window function, where each $\tilde{h}_i$ is periodic with period L_i and has unit l^2-norm over the fundamental period. With $\mathbf{p} = [p_1 \ p_2 \ldots p_n]^T \in \mathbb{Z}^n$, $\mathbf{q} = [q_1 \ q_2 \ldots q_n]^T \in \mathbb{Z}^n$, and $\mathbf{m} = [\mathbf{p}^T | \mathbf{q}^T]^T$, the discrete elementary function is given by $\tilde{h}_{\mathbf{m}}(\mathbf{k}) = \tilde{h}(\mathbf{k} - \tilde{\mathbf{T}}\mathbf{p}) \exp[j(2\pi)^n \mathbf{q}^T \tilde{\boldsymbol{\Omega}} \mathcal{L}^{-1} \mathbf{k}]$, where $0 \leq p_i < P_i$ and $0 \leq q_i < Q_i$. Let $\mathbf{1}_{2n} \in \mathbb{Z}^{2n}$ be the vector having all $2n$ entries equal to 1. Then, the finite discrete Gabor expansion for $\tilde{s}(\mathbf{k})$ is

$$\tilde{s}(\mathbf{k}) = \sum_{\mathbf{m}=0}^{\mathbf{N}-\mathbf{1}_{2n}} C_{\mathbf{m}} \tilde{h}_{\mathbf{m}}(\mathbf{k}) \tag{11}$$

The Gabor coefficients $C_{\mathbf{m}}$ are given by

$$C_{\mathbf{m}} = \langle \tilde{s}, \tilde{\gamma}_{\mathbf{m}} \rangle = \sum_{\mathbf{k}=\mathbf{0}}^{\mathbf{L}-\mathbf{1}_n} \tilde{s}(\mathbf{k}) \tilde{\gamma}_{\mathbf{m}}^*(\mathbf{k}) \tag{12}$$

where $\mathbf{1}_n \in \mathbb{Z}^n$ is the vector having all entries equal to 1. Equation (12) defines the multidimensional discrete Gabor transform of $\tilde{s}(\mathbf{k})$ with respect to $\tilde{h}(\mathbf{k})$. The frequency-modulated translate $\tilde{\gamma}_{\mathbf{m}}$ is given by $\tilde{\gamma}_{\mathbf{m}}(\mathbf{k}) = \tilde{\gamma}(\mathbf{k} - \tilde{\mathbf{T}}\mathbf{p}) \exp[j(2\pi)^n \mathbf{q}^T \tilde{\boldsymbol{\Omega}} \mathcal{L}^{-1} \mathbf{k}]$, where $0 \leq p_i < P_i$, $0 \leq q_i < Q_i$, and where the auxiliary function $\tilde{\gamma}(\mathbf{k}) = \prod_{i=1}^{n} \tilde{\gamma}_i(k_i)$ is separable. Each $\tilde{\gamma}_i$ is periodic with period L_i and may be obtained independently by solving the separated discrete biorthogonality constraint equation

$$\sum_{k=0}^{L_i-1} \tilde{h}_i(k + pQ_i) \exp(-j2\pi q P_i L_i^{-1} k) \tilde{\gamma}_i^*(k) = \delta(p)\delta(q), \quad 0 \leq p < \tilde{\Omega}_i,$$
$$0 \leq q < \tilde{T}_i \tag{13}$$

Let $\tilde{\boldsymbol{\gamma}}_i = [\tilde{\gamma}_i(0) \ \tilde{\gamma}_i(1) \ldots \tilde{\gamma}_i(L_i - 1)]^T$. For $0 \leq p < \tilde{\Omega}_i$, $0 \leq q < \tilde{T}_i$, and $0 \leq k < L_i$, define a $\tilde{\Omega}_i \tilde{T}_i \times L_i$ matrix $\mathbf{H}_i$ by $\mathbf{H}_i(p\tilde{T}_i + q, k) = \tilde{h}_i(k + pQ_i) \exp(-j2\pi q P_i L_i^{-1} k)$. Let $\boldsymbol{\mu}_i \in \mathbb{Z}^{\tilde{\Omega}_i \tilde{T}_i}$ be defined by $\boldsymbol{\mu}_i = [1 \ 0 \ 0 \ldots 0]^T$. Then, Eq. (13) may be written as

(78,79,81)

$$\mathbf{H}_i \tilde{\boldsymbol{\gamma}}_i^* = \boldsymbol{\mu}_i \tag{14}$$

If $\tilde{T}_i \tilde{\Omega}_i = L_i$, then the solution $\tilde{\gamma}_i$ of Eq. (14) is unique if it exists. If it does not exist, then the expansion in Eq. (11) also fails to exist. Alternatively, if $\tilde{T}_i \tilde{\Omega}_i < L_i$ for each $i \in [0, n]$, then the solution of Eq. (14) is not unique. Qian et al. (79–81) proposed selecting $\tilde{\gamma}_i$ according to

$$\tilde{\boldsymbol{\gamma}}_i = \underset{\tilde{\boldsymbol{\gamma}} : \mathbf{H}_i \tilde{\boldsymbol{\gamma}}^* = \boldsymbol{\mu}_i}{\arg \min} \left\| \frac{\tilde{\boldsymbol{\gamma}}}{\|\tilde{\boldsymbol{\gamma}}\|} - \tilde{h}_i \right\|^2 \tag{15}$$

where $\tilde{\boldsymbol{\gamma}} = [\tilde{\gamma}(0)\ \tilde{\gamma}(1) \ldots \tilde{\gamma}(L_i - 1)]^T$. In this case, the auxiliary function $\tilde{\gamma}(\mathbf{k})$ is optimally similar to the window function $\tilde{h}(\mathbf{k})$; the expansion of Eq. (11) is then called the *discrete quasi-orthogonal Gabor expansion,* and Eq. (12) is called the *discrete quasi-orthogonal Gabor transform* of $\tilde{s}(\mathbf{k})$ with respect to the window $\tilde{h}(\mathbf{k})$ (79–81). The $\tilde{\gamma}_i$ prescribed by Eq. (15) may be obtained via the matrix pseudo-inverse (78,79,81) $\tilde{\boldsymbol{\gamma}}_i^* = \mathbf{H}^T(\mathbf{H}\mathbf{H}^T)^{-1}\boldsymbol{\mu}$. Define

$$\Gamma_i = \underset{\tilde{\boldsymbol{\gamma}} : \mathbf{H}_i \tilde{\boldsymbol{\gamma}}^* = \boldsymbol{\mu}_i}{\min} \left\| \frac{\tilde{\boldsymbol{\gamma}}}{\|\tilde{\boldsymbol{\gamma}}\|} - \tilde{h}_i \right\|^2 \tag{16}$$

If each $\Gamma_i \approx 0$, then $\tilde{\gamma}(\mathbf{k}) \approx \alpha \tilde{h}(\mathbf{k})$, where $\alpha = \Pi_{i=1}^n \|\tilde{\gamma}_i\|$ and the Gabor coefficients $C_\mathbf{m}$ are approximately proportional to inner products $\langle \tilde{s}, h_\mathbf{m} \rangle$ between the signal and the modulated translates of the window function. The expansion (11) then becomes quasi-orthogonal and the local structure of the signal can be inferred from the Gabor coefficients and the structure of $\tilde{h}(\mathbf{k})$ (79–81).

4. THE COMPUTATION OF GENERAL AM–FM MODELS

In Sec. 3, we develop techniques for computing a special class of multicomponent AM–FM signal models (Eq. 1) where the computed components were constrained to be frequency-modulated translates of a single window function $h(\mathbf{x})$. In this section, we explore approaches for computing more general AM–FM models by allowing the modulating functions $\{a_i(\mathbf{x}), \nabla \varphi_i(\mathbf{x})\}_{i \in [1,K]}$ to be arbitrary but smoothly varying. We discuss continuous and discrete-domain algorithms for demodulating a single component of a multidimensional sophisticated signal in Sec. 4.1. In Sec. 4.2, we give the design of a biologically motivated 2-D multiband Gabor filterbank suitable for decomposing a sophisticated image into locally coherent components. We also present filtered demodulation algorithms capable of estimating the modulating functions of individual image components directly from the filterbank channel responses. An analysis paradigm called *dominant component analysis,* which estimates the dominant modulations at each point in an image, is presented in Sec. 4.3. Finally, the computation of general multicomponent AM–FM image models is addressed in Sec. 4.4.

4.1. Single-Component Demodulation

Consider an arbitrary, complex-valued n-dimensional sophisticated signal component s_i: $\mathbb{R}^n \to \mathbb{C}$ modeled as $s_i(\mathbf{x}) = a_i(\mathbf{x}) \exp[j\varphi_i(\mathbf{x})]$, where $\mathbf{x} = [x_1\ x_2 \ldots x_n]^T$. If $n = 1$, then

Re[$s_i(\mathbf{x})$] might be a component of a speech signal. In the case $n = 2$, Re[$s_i(\mathbf{x})$] might be an image component, and in the case $n = 3$, Re[$s_i(\mathbf{x})$] might be one component of a video sequence. We assume only that the modulating functions $a_i(\mathbf{x})$ and $\nabla \varphi_i(\mathbf{x})$ are locally coherent. The problem of estimating these modulating functions from the values of the component $s_i(\mathbf{x})$ is called *demodulation*. It is straightforward to obtain the amplitude-modulation function $a_i(\mathbf{x})$:

$$a_i(\mathbf{x}) = |s_i(\mathbf{x})| = |a_i(\mathbf{x}) \exp[j\varphi_i(\mathbf{x})]| \tag{17}$$

The frequency-modulation function may be obtained with the spatially local nonlinear algorithm (64)

$$\nabla \varphi_i(\mathbf{x}) = \text{Re}\left(\frac{\nabla s_i(\mathbf{x})}{j s_i(\mathbf{x})}\right) \tag{18}$$

which is applicable at all points $\mathbf{x}$ where $s_i(\mathbf{x}) \neq 0$. The joint demodulation algorithm in Eq. (17)–(18) is *exact* for *any* complex-valued n-dimensional signal interpreted as a single AM–FM component; it also gives correct signs for the components of the instantaneous frequency vector $\nabla \varphi_i(\mathbf{x})$ (60,64,86).

The algorithm just presented depends critically on the fact that the signal component $s_i(\mathbf{x})$ is *complex valued*. However, many practical engineering applications are concerned exclusively with *real-valued* signals. In Sec. 2, we stated that the problem of associating a pair of amplitude- and frequency-modulating functions with a real-valued signal is inherently ill-posed, and we now examine this fact in more detail. Let $t_i: \mathbb{R}^n \rightarrow \mathbb{R}$ be a single component of a real-valued multidimensional sophisticated signal. We seek an AM–FM model of the form $t_i(\mathbf{x}) = a_i(\mathbf{x}) \cos[\varphi_i(\mathbf{x})]$. Upon applying Eq. (17), we have $|t_i(\mathbf{x})| = a_i(\mathbf{x})|\cos[\varphi_i(\mathbf{x})]|$, which is not equal to $a_i(\mathbf{x})$ in general. Furthermore, Re[$\nabla t_i(\mathbf{x})/j t_i(\mathbf{x})$] = 0. Thus, the demodulation algorithm in Eq. (17)–(18) fails to deliver a correct interpretation when applied directly to a real-valued signal component.

Let us momentarily restrict our attention to the 1-D case. Corresponding to *any* real-valued component $t_i(x)$ there are, in fact, uncountably infinitely many pairs of modulating functions $a_i(x)$ and $\varphi_i'(x)$ for which $t_i(x) = a_i(x) \cos[\varphi_i(x)]$. To see that this is true, consider that we may take $a_i(x) = \sup_x|t_i(x)|$ and $\varphi_i(x) = \arccos(t_i(x)/\sup_x|t_i(x)|)$. In this case, we interpret the variations in $t_i(x)$ entirely as *frequency modulation*. Equally extreme, we may take $a_i(x) = |t_i(x)|$ and $\varphi_i(x) = \arccos[\text{sgn } t_i(x)]$, in which case we interpret the variations in $t_i(x)$ entirely as *amplitude modulation*. Lying between these two extremes there exists an infinite set of modulating function pairs that interpret the signal variations as combinations of amplitude and frequency modulations.

A number of approaches for associating modulating functions $a_i(x)$ and $\varphi_i'(x)$ with a real-valued signal component have been proposed in the last half century (53,54,57,87–95). Notable among these is the recently popularized Teager–Kaiser energy operator (TKEO) with its associated energy-separation algorithm (ESA) (53,54). Multidimensional versions of the TKEO and ESA have also been developed (59,62,96). In 1-D applications such as speech processing, the ESA is particularly attractive because of its computational simplicity and temporal localization. It frequently delivers modulating functions that are in close agreement with those obtained by more computationally intensive techniques such as the analytic signal approach advocated by Gabor (1), Ville (95), Vakman (87), and others. In multiple dimensions, however, the fact that the ESA delivers *unsigned* frequency estimates can be problematic, especially for signal components that are globally wide band: The relative signs of the components of the instantaneous frequency vector

$\nabla\varphi_i(\mathbf{x})$ embody the local orientation of the signal. In Sec. 2, we mentioned that our approach for analyzing real-valued multidimensional signals against the complex model in Eq. (1) is to use the directional multidimensional Hilbert transform to generate a complex extension. The demodulation algorithm of Eq. (17)—(18) can be applied directly to this extension, which we call the *analytic image* (64,65,97,98). In many respects, the analytic image may be interpreted as the multidimensional counterpart of the 1-D analytic signal.

Suppose now that $s_i\colon \mathbb{Z}^n \to \mathbb{C}$ is one locally coherent component of a discrete multidimensional sophisticated signal $s(\mathbf{k})$, where $\mathbf{k} = [k_1\ k_2 \ldots k_n]^T$. Let $\mathbf{e}_p$ be a unit vector in the k_p direction. The discrete demodulation problem is to find modulating functions $a_i\colon \mathbb{Z}^n \to [0, \infty)$ and $\nabla\varphi_i\colon \mathbb{Z}^n \to \mathbb{R}^n$ such that $s_i(\mathbf{k}) = a_i(\mathbf{k}) \exp[j\varphi_i(\mathbf{k})]$, where the notation $\nabla\varphi_i(\mathbf{k})$ indicates the samples of the continuous-domain frequency vector $\nabla\varphi_i(\mathbf{x})$. As in the continuous-domain case, the amplitude modulation function may be obtained via

$$a_i(\mathbf{k}) = |s_i(\mathbf{k})| \tag{19}$$

The discrete counterpart of the frequency algorithm (18) is

$$\mathbf{e}_p^T \nabla\varphi_i(\mathbf{k}) \approx \arcsin\left(\frac{s_i(\mathbf{k} + \mathbf{e}_p) - s_i(\mathbf{k} - \mathbf{e}_p)}{2j s_i(\mathbf{k})}\right) \tag{20}$$

$$\approx \arccos\left(\frac{s_i(\mathbf{k} + \mathbf{e}_p) + s_i(\mathbf{k} - e_p)}{2 s_i(\mathbf{k})}\right) \tag{21}$$

provided that $s_i(\mathbf{k}) \neq 0$ (99–101). Equation (20) is called the *sine algorithm;* Eq. (21) is called the *cosine algorithm.* Either algorithm may be used individually to place each component of the estimated instantaneous frequency vector to within π radians. Together, they place the estimated frequencies to within 2π radians.

In contrast to the continuous case, the discrete algorithms in Eq. (20) and Eq. (21) are not exact. They generally contain approximation errors that are tightly bounded by error functionals involving certain Sobolev norms of the continuous-domain functions $a_i(\mathbf{x})$ and $\varphi_i(\mathbf{x})$, where it is assumed that $a_i(\mathbf{k})$ contains the samples of $a_i(\mathbf{x})$ and that $\nabla\varphi_i(\mathbf{k})$ contains the samples of $\nabla\varphi_i(\mathbf{x})$ (99,100). The errors are generally small or negligible, provided that the component $s_i(\mathbf{k})$ is locally coherent.

4.2. The Decomposition of a Sophisticated Signal into Components

The nonlinear algorithms in Eq. (17)–(18) and Eq. (19)–(21) cannot be applied directly to a multicomponent signal: Prior to demodulation, components must be isolated from one another on a spatio-spectrally localized basis. Multiband linear filtering may be used to effect this separation. The decomposition of a signal into components according to Eq. (1) is not unique; in any particular computational paradigm, the structure of the filterbank drives the decomposition, and in so doing, it determines the multicomponent interpretation of the signal. By *spatio-spectrally localized isolation,* we mean that the filterbank need not isolate the signal components from one another on a global scale. Rather, what is required is that the response of each filter be dominated by at most one component at each point in the domain of the signal. In this section, we present the design of a biologically motivated 2-D filterbank suitable for decomposing images into locally coherent components.

Our goal is to emulate biological vision systems by representing the information content of an image as smooth modulations in frequency- and orientation-selective channels. Because visual cortex cell receptive fields are well described by Gabor filters, we employ a multiband Gabor filterbank. The Fourier spectra of real-valued images are conjugate symmetric, and their associated complex-valued analytic images admit spectral support only on a frequency half-plane. Hence, based on the results of Richards and Polit (9) together with those of Caelli and Bevan (10), we expect that a bank of approximately four filters at each of approximately nine orientations should suffice (as only half of the possible orientations need be explicitly considered). There is a second reason that the choice of Gabor filters is particularly advantageous. The filters must be sufficiently spectrally localized to resolve individual image components from one another, but they must simultaneously be well localized spatially if their responses are to capture the local nonstationary structure typical of sophisticated images. In this regard, it is, indeed, salutary that Gabor filters uniquely realize the multidimensional uncertainty principle lower bound on conjoint resolution.

The isotropic unity L^2-norm baseband filter is $h(\mathbf{x}) = (1/\sqrt{2\pi}) \exp(-\frac{1}{4}\mathbf{x}^T\mathbf{x})$. Upon adding scaling and frequency modulation (i.e., translation) while maintaining the unity L^2-norm and aspect ratio, we obtain for the scaled, translated channel filter with center frequency $\mathbf{\Omega}_m$

$$g_m(\mathbf{x}) = \frac{1}{\sigma_m\sqrt{2\pi}} \exp\left(-\frac{1}{4\sigma_m^2}\mathbf{x}^T\mathbf{x}\right) \exp(j\mathbf{\Omega}_m^T\mathbf{x}) \qquad (22)$$

In the frequency domain, the filter in Eq. (22) is the Gaussian

$$G_m(\mathbf{\Omega}) = \mathcal{F}[g_m(\mathbf{x})] = 2\sqrt{2\pi}\sigma_m \exp[-\sigma_m^2(\mathbf{\Omega} - \mathbf{\Omega}_m)^T(\mathbf{\Omega} - \mathbf{\Omega}_m)] \qquad (23)$$

with radial center frequency $r_m = |\mathbf{\Omega}_m|$ and orientation $\theta_m = \arg(\mathbf{\Omega}_m)$. Figure 1 depicts several quantities useful in characterizing the filter bandwidth. The circular contour in

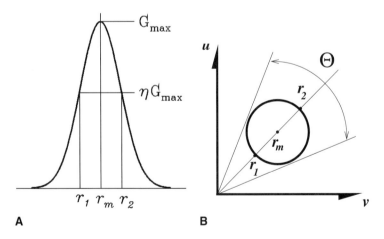

Fig. 1 Quantities used in defining the bandwidth B of a Gabor filter. (A) Radial octave bandwidth. The curve shows the filter evaluated on a line from the frequency origin through its center frequency, where $|\mathbf{\Omega}_m| = r_m$. The filter magnitude response is at a fraction $\eta < 1$ of peak response at the radial frequencies r_1 and r_2. (B) Orientation bandwidth. The circular contour with center r_m is the η-peak contour of the filter in the $\mathbf{\Omega} = [u\ v]^T$ plane.

Fig. 1b is the η-peak contour of the filter frequency response, where $\eta < 1$. A line running from the frequency origin through the filter center frequency Ω_m intercepts the η-peak contour at radial frequencies $r_1 < r_2$ given by

$$r_1 = r_m - \frac{\sqrt{-\ln \eta}}{\sigma_m} \qquad (24)$$

and

$$r_2 = r_m + \frac{\sqrt{-\ln \eta}}{\sigma_m} \qquad (25)$$

A section of the filter evaluated along this line is shown in Fig. 1a. The η-peak radial octave bandwidth is defined by $B = \log_2(r_2/r_1)$, which is a design parameter of the filterbank. The orientation bandwidth is the angle between two lines which pass through the frequency origin and are tangent to the η-peak contour. It is given by $\Theta = 2$ arc tan$(\sqrt{\gamma})$, where $\gamma = (2^B - 1)^2/(2^B + 1)^2$.

We arrange the filter center frequencies in quadrants I and IV of the spatial frequency plane along rays such that any group of four adjacent filters intersect precisely where each is at a fraction η of peak response. With this dense spacing, practically every point in the right half-plane is covered by a filter responding at the η peak or higher. Along each ray, the filter radial center frequencies progress geometrically with a common ratio R, which is a design parameter of the filterbank. The radial center frequency of the first filter on each ray is also a design parameter and is denoted by r_0. The angular spacing between rays is given by

$$\Lambda = 2 \arcsin\{(4R)^{-1/2}[(R^2 + 1)(\gamma - 1) + 2R(\gamma + 1)]^{1/2}\} \qquad (26)$$

The four free-design parameters r_0, R, η, and B completely specify the filterbank. However, all arbitrary choices of R and B are not realizable, as certain combinations lead to angular ray spacings Λ that are zero or complex. Figure 2 graphically depicts the realizable combinations by showing Λ as a function of R, where the individual curves are parameterized by B. In Fig. 2, values of R less than unity correspond to a decreasing progression of filter radial center frequencies on each ray, in which case the design parameter r_0 specifies the *greatest* radial center frequency assumed by any filter.

As an example filterbank design, let $r_0 = 9.6$ cycles per image, $R = 1.8$, $B = 1$ octave, and $\eta = \frac{1}{2}$. Then, $\gamma = \frac{1}{9}$ and $\Theta \approx 38.9424°$. In this case, the filterbank coincides with one used by Bovik in Ref. 52 and by Bovik et al. in Ref. 4. Figure 3 shows the entire filterbank in the frequency domain for these parameter choices. The frequency-domain coordinates of the figure are such that the first quadrant is located in the lower right portion of the figure. There are 40 filters arranged along 8 rays spaced equally at angles of $\Lambda \approx 20.6418°$, with five filters per ray. Hence, the filterbank comprises channels at eight orientations and at five magnitude spatial frequencies. This channel structure is in *rough* agreement with that proposed for the human visual system by the work of Richards and Polit and of Gaelli and Bevan. For display, each filter in Fig. 3 has been *independently* scaled for maximum dynamic range in the available 256 gray scales. Additional scaling has been applied to accentuate the intersections between individual filters.

In treating discrete images, it is necessary to sample the filterbank depicted in Fig. 3. Design issues concerning sampled Gabor filters were discussed in Refs 50 and 51. Design parameters for all 40 filters in a discrete filterbank applicable to 256×256 images

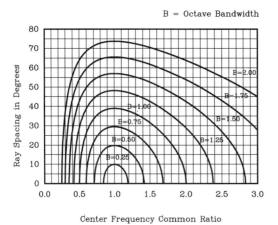

Fig. 2 Realizable combinations of the filterbank free-design parameters R, the common ratio of the geometric progression of filter radial center frequencies along each ray, and B, the filter radial octave bandwidth. The graph shows the ray spacing Λ as a function of R, parameterized by B. Combinations falling below the abscissa of the graph correspond to negative or complex values of the ray spacing Λ, which are not realizable.

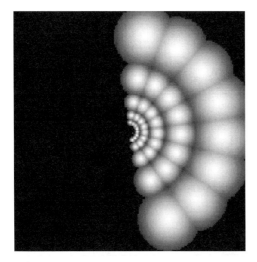

Fig. 3 Frequency-domain representation of the multiband Gabor filterbank for design parameter choices $r_0 = 9.6$ cycles per image, $R = 1.8$, $B = 1$ octave, and $\eta = \frac{1}{2}$. There are 40 filters arranged in a polar waveletlike tesselation along 8 rays with five filters per ray. Each of the 40 filters in the figure has been independently scaled for maximum dynamic range in the available gray levels.

Table 1 Design Parameters for Filters 0–19

Filter	Ray	Index on ray	Orient. θ_m (rad)	r_m $\left(\dfrac{\text{Hz}}{\text{pix}}\right)$	u_m $\left(\dfrac{\text{Hz}}{\text{pix}}\right)$	v_m $\left(\dfrac{\text{Hz}}{\text{pix}}\right)$	Space const. σ_m (pix)	max $|G(\Omega)|$
0	0	0	−1.3907	0.0375	0.0067	−0.0369	10.6004	53.1426
1	0	1	−1.3907	0.0675	0.0121	−0.0664	5.8891	29.5237
2	0	2	−1.3907	0.1215	0.0218	−0.1195	3.2717	16.4020
3	0	3	−1.3907	0.2187	0.0392	−0.2152	1.8176	9.1122
4	0	4	−1.3907	0.3937	0.0705	−0.3873	1.0098	5.0626
5	1	0	−1.0304	0.0375	0.0193	−0.0322	10.6004	53.1426
6	1	1	−1.0304	0.0675	0.0347	−0.0579	5.8891	29.5237
7	1	2	−1.0304	0.1215	0.0625	−0.1042	3.2717	16.4020
8	1	3	−1.0304	0.2187	0.1125	−0.1875	1.8176	9.1122
9	1	4	−1.0304	0.3937	0.2025	−0.3376	1.0098	5.0626
10	2	0	−0.6701	0.0375	0.0294	−0.0233	10.6004	53.1426
11	2	1	−0.6701	0.0675	0.0529	−0.0419	5.8891	29.5237
12	2	2	−0.6701	0.1215	0.0952	−0.0755	3.2717	16.4020
13	2	3	−0.6701	0.2187	0.1714	−0.1358	1.8176	9.1122
14	2	4	−0.6701	0.3937	0.3085	−0.2445	1.0098	5.0626
15	3	0	−0.3099	0.0375	0.0357	−0.0114	10.6004	53.1426
16	3	1	−0.3099	0.0675	0.0643	−0.0206	5.8891	29.5237
17	3	2	−0.3099	0.1215	0.1157	−0.0370	3.2717	16.4020
18	3	3	−0.3099	0.2187	0.2083	−0.0667	1.8176	9.1122
19	3	4	−0.3099	0.3937	0.3749	−0.1200	1.0098	5.0626

Note: See Table 3 for descriptions of the individual parameters.

are given in Tables 1 and 2. Each row of the tables gives the parameters for one filter. Descriptions of the individual parameters appear in Table 3. The real and imaginary components of the unit-pulse response of filter 10 are depicted in Fig. 4, both as surfaces and as gray-scale images.

Although multiband Gabor filtering is a tremendously effective technique for achieving the spatio-spectrally localized decomposition of a sophisticated image into components, it also complicates the component-demodulation problem. Subsequent to filtering, the modulating functions of each individual image component must be estimated from the filterbank channel responses. A family of novel approximation theorems known collectively as *quasi-eigenfunction approximations* may be used to establish the validity of applying the frequency-demodulation algorithms in Eq. (18)–(20), and Eq. (21) directly to the filterbank channel responses (as a consequence of the filtering, the continuous-domain frequency algorithm of Eq. (18) is no longer exact but contains an error that is generally small or negligible for locally coherent image components) (4,60,68,99–101). The amplitude-demodulation algorithms in Eq. (17) and Eq. (19) require modification to compensate for the scaling incurred during multiband filtering, however.

For the continuous-domain case, assume that $y_m(\mathbf{x})$ is the response of filterbank channel m and that image component $s_i(\mathbf{x})$ dominates $y_m(\mathbf{x})$ at the point $\mathbf{x}$. Application of Eq. (18) directly to $y_m(\mathbf{x})$ produces an estimate $\nabla\hat{\varphi}_i(\mathbf{x})$ of the instantaneous frequency vector $\nabla\varphi_i(\mathbf{x})$. The amplitude-modulation function $a_i(\mathbf{x})$ may then be estimated using the

Table 2 Design Parameters for Filters 20–39

Filter	Ray	Index on ray	Orient. θ_m (rad)	r_m $\left(\dfrac{\text{Hz}}{\text{pix}}\right)$	u_m $\left(\dfrac{\text{Hz}}{\text{pix}}\right)$	v_m $\left(\dfrac{\text{Hz}}{\text{pix}}\right)$	Space const. σ_m (pix)	max $|G(\Omega)|$
20	4	0	0.0504	0.0375	0.0375	0.0019	10.6004	53.1426
21	4	1	0.0504	0.0675	0.0674	0.0034	5.8891	29.5237
22	4	2	0.0504	0.1215	0.1213	0.0061	3.2717	16.4020
23	4	3	0.0504	0.2187	0.2184	0.0110	1.8176	9.1122
24	4	4	0.0504	0.3937	0.3932	0.0198	1.0098	5.0626
25	5	0	0.4107	0.0375	0.0344	0.0150	10.6004	53.1426
26	5	1	0.4107	0.0675	0.0619	0.0269	5.8891	29.5237
27	5	2	0.4107	0.1215	0.1114	0.0485	3.2717	16.4020
28	5	3	0.4107	0.2187	0.2005	0.0873	1.8176	9.1122
29	5	4	0.4107	0.3937	0.3609	0.1572	1.0098	5.0626
30	6	0	0.7709	0.0375	0.0269	0.0261	10.6004	53.1426
31	6	1	0.7709	0.0675	0.0484	0.0470	5.8891	29.5237
32	6	2	0.7709	0.1215	0.0871	0.0847	3.2717	16.4020
33	6	3	0.7709	0.2187	0.1569	0.1524	1.8176	9.1122
34	6	4	0.7709	0.3937	0.2824	0.2743	1.0098	5.0626
35	7	0	1.1312	0.0375	0.0160	0.0339	10.6004	53.1426
36	7	1	1.1312	0.0675	0.0287	0.0611	5.8891	29.5237
37	7	2	1.1312	0.1215	0.0517	0.1100	3.2717	16.4020
38	7	3	1.1312	0.2187	0.0931	0.1979	1.8176	9.1122
39	7	4	1.1312	0.3937	0.1675	0.3562	1.0098	5.0626

Note: See Table 3 for descriptions of the individual parameters.

Table 3 Description of Filter Parameters Appearing in Tables 1 and 2

Parameter name	Description
Filter	Filter number. Along each ray, filter numbers increase with increasing radial center frequency.
Ray	Number of the ray on which the filter is located. Ray zero is at the top of Fig. 3, and rays are numbered sequentially proceeding clockwise around the filterbank.
Index on ray	Index of the filter on its respective ray. The filter with index 0 on each ray is the one with lowest radial center frequency.
θ_m	Orientation of the filter center frequency Ω_m in radians; $\theta_m = \arg[\Omega_m]$.
r_m	Radial center frequency of the filter in cycles per pixel; $r_m = |\Omega_m|$.
u_m	Horizontal center frequency of the filter in cycles per pixel.
v_m	Vertical center frequency of the filter in cycles per pixel.
σ_m	Filter space constant in pixels.
$\max|G(\Omega)|$	Peak magnitude value of the filter frequency response.

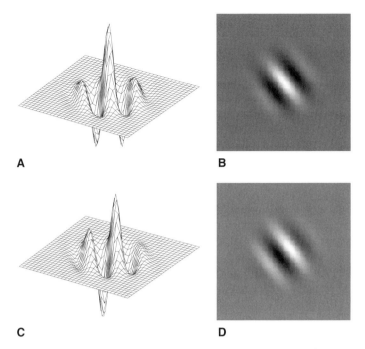

Fig. 4 Space-domain representation of Filter 10. (A) Real part plotted as a surface. (B) Real part depicted as a gray-scale image. (C) Imaginary part plotted as a surface. (D) Imaginary part depicted as a gray-scale image.

algorithm

$$a_i(\mathbf{x}) \approx \hat{a}_i(\mathbf{x}) = \left| \frac{y_m(\mathbf{x})}{G[\nabla \hat{\varphi}_i(\mathbf{x})]} \right| \qquad (27)$$

Modification of the discrete-domain amplitude-demodulation algorithm (19) proceeds similarly. Suppose that $y_m(\mathbf{k})$ is the response of filterbank channel m and that component $s_i(\mathbf{k})$ dominates $y_m(\mathbf{k})$ at the point $\mathbf{k}$. Once the algorithms in Eq. (20) and Eq. (21) have been used to obtain an estimate $\nabla \hat{\varphi}_i(\mathbf{k})$ of the frequency vector $\nabla \varphi_i(\mathbf{k})$, the amplitude-modulation function of component $s_i(\mathbf{k})$ may be estimated at the point $\mathbf{k}$ by

$$a_i(\mathbf{k}) \approx \hat{a}_\mathbf{i}(\mathbf{k}) = \left| \frac{y_m(\mathbf{k})}{G[\nabla \hat{\varphi}_i(\mathbf{k})]} \right| \qquad (28)$$

For locally coherent image components, the approximation errors inherent in the algorithms of Eq. (27) and Eq. (28) are generally small or negligible (4,60,68,99–101).

4.3. Dominant-Component Image Analysis

Locally coherent evolutionary, granular, and oriented textures occur frequently in natural images. The local coarseness, flow, and orientation of such textures can be succinctly characterized by the nonstationary instantaneous frequencies that dominate the image spectrum on a spatially local basis. These frequencies, which are termed *emergent*, are given at each image point **x** by the frequency vector $\nabla\varphi_i(\mathbf{x})$ corresponding to the component $s_i(\mathbf{x})$ that is locally dominant at the point **x**. Often, the emergent frequencies carry a rich description of the local texture structure (4). Consequently, estimates of the emergent frequencies can be extremely useful in a wide variety of machine vision tasks. In this section, we describe an analysis paradigm called *dominant-component analysis* which, for an image modeled according to Eq. (1), estimates the modulating functions of the locally dominant image component on a pointwise basis. The estimated dominant-component amplitude modulations $a_D(\mathbf{x})$ correspond to contrast in the dominant texture structure, whereas the computed dominant-component frequency modulations $\nabla\varphi_D(\mathbf{x})$ are estimates of the emergent frequencies.

A block diagram of the dominant-component approach is shown in Fig. 5, where the multicomponent sophisticated image $s(\mathbf{x})$ is analyzed with an M-channel multiband filterbank to isolate components from one another on a spatio-spectrally localized basis. At each point in the domain, the response of each filterbank channel is demodulated in the blocks marked DEMOD in Fig. 5. Amplitude and frequency estimates are computed from each filterbank channel response at each point in the domain. The dominant-component demodulation problem is that of determining which channel estimates correspond to the locally dominant image component at each point. This determination cannot be based solely on the magnitudes of the channel responses, as all of the filters have identical octave bandwidths and unity L^2-norms. Hence, the peak magnitudes of the frequency responses of low-frequency channels are much greater than those of high-frequency channels.

For each channel, a metric called the *filter selection criterion* is computed at each point in the domain, and estimates of the dominant-component modulating functions at a

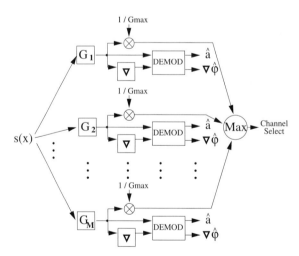

Fig. 5 Block diagram of the dominant-component approach. The filtered demodulation algorithm is performed in the blocks marked DEMOD. The magnitude of each channel response is divided by the peak value of the channel filter frequency response magnitude to obtain the filter selection criterion. On a pixel-by-pixel basis, the estimation is performed using the demodulation results from the channel which maximizes this quantity.

given point are taken from the filterbank channel which maximizes the filter selection criterion at that point. The filter selection criterion $\Psi_m(\mathbf{x})$ for channel m is defined by

$$\Psi_m(\mathbf{x}) = \frac{|y_m(\mathbf{x})|}{\max_{\Omega}|G_m(\mathbf{\Omega})|} \tag{29}$$

where $y_m(\mathbf{x})$ is the response of filterbank channel m and where $G_m(\mathbf{\Omega})$ is the frequency response of the mth channel filter. In view of the algorithms in Eq. (27) and Eq. (28), the quantity $\Psi_m(\mathbf{x})$ may be regarded as a crude estimate of the true amplitude-modulation function of the image component which dominates the response of channel m at the point $\mathbf{x}$. Although reasonable estimates of the dominant-component modulating functions could, in many cases, be obtained from any one of several channel filters having center frequencies near the dominant instantaneous frequency vector, the technique based on Eq. (29) tends to select the channel with the best signal-to-noise ratio. Hence, it affords maximal rejection of cross-component interference and other out-of-band information such as random noise and subemergent components.

4.4. General Multicomponent AM–FM Representations

It is well known that the Fourier spectra of natural images tend to be rich in low frequencies. Furthermore, low-frequency information plays an important role in visual perception. Therefore, in performing multicomponent AM–FM image modeling, it is advantageous to supplement the multiband filterbank described in Sec. 4.2 with a baseband channel to capture visually important low-frequency information. The impulse response $g_b(\mathbf{x})$ of the baseband channel filter is a Gaussian with space constant σ_b and takes the form

$$g_b(\mathbf{x}) = \frac{1}{\sigma_b\sqrt{2\pi}} \exp\left(-\frac{1}{4\sigma_b^2}\mathbf{x}^T\mathbf{x}\right) \tag{30}$$

The filter frequency response is

$$G_b(\mathbf{\Omega}) = \mathcal{F}[g_b(\mathbf{x})] = 2\sqrt{2\pi}\sigma_b \exp(-\sigma_b^2\mathbf{\Omega}^T\mathbf{\Omega}) \tag{31}$$

Let $G_m(\mathbf{\Omega})$ and $G_n(\mathbf{\Omega})$ be the first filters on two adjacent rays of the filterbank shown in Fig. 3. The baseband filter space constant σ_b is designed so that the η-peak contour of $G_b(\mathbf{\Omega})$ intercepts the η-peak contours of $G_m(\mathbf{\Omega})$ and $G_n(\mathbf{\Omega})$ at a single frequency, as depicted in Fig. 6a. Thus, the η-peak radial frequency of G_b is given by

$$r_{b,\eta} = r_0 \cos\left(\frac{\Lambda}{2}\right) - \sqrt{\frac{-\ln \eta}{\sigma_m^2} - r_0^2 \sin^2\left(\frac{\Lambda}{2}\right)} \tag{32}$$

The baseband filter space constant is then

$$\sigma_b = \frac{\sqrt{-\ln \eta}}{r_{b,\eta}} \tag{33}$$

where the filterbank parameters r_0 and Λ were described in Sec. 4.2.

We further supplement the filterbank of Fig. 3 with two high-frequency *corner filters* placed near the corners of the spatial-frequency plane. The corner filter center frequencies have orientations $\theta_m = -\pi/4$ radians and $\theta_m = \pi/4$ radians. Their radial center

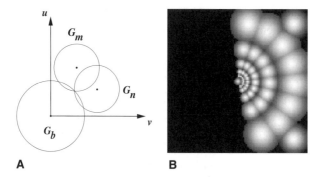

Fig. 6 Multiband Gabor filterbank with baseband filter and two high-frequency corner filters. (A) η-Peak contours of the baseband filter and two adjacent filters from the innermost ring of the regular filter tesselation. The η-peak contours of these three filters intersect in a single point. (B) Augmented filterbank including baseband filter and two corner filters. The filters appearing in the regular tesselation are identical to those of Fig. 3. The two high-frequency corner filters are at orientations $\pm \pi/4$ radians and have space constants equal to the geometric mean of the space constants of the filters in the ultimate and penultimate rings of the regular tesselation.

frequencies are

$$r_m = r_M + \frac{1}{3}\left(\frac{\pi}{\sqrt{2}} - r_M\right) \tag{34}$$

where r_M is the greatest radial center frequency assumed by any filter in the filterbank of Fig. 3. Thus, the corner filter radial center frequencies lie one-third of the way between the radial center frequencies of the filters in the outside ring of Fig. 3 and the radial frequency at the right-hand corners of the figure. The corner filter space constants σ_m are set to the geometric mean of the space constants of the filters in the ultimate ring of Fig. 3 and those of the filters in the penultimate ring of Fig. 3. The augmented filterbank including the baseband filter and both corner filters has 43 channels in total and is depicted in Fig. 6b.

4.4.1. The Channelized-Components Approach

The simplest and most straightforward approach to the computation of general multicomponent AM–FM image models is to estimate modulating functions for one image component from each channel of the multiband filterbank shown in Fig. 6b. We call this computational paradigm the *channelized-components* approach. Consistent with what is known about the processing that occurs in biological vision systems, the channelized-components approach represents the information content of an image by smoothly varying modulations occurring in a small number of frequency- and orientation-selective linear channels.

Although it is simple to compute, the main drawback of the channelized-components approach is that, for an M-channel filterbank, it necessarily leads to representations admitting M components. Thus, the channelized approach does not fulfill the goal of capturing the visually important information in an image using the smallest possible number of components. For example, consider the multicomponent sophisticated image shown in

AM–FM Image Modeling and Gabor Analysis

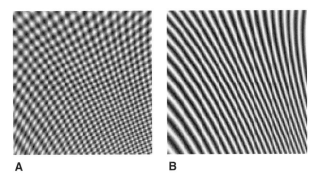

Fig. 7 Example of a multicomponent sophisticated image that admits an obvious decomposition into two locally coherent components: (A) image; (B) one of the locally coherent components in the obvious two-component interpretation.

Fig. 7a. This image admits an obvious interpretation as the sum of two highly locally coherent components, one of which is shown in Fig. 7b. Using the filterbank of Fig. 6b, however, the channelized components approach would result in a 43-component interpretation for this image. Thus, channelized-components representations are inherently inefficient.

The other main disadvantage of the channelized components approach relates to the recovery of an image from its computed multicomponent AM–FM representation. Because the Gabor filters in the multiband filterbank of Fig. 6b are not orthogonal, certain features of the image structure are generally manifest in multiple channelized components. Multiple redundant copies of this structure would be included in a direct reconstruction of the image from Eq. (1). Despite these drawbacks, however, channelized-components multicomponent AM–FM image representations often deliver reconstructed images of remarkably high quality.

4.4.2. Toward More Efficient Representations

Both of the main disadvantages to the channelized-components approach discussed in Sec. 4.4.1 were related to the fact that the number of components in a computed channelized components AM–FM image representation is completely determined by the structure of the multiband filterbank and does not depend in any way on the structure of the particular image under analysis. In this section, we outline a practical approach called *tracked multicomponent analysis* for computing the multicomponent AM–FM representation

$$\{\tilde{a}_i(\mathbf{k}), \nabla\tilde{\varphi}_i(\mathbf{k})\}_{i\in[1,K]} \tag{35}$$

of a K-component discrete-domain sophisticated image modeled as samples of Eq. (1). Although the nature of the decomposition of the image into components is still driven by the filterbank structure with this approach, the number of components in the computed representation in this case depends on the multipartite character of the image being represented. In addition to determining K, computation of the representation requires simultaneously estimating the modulating functions of all K components. Figure 8 shows a block diagram of the computational paradigm. As before, components are isolated on a spatiospectrally localized basis using a multiband Gabor filterbank of the type depicted in Fig.

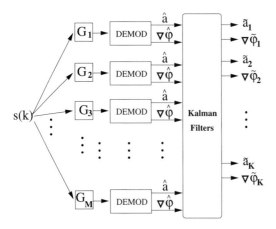

Fig. 8 Block diagram of the tracked multicomponent paradigm.

6b. Subsequent to filtering, the filtered demodulation algorithm is applied to each channel response. The track processor depicted in Fig. 8 assimilates the estimates obtained from each channel into estimates for each component to obtain the representation in Eq. (35).

The track processor is a Kalman filter based on a novel statistical state-space image-component model. Let $\varphi_i^x(\mathbf{k})$ and $\varphi_i^y(\mathbf{k})$ denote, respectively, the samples of the horizontal and vertical components of the continuous-domain frequency-modulation function $\nabla \varphi_i(\mathbf{x})$ of image component $s_i(\mathbf{x})$ in the model (1): $\nabla \varphi_i(\mathbf{k}) = [\varphi_i^x(\mathbf{k}) \ \varphi_i^y(\mathbf{k})]^T$ [the samples of $s_i(\mathbf{x})$ are denoted by $s_i(\mathbf{k})$]. Introduce an artificial temporal causality relationship between points in the sampled image domain by mapping them to a 1-D lattice according to a path function $\mathcal{O}: \mathbf{k} \mapsto k, k \in \mathbb{N}$. The mapping $\mathcal{O}$ is such that pixels which are adjacent on the 1-D lattice are also adjacent in the image. $\mathcal{O}$ maps the modulating functions of a component according to, for example, $a_i(\mathbf{k}) \stackrel{\mathcal{O}}{=} a_i(k)$. Let ρ denote continuous arc length along $\mathcal{O}$ and use the notation

$$\varphi_i^{x'}(k) = \left.\frac{\partial}{\partial \rho} \varphi_i^x(\rho)\right|_{\rho=k} \tag{36}$$

to indicate the restriction to the discrete 1-D lattice of the derivative of a modulating function taken with respect to ρ. Then, each modulating function may be expanded in a first-order Taylor series about a lattice point k; for example,

$$\varphi_i^x(k+1) = \varphi_i^x(k) + \varphi_i^{x'}(k) + \int_k^{k+1} (k+1-\rho)\varphi_i^{x''}(\rho)d\rho \tag{37}$$

Likewise, the first-order derivatives of the modulating functions may be expanded in zeroth-order Taylor series:

$$\varphi_i^{x'}(k+1) = \varphi_i^{x'}(k) + \int_k^{k+1} \varphi_i^{x''}(\rho)\,d\rho \tag{38}$$

Now, we model $a_i(\mathbf{x})$ and $\varphi_i(\mathbf{x})$ as independent, homogeneous mean-square (m.s.) differentiable random fields and assume that $\varphi_i(\mathbf{x})$ is *quadrant symmetric* (102). We model the Taylor series integrals associated with Eq. (37) as three noise processes $u_a(k)$, $u_{\varphi_x}(k)$, and $u_{\varphi_y}(k)$. Similarly, we model the Taylor series integrals associated with Eq. (38) by noise processes $v_a(k)$, $v_{\varphi_x}(k)$, and $v_{\varphi_y}(k)$. Collectively, we refer to these six noise processes as the *modulation accelerations,* or MAs as they involve local averages of the second derivatives of the modulating functions. Then, the series in Eq. (37)–(38) for all three modulating functions may be written together in a canonical state-variable form to obtain the statistical state-space component model

$$\begin{bmatrix} a_i(k+1) \\ a_i'(k+1) \\ \varphi_i^x(k+1) \\ \varphi_i^{x'}(k+1) \\ \varphi_i^y(k+1) \\ \varphi_i^{y'}(k+1) \end{bmatrix} = \begin{bmatrix} 1 & 1 & 0 & 0 & 0 & 0 \\ 0 & 1 & 0 & 0 & 0 & 0 \\ 0 & 0 & 1 & 1 & 0 & 0 \\ 0 & 0 & 0 & 1 & 0 & 0 \\ 0 & 0 & 0 & 0 & 1 & 1 \\ 0 & 0 & 0 & 0 & 0 & 1 \end{bmatrix} \begin{bmatrix} a_i(k) \\ a_i'(k) \\ \varphi_i^x(k) \\ \varphi_i^{x'}(k) \\ \varphi_i^y(k) \\ \varphi_i^{y'}(k) \end{bmatrix} + \begin{bmatrix} u_a(k) \\ v_a(k) \\ u_{\varphi_x}(k) \\ v_{\varphi_x}(k) \\ u_{\varphi_y}(k) \\ v_{\varphi_y}(k) \end{bmatrix} \quad (39)$$

with output vector

$$\mathbf{Y}(k) = \begin{bmatrix} a_i(k) \\ \varphi_i^x(k) \\ \varphi_i^y(k) \end{bmatrix} \quad (40)$$

We model the estimation errors inherent in the filtered demodulation algorithm by noise processes $n_a(k)$, $n_{\varphi_x}(k)$, and $n_{\varphi_y}(k)$, called the *measurement noises,* or MNs. This results in the observation equation

$$\begin{bmatrix} \hat{a}(k) \\ \hat{\varphi}^x(k) \\ \hat{\varphi}^y(k) \end{bmatrix} = \begin{bmatrix} a_i(k) \\ \varphi_i^x(k) \\ \varphi_i^y(k) \end{bmatrix} + \begin{bmatrix} n_a(k) \\ n_{\varphi_x}(k) \\ n_{\varphi_y}(k) \end{bmatrix} \quad (41)$$

We assume that the covariance structures of the second derivatives of the modulating functions are highly spatially localized, so that they behave impulsively in both the vertical and horizontal directions when viewed at the scale of the spatial sampling lattice. Then, it is straightforward to compute the following covariances between the MAs:

$$E[u_a(k)u_a(j)] = \tfrac{1}{3}\sigma_{a''}^2 \delta(k-j) \quad (42)$$

$$E[v_a(k)v_a(j)] = \sigma_{a''}^2 \delta(k-j) \quad (43)$$

$$E[u_a(k)v_a(j)] = \tfrac{1}{2}\sigma_{a''}^2 \delta(k-j) \quad (44)$$

Expressions for the MA covariances $E[u_{\varphi_x}(k)u_{\varphi_x}(j)]$, $E[u_{\varphi_y}(k)u_{\varphi_y}(j)]$, $E[v_{\varphi_x}(k)v_{\varphi_x}(j)]$, $E[v_{\varphi_y}(k)v_{\varphi_y}(j)]$, $E[u_{\varphi_x}(k)v_{\varphi_x}(j)]$, and $E[u_{\varphi_y}(k)v_{\varphi_y}(j)]$ are analogous in form. The remaining 12 MA covariance functions are all identically zero.

Analytical characterization of the MN moments and of the joint moments between the MNs and MAs is difficult. However, in practice, we have found that when dealing with locally coherent images, the MN moments are normally quite small with relation to the magnitudes of the modulating functions themselves. Therefore, within the scope of this chapter, we shall assume that the MNs are jointly uncorrelated and also uncorrelated with the MAs.

In the preceding, we have modeled the estimated modulating functions of a nonstationary image component as noisy observations of an affine function of the state vector of a finite-dimensional linear system driven by uncorrelated noise. Hence, the minimum mean-squared error (MMSE) optimal track processor involves Kalman filters. Due to the block-diagonal structure of the state transition matrix in Eq. (39), the system modes corresponding to the amplitude modulation and the two components of the frequency modulation may be decoupled, resulting in three independent second-order systems. We use independent second-order Kalman filters to track each one.

The explicit recursive formulation for the optimal amplitude estimates $\tilde{a}_i(k|k)$ is

$$\tilde{a}_i(k|k) = \tilde{a}_i(k|k-1) + \alpha_a(k)[\hat{a}(k) - \tilde{a}_i(k|k-1)] \qquad (45)$$

$$\tilde{a}'_i(k+1|k) = \tilde{a}'_i(k|k-1) + \beta_a(k)[\hat{a}(k) - \tilde{a}_i(k|k-1)] \qquad (46)$$

$$\tilde{a}_i(k+1|k) = \tilde{a}_i(k|k) + \tilde{a}'_i(k+1|k) \qquad (47)$$

The optimal frequency estimators are analogous in form. Formulations for the Kalman filter gain sequences $\alpha_a(k)$ and $\beta_a(k)$ follow from recursive expressions for the state vector error-covariance matrices associated with the decoupled systems. The gain sequences typically converge rapidly. For images that are reasonably locally coherent, more than 20 gains are rarely required to achieve absolute convergence to within 10^{-14}. A discussion of issues related to track initialization may be found in Ref. 64.

The Kalman filters in Eq. (45)–(47) work well for images which contain only a few components that are well separated in frequency. However, many natural images contain phase discontinuities arising from occlusions, surface defects, or surface discontinuities. The frequency modulating function $\nabla\varphi_i(\mathbf{x})$ of an image component $s_i(\mathbf{x})$ can generally contain *unbounded* excursions of either sign near such discontinuities. If the Kalman filters were permitted to track such frequency excursions in an image comprising several components, then the track corresponding to a particular component $s_i(\mathbf{k})$ containing a wide-band frequency excursion would typically take observations from a filterbank channel that was dominated by another tracked component $s_q(\mathbf{k})$. Typically, the tracks corresponding to these two components would then become coincident and continue updating from the same filterbank channels indefinitely.

Therefore, in computing representations for multipartite natural images, it is generally necessary to postfilter the estimated modulating functions computed from the filterbank channel responses in order to compensate for frequency excursions. We use lowpass Gaussian postfilters for this purpose, as their envelopes and bandwidths can be simply related to the channel filters. With $\mathbf{x} = [x\ y]^T$, the impulse response of the continuous-domain unity L^1-norm postfilter for smoothing the estimated modulating functions delivered by filterbank channel m is

$$p_m(\mathbf{x}) = \frac{1}{4\pi\kappa_\theta\kappa_\Omega\sigma_m^2} \exp\left[-\frac{1}{4\sigma_m^2}\left(\frac{\zeta^2}{\kappa_\theta^2} + \frac{\xi^2}{\kappa_\Omega^2}\right)\right] \qquad (48)$$

where σ_m is the space constant of the Gabor channel filter $g_m(\mathbf{x})$, κ_θ and κ_Ω are *postfilter space constant scaling factors*,

$$\begin{bmatrix}\zeta \\ \xi\end{bmatrix} = \begin{bmatrix}\cos\vartheta & \sin\vartheta \\ -\sin\vartheta & \cos\vartheta\end{bmatrix}\begin{bmatrix}x \\ y\end{bmatrix} \qquad (49)$$

$\vartheta = \theta_m - \pi/2$, and θ_m is the orientation of channel filter $g_m(\mathbf{x})$.

Fig. 9 Postfiltered channel model.

Thus, the orientation bandwidth of the filter in Eq. (48) is governed by the scaling factor κ_θ and the magnitude-frequency bandwidth is governed by κ_Ω. If $\kappa_\theta = \kappa_\Omega = 1.0$, then the linear bandwidth of $p_m(\mathbf{x})$ is identical to that of $g_m(\mathbf{x})$. Larger postfilter space constant scaling factors yield a filter with a narrower bandwidth, which performs more smoothing. Taking κ_θ small facilitates the movement of tracked image components between channels which differ in orientation, whereas taking κ_Ω small facilitates the movement of tracked components between channels which differ in magnitude frequency. With the postfilters, initial frequency estimates are computed using the frequency algorithms in Eq. (18) or Eq. (20)–(21). These estimates are then postfiltered, and the postfiltered frequency estimates are used in the filtered amplitude-estimation algorithms of Eq. (27) or Eq. (28). The amplitude estimates themselves are then postfiltered. This postfiltered channel model is shown in Fig. 9.

5. APPLICATIONS AND EXAMPLES

Up to this point, we have observed that what is known about the function of biological vision systems strongly suggests that they represent the visual information in images as smoothly varying modulations in frequency- and orientation-selective linear channels. Furthermore, we have claimed that computed modulations are extremely useful in performing a wide variety of image-processing and machine vision tasks. We have also described several recently developed techniques for computing multicomponent AM–FM image models. Gabor analysis played a central role in each technique. In this section, we describe how these techniques may be applied in real machine vision applications and present a number of concrete examples.

5.1. Gabor Transform

As we mentioned in Sec. 3.1, an analytical technique for computing the coefficients in the 2-D Gabor expansion was developed by Zeevi and Porat, who used the expansion for image representations in general machine vision and image processing applications (82). Porat and Zeevi developed the mathematics to compute 2-D Gabor expansions with nonuniform spatial and spectral sampling intervals (2). They also introduced the *Gaborian Pyramid*, a special nonuniformly sampled Gabor expansion wherein the modulation frequencies of the Gaussian elementary functions are harmonically related. They demonstrated high-quality reconstructions of textured images using Gabor expansions truncated to a few

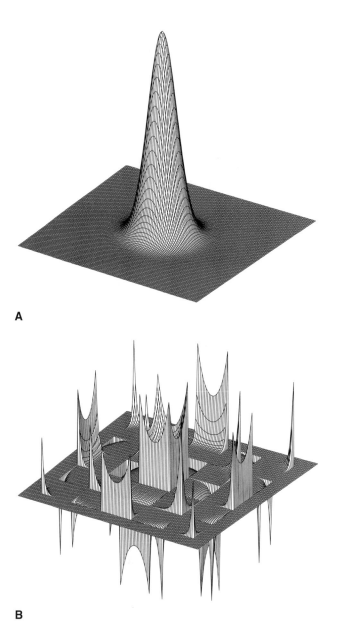

Fig. 10 Gaussian analysis window and auxiliary function for 2-D Gabor transform. (A) Gaussian window function $h(\mathbf{x})$; $\sigma = 1.0$. (B) Auxiliary function $\gamma(\mathbf{x})$.

thousand terms, and used simple statistics of the Gabor coefficients as features to effectively segment juxtaposed Brodatz textures (83). Also in Ref. 83, they used 2-D Gabor functions to add natural-looking synthesized textures to simple line drawings.

A separable Gaussian window function $h(\mathbf{x})$ is shown in Fig. 10a. The corresponding 1-D window functions $h_1(x_1)$ and $h_2(x_2)$ are each defined according to

AM–FM Image Modeling and Gabor Analysis

$$h_i(x_i) = \left(\frac{2}{\sigma}\right)^{1/4} \exp\left(-\frac{\pi}{\sigma}x_i^2\right) \tag{50}$$

where $\sigma = 1.0$. The biorthogonal auxiliary function $\gamma(\mathbf{x})$ associated with this window function is separable into the n-fold product of

$$\gamma_i(x_i) = (2\sigma)^{-1/4}\left(\frac{K_0}{\pi}\right)^{-3/2} \exp\left(\frac{\pi}{\sigma}x_i^2\right) \sum_{n > x_i/\sqrt{\sigma}-1/2} (-1)^n \exp\left[-\pi\left(n+\frac{1}{2}\right)^2\right] \tag{51}$$

where $K_0 \approx 1.85$ is the complete elliptic integral with modulus $1/\sqrt{2}$. The marked dissimilarity between the synthesis window function of Fig. 10a and the auxiliary function of Fig. 10b makes it strikingly clear why oversampled versions of the quasi-orthogonal discrete Gabor transform developed by Qian and Chen (79–81) are of such great practical interest for nonstationary spatio-spectral analysis.

For comparison, a 64×64-point one-sided exponential synthesis window function is shown in Fig. 11a. This window function is separable into the n-fold product of

$$\tilde{h}_i(k_i) = A_0 \exp\left(-\frac{3}{20}(k-16)\right) \tag{52}$$

where $A_0 \approx 0.0462$. For the parameter choices indicated in Fig. 11, the optimal quasi-orthogonal auxiliary function $\tilde{\gamma}(\mathbf{k})$ in Eq. (15) is shown in Fig. 11b, where it is indeed seen to bear strong similarity to the synthesis window of Fig. 11a. Qian and Chen also

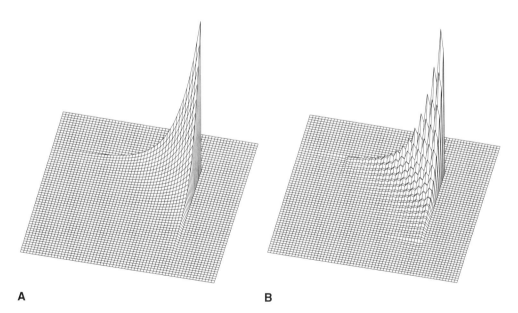

Fig. 11 One-sided exponential window function and optimal auxiliary function for 2-D finite quasi-orthogonal discrete Gabor transform. $L_1 = L_2 = 64$, $Q_1 = Q_2 = 32$, $P_1 = P_2 = 32$, $\tilde{\Omega}_1 = \tilde{\Omega}_2 = 2$, and $\tilde{T}_1 = \tilde{T}_2 = 2$. (A) One-sided exponential window function $h(\mathbf{k})$; (B) optimal auxiliary function $\tilde{\gamma}(\mathbf{k})$.

formulated a generalization of the quasi-orthogonal discrete Gabor transform wherein significant computational savings can be realized by employing a window function $h(\mathbf{x})$ with spatial support much smaller than the support of the image under analysis (79,81).

5.2. Dominant-Component Analysis

Numerous authors have used dominant-component amplitude and frequency modulations computed by one method or another to perform texture segmentation and classification (2, 4, 40, 48–50, 82, 83). Yu et al. employed dominant-component modulations computed using the 2-D discrete Teager–Kaiser operator to perform edge detection in images (59). Mitra and others also used dominant modulations computed by the Teager–Kaiser operator for image enhancement (61,103–106) and for noise removal (107). The Teager–Kaiser operator was used to compute dominant-component modulations for performing nonstationary spatio-spectral analysis by Maragos et al. (96) and by Maragos and Bovik (62,108). The specific techniques presented in Sec. 4.3 were used for nonstationary spatio-spectral analysis by Havlicek et al. (60,100,109). Super and Bovik (110–112) and Bovik et al. (4) used computed dominant-component modulations to obtain excellent 3-D surface shape reconstructions from image texture. Estimates of the dominant-component modulating functions were also used by Chen et al. to perform texture-based computational stereopsis (113,114) and by Pattichis et al. in the computation of image flow lines (109,115).

An example dominant-component analysis of the Brodatz texture image Tree is shown in Fig. 12. The original image appears in Fig. 12a. A needle diagram depicting the the emergent frequency estimates is given in Fig. 12b, where one needle is shown for each block of 10 × 10 pixels. The needle lengths in Fig. 12b are inversely proportional to the magnitude of the instantaneous frequency vector. Thus, they are proportional to the dominant *instantaneous period*. The needles are oriented along the emergent frequency vector. Note that the needles are normal to dominant texture features, as expected. The estimated dominant-component amplitude-modulation function is shown in Fig. 12c and may generally be interpreted as the *contrast* of the dominant component: Bright areas in Fig. 12c generally correspond to regions of high contrast in Fig. 12a. A reconstruction of the dominant image component from the estimated dominant-component modulating functions is given in Fig. 12d. Although highly nonstationary, this image possess significant locally coherent texture structure. The similarity between the dominant-component reconstruction and the original image is striking and suggests that, over much of the image, a substantial fraction of the total texture structure has been captured in only a single computed AM–FM component.

In the next set of examples, we illustrate one approach by which texture segmentation may be performed by detecting edges in the estimated dominant-component modulating functions. The *Laplacian of Gaussian* (or LoG) *filter* is an edge-detection filter with a circularly symmetric impulse response (34). The filter space constant σ regulates the scale at which the filter detects edges. Let $t(\mathbf{k})$ be a real-valued discrete image containing computed estimates of one of the dominant-component modulating functions of a discrete sophisticated image, $\nabla^2 g(\mathbf{k})$ be the unit pulse response of a sampled LoG filter, and $y(\mathbf{k}) = t(\mathbf{k}) * \nabla^2 g(\mathbf{k})$. Then, $y(\mathbf{k})$ will contain zero crossings where there are edges present in $t(\mathbf{k})$ at the scale σ. One constructs an edge map $\mathcal{Y}(\mathbf{k})$ by setting $\mathcal{Y}(\mathbf{k}) = 1$, where $y(\mathbf{k})$ contains zero crossings. Often, it is advantageous to suppress weak edges in $t(\mathbf{k})$ from appearing in the edge map. This may be accomplished by thresholding the gradient magni-

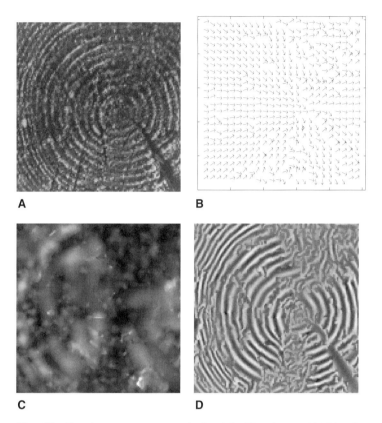

Fig. 12 Dominant-component analysis of the Tree image: (A) Tree image; (B) needle diagram depicting estimated emergent frequencies; (C) estimated dominant-component amplitude-modulation function; (D) reconstruction of dominant component from estimated modulating functions.

tude of $y(\mathbf{k})$; one sets $\mathcal{Y}(\mathbf{k}) = 1$ only if $y(\mathbf{k})$ contains a zero crossing at $\mathbf{k}$ *and* $|\nabla y(\mathbf{k})| > \tau$ for some threshold value τ.

Figure 13a shows the 256×256 image Mica-Burlap. A reconstruction of the dominant image component is shown in Fig. 13b. The estimated dominant-component amplitude-modulation function is shown in Fig. 13c, and the estimated emergent frequencies are depicted in the needle diagram of Fig. 13d. The arrows in the needle diagram are proportional to the instantaneous period, and their lengths have been squared to accentuate the differences between frequency estimates in different image regions. Figure 13e presents the magnitudes of the estimated emergent frequency vectors as a gray-scale image. Thus, arrow lengths in Fig. 13d are reciprocally related to the gray scales in Fig. 13e. A LoG filter with space constant $\sigma = 15$ pixels was applied to the magnitude frequency estimates. The resulting edge map contained several contours. However, after thresholding the gradient magnitude of the LoG response image with $\tau = 1.5$, only one closed contour remained. This contour is shown overlaid on the image in Fig. 13f.

The 256×256 image Pellets-Beans is shown in Fig. 14a. The dominant-component reconstruction is given in Fig. 14b, and the dominant-component estimated amplitude-

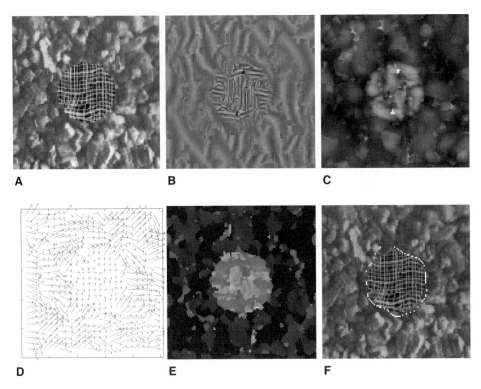

Fig. 13 Dominant-component texture segmentation of Mica-Burlap image: (A) Mica-Burlap image; (B) dominant-component reconstruction; (C) estimated amplitude-modulation function; (D) estimated emergent frequencies; (E) estimated emergent frequency magnitudes displayed as a gray-scale image; (F) texture segmentation computed by applying a $\nabla^2 g$ filter to the estimated emergent frequency magnitudes.

modulation function is shown in Fig. 14c. The estimated emergent frequencies are depicted in Fig. 14d, where arrow length is proportional to the instantaneous period and the arrow lengths have been squared for display. As in the example of Fig. 13, texture segmentation was performed by applying a LoG filter to the emergent frequency magnitudes given in the gray-scale image of Fig. 14e. However, the filter space constant was $\sigma = 49$ pixels in this case. The resulting edge map contained only one contour; therefore, thresholding was not required. This contour is shown overlaid on the image in Fig. 14f.

The other quantities delivered by dominant-component analysis may also be used effectively for texture segmentation. The image Wood-Wood is shown in Fig. 15a. This 256×256 image was generated by rotating the original image counterclockwise by 45° and subsequently replacing pixels in the central diamond-shaped region of the original image with their counterparts in the rotated image. The dominant-component reconstruction is shown in Fig. 15b. The estimated dominant-component amplitude-modulation function is depicted in Fig. 15c and the estimated emergent frequencies are given in Fig. 15d. Figure 15e presents the *orientations* of the estimated emergent frequency vectors as a gray-scale image. A LoG filter with space constant $\sigma = 14$ pixels was applied to the orientations in Fig. 15e and the LoG response gradient magnitude was thresholded at τ

AM–FM Image Modeling and Gabor Analysis

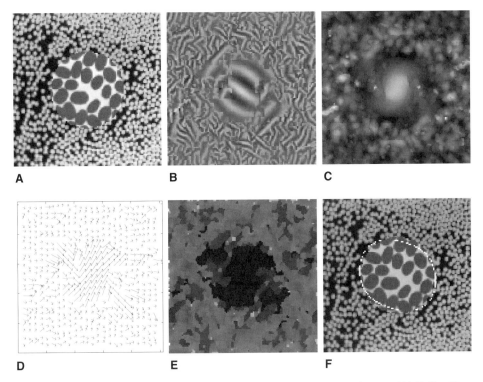

Fig. 14 Dominant-component texture segmentation of Pellets-Beans image: (A) Pellets-Beans image; (B) dominant-component reconstruction; (C) estimated amplitude-modulation function; (D) estimated emergent frequencies; (E) estimated emergent frequency magnitudes displayed as a gray-scale image; (F) texture segmentation computed by applying a $\nabla^2 g$ filter to the estimated emergent frequency magnitudes.

$= 11$. The resulting edge map contained only one closed contour, which is shown overlaid on the image in Fig. 15f.

Finally, Fig. 16a shows the 256×256 image Paper-Burlap. The dominant-component reconstruction is given in Fig. 16b. The dominant-component estimated amplitude-modulation function is given in Fig. 16c and the emergent frequency estimates are depicted in Fig. 16d. In this case, a LoG filter with space constant $\sigma = 46.5$ pixels was applied to the amplitude estimates of Fig. 16c. The resulting edge map contained only a single contour, which is shown overlaid on the image in Fig. 16e.

The examples of this section establish that all of the estimated quantities delivered by a dominant-component image analysis—namely the estimated dominant-component amplitude modulation function, the estimated emergent frequency magnitudes, and the estimated emergent frequency orientations—are useful in solving the classical texture-segmentation problem.

5.3. General AM–FM Representations

The techniques presented in Sec. 4 for computing general multicomponent AM–FM image representations are quite new and exciting. We believe that once they reach full maturity,

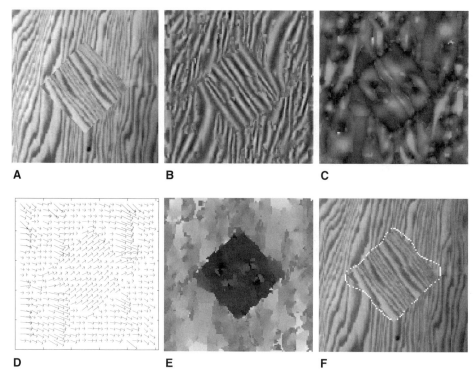

Fig. 15 Dominant-component texture segmentation of Wood-Wood image: (A) Wood-Wood image; (B) dominant-component reconstruction; (C) estimated amplitude-modulation function; (D) estimated emergent frequencies; (E) estimated emergent frequency orientations displayed as a grayscale image; (F) texture segmentation computed by applying a $\nabla^2 g$ filter to the estimated emergent frequency orientations.

they will find significant future applications in AM–FM-based image and video coding for telecomputing, telemedicine, multimedia communications, and CD-ROM mass storage systems. Figures 17 and 18 show eight natural images along with reconstructions obtained using the channelized-component approach. In each case, the reconstructions are of high quality. The images in Figs. 17, 18a, and 18c each portray a single, homogeneous Brodatz texture, whereas the images in Figs. 17e and 17g are more complicated in that they contain multiple objects with textures exhibiting significantly differing characteristics. In each example, the image was processed with the filterbank of Fig. 6b. The postfiltered channel model of Fig. 9 was employed using isotropic postfilters with space constant scaling factors $\kappa_\theta = \kappa_\Omega = 1.0$. One component was reconstructed from the estimated modulating functions delivered by each filterbank channel. These reconstructed components were summed to produce the image reconstructions shown in Figs. 17 and 18.

Agreement between the reconstructed and original images in these examples is excellent; in several cases, they are virtually indistinguishable visually. Of particular interest are the Pebbles, Beach, and Celebrity images of Figs. 18c, 18e, and 18g, respetively. The Beach image is an aerial photograph of Ocean City, NJ. On cursory inspection, none of these images seems particularly well suited to AM–FM modeling. On a global scale, the

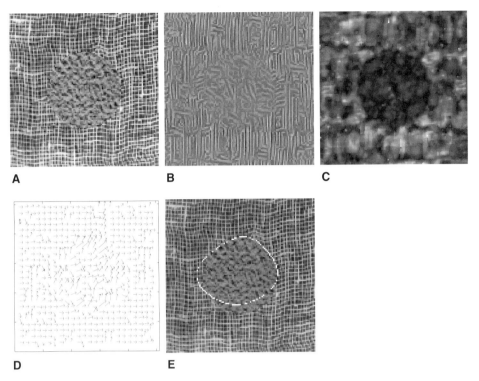

Fig. 16 Dominant-component texture segmentation of Paper-Burlap image: (A) Paper-Burlap image; (B) dominant-component reconstruction; (C) estimated amplitude-modulation function; (D) estimated emergent frequencies; (E) texture segmentation computed by applying a $\nabla^2 g$ filter to the estimated amplitude-modulation function.

Pebbles image appears almost devoid of coherent structure. The Beach and Celebrity images each have significant regions dominated by very low-frequency information and also contain multiple complicated textures with disparate characteristics supported over small, irregularly shaped regions. The quality of the reconstructions shown in these examples attests to the considerable power of AM–FM modeling techniques for nonstationary image analysis and representation.

Our final two examples demonstrate the tracked, postfiltered paradigm depicted in Figs. 8 and 9. The approach was applied to two 256×256 Brodatz texture images using the filterbank of Fig. 6b. In each case, modulating functions for a baseband image component were extracted from the baseband channel of the filterbank, and these estimates were not considered by the track processor of Fig. 8. Consequently, the baseband channel postfilter space constant scaling factors were set to the relatively small values $\kappa_\theta = \kappa_\Omega = 1.0$. More pronounced postfiltering was required for the tracked channels to ensure that the track processor could maintain component tracking near pixels where phase discontinuities and their associated frequency excursions occurred. The postfilter space constant scaling factors for the tracked channels were set to $\kappa_\theta = 3.0$ and $\kappa_\Omega = 3.25$. The small difference between these values *slightly* favors the movement of tracks between adjacent channels with the same radial center frequencies but different orientations over the move-

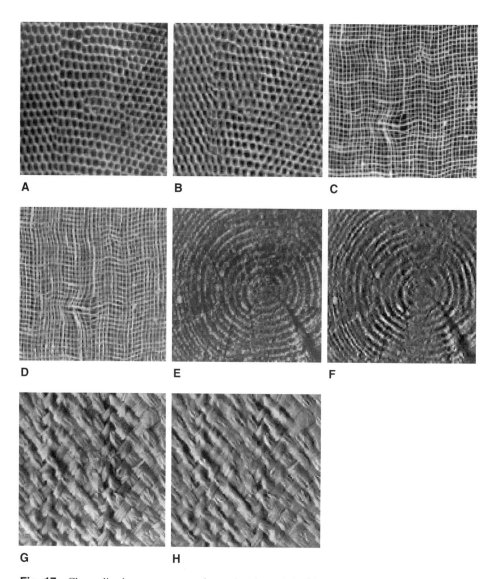

Fig. 17 Channelized-components analyses showing original images and reconstructions from 43 channelized components: (A) Reptile image and (B) reconstruction; (C) Burlap image and (D) reconstruction; (E) Tree image and (F) reconstruction; (G) Raffia image and (H) reconstruction.

ment of tracks between adjacent channels with the same orientation but different radial center frequencies.

The image Reptile is shown again in Fig. 19a. The track processor detected the presence of five locally coherent AM–FM image components in this image. A reconstruction of the image from the modulating functions of these five components and the baseband component is shown in Fig. 19b. It is of remarkable quality for such a small number of components. The baseband component reconstruction appears in Fig. 19c, whereas

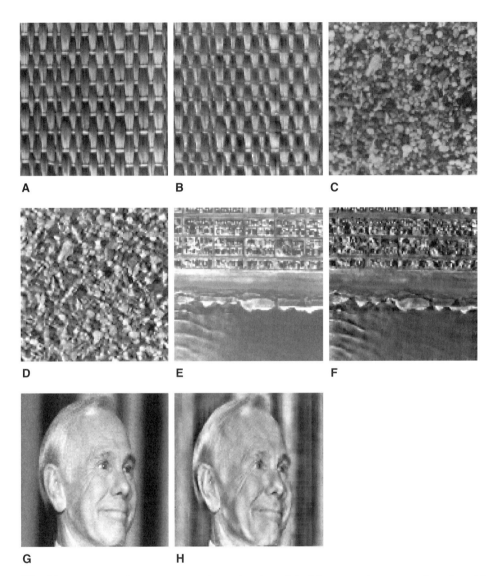

Fig. 18 Channelized-components analyses showing original images and reconstructions from 43 channelized components: (A) Straw image and (B) reconstruction; (C) Pebbles image and (D) reconstruction; (E) Beach image and (F) reconstruction; (G) Celebrity image and (H) reconstruction.

reconstructions of tracked components one through five appear in Figs. 19d–19h, respectively. Considerable nonstationary amplitude and frequency modulation is visible in each reconstructed component.

The baseband component primarily captures low-frequency information. The low-frequency structure of this image embodies smooth shading and contrast variations occurring at large spatial scales. Components one through three and component five each capture elements of the perceptually dominant hexagonal texture structure of the image. Note that

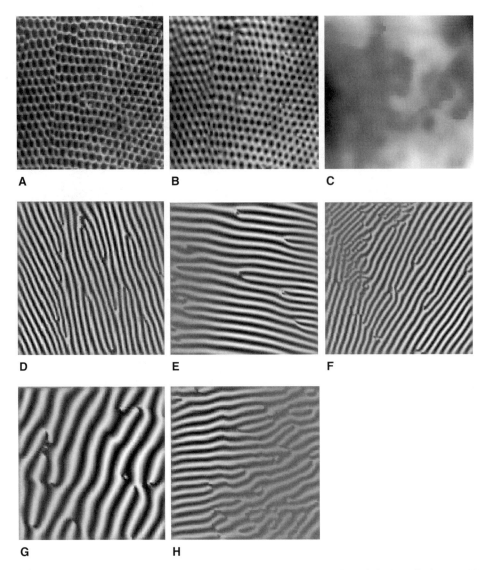

Fig. 19 Multicomponent AM–FM representation and reconstruction of the Reptile image: (A) Image; (B) reconstruction from five tracked components and baseband component; (C) baseband component reconstruction; (D)–(H) reconstructions of components one through five.

the magnitude spatial frequency of this structure is approximately 16–20 cycles per image and varies nonstationarily. The perceptually dominant orientation of this structure also varies nonstationarily and is quite different (e.g., in the leftmost and rightmost regions of the image). It is interesting to observe how these nonstationary frequency variations are manifest in the individual computed AM–FM components. Components one and three are each dominated by different orientations in different image regions. The structure of the amplitude modulations in components two and five is complementary in character:

component two has lower contrast on the left and higher contrast on the right, whereas component five has higher contrast on the left and lower contrast on the right. Finally, component four exhibits structure of relatively lower frequency and may be interpreted loosely as an AM–FM *harmonic* of component four.

The Burlap image appears again in Fig. 20a. In this case, the track processor identified the presence of eight locally coherent AM–FM image components. A reconstruction of the image from these eight tracked components and the baseband component is given in Fig. 20b. This reconstruction is again in remarkable agreement with the original. As with the Reptile image, the essential structure of the Burlap image has been effectively captured by the computed multicomponent AM–FM representation and faithfully reproduced in the reconstruction of Fig. 20b. However, noticeable high-frequency information is missing throughout the reconstruction, particularly from the highly nonstationary structure appearing just to the left and below the center of the image. Furthermore, many of the sharp edges which appear in the original image appear blurred to varying degrees in the reconstruction.

These examples demonstrate that the postfiltered, tracked approach for computing multicomponent AM–FM image representations is extremely powerful. In both cases, the essential structure of the image was captured in a remarkably small number of components. The structure that was missing from the reconstructed images tended to be locally supported and high frequency in character. Note that both of the images analyzed in these examples consisted of a single nonstationary texture. The postfiltered, tracked approach tends to have difficulty with texture structure that is supported only locally and with images that contain multiple objects or textured regions with disparate characteristics. This is because, in processing the image pixels along the 1-D path function $\mathcal{O}$, the track processor cannot remain locked onto coherent structure that is supported on irregularly shaped subregions of the image. The state-space model in Eq. (39)–(41) assumes that a component is supported on all tracked pixels. One possible approach for dealing with images containing multiple textured objects is to perform a texture segmentation using the dominant-component analysis paradigm prior to application of the track processor. Tracking could then be carried out independently over each segmented region. This approach merits considerable future investigation.

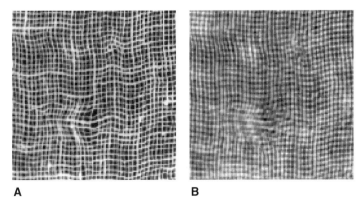

Fig. 20 Multicomponent AM–FM representation of Burlap: (A) Image; (B) reconstruction from eight tracked components and baseband component.

6. CONCLUSION

In this chapter, we have examined a number of recently developed techniques that use Gabor analysis to model images as sums of AM–FM functions. Because AM–FM functions generalize the 2-D Fourier kernel, most natural images can be modeled reasonably well using substantially fewer than N^2 AM–FM components. Because Gabor analysis is inherently nonstationary, computed AM–FM models provide image representations that are simultaneously localized in both space and spatial frequency.

Such AM–FM image models are of practical engineering interest for at least three reasons. First, they are convenient and intuitive tools for performing time–frequency and spatio-spectral analysis: The quasi-orthogonal Gabor transform as well as the representations shown in Figs. 12 and 19 naturally facilitate signal analysis *in terms of* the nonstationarities present in an image. Second, the estimated modulating functions delivered by the dominant-component analysis paradigm presented in Sec. 4.3 have demonstrable utility in solving a number of classical machine vision problems. Most notably, they are useful in the 3-D shape from the texture problem, the stereo correspondence problem, and the texture segmentation problem, all of which are germane to robotic control, autonomous vehicle navigation, automatic target recognition, and a host of other applications. A particularly satisfying feature of computed AM–FM image models is that, like biological vision systems, they represent the information in an image as locally coherent modulations in frequency- and orientation-selective channels. Third, we are confident that the multicomponent AM–FM representations presented in Secs. 3 and 4 hold great promise as the bases of future AM–FM signal, image, and video coding schemes. Such coding schemes would be of great utility in multimedia communications, telecomputing, telemedicine, and CD-ROM mass storage system applications.

Gabor transforms represent an image as a sum of AM–FM functions that are all frequency-modulated translates of the synthesis window. The Gabor expansion coefficients are inner products between the image and frequency-modulated translates of the biorthogonal auxiliary function. In theory, the multidimensional Gabor transform of Sec. 3.1 is invertible. In practice, however, the Gabor expansion must be truncated to a finite number of terms, and this generally leads to losses in an image reconstructed from its truncated Gabor expansion. However, both Porat and Zeevi (2,83) and Qian et al. (79,81) have demonstrated excellent reconstructions from finite Gabor expansions admitting only small or negligible losses.

For discrete signals and periodic signals, the Gabor transform presented in Sec. 3.2 is both finite and discrete. In the oversampled case, the auxiliary function obtained by solving the multidimensional biorthogonality constraint equation is not unique. Thus, in the oversampled case, it is possible to construct a quasi-orthogonal discrete Gabor transform in which the auxiliary function is optimally similar to the synthesis window (79–81). Then, the Gabor coefficients reflect the local similarity between the signal and the translated and frequency-modulated elementary functions.

In Sec. 4, we presented practical techniques for computing estimates of the component-modulating functions in the general multicomponent AM–FM representation in Eq. (35). The individual components embody physically meaningful, smooth, nonstationary, or evolutionary structure within an image. The AM–FM modeling problem is decidedly ill-posed; any signal or image may be exactly represented by a single AM–FM function in uncountably infinitely many ways and also by multiple AM–FM functions in uncountably infinitely many ways. The most useful representations are those wherein each AM–FM component admits only smooth modulating functions. We have observed that, for many natural images, the computed modulating functions tend to be extraordinarily smooth. In raw form, the storage required for the representation in Eq. (35) is $3K/2$ times the size of the DFT. We are currently investigating linear prediction, quantization, and other techniques for coding the computed modulations. We expect that such techniques will deliver considerable coding gains when applied to computed AM–FM models for locally coherent images and video. From the reconstructed images presented in Figs. 17–20 of Sec. 5.3, it is clear that the computed AM–FM image representations described in Sec. 4.4 are lossy. These new techniques are still evolving, and have yet to reach full maturity. A wealth of exciting research problems remain open in the area.

ACKNOWLEDGMENTS

This research was supported in part by the Army Research Office under contract DAAH 049510494 and by the Air Force Office of Scientific Research, Air Force Systems Command, USAF, under grant number F49620-93-1-0307.

REFERENCES

1. D Gabor. Theory of communication. J Inst Elect Eng London 93(III):429–457, 1946.
2. M Porat, YY Zeevi. The generalized Gabor scheme of image representation in biological and machine vision. IEEE Trans Pattern Anal Machine Intell PAMI-10(4):452–468, 1988.
3. A Grossmann, J Morlet. Decomposition of Hardy functions into square integrable wavelets of constant shape. SIAM J Math Anal 15(4):723–736, 1984.
4. AC Bovik, N Gopal, T Emmoth, A Restrepo. Localized measurement of emergent image frequencies by Gabor wavelets. IEEE Trans Inform Theory IT-38(2):691–712, 1992.
5. P Goupillaud, A Grossmann, J Morlet. Cycle-octave and related transforms in seismic signal analysis. Geoexploration 23:85–102, 1984.
6. FW Campbell, JG Robson. Applications of Fourier analysis of the visibility of gratings. J Physiol London 197:551–556, 1968.
7. N Graham, J Nachmias. Detection of grating patterns containing two spatial frequencies: A comparison of single-channel and multiple-channels models. Vision Res 11:251–259, 1971.
8. C Blakemore, FW Campbell. On the existence of neurones in the human visual system selectively sensitive to the orientation and size of retinal images. J Physiol London 203: 237–260, 1969.
9. W Richards, A Polit. Texture matching. Kybernetic 16(3):155–162, 1974.
10. T Caelli, P Bevan. Probing the spatial frequency spectrum for orientation sensitivity in stochastic textures. Vision Res 23:39–45, 1983.
11. L Maffei, A Fiorentini. The visual cortex as a spatial frequency analyser. Vision Res 13: 1255–1267, 1973.
12. VD Glezer, VA Ivanoff, TA Tscherbach. Investigation of complex and hypercomplex receptive fields of visual cortex of the cat as spatial frequency filters. Vision Res. 13:1875–1904, 1973.
13. N Graham. Visual detection of aperiodic spatial stimuli by probability summation among narrowband channels. Vision Res. 17:637–652, 1977.
14. VD Glezer, AM Cooperman. Local spectral analysis in the visual cortex. Biol Cybern 28: 101–108, 1977.
15. L Maffei, MC Morrone, M Pirchio, G Sandini. Responses of visual cortical cells to periodic and non-periodic stimuli. J Physiol London 296:27–47, 1979.

16. DJ Tolhurst, ID Thompson. On the variety of spatial frequency selectivities shown by neurons in area 17 of the cat. Proc Roy Soc London Series B 213:183–199, 1981.
17. JJ Kulikowski, PO Bishop. Linear analysis of the responses of simple cells in the cat visual cortex. Exp Brain Res 44:386–400, 1981.
18. RL De Valois, DG Albrecht, LG Thorell. Spatial frequency selectivity of cells in macaque visual cortex. Vision Res 22:545–559, 1982.
19. JA Movshon, DJ Tolhurst. On the response linearity of neurons in cat visual cortex. J Physiol London 249:56–57, 1975.
20. JA Movshon, ID Thompson, DJ Tolhurst. Spatial summation in the receptive fields of simple cells in the cat's striate cortex. J Physiol London 283:53–77, 1978.
21. KK De Valois, RL De Valois, EW Yund. Responses of striate cortical cells to grating and checkerboard patterns. J Physiol London 291:483–505, 1979.
22. BW Andrews, DA Pollen. Relationship between spatial frequency selectivity and receptive field profile of simple cells. J Physiol London 287:163–176, 1979.
23. JJ Kulikowski, PO Bishop. Fourier analysis and spatial representation in the visual cortex. Experientia 37:160–163, 1981.
24. JP Jones, LA Palmer. An evaluation of the two-dimensional Gabor model of simple receptive fields in cat striate cortex. J Neurophysiol 58(6):1233–1258, 1987.
25. S Marčelja. Mathematical description of the responses of simple cortical cells. J Opt Soc Am 70(11):1297–1300, 1982.
26. JJ Kulikowski, S Marčelja, PO Bishop. Theory of spatial position and spatial frequency relations in the receptive fields of simple cells in the visual cortex. Biol Cybern 43(3): 187–198, 1982.
27. B Sakitt, HB Barlow. A model for the economical encoding of the visual image in cerebral cortex. Biol Cybern 43:97–108, 1982.
28. DA Pollen, SF Ronner. Visual cortical neurons as localized spatial frequency filters. IEEE Trans Syst Man Cybern SMC-13(5):907–916, 1983.
29. T Caelli, G Moraglia. On the detection of Gabor signals and discrimination of Gabor textures. Vision Res 25(5):671–684, 1985.
30. JP Jones, LA Palmer. The two-dimensional structure of simple receptive fields in cat striate cortex. J Neurophysiol 58(6):1187–1211, 1987.
31. JP Jones, A Stepnoski, LA Palmer. The two-dimensional spectral structure of simple receptive fields in cat striate cortex. J Neurophysiol 58(6):1212–1232, 1987.
32. DG Stork, HR Wilson. Analysis of Gabor function descriptions of visual receptive fields. Proc. ARVO Ann Meeting Investig Opth Visual Sci, Sarasota, FL, 1988, Vol 29, p 398.
33. HR Wilson, D Levi, L Maffei, J Rovamo, R De Valois. The perception of form: Retina to striate cortex. In: L Spillmann, JJS Werner, eds. Visual Perception: the Neurophysiological Foundations. San Diego, CA: Academic Press, 1990, pp 232–241.
34. D Marr. Vision. New York: WH Freeman, 1982.
35. RC Gonzalez, RE Woods. Digital Image Processing. Reading, MA: Addison-Wesley, 1992.
36. D H Ballard, C M Brown. Computer Vision. Englewood Cliffs, NJ: Prentice-Hall, 1982.
37. H Voorhees, T Poggio. Detecting textons and texture boundaries in natural images. Proc First Intl Conf Comput Vision, London, 1987.
38. B Julesz, JR Bergen. Textons, the fundamental elements in preattentive vision and perception of textures. Bell Syst Tech J 62:1619–1645, 1983.
39. JEW Mayhew, JP Frisby. Texture discrimination and Fourier analysis in human vision. Nature, 275:438–439, 1978.
40. MR Turner. Texture discrimination by Gabor functions. Biol Cybern 55(2):71–82, 1986.
41. A Witkin. Recovering surface shape and orientation from texture. Artif Intell 17:17–45, 1981.
42. M Kass, A Witkin. Analyzing oriented patterns. Comput Vision Graphics Image Process 37: 362–385, 1987.

43. JM Coggins, AK Jain. A spatial filtering approach to texture analysis. Pattern Recogn Lett 3(3):195–203, 1985.
44. AR Rao, BG Schunck. Computing oriented texture fields. Proc IEEE Comput Soc Conf Comput Vision Pattern Recog, San Diego, CA, 1989, pp 61–68.
45. AR Rao, RC Jain. Computerized flow field analysis: Oriented texture fields. IEEE Trans Pattern Anal Machine Intell PAMI-14(7):693–709, 1992.
46. J Malik, P Perona. Preattentive texture discrimination with early vision mechanisms. J Opt Soc Am A 7(5):923–932, 1990.
47. B Schachter. Long created wave models. Comput Vision Graphics Image Proc 12(2):187–201, 1980.
48. M Clark, AC Bovik. Texture discrimination using a model of visual cortex. Proc IEEE Intl Conf Syst Man Cybern, Atlanta, GA, 1986.
49. AC Bovik, M Clark, WS Geisler. Computational texture analysis using localized spatial filtering. Proc IEEE Comput Soc Workshop Comput Vision, Miami Beach, FL, 1987.
50. AC Bovik, M Clark, WS Geisler. Multichannel texture analysis using localized spatial filters. IEEE Trans Pattern Anal Machine Intell PAMI-12(1):55–73, 1990.
51. AC Bovik. Analysis of multichannel narrow-band filters for image texture segmentation. IEEE Trans Signal Process SP-39(9):2025–2043, 1991.
52. AC Bovik. Variational pattern analysis using Gabor wavelets. Proc IEEE Intl Conf Acoust Speech Single Process, San Francisco, CA, 1992, vol IV, pp 669–672.
53. P Maragos, JF Kaiser, TF Quatieri. On amplitude and frequency demodulation using energy operators. IEEE Trans Signal Process SP-41(4):1532–1550, 1993.
54. P Maragos, JF Kaiser, TF Quatieri. Energy separation in signal modulations with applications to speech analysis. IEEE Trans Signal Process SP-41(10):3024–3051, 1993.
55. AC Bovik, P Maragos, TF Quatieri. AM–FM energy detection and separation in noise using multiband energy operators. IEEE Trans Signal Process SP-41(12):3245–3265, 1993.
56. HM Hanson, P Maragos, A Potamianos. A system for finding speech formants and modulations via energy separation. IEEE Trans Speech Audio Process SAP-2(3):436–443, 1994.
57. S Lu, PC Doerschuk. Nonlinear modeling and processing of speech based on sums of AM-FM formant models. IEEE Trans Signal Process SP-44(4):773–782, 1996.
58. CS Ramalingam. On the equivalence of DESA-1a and Prony's method when the signal is a sinusoid. IEEE Signal Process Lett 3(5):141–143, 1996.
59. T-H Yu, SK Mitra, JF Kaiser. Novel algorithm for image enhancement. Proc SPIE/SPSE Conference on Image Processing Algorithms and Techniques II, San Jose, CA, 1991.
60. JP Havlicek, AC Bovik, P Maragos. Modulation models for image processing and wavelet-based image demodulation. Proc 26th IEEE Asilomar Conf Signals, Syst Comput Pacific Grove, CA, 1992, pp 805–810.
61. SK Mitra, S Thurnhofer, M Lightstone, N Strobel. Two-dimensional Teager operators and their image processing applications. Proc 1995 IEEE Workshop Nonlin Signal and Image Proc, Neos Marmaras, Halkidiki, Greece, 1995, pp 959–962.
62. P Maragos, AC Bovik. Image demodulation using multidimensional energy separation. J Opt Soc Am A 12(9):1867–1876, 1995.
63. B Friedlander, JM Francos. An estimation algorithm for 2-D polynomial phase signals. IEEE Trans Image Process IP-5(6):1084–1087, 1996.
64. JP Havlicek, DS Harding, AC Bovik. The multi-component AM-FM image representation. IEEE Trans Image Process 5(6):1094–1100, 1996.
65. JP Havlicek, JW Havlicek, AC Bovik. The analytic image. Proc IEEE Intl Conf Image Process Santa Barbara, CA, 1997.
66. EM Stein. Singular Integrals and Differentiability Properties of Functions. NJ: Princeton University Press, Princeton, 1970.
67. EM Stein, G Weiss. Introduction to Fourier Analysis on Euclidean Spaces. Princeton, NJ: Princeton University Press, 1971.

68. JP Havlicek, DS Harding, AC Bovik. Computation of the multi-component AM-FM image representation. Technical Report TR-96-001, Center for Vision and Image Sciences, The University of Texas at Austin, 1996.
69. AC Bovik. Integral inequality bounding the weighted absolute deviation of an n-dimensional function. IEEE Trans Signal Process 40(4):973–975, 1992.
70. AC Bovik. A bound involving n-dimensional instantaneous frequency. IEEE Trans Circuit Syst CS-38(11):1389–1390, 1991.
71. DE Vakman. Sophisticated Signals and the Uncertainty Principle in Radar. New York: Springer-Verlag, 1968.
72. A Papoulis. Signal Analysis. New York: McGraw-Hill, 1977.
73. MJ Bastiaans. Gabor's expansion of a signal into Gaussian elementary signals. Proc IEEE 68:538–539, 1980.
74. T Genossar, M Porat. Can one evaluate the Gabor expansion using Gabor's iterative algorithm? IEEE Trans Signal Process SP-40(8):1852–1861, 1992.
75. R Orr. The order of computation for finite discrete Gabor transforms. IEEE Trans Signal Process SP-41(1):122–130, 1993.
76. MJ Basstiaans. A sampling theorem for the complex spectrogram and Gabor expansion of a signal into Gaussian elementary signals. Opt Eng 20(4):594–598, 1981.
77. MJ Basstiaans. Gabor's signal expansion and degrees of freedom of a signal. Opt Acta 29(9): 1223–1229, 1982.
78. J Wexler, S Raz. Discrete Gabor expansions. Signal Process 21(3):207–220, 1990.
79. S Qian, D Chen. Discrete Gabor transform. IEEE Trans Signal Process SP-41(7):2429–2438, 1993.
80. S Qian, K Chen, S Li. Optimal biorthogonal functions for finite discrete-time Gabor expansions. Signal Process 27(2):177–185, 1992.
81. S Qian, D Chen. Optimal biorthogonal analysis window function for discrete Gabor transform. IEEE Trans Signal Process SP-42(3):694–697, 1994.
82. YY Zeevi, M Porat. Combined frequency-position scheme of image representation in vision. J Opt Soc Am A 1(12):1248, 1984.
83. M Porat, YY Zeevi. Localized texture processing in vision: Analysis and synthesis in the Gaborian space. IEEE Trans Biomed Eng 36(1):115–129, 1989.
84. DM MacKay. Strife over visual cortical function. Nature 289:176–218, 1981.
85. JG Daugman. Uncertainty relation for resolution in space, spatial frequency, and orientation optimized by two-dimensional visual cortical filters. J Opt Soc Am A 2(7):1160–1169, 1985.
86. JP Havlicek, AC Bovik. Multi-component AM–FM image models and wavelet-based demodulation with component tracking. Proc IEEE Intl Conf Image Processing, Austin, TX, 1994, vol I, pp 41–45.
87. D Vakman. On the analytic signal, the Teager–Kaiser energy algorithm, and other methods for defining amplitude and frequency. IEEE Trans Signal Process SP-44(4):791–797, 1996.
88. L Cohen. Time-Frequency Analysis. Englewood Cliffs, Prentice-Hall NJ: 1995.
89. MA Poletti. Instantaneous frequency and conditional moments in the time-frequency plane. IEEE Trans Signal Process SP-39(3):755–756, 1991.
90. H Broman. The instantaneous frequency of a Gaussian signal: The one-dimensional density function. IEEE Trans Acoust Speech Signal Process ASSP-29(1):108–111, 1981.
91. MS Gupta. Definition of instantaneous frequency and frequency measurability. Am J Phys 43(12):1087–1088, 1975.
92. L Mandel. Interpretation of instantaneous frequencies. Am J Phys 42:840–846, 1974.
93. LM Fink. Relations between the spectrum and instantaneous frequency of a signal. Probl Inform Transmiss 2:11–21, 1966.
94. J Shekel. Instantaneous' frequency. Proc Inst Radio Eng 41:548, 1953.
95. J Ville. Théorie et applications de la notation de signal analytique. Cables Transmiss 2A: 61–74, 1948; translated from the French in I Selin. Theory and applications of the Notion

of Complex Signal. Technical Report T-92, The RAND Corporation, Santa Monica, CA, 1958.
96. P Maragos, AC Bovik, TF Quatieri. A multidimensional energy operator for image processing. Proc SPIE Symp Visual Commun Image Process Boston, MA, 1992, pp 177–186.
97. JP Havlicek, DS Harding, AC Bovik. Reconstruction from the multi-component AM-FM image representation. Proc IEEE Intl Conf Image Process Washington, DC, 1995, vol II, pp 280–283.
98. JP Havlicek, AC Bovik. AM–FM models, the analytic image, and nonlinear demodulation techniques. Technical Report TR-95-001, Center for Vision and Image Sciences, The University of Texas at Austin, 1995.
99. AC Bovik, JP Havlicek, DS Harding, MD Desai. Limits on discrete modulated signals. IEEE Trans Signal Process SP-45(4):867–879, 1997.
100. JP Havlicek, DS Harding, AC Bovik. Discrete quasi-eigenfunction approximation for AM-FM image analysis. Proc IEEE Intl Conf Image Process, Lausanne, Switzerland, 1996, pp 633–636.
101. JP Havlicek, AC Bovik, MD Desai, DS Harding. The discrete quasi-eigenfunction approximation. Proc Intl Conf Digital Signal Process, Limassol, Cyprus, 1995, pp 747–752.
102. E Vanmarcke. Random Fields, Analysis and Synthesis. Cambridge, MA: MIT Press, 1983.
103. SK Mitra, H Li, IS Lin, T-H Yu. A new class of nonlinear filters for image enhancement. Proc IEEE Intl Conf Acoust Speech Signal Process, Toronto, Canada, 1991, pp. 2525–2528.
104. S Thurnhofer, SK Mitra. Nonlinear detail enhancement of error-diffused images. Proc IS&T/SPIE Symp Elect Imaging: Science & Tech, San Jose, CA, 1994, SPIE 2179, pp 170–181.
105. T-H Yu, SK Mitra. Unsharp masking with nonlinear filters. Signal Processing VII: Theories and Applications. M Holt, C Cowan, P Grant, W Sandham, eds. European Association for Signal Processing, 1994.
106. N Strobel, SK Mitra. Quadratic filtes for image contrast enhancement. Proc 28th Annual Asilomar Conf Signals Syst Comput, Pacific Grove, CA, 1994, pp 208–212.
107. SK Mitra, T-H Yu. A new nonlinear algorithm for the removal of impulse noise from highly corrupted images. Proc 1994 IEEE Int Symp Circuit Syst. London, 1994, pp 17–20.
108. P Maragos, AC Bovik. Demodulation of images modeled by amplitude-frequency modulation using multidimensional energy separation. Proc IEEE Intl Conf Image Process, Austin, TX, 1994, Vol III, pp 421–425.
109. JP Havlicek, MS Pattichis, DS Harding, AC Christofides, AC Bovik. AM–FM image analysis techniques. Proc IEEE Southwest Symp Image Anal Interpretation, San Antonio, TX, 1996, pp 195–199.
110. BJ Super, AC Bovik. Solution to shape-from-texture for wavelet-based measurement of local spectral moments. Technical Report TR-92-5-80, Computer and Vision Research Center, The University of Texas at Austin, 1991.
111. BJ Super, AC Bovik. Shape from texture using local spectral moments. IEEE Trans Pattern Anal Machine Intell PAMI-17(4): 333–343, 1995.
112. BJ Super, AC Bovik. Planar surface orientation from texture spatial frequencies. Pattern Recogn 28(5):728–743, 1995.
113. T-Y Chen, AC Bovik, BJ Super. Multiscale stereopsis via Gabor filter phase response. Proc IEEE Intl Conf Syst, Man, and Cybernetics, SanAntonio, TX, 1994, pp 55–60.
114. T-Y Chen, AC Bovik. Stereo disparity from multiscale processing of local image phase. Proc. IEEE Intl Symp Comput Vision, Coral Gables, FL, 1995.
115. MS Pattichis, AC Bovik. Multi-dimensional frequency modulation in texture images. Proc Intl Conf on Digital Signal Process, Limassol, Cyprus, 1995, pp 753–758.

13
Electronic Digital Image Stabilization and Mosaicking

Carlos Morimoto
IBM Almaden Research Center, San Jose, California

Rama Chellappa
University of Maryland, College Park, Maryland

1. INTRODUCTION

Recently, electronic digital image stabilization (EDIS) has become an integral component of several imaging devices, such as binoculars and commercial portable video recorders. Before the introduction of such a technique, mechanical gyroscopic systems were the most commonly used approach to achieve image stabilization. These systems are unattractive for general applications due to size, complexity, and cost considerations. EDIS offers low cost and more flexibility than mechanical systems, as the user can specify which components of the motion are to be stabilized (''on-demand stabilization'').

Image frames of a sequence taken from a moving imaging system differ due to the motion of the camera, changes in the scene, and noise corruption during the acquisition process. The purpose of image stabilization is to remove the unwanted or unintended motion of the imaging device, so that the remaining contents of the sequence can be more easily viewed or subsequently processed. In many cases, it is possible to segment the image into regions which describe different motion fields, in which case the system can select one of them to be stabilized in order to create an output sequence where that particular region always appears at the same position. For most applications though, the largest or dominant region is stabilized (the background).

Electronic digital image stabilization can be used as a front-end system in a variety of image analysis applications or simply as a visualization tool. Typically, the application defines the ''unwanted'' components of the motion. For example, for tele-operation of robotic vehicles, it is desirable to provide the tele-operator with a smoothed sequence where the high-frequency or oscillatory motion components are removed from the original sequence. Other applications like tracking independently moving objects from moving platforms can be simplified by removing the motion due to the camera, so that the scene background is relatively still.

The main component of an EDIS system is an appropriate image registration technique. Image registration algorithms compute a global transformation between two image frames which minimizes their spatial misalignment (it is assumed that the scene is rigid and movement of the camera causes the misalignments). Assuming an arbitrary frame of the sequence as a reference, the transformations can be filtered or directly applied to all the other frames in order to generate the stabilized output. When no filtering is performed, images can be composed by superimposing their overlapping regions. For example, if the camera is simply panning, a panoramic view of the scene can be constructed. The resulting composite image is referred to as a mosaic, and the process of building mosaics is called mosaicking.

2. IMAGE STABILIZATION AND MOSAICKING

The methods proposed for electronic digital image stabilization can be distinguished by the models adopted to estimate the camera motion. Several two-dimensional (2-D) and three-dimensional (3-D) stabilization schemes are described by Davis et al. (1). For 2-D models, in general all the estimated motion parameters are compensated for (i.e., all motion is removed from the input sequence) (2–4). Stabilization in three dimensions is achieved by derotating the frames, generating a translation-only sequence or a sequence containing translation and low-frequency rotation (smoothed rotation). Yao et al. (5) compensate for 3-D rotation by tracking multiple visual cues like distant points and horizon lines, using an extended Kalman filter for the estimation of the 3-D-motion parameters of interest. Both kinematic and kinetic models suitable for determining the smoothed and oscillatory rotational-motion components are considered, so that images obeying a smoothed rotation model can be obtained. Durić and Rosenfeld (6) also use a vehicle model to filter the high-frequency components of the rotational parameters. A flow-based motion estimator applied to points close to the horizon (distant points) is used to estimate the rotational parameters, and the solution is recursively refined. Rousso et al. (7) propose a feature-based method that uses three frames to estimate the 3-D rotation, which is then used for stabilization.

Two-dimensional models are used by Viéville et al. (8), Irani et al. (9), Kwon et al. (10), and Balakirsky (11). Viéville et al. (8) use linear segments from the input images and align them with the absolute vertical direction, which can be provided by an inertial sensor, eliminating the need to estimate the rotation around the optical axis. Stabilization is achieved by compensating for 2-D linear translation, which minimizes the disparity between two successive frames. The focus of expansion (FOE) and 3-D structure are then computed from the stabilized frames. Irani et al. (9) also compute egomotion from 2-D stabilization by estimating the 2-D affine motion of a single image region, which is used to cancel the 3-D rotation of the camera motion. Then, the 3-D camera translation is computed by finding the FOE in the translation-only sequence. Stabilization by derotating the 3-D parameters is then possible. Later, Irani and Anandan extended the 2-D model to include layers of parametric transformations (12). Kwon et al. (10) suggest the use of image stabilization for subband video coding, and Balakirsky and Chellappa (13) compare the performance of several stabilization systems by using them as input to an automatic object tracker. The quality of stabilization improves the ability to segment and track the moving objects and the false alarm rate is used for evaluation.

2.1. Fast Implementations of EDIS Systems

Fast implementations of 2-D stabilization algorithms are presented in Hansen et al. (3) and Morimoto and Chellappa (14,15). Hansen et al. (3) describe the implementation of an image stabilization system used in the ARPA UGV Demo II program to perform reconnaissance, surveillance, and target acquisition from a moving unmanned vehicle (2). The system uses a mosaic-based registration technique and was implemented on a pyramidal hardware (VFE-100). The interframe motion is computed using a multiresolution, iterative process that estimates affine motion parameters between levels of Laplacian pyramid images. From coarse to fine levels, the optical flow of local patches of the image is computed using a cross-correlation scheme. The motion parameters are then computed by fitting an affine motion model to the flow. These parameters are used to warp the previous image (on the next finer pyramid level) to the current image, and the refinement process continues until the desired precision is achieved. This scheme, combined with the construction of a mosaic image, allows the system to cope with large image displacements. One of the earlier VFE implementations (3) is capable of stabilizing images of size 128×120 pixels, with image displacements ranging from ± 32 to ± 64 pixels, at a rate of approximately 10 fps (frames per second).

The system developed by Morimoto and Chellappa (14) was implemented on a Datacube board MV200, which is a parallel-pipeline machine commonly used for real-time image-processing applications and to detect independently moving objects from a moving vehicle (4). The system is based on a similarity model-based multiresolution motion-estimation technique which is able to stabilize 18 fps using images of resolution 128×120 pixels, with image displacements of up to ± 21 pixels between consecutive frames. The motion is estimated from the displacement of a small set of feature displacements, which are automatically detected and tracked between consecutive frames. Morimoto and Chellappa also describe the implementation of a fast 3-D derotation stabilization system in Ref. 15, based on the same hardware platform. An extended Kalman filter (EKF) is employed to estimate 3-D rotation represented by unit quaternions. A small set of feature points is hierarchically tracked using Gaussian and Laplacian pyramids, and the interframe displacements are used as measurements for the EKF.

2.2. Mosaicking

Mosaicking, or the process of building mosaic images, is achieved by aligning consecutive frames using global motion transformations to create a panoramic view of the scene. In principle, any technique used to estimate global camera motion can be directly used for mosaicking. Recently, many applications have been implemented which explore the compactness, efficiency, and large field of view of mosaic representations. Several mosaic representations and applications can be found in Ref. 16.

As mentioned earlier Burt and Anandan (2) and Hansen et al. (3) use mosaics in order to stabilize image sequences with large interframe displacements. To register a new frame to the mosaic, an initial estimate of the position of the new frame (indexing) is determined by correlating small ''representative landmark'' regions of the new frame with the mosaic. After indexing and registration, the mosaic is updated before the next frame is processed. Whenever a frame gets warped outside the mosaic, a new mosaic ''tile'' is created.

Morimoto and Chellappa (15) suggest the use of inverse mosaics for image stabilization. Inverse mosaics can be obtained by mosaicking the reverse input sequence, or simply by an inverse global transformation of the current mosaic; inverse mosaics have the property of keeping the last frame always at the same position and they do not introduce distortions when the images are warped into the mosaic. This increases the accuracy of the registration process. A fast implementation of a mosaicking system is described in Ref. 17.

For the purpose of video coding, Irani et al. (18) describe two techniques, one suitable for storage and the other suitable for transmission. For storage, the entire sequence can be used to build the mosaic which is then encoded; but for transmission, each new frame is encoded separately. Sawhney et al. (19), and Dufaux and Moscheni (20) also suggest the use of mosaics for video coding, particularly stressing the advantages of background mosaics, where moving objects in the foreground are segmented out. Wang and Adelson (21) describe a layered representation where the sequence is decomposed into layers. Each layer is a mosaic, composed by a region describing affine motion. The layers are ordered by depth, and each one is represented by an intensity or texture map, a binary transparency map, and a set of affine motion parameters used by the decoder to reconstruct the motion described by the layer. At the decoder, the sequence is synthesized by warping each layer onto the viewing window, from the deepest to the closest layer. The transparency maps guarantee that only the useful pixels are drawn onto the viewing window.

Mosaics have also been used for image enhancement (22,23) and to estimate structure information. Ishiguro et al. (24) construct 360° mosaics (which they call omni-directional views) mechanically and extract range information during the mosaicking process. The range map is then further refined using pairs of omni-directional views, as two "wide-angle" images, for stereo correspondence. A similar idea was implemented by Kang and Szeliski (25), who use multiple cylindrical images (360° mosaics) to extract 3-D structure data. To obtain the panoramic image, the system is first calibrated, and each frame of the 360° sequence is corrected using the intrinsic camera parameters, before being converted to cylindrical coordinates. The image composition is done by phase correlation and local image registration. Several such cylindrical panoramic images are used by a multibaseline stereo technique, based on the eight-point structure from the motion algorithm (26,22). The "large" field of view of the cylindrical images provide the necessary robustness to noise, making the eight-point algorithm useful. To improve accuracy, an iterative least squares minimization is simultaneously applied to refine both camera motion and 3-D positions.

Finally, it is important to note that mosaics have recently been included in commercial products such as VideoBrush© and QuickTime© VR and used as tools for visualization, computer graphics, virtual reality applications, and so forth. These topics are further discussed in Refs. 28–31.

3. AN IMAGE STABILIZATION ALGORITHM

A general EDIS system can be decomposed into three main modules, as shown in Fig. 1. From the video input, the first module estimates the interframe global transformation

Fig. 1 Modules of a general electronic image stabilization system.

parameters, which are combined and filtered by the motion-compensation module to update the global transformation used to bring the current frame into alignment with the reference frame. The third module uses the global transformation to generate the stabilized output sequence, and possibly mosaics.

The primary component of an EDIS system is the motion estimation or image registration module. It computes the interframe transformation parameters, which brings two consecutive frames into alignment, according to some global image transformation model.

3.1. Image Transformation Models

A global transformation which maps an image I_0 into an image I_1 can be defined as follows (32):

$$I_1(x, y) = \gamma(I_0(\psi(x, y))) \tag{1}$$

where $I_i(x, y)$ represents the intensity of pixel (x, y) of image I_i, γ is a one-dimensional intensity transformation function, and $\psi(x, y)$ is a two-dimensional spatial coordinate transformation which maps pixel coordinates $\mathbf{p}_0 = (x_0, y_0)$, to new pixel coordinates $\mathbf{p}'_0 = (x'_0, y'_0)$ such that $\mathbf{p}'_0 = \psi(\mathbf{p}_0)$. For stabilization purposes, this problem can be reduced to the computation of the optimal spatial transformation ψ which properly aligns the two frames.

To facilitate the description of the estimation techniques used to compute the coordinate transformation among frames, their composition, and other operations necessary for stabilization, we will restrict our analysis to subsets of the affine group of transformations. The usual first step to recover the camera motion or a global image transformation which aligns two images is to determine image point correspondences. Reviews and comparisons of such techniques can be found in Refs. 32–35.

A simple transformation which is sufficient to register two images taken from the same viewing angle but from a different position can be defined using four parameters, such as

$$\mathbf{p}_1 = s\mathbf{R}\mathbf{p}_0 + \mathbf{T} \tag{2}$$

where $\mathbf{T}$ is a translation vector, s is a scalar corresponding to the scaling factor, and $\mathbf{R}$ is an orthogonal rotational matrix defined by

$$\mathbf{R} = \begin{pmatrix} \cos\Theta & -\sin\Theta \\ \sin\Theta & \cos\Theta \end{pmatrix} \tag{3}$$

where the parameter Θ defines a rotation around the optical axis. These transformations preserve angles and relative lengths and belong to the *similarity* group.

Rigid transformations where scaling is not allowed are also described by Eq. (2) when s is set to be of unit value. The set of all such three-parameter transformations forms the *Euclidean* group.

A six-parameter *affine* transformation is obtained by relaxing the constraints on the rotational matrix $\mathbf{R}$; it is defined by

$$\mathbf{p}_1 = \begin{pmatrix} r_{11} & r_{12} \\ r_{21} & r_{22} \end{pmatrix}\mathbf{p}_0 + \begin{pmatrix} t_x \\ t_y \end{pmatrix} = \mathbf{R}\mathbf{p}_0 + \mathbf{T} \tag{4}$$

Angles and lengths are no longer preserved in this transformation, although parallel lines remain parallel. The affine transformation allows for skewing, and for change in aspect

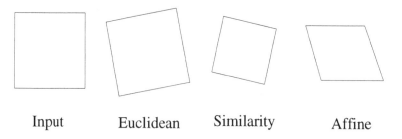

Fig. 2 Image transformation models.

ratio due to nonuniform scaling in x and y. Figure 2 illustrates the possible effects of these image transformations when applied to the square shown on the left-hand size.

The family of transformations described so far cannot account for some distortions which appear in more general 3-D motion, such as those caused by pan-tilt movement. The eight-parameter transformation of the *projective* group, which is able to describe general 3-D camera motion, requires more complex estimation techniques not discussed in this chapter. The use of more complex models is discussed in Refs. 36 and 37.

3.2. Motion Estimation

This section presents a fast, accurate, and robust feature-based motion-estimation technique, selected for implementation in a real-time platform as described in Ref. 14. The algorithm tracks a small set of feature points, and the interframe displacements are used to fit one of the motion models using least squares. Multiresolution is used to increase the robustness to large image displacements.

The process starts with the construction of Gaussian and Laplacian pyramids for both images to be registered, I_t and I_{t-1}. A Gaussian pyramid G_t is formed by combining several reduced-resolution Gaussian images of I_t. The image at level l of the pyramid is denoted by $G_t[l]$, where $G_t[0]$ is the highest resolution image, which might be a copy of I_t. An image at level l has resolution $R[l] = R[0]/2^l$, where $R[0]$ is the resolution of $G_t[0]$. The Laplacian pyramid L_t is obtained by convolving G_t with a Laplacian kernel operator. The levels of the pyramids G_{t-1} and G_t are used from coarse to fine resolutions; each new processed level contributes to refining the motion estimates.

Features are selected from the Laplacian pyramids. For an arbitrary level l of the pyramid, N features are chosen by dividing $L_{t-1}[l]$ into N nonoverlapping regions. To speed up the feature-detection process, only the pixel with maximum intensity and uncorrelated with its eight immediate neighbors is selected for tracking. Confidence measures are computed before estimation to increase robustness to noise and discard inappropriate features. This topic is further discussed in Sec. 3.2.3.

3.2.1. Feature Tracking

Once features are selected for an arbitrary pyramid level l, the corresponding Gaussian pyramids are used for tracking. A match for the corresponding feature from $G_{t-1}[l]$ is obtained by minimizing the sum of squared differences (SSD) over a neighborhood (search window) around the candidate matches in $G_t[l]$.

Image Stabilization and Mosaicking

The SSD between two windows of size $W = (2w + 1) \times (2w + 1)$ centered at the image coordinate $G_{t-1}[l](x, y)$ and its corresponding candidate location $G_t[l](u, v)$ is given by

$$\text{SSD} = \sum_{i=-w}^{+w} \sum_{j=-w}^{+w} [(G_{t-1}[l](x + i, y + j) - G_t[l](u + i, v + j))^2] \quad (5)$$

For a feature $G_{t-1}[l](x, y)$, a search for the minimum SSD is performed in a window of size $S = (2s + 1) \times (2s + 1)$ centered at $G_t[l](x, y)$. The values of s and w have great influence in the performance of the system and must be kept small. Small correlation windows are more susceptible to false matches though, and a trade-off between performance and robustness must be made. Experiments concerning performance evaluation are presented in Sec. 6.3. After the grid-to-grid matches are obtained, displacements with subpixel accuracy are computed using a differential method (38). Subpixel accuracy is necessary to reduce the quantization error introduced when the images are digitized.

3.2.2. Subpixel Matching

If a feature $P_t^0(u, v)$ has offset $(\delta x, \delta y)$ relative to $P_{t-1}^0(u, v)$ [assume they are tracked and registered so that the translation $(\delta x, \delta y)$ is very small] [i.e., $P_t^0(u, v) = P_{t-1}^0(u - \delta x, v - \delta y)$], the frame difference can be expanded as

$$\begin{aligned} d(u, v) &= P_{t-1}^0(u, v) - P_t^0(u, v) \\ &= P_{t-1}^0(u, v) - P_{t-1}^0(u, -\delta x, v - \delta y) \\ &\approx \frac{\partial P_{t-1}^0(u, v)}{\partial x} \delta x + \frac{\partial P_{t-1}^0(u, v)}{\partial y} \delta y \end{aligned} \quad (6)$$

The terms $\partial P_{t-1}^0(u, v)/\partial x$ and $\partial P_{t-1}^0(u, v)/\partial y$ can be approximated by forward differences, so that for a small neighborhood around $P_{t-1}^0(u, v)$ of size $W = (2w + 1) \times (2w + 1)$, there will be W simultaneous equations

$$\mathbf{D} = \mathbf{G}\mathbf{\Delta} \quad (7)$$

where $\mathbf{D}$ is the known difference of intensities vector $(P_{t-1}^0(u, v) - P_t^0(u, v))$, $\mathbf{G}$ is the known gradient matrix, and $\mathbf{\Delta}$ is the unknown vector of translations, so that

$$\mathbf{D} = \begin{pmatrix} d(-w, -w) \\ \vdots \\ d(w, w) \end{pmatrix};$$

$$\mathbf{G} = \begin{pmatrix} \frac{\partial P_{t-1}^0(-w, -w)}{\partial x} & \frac{\partial P_{t-1}^0(-w, -w)}{\partial y} \\ \vdots & \vdots \\ \frac{\partial P_{t-1}^0(w, w)}{\partial x} & \frac{\partial P_{t-1}^0(w, w)}{\partial y} \end{pmatrix}; \quad \mathbf{\Delta} = \begin{pmatrix} \delta x \\ \delta y \end{pmatrix} \quad (8)$$

The system of equations in (7) can be solved using $\mathbf{\Delta} = (\mathbf{G}^t\mathbf{G})^{-1}\mathbf{G}^t\mathbf{D}$.

3.2.3. Confidence Measures

Several feature-tracking algorithms stress the importance of evaluating how well features are tracked (34,39). Good features to track can be selected on the basis of their texturedness and cornerness. Unfortunately, these features can, for example, become occluded, lie on depth discontinuities, or be part of reflection highlights. In each case, tracking would probably be very poor. In this section, we describe three confidence measures used to reject these features and increase the robustness and accuracy of the motion estimates.

The first two confidence measures are given by the eigenvalues of $\mathbf{G}^t\mathbf{G}$ from Eqs. (8). These eigenvalues are a good measure of the texturedness around the feature and were also used in Refs. 34, 39, and 40, Because $\mathbf{G}^t\mathbf{G}$ is a 2×2 symmetric matrix, it has two real eigenvalues. Small eigenvalues correspond to regions of approximately constant intensity profile, whereas one small and one large eigenvalue correspond to a unidirectional texture pattern, where only the normal flow can be reliably estimated. Features are suitable for tracking when both eigenvalues are larger than a certain threshold, which, in practice, also guarantees that $\mathbf{G}^t\mathbf{G}$ is well conditioned, and when the ratio between the largest and smallest eigenvalues is not greater than one magnitude (typically). The thresholds are determined empirically and depend on the sizes of the correlation and subpixel windows. When one of these criteria is not satisfied, the feature is dropped and is not used for estimation. This is more restrictive than the criteria used in Refs. 34, 39, and 40. Simoncelli et al. (40) use the sum of both eigenvalues for thresholding, whereas Barron et al. (34) and Shi and Tomasi (39) select features based only on the smallest eigenvalue.

The third confidence measure is given by the minimum SSD used for determining the best match. First, when there is more than one candidate for the minimum within the search window, the feature is dropped. When the minimum is unique but high, this might be an indication that it is on a depth discontinuity, or it is occluded, or the search window is not large enough. In the current implementation, only features with residues below a certain threshold are used for estimation. A robust estimation framework (19,41) could also be applied using the reduced set of features to reject remaining outliers.

3.2.4. Multiresolution Estimation

The algorithm for multiresolution estimation is very similar to the one described in Ref. 42. Starting from the coarsest resolution, features are detected using L_{t-1} and tracked between G_t and G_{t-1}. The transformation parameters are computed from the feature displacements as follows: Each tracked feature contributes two equations from the x and y coordinates of Eqs. (2) and (4). Because in general we have $2N > P$, where P is the number of parameters, the overdetermined system with $2N$ equations and P unknowns can be solved using a least squares method.

For an arbitrary higher-resolution level l, the transformation estimated up to level $l + 1$ must be properly scaled to level l and used to warp $G_t[l]$. The warping brings the images at level l into alignment, and the refinement process continues by scaling the features detected from $L_{t-1}[l + 1]$ or computing new features in $L_{t-1}[l]$, and tracking the features from $G_{t-1}[l]$ to the warped image of $G_t[l]$. The transformation computed from the feature displacements at level l must be combined with the estimate from the previous level to produce the correct interframe transformation used to warp $G_t[l - 1]$. This process is repeated until the finest resolution level is reached. Note that the displacement is doubled after every level; hence, the total displacement that this algorithm can handle can be very large even for small search window sizes.

3.3. Motion Compensation

To generate a stabilized sequence, it is necessary to determine the transformation which brings the current frame into alignment with the reference frame. The motion-compensation module computes this global transformation from the interframe estimates. Let ψ_t be the global transformation which aligns image frame I_t with the reference frame I_0 [i.e., $I_0(x, y) = I_t(\psi_t(x, y))$]. When the interframe estimate ψ which aligns I_{t+1} with I_t is available [$I_t(x, y) = I_{t+1}(\psi(x, y))$], the new global transformation ψ_{t+1} must be updated using the composition rule

$$\psi_{t+1}(x, y) = \psi_t(\psi(x, y)) \qquad (9)$$

If a smoothed output is desired instead of a completely stable one, the motion parameters can be filtered independently, or altogether. For example, a buffer can be created to store the motion from the last few frames, and a polynomial or spline interpolation could be used to warp the current, or some previous frame. Warping some previous frame introduces a constant delay on the system but might produce better smoothing. These filters can also be applied dynamically; for example, for a camera mounted on a ground vehicle, when the vehicle makes a turn or starts going up or down a hill, the horizontal or vertical translations might be compensated differently, or not compensated at all.

3.4. Image Composition

The stabilized output sequence is generated by warping the current frame according to the global transformation computed by the motion-compensation module. Warping is performed by scanning the output frame and determining the intensity at the corresponding pixel of the input frame (43), which is then copied onto the output in case the transformed

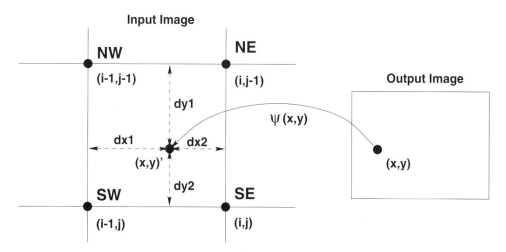

Fig. 3 Bilinear interpolation.

pixel lies inside the input image; otherwise, some background default intensity is placed at the current output pixel. In general, the transformed coordinates will have noninteger values. To obtain the intensity at a nongrid point $(x, y)' = \psi(x, y)$, bilinear interpolation using the nearest four grid points (NW, NE, SW, SE) is performed

$$\text{Out}(x, y) = \text{NW} \times dx_2 \times dy_2 + \text{SW} \times dx_2 \times dy_1 + \text{NE} \times dx_1 \times dy_2 \quad (10)$$
$$+ \text{SE} \times dx_1 \times dy_1$$

where dx_1, dx_2, dy_1, and dy_2 are shown in Fig. 3.

Mosaics are constructed in a similar way. To avoid scanning the whole mosaic, which might be considerably larger than the input frame, the four corners of the input image are warped first, to determine the bounding rectangle in the mosaic which contains the warped input frame. Only this output rectangle needs to be scanned. Temporal filtering can be used to segment the background image (18,19). For the fast implementation described in Ref. 17, no temporal filtering is applied, and the mosaic is treated as a write-once memory, so that only new parts of the scene are drawn on top of the mosaic. Whenever the new frame is warped outside the bounds of the mosaic, the mosaic is scrolled to maintain the current view on the screen. Areas which scroll off the edge of the display are lost.

4. STABILIZATION AND MOSAICKING EXAMPLES

Because stabilization is supposed to keep the reference image still, the difference between two stabilized frames should be close to zero almost everywhere; this is not true in the case of the original sequence due to the motion of the camera. The following examples show the results of the EDIS algorithm describe in Sec. 3.2 based on the differences of input and stabilized frames. The residue from stabilized frames is considerably smaller than the residue between input frames.

Figure 4 shows the results of the similarity model-based stabilization system for an off-road sequence provided by NIST. Processing off-road sequences is a big challenge due to the irregularities of the terrain, which might cause very fast and large image displacements. Just one frame of the input sequence is shown, on the top row at the left-hand side, along with its corresponding stabilized frame on the right. The bottom row shows the image differences between two consecutive input and stabilized frames. Observe that the error is larger for those regions which are closer to the camera, basically due to perspective distortions and parallax. Because the vehicle was moving forward, the stabilized frame is scaled down. Homogeneous black regions around the difference between stabilized frames are due to the compensation of motion (the warping performed to register the frames).

Figure 5 shows the results for another off-road sequence, provided by Martin Marietta, which was processed using the Euclidean model-based stabilization system. For this sequence, the camera was pointing sideways (i.e., almost perpendicular to the direction of motion), so that the changes in scale are not relevant. To improve visualization, the residual image was thresholded. Note that the other vehicle moving at a distance in this scene is segmented by the simple difference of stabilized frames.

Although stabilization produces nice results for visualization, segmenting independently moving objects directly from the difference of stabilized frames is not straightfor-

Image Stabilization and Mosaicking

Fig. 4 Stabilization of an off-road sequence. (Courtesy of NIST.)

ward, mainly due to regions which do not fit the 2-D motion assumption, because they are also segmented after the difference image is thresholded. Temporal median filters combined with velocity-tuned filters are used in Ref. 4 to detect independently moving objects (IMOs). Qualitative approaches (44) to the detection of IMOs could also benefit from electronic image stabilization. A simple difference between stabilized frames with normalization using spatial gradients is used in Ref. 45.

The UGV sequence will be used to illustrate mosaicking. The sequence starts zooming out of a vehicle, then pans from right to left, and finally zooms in again. The similarity-model-based system was used in order to compensate for the zooming. The top-left and top-right images of Fig. 6 show input and stabilized frames respectively, during simultaneous panning and zooming out. Again, the bottom images show the residue between consecutive input and stabilized frames (the difference between stabilized frames is shown here as the difference between the previous frame and the warped current frame using the interframe motion estimate).

Due to zooming and fast panning, the residue from the difference between input frames is large everywhere (see image at the bottom left), and the stabilized frames are warped completely out of the display window after a few frames. For better visualization, the motion estimates are used to align the input image frames and compose the mosaic

Fig. 5 Stabilization results for an off-road sequence. (Courtesy of Martin Marietta.)

Fig. 6 Stabilization results for the UGV sequence. The zooming out and panning of the sequence is noticeable from the stabilized frame (top right).

Fig. 7 Mosaicking. The top row shows the mosaic constructed for the UGV sequence up to the 90th frames. The bottom row shows, from right to left, the 1st, 45th, and 90th frames. Observe the change in scale between the 1st and 45th frames, which corresponds to the zooming out part of the sequence.

shown in Fig. 7. The top row of this figure shows the mosaic composed by 90 frames of the sequence. The bottom row shows the 90th, 45th, and 1st frames of the input sequence, from left to right (remember that the camera is panning from right to left). The zooming ends approximately after the 30th frame. Observe the scale change between the 1st and 45th frames and how they appear in the mosaic. Figure 8 shows the mosaic after 190

Fig. 8 Mosaic for the whole UGV sequence. Shortly after this frame, the sequence zooms in. This cannot be seen because the mosaic is implemented as write-once.

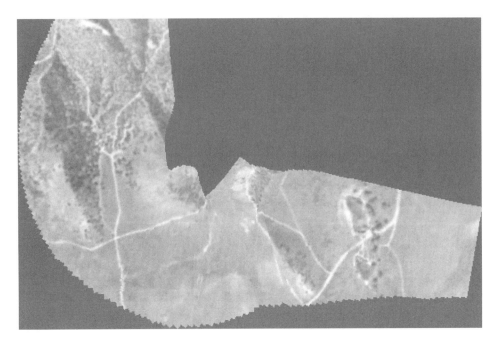

Fig. 9 Mosaic for an aerial sequence.

frames were processed. Shortly after that frame, the sequence starts zooming in, and because the mosaic is implemented as write-once, the rest of the sequence cannot be seen.

Figure 9 shows the results of the Euclidean model-based mosaicking algorithm for an aerial image sequence provided by the David Sarnoff Research Center. Observe that the path followed by the aircraft is reconstructed. No significant differences in performance are noticeable when the affine or similarity models are applied.

5. EVALUATION OF IMAGE STABILIZATION SYSTEMS

In this section, the performance of the 2-D global parametric motion models is compared using the image stabilization algorithm described previously. Evaluation methods for image stabilization systems have been presented in Refs. 43 and 14. Reference 13 compares the performance of different stabilization algorithms based on the accuracy of a real-time object tracker, and Reference 14 considers the maximum displacement velocity in pixels per second, computed as the product of the frame rate and the maximum image displacement between frames.

As motion estimation is the main component of an image stabilization system, the evaluation of the system could be based on the performance of the motion-estimation module alone, in which case one could use synthetic or calibrated sequences where the interframe motions are known, such as in Refs. 34 and 46. Aside from the issue of generating sequences with known motion, the stabilization system is based on approximate parametric global transformations, which creates the problem of finding the optimal transformation from the ground-truth data, so that the motion estimates can be evaluated in terms of a distance measure from these optimal parameters. Another important issue is how to compare the performance of systems based on different motion models, as distance measures might be model dependent.

The peak signal-to-noise ratio (PSNR) is used to evaluate the fidelity of an EDIS system. This method does not require the use of calibrated sequences and provides a simple way of comparing systems based on different motion models. Synthetic sequences are used to measure other properties, such as the range of displacements within which they operate.

Intuitively, because all motion is compensated for after stabilization, the difference between two stabilized frames should be, ideally, zero in the overlapping regions. Several factors contribute to this measure not being zero, such as noise, changes in the structure of the scene, and limitations of the motion model. For stabilization purposes, the PSNR can be considered as a measure of departure from the optimal case (i.e., two identical images) or as a measure of the overlap between two images, which is maximized when the images are the same. When two images do not overlap, stabilization is not possible, and the PSNR is meaningless. But if pixels from nonoverlapping regions are replaced by the corresponding pixels from the original frame before the PSNR is computed, a lower bound (LB) for the fidelity measure is created, which is given by the PSNR between the corresponding frames from the original sequence. We also assume that the stabilization system has produced a valid output whenever the PSNR is higher than LB. Erroneous motion estimates can, in fact, produce PSNRs below LB.

5.1. Evaluation Tests

The following experiments were designed to evaluate three characteristics of stabilization systems: fidelity, displacement range, and performance. *Fidelity* is a measure of how well stabilization is compensating the motion of the camera (i.e., how precisely the motion model fits the actual camera motion). Because motion must be estimated between frames, it is also directly dependent on the estimation process. *Displacement range* is defined by the minimum and maximum image displacements supported by the stabilization system, and *performance* is defined by the maximum displacement velocity which the system can compensate for, in pixels per second, given by the product of the frame rate and the maximum interframe translational displacement.

5.1.1. The Fidelity Measure

The PSNR between stabilized frames can be used to measure the fidelity of a system. Intuitively, when all motion is compensated for, there should be no residual motion after stabilization, which means that the "same" frame should be obtained continuously. Because the images are similar, the difference between two stabilized images should be zero almost everywhere. Many factors contribute to this difference being nonzero, such as noise, estimation errors, distortions caused by departures from the motion model and by the interpolation during warping, and so forth. The fidelity measure integrates the imprecision of the system due to all these factors.

The mean squared error (MSE) between images I_1 and I_0 of sizes $N \times M$ pixels is given by

$$\text{MSE}(I_1, I_0) = \frac{1}{(N \times M)} \sum_{i=1}^{N} \sum_{j=1}^{M} (I_1(i,j) - I_0(i,j))^2 \quad (11)$$

The MSE is a measure of the average departure per pixel from the desired stabilized result. The PSNR between I_1 and I_0 is

$$\text{PSNR}(I_1, I_0) = 10 \log\left(\frac{(255)^2}{\text{MSE}(I_1, I_0)}\right) \quad (12)$$

The PSNR gives a relation between the desired output and the residual image, in terms of their powers (for gray images with a maximum intensity of 255). The higher the PSNR between two stabilized frames, the better the fidelity of the system.

The above formulation does not account for the fact that when a camera moves, it probably produces nonoverlapping regions where compensation cannot be done. If the PSNR were computed just for the overlapping areas, the fidelity measure would not be meaningful when the overlapping areas are small. In order to handle these regions, we propose that every pixel belonging to a compensated frame and which does not overlap with the reference frame be copied from the current frame before computing the PSNR. For the case when the motion estimate warps the image completely outside the reference frame, the PSNR between the reference and the current frames without compensation are used as the LB. We assume that the system has produced a valid output whenever the PSNR between stabilized frames is higher than the LB.

To run this experiment, the warping functions were modified to account for the nonoverlapping areas. Given a particular stabilization system and an arbitrary sequence, two measures are computed. The first is a measure of the interframe transformation fidelity (ITF) and the second measures the global transformation fidelity (GTF). The ITF is defined as the PSNR between two consecutive stabilized frames, where the previous frame is considered the reference and the current frame is warped using the interframe motion parameters before the ITF is computed; the GTF corresponds to the PSNR between the reference frame and the current stabilized frame. The lower bounds on ITF and GTF will be respectively denoted by LBI and LBG. Formally, these measures are defined as

$$\text{LBI} = \text{PSNR}(I_t, I_{t-1}) \quad (13)$$

$$\text{LBG} = \text{PSNR}(I_t, I_{\text{ref}}) \quad (14)$$

$$\text{ITF} = \text{PSNR}(S_t^i, I_{t-1}) \quad (15)$$

$$\text{ITF} = \text{PSNR}(S_t^g, I_{\text{ref}}) \quad (16)$$

$$S_t^i(x, y) = \begin{cases} I_t(\psi_i(x,y)) & \text{if } \psi_i(x,y) \text{ is within the boundaries of } I_t \\ I_t(x,y) & \text{otherwise} \end{cases} \quad (17)$$

$$S_t^g(x, y) = \begin{cases} I_t(\psi_g(x,y)) & \text{if } \psi_g(x,y) \text{ is within the boundaries of } I_t \\ I_t(x,y) & \text{otherwise} \end{cases} \quad (18)$$

where $\psi_i(x,y)$ and $\psi_g(x,y)$ are the warping functions which stabilize the current frame I_t to the previous frame I_{t-1} and to the reference frame I_{ref}, respectively.

5.1.2. The Minimum Image Displacement Measure

The second experiment determines the minimum image displacement that a system can measure. Because the estimation is based on feature tracking, this experiment could also be used to compare different tracking algorithms with subpixel precision. For this experiment, synthetic image sequences were created by the following procedure. Given a single image I of large dimensions, a window of smaller size (e.g., 128 × 128 pixels) is first placed at a fixed position on the image. This window is used to compose the output sequence. The first frame is defined by the window itself, and the displacement velocity increment (acceleration) is set to zero. The following frames are then created by incrementing the displacement velocity by a small amount and warping I according to the new displacement velocity using bilinear interpolation. As a result, the contents of the window, when placed on the warped image, will change proportionally. The precision of this measurement is defined by the acceleration step between frames.

For very small displacements, the PSNR between consecutive frames is very high (i.e., LBI is very high). If the errors in the estimated parameters are larger than the true transformation parameters, the ITF will be lower than LBI. As the displacement increases, LBI decreases and ITF increases if the displacement is large enough to be estimated. Therefore, one curve eventually crosses the other. This crossing point is used to define the minimum image displacement compensated by the system.

5.1.3. The Maximum Image Displacement Measure

This experiment determines the upper bound on the range of displacements that can be handled by an EDIS system. Synthetic image sequences were created using the method described above, using larger acceleration steps between frames.

It is expected that when the system is working properly, the ITF remains higher than the LBI, which is low for large interframe displacements. When the displacement is too large, the system produces invalid motion estimates, causing the ITF to drop and possibly become lower than the LBI. The maximum image displacement is defined to be the point where the ITF curve reaches or crosses the LBI curve.

5.1.4. The Performance Measure

The frame rate is an important feature of any stabilization system, but this measure alone might be misleading because the robustness and accuracy of the system can be easily sacrificed to increase speed. For example, by reducing the sizes of the search and correlation windows described in Sec. 3.2 or reducing the number of feature points or pyramid levels to be processed, the system can be made to run considerably faster, but any of these changes have a large negative impact on the accuracy.

Performance will be measured by the maximum displacement velocity supported by the system, which is defined by the product of the frame rate and the maximum translational displacement, in pixels per second.

6. RESULTS FROM THE EVALUATION TESTS

The experiments presented in this section have the purpose of comparing the performance of the three transformation groups presented in Sec. 3.1, when applied to stabilization systems based on the algorithm described in Sec. 3. We are also interested in how the behavior of the system changes when system parameters, other than the motion model, are modified.

All experiments were run using the same settings for all parameters (i.e., the same number of feature points, pyramid levels, search window sizes, etc.). Sixteen features arranged in two rows and eight columns were tracked using search windows of size 5 pixels per pyramid level. Two pyramid levels were constructed for images of resolution 128×128 pixels, and three levels were used for images of higher resolution. All synthetic sequences were of resolution 128×128 pixels, and all real uncalibrated sequences were of resolution 320×240 pixels.

The following notation is used in the graphs presented in this section. The curves for the affine transformations are drawn using dotted lines with stars ($\cdot * \cdot$); the curves for the similarity transformations are drawn using solid lines (—); and the curves for Euclidean transformations, with dash-dotted lines (–·). Dotted lines with circles ($\cdot \circ \cdot$) are used for the measurements of the lower bounds, and other curves (if any) are drawn with dotted lines with crosses ($\cdot + \cdot$).

6.1. Fidelity

Two uncalibrated real sequences were used to evaluate the distortion after stabilization. Figure 10 shows two arbitrary frames from each sequence. The images in the top row correspond to the UGV sequence, which was also used in Sec. 4. The bottom images are from the Building sequence, which shows a left to right panning movement. Figures 11 and 12 show the results of the distortion tests for 30 frames of each sequence.

Fig. 10 Two frames from the UGV (top row) and Building (bottom row) sequences used for the distortion tests.

Image Stabilization and Mosaicking

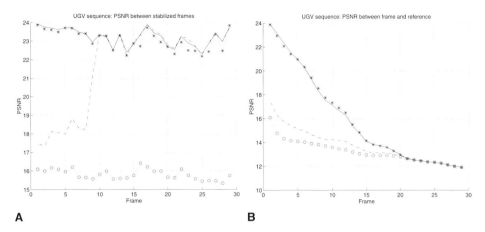

Fig. 11 (A) ITF results and (B) GTF results for the UGV sequence.

Figure 11a shows the ITF and LBI for 30 frames of the UGV sequence. Observe that the affine- and similarity-based systems have very similar curves, whereas the Euclidean system performs poorly during the first 10 frames, which correspond to the zooming part of the sequence. This result is expected because the Euclidean group does not model scale.

After the 20th frame, the sequence no longer overlaps the reference frame. This can be observed from the GTF curves shown in Fig. 11b. The GTF drops from frame to frame because each new frame has less overlap with the reference frame. The GTF of the Euclidean system is considerably smaller due to the lack of scaling compensation.

The ITF and GTF for the Building sequence are shown in Figs. 12a and 12b, respectively. In this case, as there is no change in scale, all systems perform equally well. It is important to notice from the ITF graph that feature outliers have more influence on the

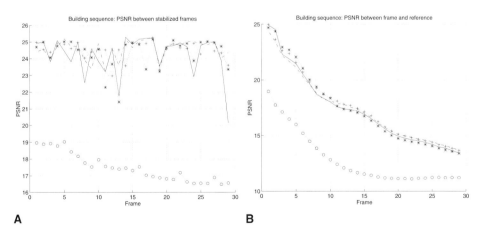

Fig. 12 (A) ITF results and (B) GTF for the Building sequence.

performance of higher-order models. To test this hypothesis, the affine system was reconfigured to use 20 features instead of 16 [shown by the $(\cdot + \cdot)$ curve]; the resulting performance improvement can be seen from the ITF and GTF curves. Since Because the same 16 features are used for all systems, the least squares fit seems to be more robust for the lower-complexity models.

6.2. Range of Displacements

Two synthetic sequences composed of 19 frames each were created to measure the minimum displacement of each system. Figure 13a and 13b show one frame from the Bahia and Boat sequences, respectively. The interframe acceleration step was set to 1/10th of a pixel/frame^2 (i.e., frame F_n has a displacement of $n/10$ pixel from F_{n-1}, for $n > 0$).

For these experiments, a fourth system was introduced. It is based on the Euclidean group, but with a simpler grid-to-grid (no subpixel precision) feature tracker. The measurements for this system are shown by the dotted line with crosses $(\cdot + \cdot)$.

Figures 14a and 14c show the ITF for the Bahia and Boat sequences, respectively. For the first three systems, the ITF becomes larger than LBI after the second frame (i.e., the minimum displacement is below 0.3 pixel/frame). For the system without subpixel precision, the minimum displacement is below 0.7 pixel/frame for the Bahia sequence and below 0.6 pixel/frame for the Boat sequence.

Figures 14b and 14d show the average ITF for these sequences. It is easier to compare the performance from these graphs. The poor performance of the system without subpixel precision is obvious, and it is also clear that the simpler models (similarity and Euclidean) outperform the more complex six-parameter affine model.

Similar sequences were created to determine maximum displacements, now with an acceleration of 1 pixel/frame^2. The maximum displacement is a property of the system related to the search sizes and number of pyramid levels. The search sizes were set to ± 5 pixels on each level, and only two pyramid levels were constructed (the highest resolution level of the pyramids is 320 × 240 pixels. A fourth system with the same search window sizes but only one pyramid level (without multiresolution) was also tested; its results are shown by the curves with crosses $(\cdot + \cdot)$.

Figure 15a shows the ITF measures for the Boat sequence. Observe that the maximum displacement of the first three systems lies between the 13th and 14th frame, so that

A   B

Fig. 13 An example from the (A) Bahia and (B) Boat sequences.

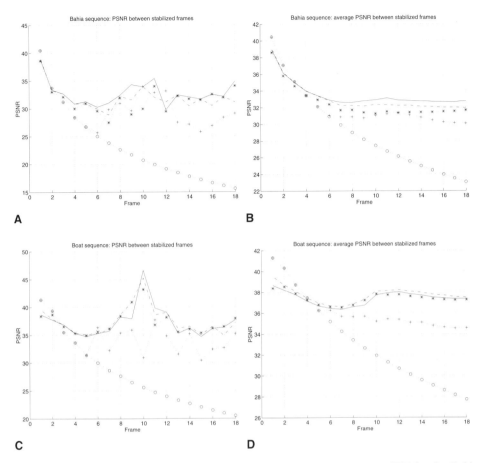

Fig. 14 Determination of minimum translations: (A) ITF and (B) average ITF for the Bahia sequence; (C) ITF and (D) average ITF for the Boat sequence.

it is safe to say that it is above 13 pixels (due to lack of subpixel information). For the Bahia sequence shown in Fig. 15c, this also seems to be the case, although there is a considerable drop after the 10th frame. For the system using only one pyramid level, the maximum displacement lies around five pixels, as this value corresponds to the size of the search window.

The average ITF for the Boat sequence is shown in Fig. 15b. It is clear from these graphs that the performance of all systems is very similar and that their performances drop after the 11th frame. Figure 15d shows the average ITF for the Bahia sequence. The irregularities on the first few frames, which can be seen from Fig. 15c, create a small difference in the average performance of the systems. The clear winner seems to be the similarity model, and the affine model seems to be performing slightly better than the Euclidean-based system for this particular sequence. All the performances drop after the 10th frame.

The analysis of minimum and maximum displacements is not limited to translations. It is simple to create a synthetic sequence by varying any parameter of a transformation

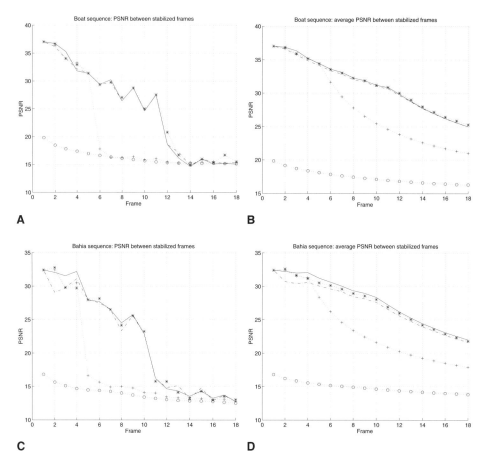

Fig. 15 Determination of maximum translations: (A) ITF and (B) average ITF for the Boat sequence; (C) ITF and (D) average ITF for the Bahia sequence.

group, and then empirically determining the system's operating range for the sequence. For example, Fig. 16 shows the lower and upper bounds for rotation sequences created by varying the rotational parameter of the Euclidean transformation.

The graph in Fig. 16a shows the ITF curves for a small interframe rotation sequence. The sequence, based on the Bahia image, was created using rotation increments of 0.1 degree. The minimum rotation for all systems lies below 0.3 degree. The bottom graph of Fig. 16b shows the ITF curves for a large interframe rotation sequence, which was created using rotation increments of 1 degree, but starting from 5 degrees. The ITF of all systems seems to break down after the 19th frame (i.e., the maximum rotation is above 23 degrees).

6.3. Performance

The algorithm described in Sec. 3 using the similarity model transformation was implemented in a real-time imaging platform, a Datacube MV200 board connected to a Sun

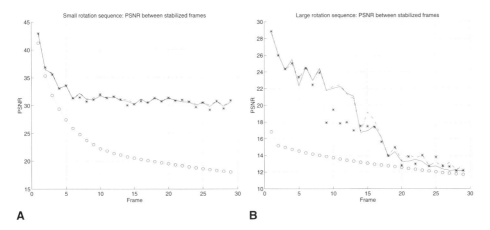

Fig. 16 (A) Minimum and (B) maximum rotations for the Bahia sequence.

Sparc 20. The Datacube is responsible for image acquisition, construction of the Gaussian and Laplacian pyramids, and for the generation of the stabilized output video (warping and digital-to-analog conversion). The Sparc station receives the pyramids from the Datacube, detects and tracks a small set of feature points which are used for motion estimation and compensation, and the global motion parameters are sent back to the Datacube, which warps the current frame and displays the stabilized sequence. The highest pyramid resolution is 128 × 120, and to achieve robustness for large image displacements, four pyramid levels are processed. The settings for the search and SSD window sizes are shown in Table 1.

Figure 17a shows the ITF curves for a sequence with increasingly large interframe translations, using the Bahia image of Fig. 13a. The continuous line at the bottom corresponds to the LBI of the sequence. The dotted line curve with stars (·*·) corresponds to the system with all search windows set to one pixel (Sys1); the dotted curve with circles (·○·) shows the results of the system with search window size set to two pixels (Sys2); the solid curve shows the results for windows of size three pixels (Sys3); and the dash-dotted curve shows the ITF for search windows of size four pixels (Sys4). From these curves, it can be seen that Sys1 can stabilize displacements up to 15 pixels, Sys2 up to 29, Sys3 up to 35, and Sys4 up to 40. Similar results can also be seen from the average ITF shown in Fig. 17b.

Table 1 Settings Used for the SSD and Search Window Sizes

Pyramid level	Level resolution	Search size	SSD size
3	16 × 15	4	5
2	32 × 30	3	5
1	54 × 60	3	5
0	128 × 120	3	7

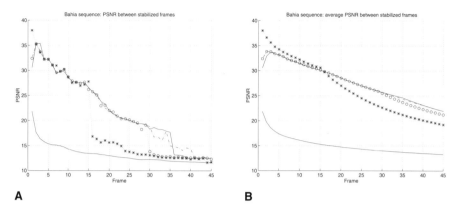

Fig. 17 (A) ITF and (B) average ITF graph for the Bahia sequence, under uniform search window sizes for all pyramid levels.

We notice that the ITF curves do not change much when the search sizes for only the higher-resolution levels are changed. Because refinement is done between pyramid levels, the search windows can be reduced provided that a good initial estimate is given by the coarser levels. Figure 18a shows the ITF for Sys4 (dash-dotted curve) and the system configured as described in Table 1 (Sys43), shown by the continuous line. Observe that the performance of these systems is remarkably similar, which can also be seen from the average ITF graphs presented in Fig. 18b.

From reducing the size of the search windows, improvement in performance is obtained. Table 2 shows how the frame rate for the stabilization system degrades when the search window is increased and the SSD windows are kept constant, for Sys1 to Sys4, and Sys43. All systems were set to detect and track 10 features between frames. The last row of Table 2 shows the performance of the stabilization system presented by Hansen

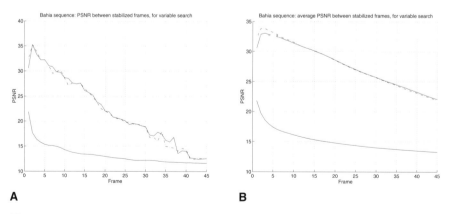

Fig. 18 (A) ITF and (B) average ITF graph for the Bahia sequence, under variable search window sizes.

Table 2 Performance Evaluation of Search Window Sizes for the Stabilization and Mosaicking Systems

System	Local search (pixels)	Max. disp. (pixels)	Frame rate (frames/s)	Image velocity (pixels/s)
Sys1	±1	±15	17.3	259.5
Sys2	±2	±30	13.9	417.0
Sys3	±3	±35	10.8	378.0
Sys4	±4	±40	8.3	332.0
Sys43		±40	10.5	420.0
VFE-100		±32	10	320

et al. (3) using the VFE-100 pyramid hardware. Their system is able to stabilize images with velocities of 320 pixels per second, running at approximately 10 frames per second. Despite the better performance figures of Sys43, we expect the system based on the VFE-100 to produce more accurate and robust motion estimates because several local patches contribute to the refinement and estimation of the parameters, whereas Sys43 uses a very limited set of feature points. On the other hand, Sys43 can handle sequences with regions that do not fit the affine model if the set of features is constrained to distant points in the scene. Also, for real-time applications, faster frame rates (closer to video frame rate) might be more appropriate.

7. CONCLUSION

We have presented several techniques for image stabilization and mosaicking and have proposed a feature-based multiresolution algorithm which is suitable for real-time implementation, yet robust for large image displacements and accurate. The algorithm was used to fit subsets of the affine group of transformations and can be easily extended to estimate higher-order models. Stabilization is achieved by combining all motion from a reference frame and generating a motion-compensated video output. Mosaics can also be constructed by directly aligning the input frames using the same motion estimates.

We have also described simple ways of evaluating the fidelity, range of displacements, and performance of image stabilization systems. Fidelity is a measure of how well EDIS compensates for the motion of the camera, the range of displacements characterizes the minimum and maximum displacements that a system can handle, and performance measures the maximum image velocity for which the system can compensate for. Although these measures are not absolute because they depend on the sequence being stabilized, they can be used to compare different systems, even those based on different transformation models. They can also be used as development tools, to easily evaluate performance as a function of different system parameters and modules.

Electronic digital image stabilization and mosaicking can also be applied for other video applications, such as video coding, indexing, and browsing; animation (such as for games and virtual-reality applications); image and video segmentation; and basically any image-processing algorithm based on correlation and matching, such as stereo and object-tracking algorithms.

REFERENCES

1. LS Davis, R Bajcsy, R Nelson, M Herman. RSTA on the move. Proc DARPA Image Understanding Workshop, Monterey, CA, 1994, pp. 435–456.
2. P Burt, P Anandan. Image stabilization by registration to a reference mosaic. Proc DARPA Image Understanding Workshop, Monterey, CA, 1994, pp 425–434.
3. M Hansen, P Anandan, K Dana, G van der Wal, PJ Burt. Real-time scene stabilization and mosaic construction. Proc DARPA Image Understanding Workshop, Monterey, CA, 1994, pp 457–465.
4. CH Morimoto, D DeMenthon, LS Davis, R Chellappa, R Nelson. Detection of independently moving objects in passive video. Proc of Intelligent Vehicles Workshop, Detroit, MI, 1995, pp 270–275.
5. YS Yao, P Burlina, R Chellappa. Electronic image stabilization using multiple visual cues. Proc International Conference on Image Processing, Washington, DC, 1995, pp 191–194.
6. Z Durić, A Rosenfeld. Stabilization of image sequences. Technical Report CAR-TR-778, Center for Automation Research, University of Maryland, College Park, 1995.
7. B Rousso, S Avidan, A Shashua, S Peleg. Robust recovery of camera rotation from three frames. Proc. DARPA Image Understanding Workshop, Palm Springs, CA, 1996, pp. 851–856.
8. T Viéville, E Clergue, PEDS Facao. Computation of ego-motion and structure from visual and internal sensors using the vertical cue. Proc International Conference on Computer Vision, Berlin, 1993, pp 591–598.
9. M Irani, B Rousso, S Peleg. Recovery of ego-motion using image stabilization. Proc IEEE Conference on Computer Vision and Pattern Recognition, Seattle, WA, 1994, pp 454–460.
10. OJ Kwon, R Chellappa, CH Morimoto. Motion compensated subband coding of video acquired from a moving platform. Proc of IEEE International Conference on Acoustics, Speech, and Signal Processing, Detroit, MI, 1995, pp 2185–2188.
11. S Balakirsky. Comparison of electronic image stabilization systems. Master's thesis, Department of Electrical Engineering, University of Maryland, College Park, 1995.
12. M Irani, P Anandan. A unified approach to moving object detection in 2d and 3d scenes. Proc DARPA Image Understanding Workshop, Palm Springs, CA, 1996, pp 707–718.
13. S Balakirsky, R Chellappa. Performance characterization of image stabilization algorithms. Technical Report CAR-TR-822, Center for Automation Research, 1996.
14. CH Morimoto, R Chellappa. Fast electronic digital image stabilization. Proc International Conference on Pattern Recognition, Vienna, 1996.
15. CH Morimoto, R Chellappa. Fast 3d stabilization and mosaicking. Proc IEEE Conference on Computer Vision and Pattern Recognition, Puerto Rico, 1997.
16. M Irani, P Anandan, S Hsu. Mosaic based representations of video sequences and their applications. Proc International Conference on Computer Vision, Cambridge, MA, 1995, pp 605–611.
17. CH Morimoto, S Balakirsky, R Chellappa. Fast image stabilization and mosaicking for predator data. Proc DARPA Image Understanding Workshop, New Orleans, LA, 1997.
18. M Irani, S Hsu, P Anandan. Mosaic based video compression. SPIE Conference on Electronic Imaging, Digital Image Compression: Algorithms and Techniques, San Jose, CA, 1995, vol 2419, pp 242-253.
19. H Sawhney, S Ayer, M Gorkani. Model-based 2d and 3d dominant motion estimation for mosaicing and video representation. Proc International Conference on Computer Vision, Cambridge, MA, 1995, pp 583–590.
20. F Dufaux, F Moscheni. Background mosaicking for low bit rate video coding. Proc International Conference on Image Processing, Lausanne, 1996, vol I, pp 673–676.
21. JYA Wang, EH Adelson. Representing moving images with layers. IEEE Trans Image Process IR 3(5):625–638, 1994.
22. S Mann, RW Picard. Virtual bellows: Constructing high quality stills from video. Proc International Conference on Image Processing, Austin, TX, 1994, vol I, pp 363–367.

23. L Teodosio, W Bender. Salient video stills, content and context preserved. Proc ACM Multimedia Conference, 1993.
24. H Ishiguro, M Yamamoto, S Tsuji. Omni-directional stereo. IEEE Trans Pattern Anal Machine Intell 14(2):257–262, 1992.
25. SB Kang, R Szeliski. 3d scene data recovery using omnidirectional multibaseline stereo. Proc. IEEE Conference on Computer Vision and Pattern Recognition, San Francisco, CA, 1996, pp 364–370.
26. RI Hartley. In defence of the 8-point algorithm. Proc International Conference on Computer Vision, Cambridge, MA, 1995, pp 1064–1070.
27. HC Longuet-Higgins. A computer algorithm for reconstructing a scene from two projections. Nature 293:133–135, 1981.
28. SE Chen. Quicktime VR—An image-based approach to virtual environment navigation. Proc of ACM SIGGRAPH, Los Angeles, CA, 1995, pp 29–38.
29. L McMillan, G Bishop. Plenoptic modeling: An image-based rendering system. Proc of ACM SIGGRAPH, Los Angeles, CA, 1995, pp 39–46.
30. S Peleg, J Herman. Panoramic mosaics by manifold projection. Proc IEEE Conference on Computer Vision and Pattern Recognition, Puerto Rico, 1997.
31. R Szeliski. Image mosaicking for tele-reality applications. Technical Report CRL 94/2, DEC Cambridge Research Laboratory, 1994.
32. LG Brown. A survey of image registration techniques. Computer Surveys 24(4):325–376, 1992.
33. JK Aggarwal, N Nandhakumar. On the computation of motion from sequences of images—A review. Proc IEEE 76(8):917–935, August 1988.
34. JL Barron, DJ Fleet, SS Beauchemin. Performance of optical flow techniques. Int J Computer Vision 12(1):43–77, 1994.
35. TS Huang, AN Netravali. Motion and structure from feature correspondences: A review. Proc IEEE 82(2):252–268, 1994.
36. G Adiv. Determining three-dimensional motion and structure from optical flow generated by several moving objects. IEEE Trans Pattern Anal Machine Intell 7(4):384–401, 1985.
37. S Mann, RW Picard. Video orbits: Characterizing the coordinate transformation between two images using the projective group. Technical Report 278, MIT Media Lab, Cambridge, MA, 1995.
38. Q Tian, MN Huhns. Algorithms for subpixel registration. Computer Vision Graphics Image Process 35:220–233, 1986.
39. J Shi, C Tomasi. Good features to track. Proc IEEE Conference on Computer Vision and Pattern Recognition, Seattle, WA, 1994, pp 593–600.
40. EP Simoncelli, EH Adelson, DJ Heeger. Probability distribution of optical flow. Proc IEEE Conference on Computer Vision and Pattern Recognition, Maui, HI, 1991, pp 310–315.
41. MJ Black, P Anandan. A framework for the robust estimation of optical flow. Proc International Conference on Computer Vision, Berlin, 1993, pp 231–236.
42. Q Zheng, R Chellappa. A computational vision approach to image registration. IEEE Trans Image Process IP-2:311–326, 1993.
43. G Woldberg. Digital Image Warping. Los Alamitos, CA: IEEE Computer Society Press, 1990.
44. R Nelson. Qualitative detection of motion by a moving observer. Int J Computer Vision 7:33–46, 1991.
45. P Anandan, P Burt, J Pearson. Spatial and temporal mechanisms in target cueing. Proc DARPA Image Understanding Workshop, Palm Springs, CA, 1996, pp 525–529.
46. TY Tian, C Tomasi, DJ Heeger. Comparison of approaches to egomotion computation. Proc IEEE Conference on Computer Vision and Pattern Recognition, San Francisco, CA, 1996, pp 315–320.

14
Highly Robust Statistical Estimates Based on Minimum-Error Bayesian Classification

Xinhua Zhuang and Kannappan Palaniappan
University of Missouri–Columbia, Columbia, Missouri

Robert M. Haralick
University of Washington, Seattle, Washington

1. INTRODUCTION

One of the most interesting and challenging areas of artificial intelligence is computer vision. In Refs. 15 and 21, it was argued that computer vision tasks algorithms need to be made more robust. All computer vision feature detectors, recognizers, matchers, groupers, trackers and localizers developed so far are unavoidably error-prone and seem to make occasional errors which, indeed, are blunders. Consequently, a realistic assumption for errors in data must be used to improve the performance of computer vision algorithms. Some vision applications for which robust statistical methods have been studied include: fitting surface patches to range data (16,22), stereo analysis (17), pose estimation (18), motion analysis (19), optic flow (20), and shape and motion-based scene segmentation (23). One choice, which is most often used, is a contaminated Gaussian noise model, which is defined as a regular white Gaussian noise model with probability $1 - \epsilon$ plus an outlier process with probability ϵ (2,3). The least squares (LS)-estimator that we are most familiar with is based on a pure regular white Gaussian noise model assumption and has been used for many years. The problem with the LS-estimator is its sensitivity to minor deviations from the Gaussian noise model assumption. It is well known that outliers have an unusually large influence on the LS-estimator (3). The outliers may come from bad data points caused by errors in measurements or may simply arise from the nature of the physical data; that is, the underlying distribution is not represented by a pure Gaussian distribution. Because outliers can deviate the LS fit a great deal, the resulting residuals are often misleading. In order to reduce the influence of outliers on the final estimates, robust procedures have been developed to modify least squares schemes. The best known robust procedure is the class of maximum-likelihood-type estimators, or simply the M-estimator, which minimizes a robust-loss objective function that is symmetric and positive-definite. Other robust estimators such as the L-estimator (linear combinations of order statistics) and the R-estimator (rank-based estimates) also arise in various applications (3,9).

The theory of the M-estimator was first developed by Huber in 1964 for the estimation of the location and scale parameters from a sequence of independent and identically distributed (iid) observations. It has been successfully generalized for robust regression (3). The essence of the M-estimator can be illustrated in the following discussion on location parameter estimation. Given a sequence of observations $x_1, \ldots, x_N$ such that $x_k = \theta + e_k$, the problem is to estimate the location parameter θ. The distribution of errors, e_k's, is not assumed to be known exactly; the only assumption is that $e_1, \ldots, e_N$ are iid and the distribution is symmetric.

Let $\rho(x)$ be a cost or influence function. Then, the M-estimator of the location parameter θ is the solution of the following minimization problem:

$$\min_{\theta} \sum_k \rho(x_k - \theta) \tag{1}$$

Let the derivative of $\rho(x)$ be $\psi(x)$. Then the minimum of problem (1) is given by

$$\sum_k \psi(x_k - \theta) = 0. \tag{2}$$

If the probability density function $f(x)$ of errors is known, then the cost function in problem (1) can be favorably chosen as $\rho(x) = -\log f(x)$; thus, $\psi(x) = -f'(x)/f(x)$, and the solution to Eq. (2) coincides with the maximum-likelihood estimate of the location parameter θ (3,9).

For the LS-estimator, the cost function is chosen as the quadratic $\rho(x) = x^2/2$; thus, $\psi(x)$ is equal to x. The LS-estimator, also known as the sample mean, coincides with the maximum-likelihood estimate if the error density function is modeled as a Gaussian. For the least absolute deviation estimator, the cost function is chosen as $\rho(x) = |x|$; thus, $\psi(x)$ is equal to $\text{sgn}(x)$. The resulting estimate is known as the sample median, which coincides with the maximum-likelihood estimate if the error distribution is Laplacian. The sample mean is very sensitive to the tail behavior of the error distribution, in the sense that even a single outlier can alter the estimate drastically. On the other hand, the sample median is sensitive to the behavior of the error distribution around its median. The breakdown point of the highest fraction of arbitrary outliers that can be tolerated is zero for the LS-estimator and is 0.5 for least absolute derivation estimator. A class of min-max estimators suggested by Huber, which lie between the sample mean and the sample median, use the following cost function:

$$\rho_0(x) = \frac{x^2}{2} \quad \text{if } |x| < a \tag{3}$$

$$= a|x| - \frac{a^2}{2} \quad \text{otherwise}$$

whose derivative is given by

$$\psi_0(x) = \min[a, \max(x, -a)] \tag{4}$$

where a represents an efficiency tuning parameter; supposedly a function of the contaminated percentage ϵ. $\psi_0(x)$ in Eq. (4) is bounded, nondecreasing, and continuous. Monotonicity assures uniqueness of the M-estimator solution. The effect of using $\psi_0(x)$ is to assign less weight to the small portion of large residuals and a unit weight to the bulk of small to moderate residuals, so that large residuals will not drastically influence the final estimate.

Huber's M-estimator coincides with the maximum likelihood estimate if the error distribution follows the so-called least favorable density function:

$$f_0(x) = \frac{1-\epsilon}{\sqrt{2\pi}} \exp[-\rho_0(x)] \qquad (5)$$

which represents a combinatorial distribution, that is Gaussian in the middle and Laplacian at the tails, with much larger variance. The least favorable density function was derived based on minimizing its maximum asymptotic variance for all possible densities $f \in P_\epsilon$, where $P_\epsilon(\phi) = \{f | f = (1-\epsilon)\phi + \epsilon h, h \in W\}$ with ϕ and W representing the standard Gaussian density and the set of unknown densities, respectively.

The function $f_0(x)$ or any other single robust density function is unlikely to provide high robustness for the following reason. As the outlier proportion ϵ increases, a significant number of outlier residuals are generated that may show a strong statistical tendency against the assumed density model $f_0(x)$. It seems that if we completely model each possible density function f in the neighborhood of $P_\epsilon(\phi)$ achieved using a single robust density function, then a high degree of robustness can hardly be achieved.

It is easy to understand then that robust estimation actually amounts to a classification between inliers and outliers. In this Chapter, we present a highly robust estimator called the MF (model fitting)-estimator (4,5), which uses the minimum-error Bayesian classification of inliers and outliers. The approach is optimal in the sense that inliers are separated from outliers with minimum misclassification errors. This naturally leads to a set of *partial* density models, each of which models the underlying unknown density function only at observed samples. There exists a subset of *partial* density models, each of which provides the same desired minimum-error Bayesian classification between inliers and outliers. The high robustness of the MF-estimator is demonstrated using computer experiments and real data.

This chapter is organized as follows. In Sec. 2, the MF-estimator is presented. In Sec. 3, the algorithm is applied to data clustering analysis. In Sec. 4, the same algorithm is used to estimate an unknown probability density in the practical context of a biomedical application cervical smear cell classification. Following the conclusion is an appendix that briefly describes the Kolmogorov–Smirnov normality test used to determine the validity of extracted Gaussian components.

2. MINIMUM-ERROR BAYESIAN CLASSIFICATION AND THE MF-ESTIMATOR

2.1. Minimum-Error Bayesian Classification and Partial Modeling

As stated in Ref. 3, the basic problem involved in the estimation of the location and scale parameter is to discover a valid Gaussian component G, characterized by $N(m, C)$ with mean vector m and covariance matrix C (see Appendix), in the given data set X. Let X consist of N samples $x^1, \ldots, x^N$ from the same event space which belong to an n-dimensional Euclidean space R_n. It is assumed that each sample $x^k \in X$ is generated by a mixture of an unknown Gaussian density $N(m, C)$ with probability $1 - \epsilon$ and an unknown outlier density $h(\cdot)$ with probability ϵ. These data samples are assumed independently and identically distributed with a common probability density $f(x)$:

$$f(x^k) = \frac{1-\epsilon}{(\sqrt{2\pi})^n \sqrt{|C|}} \exp\left(-\frac{1}{2} d^2(x^k)\right) + \epsilon h(x^k) \qquad (6)$$

where $d^2(x^k)$ represents the squared Mahalanobis distance of x^k from the unknown mean vector m, namely

$$d^2(x^k) = (x^k - m)^T C^{-1}(x^k - m) \qquad (7)$$

where the superscript T denotes the transpose operator.

The density $f(\cdot)$ becomes a pure Gaussian density when $\epsilon = 0$. Accordingly, $f(\cdot)$ is called a contaminated Gaussian density when $\epsilon > 0$ (3).

The contaminated Gaussian density model $f(\cdot)$ directly incorporates uncertainty and can be justified as an appropriate model in many practical applications. Let us imagine that *supernature* (the process we are trying to model) chooses x^k from $N(m, C)$ with probability $1 - \epsilon$ or from $h(\cdot)$ with probability ϵ. But the experimenter is only able to observe x^k. Ideally, a sample x^k is classified as an *inlier* if it is realized from $N(m, C)$, or as an outlier if otherwise. Let G be the subset of all observations from the Gaussian component comprising all inliers and let B be its complement. The subset, $B = X - G$, then contains all outliers. This can be stated as

$$G = \{x^i : x^i \text{ generated by } N(m, C)\}$$
$$B = \{x^j : x^j \text{ not generated by } N(m, C)\} \qquad (8)$$

To proceed in a fruitful direction, sound constraints for the distributions shall be used. By assuming that *supernature* is impartial, the ideal classification (8) would imply that the likelihood of any inlier being generated by $N(m, C)$ is greater than the likelihood of any outlier being generated by $N(m, C)$. Mathematically, the "impartial" assumption states that for any $x^i \in G$ and $x^j \in B$, it holds that

$$g_i > g_j \qquad (9)$$

where g_k $1/((\sqrt{2\pi})^n \sqrt{|C|}) \exp(-\frac{1}{2}d^2(x^k))$ is the likelihood of observing x^k. By defining

$$g = \min\{g_i : x^i \in G\}$$
$$b = \max\{g_j : x^j \in B\} \qquad (10)$$

then, by inequality (9), the minimum likelihood of any inlier being generated by $N(m, C)$ must be greater than the maximum likelihood of any outlier being generated by $N(m, C)$:

$$g > b \qquad (11)$$

The discovery of a valid Gaussian component in X represents a process which classifies observed samples into inliers and outliers. The ideal classification (8) practised by *supernature* cannot be exactly retrieved by the experimenter even if all the parameters m, C, ϵ, $h(x^1), \ldots, h(x^N)$ are given to the experimenter. It is clear that the best feasible classification will be the one that attains the minimum misclassification error.

Using Eq. (6), the probability of a sample x^k being an inlier is given by

$$\lambda_k = \frac{(1 - \epsilon)g_k}{f_k} \qquad (12)$$

where f_k stands for $f(x^k)$. According to Ref. 6, the misclassification error probability will be minimal if the experimenter uses the Bayesian classification

$$G = \{x^i : \lambda_i > 0.5\} = \left\{x^i : g_i > \frac{\epsilon h(x^i)}{1 - \epsilon}\right\}$$
$$B = \{x^j : \lambda_j \leq 0.5\} = \left\{x^j : g_j \leq \frac{\epsilon h(x^j)}{1 - \epsilon}\right\} \qquad (13)$$

to approximate the ideal classification (8). The minimum-error Bayesian classification (13) states that an inlier is more likely to have been generated by $N(m, C)$ than by the unknown outlier distribution $h(\cdot)$ and that an outlier is less likely to have been generated by $N(m, C)$ than by $h(\cdot)$. Due to the constraint (11), the Bayesian classification (13) enforces those N unknown outlier density values $h(x^k)$'s to fall in the half-open interval $[(1 - \epsilon)b/\epsilon, (1 - \epsilon)g/\epsilon)$. Interestingly enough, any combination of N values from that interval would realize the same Bayesian classification, as can be easily verified. One possible combination which assumes the least configurational information about the unknown outlier distribution would have all N unknowns be identically distributed, that is,

$$h(x^1) = h(x^2) = \cdots = h(x^N) = \delta \qquad (14)$$

It is important to emphasize that there exists a continuum of partial models or δ's, each of which lies in the appropriate interval and provides the same satisfactory Bayesian classification (13) for the N data samples. The partial modeling assumption (14) when applied to the outlier density function transforms Eq. (6) into

$$f_k = (1 - \epsilon)g_k + \epsilon\delta \qquad (15)$$

Because the samples $x^k, k = 1, \ldots, N$, are assumed to be independent of each other, the log-likelihood function of observing $x^1, \ldots, x^N$ conditioned on m, C, ϵ, and δ can be expressed as

$$Q = \log P(x^1, \ldots, x^N | m, C, \epsilon, \delta) = \sum_k \log f_k \qquad (16)$$

Using Eq. (15), Q can be simplified as

$$Q = N \log(1 - \epsilon) + \sum_k \log\left\{g_k + \frac{\epsilon\delta}{1 - \epsilon}\right\} \qquad (17)$$

Defining the model-fitting function

$$q(m, C; t) = \sum_k \log\{g_k + t\} \qquad (18)$$

it is easy to verify that the maximization of Q at δ with respect to m and C is equivalent to maximizing $q(m, C; t)$ at t with respect to m and C, provided that $t = \epsilon\delta/(1 - \epsilon)$. Without confusion, we shall refer to each t (≥ 0) as a partial model henceforth.

2.2. The MF-Estimator

In the following, the MF-estimator is presented within the context of detecting a valid Gaussian component. The extension to the general regression domain is straightforward and will not be presented.

A sequence of partial models is set up ($t_0 = 0 < t_1 < \cdots < t_L$, where t_L denotes an upper bound for all potentially desirable partial models). The determination of the upper bound depends on the spatial arrangements of data samples and can be estimated. At each selected partial model $t_s, s = 0, 1, \ldots, L$, $q(m, C; t_s)$ is iteratively maximized with respect to m and C by using the gradient ascent rule beginning with a randomly chosen initial mean $m^{(0)}$. Having solved $\max_{m,C} q(m, C; t_s)$ for $m(t_s)$ and $C(t_s)$ the mean and covariance, respectively, the inlier set, $G(t_s) = \{x^i : g_i(m(t_s), C(t_s)) > t_s\}$, is calculated, followed by the Kolmogorov–Smirnov (K–S) normality test on $G(t_s)$. If the test succeeds, then the valid Gaussian component, $G(t_s)$, has been determined. Otherwise, we proceed to the next partial model if the upper bound t_L has not been reached. This provides a brief description of the MF-estimator. The details for solving each $\max_{m,C} q(m, C; t_s)$ is given in Sec. 2.3.

2.3. Solving $\max_{m,C} q(m, C; t_s)$

The gradient ascent rule is used to solve each maximization of $\max_{m,C} q(m, C; t_s)$ for a specified t_s. The gradients $\nabla_m q(m, C; t_s)$ and $\nabla_C q(m, C; t_s)$ can be derived as

$$\nabla_m q(m, C; t_s) = \nabla_m \sum_k \log(g_k + t_s)$$

$$= \sum_k \frac{1}{g_k + t_s} \nabla_m g_k$$

$$= \sum_k \frac{g_k}{g_k + t_s} \nabla_m (\log g_k)$$

$$= \sum_k \lambda_k \nabla_m (\log g_k)$$

$$= \sum_k \lambda_k \left\{ \nabla_m \log\left(\frac{1}{(\sqrt{2\pi})^n \sqrt{|C|}}\right) - \frac{1}{2} \nabla_m (x_k - m)^T C^{-1}(x_k - m) \right\}$$

$$= -\frac{1}{2} \sum_k \lambda_k \nabla_m \{(x_k - m)^T C^{-1}(x_k - m)\}$$

$$= \sum_k \lambda_k C^{-1}(x_k - m)$$

$$= C^{-1} \sum_k \lambda_k (x_k - m) \tag{19}$$

$$\nabla_C q(m, C; t_s) = \sum_k \lambda_k \nabla_C (\log g_k)$$

$$= \sum_k \lambda_k \left\{ \nabla_C \log\left(\frac{1}{(\sqrt{2\pi})^n \sqrt{|C|}}\right) - \frac{1}{2} \nabla_C (x_k - m)^T C^{-1}(x_k - m) \right\}$$

$$= -\frac{1}{2} \sum_k \lambda_k \{\nabla_C \log|C| + \nabla_C (x_k - m)^T C^{-1}(x_k - m)\}$$

$$= -\frac{1}{2} \sum_k \lambda_k \{C^{-1} - C^{-2}(x_k - m)(x_k - m)^T\}$$

$$= -\frac{1}{2} C^{-1} \left\{ \sum_k \lambda_k - C^{-1} \sum_k \lambda_k (x_k - m)(x_k - m)^T \right\} \tag{20}$$

where

$$\lambda_k = \frac{g_k}{g_k + t_s}, \quad g_k = \frac{1}{(\sqrt{2\pi})^n \sqrt{|C|}} \exp[-\tfrac{1}{2} d^2(x^k)].$$

Motivated by the gradient expressions (19) and (20), the following fast iterative scheme is used to solve each $\max_{m,C} q(m, C; t_s)$ for $m(t_s)$ and $C(t_s)$:

Step 1. Randomly choose $m^{(0)}$, then form $C^{(0)}$ by

$$C^{(0)} = \frac{1}{N} \sum_k (x^k - m^{(0)})(x^k - m^{(0)})^\text{T}$$

Step 2. Given the jth estimates $m^{(j)}$ and $C^{(j)}$ at the jth iterative step, $g_k^{(j)}$ and $\lambda_k^{(j)}$, $k = 1, \ldots, N$, are first calculated and the $(j + 1)$st estimates $m^{(j+1)}$ and $C^{(j+1)}$ are then calculated as the solution to the following weighted least squares problem:

$$\max_{m,C} \sum_k \lambda_k^{(j)} \log g_k \rightarrow m^{(j+1)}, C^{(j+1)} \qquad (21)$$

From Eqs. (19) and (20), $m^{(j+1)}$ and $C^{(j+1)}$ can be derived as

$$m^{(j+1)} = \frac{1}{\lambda^{(j)}} \sum_k \lambda_k^{(j)} x^k$$

$$C^{(j+1)} = \frac{1}{\lambda^{(j)}} \sum_k \lambda_k^{(j)} (x^k - m^{(j+1)})(x^k - m^{(j+1)})^\text{T} \qquad (22)$$

where $\lambda_k^{(j)} = g_k^{(j)}/(g_k^{(j)} + t_s)$ and $\lambda^{(j)} = \sum_k \lambda_k^{(j)}$.

3. DATA CLUSTERING ANALYSIS

An important application of the algorithm is to the problem of identifying clusters or classes in multidimensional datasets. Clustering analysis plays a central role in pattern recognition, particularly when only unlabeled training data samples are available (7,8). A cluster can be loosely defined as a set of samples whose density is larger than the density of the surrounding volume. Clustering methods attempt to partition a set of observations by grouping them into a number of statistical classes. The objective of clustering unlabeled data is to obtain an organization and description of the data that is both meaningful and compact. The general class of methods for extracting information from unlabeled samples are known as *unsupervised learning* algorithms. Once data samples are labeled using a clustering algorithm, the resulting classes can be used to design a pattern classifier for classifying unknown samples. Consequently, the effectiveness of clustering analysis is crucial to the performance of the classifier in identifying the class to which noisy observations belong. There are several major difficulties encountered in clustering analysis, which are also common to mixture data analysis:

- The characteristics and locations of clusters are usually not known a priori; thus, a parameterization of the clusters is needed.
- The number of clusters in the sample space is rarely known a priori.
- In real applications, well-defined clusters are atypical. Contamination of the data due to non-Gaussian noise and mixing with outliers makes the statistical identification and estimation of cluster parameters a difficult problem not only to solve but to formulate in a proper framework without using simplifying assumptions. However, Duda and Hart (7) convincingly argue that modeling observed data samples as being realizations from a Gaussian mixture distribution is an effective approximation for a variety of complex probability distributions.

A number of clustering algorithms have been developed and modified over the past several decades by researchers in the statistical and pattern-recognition fields as well as applied areas like medicine, pharmacology, remote sensing, astronomy, psychology, manufacturing, marketing, and finance (8). One of the most widely used clustering algorithms is the K-means algorithm, along with the variation known as ISODATA (6,8). Recently, the minimum volume ellipsoid (MVE) robust estimator was introduced in statistics (9). The MVE-estimator was extended in Ref. 10 to a clustering method known as the general minimum volume ellipsoid (GMVE) clustering algorithm and was successfully applied to several computer vision problems. The GMVE algorithm uses a random-sampling approach in order to avoid combinatorial intractability. The reliability of the initial guess for the GMVE algorithm is also extremely important. Due to these two limitations of the GMVE algorithm, results of experiments for clustering analysis are compared only with the performance of the K-means algorithm.

Clustering analysis in a general sense can be considered to be a special case of mixture analysis. As argued by Duda and Hart (7), a great variety of data clusters can be approximated by assuming that data samples are realized from a Gaussian mixture density. The cluster identification problem, however, does have certain additional requirements. For example, large clusters are preferred over many small ones, and clusters with a large degree of overlap should be merged into a single cluster. Such perceptual constraints on the geometry of clusters can usually be translated into an appropriate statistical significance level in a normality test.

The application of the MF-estimator algorithm to clustering analysis is straightforward. The output of the algorithm will be a number of clusters and possibly an *unassigned set* containing samples that do not belong to any of the detected clusters. Depending on the application, the unassigned set of observations can be discarded or each sample from this set can be assigned to a cluster based on a criterion such as the k nearest-neighbor rule. The application of the MF-estimator to other vision tasks can be found in Ref. 17, 18, 19, 20.

3.1. Clustering Analysis Using Simulated Data

Two-dimensional mixture data sets are used for computer simulation experiments in order to illustrate the results clearly. Three experiments using three different mixture models are performed, as discussed below.

3.1.1. Gaussian Mixture Densities

The first case compares the performance of the MF-estimator and K-means algorithm in reliably segmenting data generated from *noiseless* Gaussian mixture distributions with known parameters. The details of the Gaussian mixture density decomposition (GMDD) algorithm using the MF-estimator can be found in Ref. 5. Figures 1 and 2 show the clustering results using noiseless data generated from mixtures with three and six distinct Gaussian components, respectively. The K-means algorithm requires an a priori estimate of the number of clusters to be identified from the data. Even with the additional information on the correct number of mixture components and noiseless samples, the K-means algorithm is able to identify the clusters well only when the intercluster distance is large, as shown in Fig. 1. The three clusters making up classes B, C, and F are particularly poorly resolved by the K-means algorithm, as shown in Fig. 2. The MF-estimator is successful even when the clusters are not well separated, as shown in Fig. 2. The initializations used in the K-means algorithm were the true means in each case.

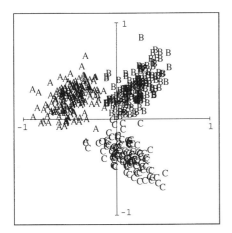 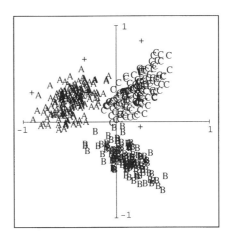

Fig. 1 Clustering by *K*-means (*left*) and GMDD algorithm (*right*): three clean Gaussian components.

3.1.2. Contaminated Gaussian Mixture Densities

The second case compares the performance of the MF-estimator and *K*-means algorithm in robustly grouping data generated from *noisy* Gaussian mixture distributions with known parameters. The contaminating noise is used to simulate the effects of outlier behavior. A very noisy background consisting of 200 data points uniformly distributed within the region defined by $[-1, 1] \times [-1, 1]$ is added to a pure Gaussian mixture with five components. The performance of both algorithms on the contaminated Gaussian mixture data can be compared in Fig. 3. The performance of the *K*-means algorithm has severely degenerated in that it tries to classify all of the corrupt data, whereas the MF-estimator is much more robust as it tries to ignore outliers. The data values that are unassigned to

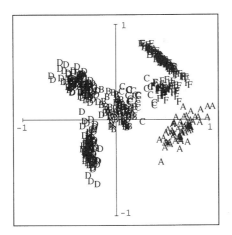 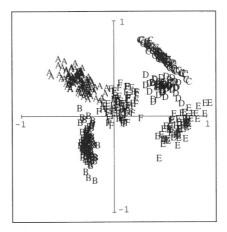

Fig. 2 Clustering by *K*-means (*left*) and GMDD algorithm (*right*): six clean Gaussian components.

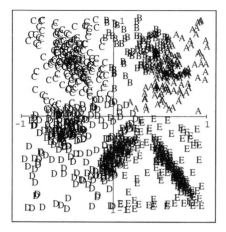 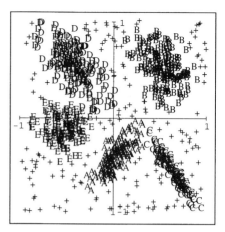

Fig. 3 Clustering by *K*-means (*left*) and GMDD algorithm (*right*): five clean Gaussian components in a uniform noise background.

any cluster are marked by the "+" symbol in the results for the MF-estimator clustering. The unassigned set matches very closely the set of outliers contaminating the mixture model. The correct number of mixture components and true means as initializations are provided to the *K*-means algorithm in this case, too.

3.1.3. Special Case

A special case is designed to examine the sensitivity of the MF-estimator to the Gaussian distribution assumption. A pair of heavily overlapping Gaussian-distributed clusters were constructed so that the two distributions form a single larger cluster; note that such a combined cluster does not quite follow a Gaussian distribution. The MF-estimator detects and estimates the combined cluster as a single cluster (see the cluster marked A in Fig. 4) at the appropriate (lower) significance level in the normality test for the detected component.

4. UNKNOWN PROBABILITY DENSITY ESTIMATION

In this section, we apply the MF-estimator to modeling and estimating an unknown probability density in the practical context of automated cervical smear cell classification.

Estimating an unknown density function given a set of observations is fundamentally important in pattern recognition (7). In practice, many pattern-recognition systems are built by acquiring knowledge from training samples. Such knowledge usually represents the overall tendency or statistics of the underlying process being studied, and, thus, many learning procedures involve estimating the probability density based on data samples. Probability density function estimation is not only an important theoretical problem but has tremendous practical applications. Currently there are only a few parametric or nonparametric techniques available. Moreover, most of the existing techniques are either computationally expensive or too simple to approximate the complex densities encountered in practice. As emphasized earlier, multimodal Gaussian mixture densities provide a realistic

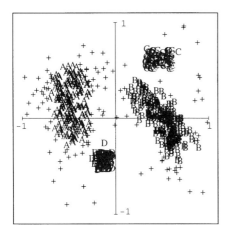

Fig. 4 The GMDD algorithm detects a combined cluster (marked by A) that actually does not quite follow a Gaussian distribution.

probability density structure for modeling or approximating an unknown density function (7,11,12). The goal of estimating an unknown probability density function is different from cluster analysis. In cluster analysis we usually want to identify distinct classes that can be used for classifier design. In probability density function estimation the composition of mixture components that is a good approximation of the time probability density is more important than the organization of components (probability densities) as reflected in the separability, size or scale of indivudual components.

An important biomedical application is automated cervical cancer slide prescreening which has been extensively studied for decades (13). The basic goal is to design a medical instrument or machine that automatically inspects patients' pap smear slides and detects and identifies the presence of abnormal or cancerous cells. The most crucial component in the system is the cell classifier algorithm, which is used to distinguish between three classes—artifacts, normal cells, and abnormal cells, denoted as C_0, C_1, and C_2, respectively. A widely accepted classification method is the Bayesian paradigm. The Bayesian method is a very flexible and powerful framework for this particular system, since it is able to integrate a variety of diagnostic information (14).

Let $\mathbf{F}$ denote an M-dimensional feature vector extracted from an object of interest in the slides. Our goal is to design a Bayesian cell classifier so as to minimize the misclassification error probability, namely,

$$\arg\max_{0 \leq j \leq 2} P(C_j|\mathbf{F}) = \arg\max_{0 \leq j \leq 2} P(C_j)P(\mathbf{F}|C_j) \tag{23}$$

where $P(C_j)$ represents the frequency or prior probability of observing cell type C_j, that is estimated in advance using clinical statistics. We can model each $P(\mathbf{F}|C_j)$, for $j = 0$, 1, 2, as a Gaussian mixture density and apply the MF-estimator to estimating the components using training samples collected and labeled by cytotechnologists from pap smear slides of a target patient group. Past experience indicates that the training sample set is contaminated with outlier noise. To account for these gross errors, the significance level of the K–S normality test is reasonably relaxed so that roughly approximating Gaus-

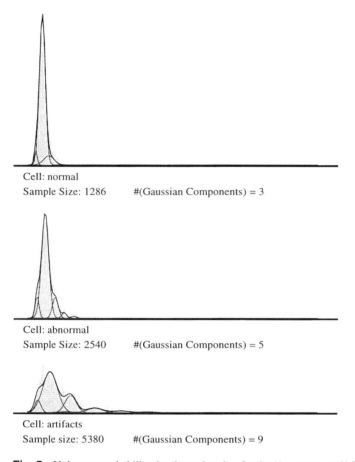

Fig. 5 Unknown probability density estimation for the "compactness" feature for three cell types.

sian densities will be accepted. For clarity, we show only the case of one-dimensional features. The actual Bayesian classifier implemented in the system uses four-dimensional or five-dimensional feature vectors. The computer experiments are arranged as follows. For each cell type C_j, $j = 0, 1, 2$, $P(\mathbf{F}|C_j)$, the class distribution of a specific feature is estimated. Figures 5 and 6 show two sets of experimental results for two different cell features: the "compactness" and the "integrated optical density." The definition of these features and the method to measure them are in Ref. 13 and 14. In the figures, each extracted Gaussian component is shown at separate densities and the overall estimate of each $P(\mathbf{F}|C_j)$ approximated by Gaussian mixture shown as the solid boundary curve. The training sample size and the detected number of Gaussian components in each case are also illustrated in Figs. 5 and 6. A comparison to the raw data histogram reveals that the MF-estimator estimates each mixture density very well, even though the statistics of the training sample sets varies considerably and is noisy.

5. CONCLUSION

Many tasks in computer vision and image understanding appear to be "mission impossible" problems. The lack of sufficient robustness in most existing computer vision algo-

Highly Robust Statistical Estimates

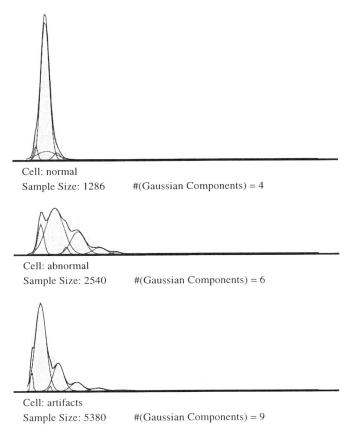

Fig. 6 Unknown probability density estimation for the "integrated optical density" feature for three cell types.

rithms is a key aspect of this challenge. In this chapter, it was argued that high robustness cannot be achieved by using a single density function. The MF-estimator based on the minimum-error Bayesian classification of inliers and outliers was shown to be a good robust statistical approach. The computer experiments demonstrated the effectiveness of the resulting partial modeling algorithm.

APPENDIX

A.1. Kolmogorov–Smirnov Normality Test

Suppose β is a random variable with observed samples β_k, $k = 1, \ldots, K$. Let $P(\overline{\beta})$ represent the known cumulative distribution function of β, namely

$$P(\overline{\beta}) = \text{Prob}(\beta \leq \overline{\beta})$$

An unbiased estimator $S_K(\overline{\beta})$ of $P(\overline{\beta})$ can be constructed based on the observed samples β_k, $k = 1, \ldots, K$, as follows:

$$S_K(\overline{\beta}) = \frac{1}{K}\#\{\beta_k: \beta_k \le \overline{\beta}\}$$

Different data sets provide different estimates of the cumulative distribution function. A number of statistics are available to measure the overall difference between a cumulative distribution function estimate and the known theoretical cumulative distribution function. The Kolmogorov–Smirnov (K–S) test statistic D is a particularly simple measure. The K–S statistic is defined as the maximum value of the absolute difference between the estimated and theoretical cumulative distributions. The K–S statistic for comparing a cumulative distribution function estimate $S_K(\overline{\beta})$ to the known cumulative distribution function $P(\overline{\beta})$ is

$$D = \max_{-\infty < \overline{\beta} < \infty} |S_K(\overline{\beta}) - P(\overline{\beta})|$$

What makes the K–S statistic attractive is that the distribution of D can be usefully approximated under the null hypothesis when the data sets are indeed drawn from the same distribution. The distribution of D can be used to assess the significance of any observed nonzero value of D.

The function that is needed for the calculation of the significance of any observed D is the limit of $\text{Prob}(\sqrt{K}D > \lambda)$ as $K \to \infty$ and can be written as

$$Q_{\text{K-S}}(\lambda) = \lim_{K \to \infty} \text{Prob}\{\sqrt{K}D > \lambda\}$$

$$= 2 \sum_{r=\infty}^{\infty} (-1)^r e^{-2r^2\lambda^2}$$

which is a monotonic function with the limiting values

$$Q_{\text{K-S}}(\infty) = 0, \quad Q_{\text{K-S}}(0) = 1$$

In terms of this function, the significance level of an observed value of D is approximately given by $Q_{\text{K-S}}(\sqrt{K}D)$. The approximation becomes asymptotically accurate as K becomes large. A good approximation to $Q_{\text{K-S}}(\sqrt{K}D)$ is given by the first term in the following series expansion:

$$Q_{\text{K-S}}(\sqrt{K}D) = 2 \sum_{r=1}^{\infty} (-1)^{r-1} e^{-2r^2 KD^2}$$

that is,

$$Q_{\text{K-S}}(\sqrt{K}D) \approx 2e^{-2KD^2}$$

due to the rapid convergence of the series. So, for a significance level of α, the K–S statistic $D \ge [(-1/2K \ln(\alpha/2)]^{1/2}$ can be used to reject the null hypothesis.

A.2. Gaussian-Distributed Cluster

Let G consist of K pattern vectors $x^1, \ldots, x^K$, each of which belongs to R_n. The question to be answered is, Are these K pattern vectors generated by a known Gaussian distribution $N(m, C)$? If the pattern vectors are x^k, then the squared Mahalanobis distances, $\beta_k = d^2(x^k)$, $k = 1, \ldots, K$, represent K observed samples from a chi-square distribution with n degrees of freedom. The theoretical cumulative distribution function $P(\overline{\beta})$ for the chi-square distribution with n degrees of freedom is calculated using the relation

$$\overline{\beta} = \chi^2_{n,P(\overline{\beta})}$$

The unbiased estimator $S_K(\overline{\beta})$ of $P(\overline{\beta})$ can be calculated by using the procedure in Sec. A.1. The K–S statistic can then be approximated by

$$D_K = \max_j \left| P(\overline{\beta}_j) - S_K(\overline{\beta}_j) \right|$$

As a result, the significance level of the null hypothesis that the data set G is generated by $N(m, C)$ is approximated by $Q_{K-S}(\sqrt{K}D_K)$.

The above testing procedure assumes that the mean vector m and the covariance matrix C are known. Quite often, the more practical question is whether the K pattern vectors were generated by a Gaussian distribution whose mean vector m and covariance matrix C each are unknown a priori. The unknown squared Mahalanobis distances $\overline{\beta}_k = d^2(x^k)$, $k = 1, \ldots, K$, which represent K samples from a chi-square distribution with n degrees of freedom need to be appropriately estimated first. The unbiased estimators of m and C are given by

$$m = \frac{1}{K} \sum_k x^k$$

$$C = \frac{1}{K - 1} \sum_k (x^k - m)(x^k - m)'$$

Using the above unbiased estimates for the mean vector and covariance matrix, each squared Mahalanobis distance $\overline{\beta}_k = d^2(x^k)$ can be estimated. Proceeding as before, the K–S statistic, D_K, and the significance level $Q_{K-S}(\sqrt{K}D)$ are each calculated appropriately. If the null hypothesis is supported using the above K–S test, then we would accept that the hypothesis that the set of observations G describes a valid Gaussian-distributed cluster.

REFERENCES

1. RM Haralick. Computer vision theory: The lack thereof. Computer Vision Graphics Image Process 36:372–386, 1986.
2. RM Haralick, H Joo, CN Lee, X Zhuang, VG Vaidya, MB Kim. Pose Estimation from Corresponding Point Data. IEEE Trans Syst Man Cybern, 19(6):1426–1446, 1989.
3. PJ Huber. Robust Statistics. New York: John Wiley & Sons, 1981.
4. X Zhuang, T Wang, P Zhang. A highly robust estimator through partially likelihood function modeling and its application in computer vision. IEEE Trans. Pattern Anal Machine Intell PAMI-14(1):19–35, 1992.
5. X Zhuang, Y Huang, K Palaniappan, Y Zhao. Gaussian mixture density modeling, decomposition and applications. IEEE Trans. Image Process IP-5(9):1293–1302, 1996.
6. JT Tou, RC Gonzalez. Pattern Recognition Principles. Reading, MA: Addison-Wesley, 1974.
7. RO Duda, PE Hart. Pattern Classification and Scene Analysis. New York: John Wiley & Sons, 1973.
8. AK Jain, RC Dubes. Algorithms for Clustering Data. Englewood Cliffs, NJ: Prentice-Hall, 1988.
9. PJ Rousseeuw, AM Leroy. Robust Regression and Outlier Detection. New York: John Wiley & Sons, 1987.

10. JM Jolion, P Meer, S Bataouche. Robust clustering with applications in computer vision. IEEE Trans Pattern Anal Machine Intell PAMI-13(8):791–802, 1991.
11. DM Titterington, AFM Smith, UE Markov. Statistical Analysis of Finite Mixture Distributions. New York: John Wiley & Sons, 1985.
12. K Fukunaga. Statistical Pattern Recognition, 2nd ed. New York: Academic Press, 1990.
13. JS Lee, W Bannister, L Kuan, P Bartels, A Nelson. A processing strategy for automated pap smear screening. Anal Quant Cytol Histol 14(5):415–425, 1992.
14. X Zhuang, JS Lee, Y Huang, A Nelson. Staining independent Bayes classifier for automated cell pattern recognition. Proc. SPIE Biomedical Image Processing and Biomedical Visualization, Ed. RS Acharya, DB Goldgof. Volume 1905, pp. 585–594, 1993.
15. BG Schunck. Robust computational vision. Proc. Int. Workshop on Robust Computer Vision. IEEE Computer Society, Seattle, Washington, 1–30 of 1990, pp. 1–18, 1990.
16. CV Stewart. Bias in robust estimation caused by discontinuities and multiple structures. IEEE Trans Pattern Anal Machine Intell, PAMI-19(8):818–833, 1997.
17. K Palaniappan, Y Huang, X Zhuang, AF Hasler, Robust stereo analysis. IEEE Int Symp Computer Vision, Coral Gables, Florida, pp. 175–181. Nov 19–21, 1995.
18. X Zhuang, Y Huang. Robust 3D-3D pose estimation. IEEE Trans Pattern Anal Machine Intell, PAMI-16(8):818–824, 1994.
19. X Zhuang, Y Zhao, TS Huang. Residual-based robust estimation and image-motion analysis. Int Journal Imaging Systems and Technology. 2:371–379, 1990.
20. Y Huang, K Palaniappan, X Zhuang, JE Cavanaugh. Optic flow field segmentation and motion estimation using a robust genetic partitioning algorithm. IEEE Trans Pattern Anal Machine Intell, PAMI-17(12):1177–1190, 1995.
21. P Meer, D Mintz, A Rosenfeld, DY Kim. Robust regression methods for computer vision: A review. Int J Computer Vision, 6(1):59–70, 1991.
22. KL Boyer, MJ Mizra, G Ganuly. The robust sequential estimator: A general approach and its application to surface organization in range data. IEEE Trans Pattern Anal Machine Intell, PAMI-16:987–1001, 1994.
23. T Darrell, A Pentland. Cooperative robust estimation using layers of support. IEEE Trans Pattern Anal Machine Intell, PAMI-17(5):474–487, 1995.

15
Learning in Computer Vision and Beyond: Development

Juyang Weng
Michigan State University, East Lansing, Michigan

1. MOTIVATIONS

Despite the power of modern computers, the principle of which was first introduced in 1936 by Alan Turing in his now celebrated paper (1), we have seen a paradoxical picture of artificial intelligence: Computers have done very well in some areas that are typically considered very difficult (by humans), such as playing chess games, but they have done poorly in other areas that are commonly considered easy (by humans), such as vision (2).

1.1. Challenges

It seems a relatively simpler task to write a program to solve a symbolic problem where the problem is well defined and the input is human-preprocessed symbolic data. However, it is very difficult for a computer to solve a problem that is not definable mathematically using the raw sensory data in their original form. For example, although there have been some limited applications in controlled settings (3), image understanding is extremely difficult, especially for tasks such as recognizing objects in a general setting. The recognition must cope with many variation factors, such as lighting, viewing angle, viewing distance, and object changes (e.g., facial expressions). The difficulties encountered in computer vision might not be surprising if one realizes that it is the most difficult sensing modality in humans. In fact, about a half of the cerebral cortex in the human brain is given over to visual information processing. As is well known, a large portion of human knowledge is acquired through vision.

    The speech-recognition field has some limited applications for recognizing words and short sentences in a controlled setting (4,5). However, further advances for general settings require understanding the situation, context, speaker's intention, speaker characteristics, language, and the meaning of what is said (6,7). In the natural language understanding field, we have seen various low-level applications in language processing from text inputs, such as spell checkers, grammar checkers, and keyword-based search (8), but these low-level applications do not require a true understanding of the text. It has been known that language understanding requires not only syntax but also semantics and ''commonsense'' knowledge. Several grandiose language knowledge bases are being developed, such as the commonsense knowledge base, CYC (9,10), and the lexical database, WordNet (11). However, it is an open question whether a machine can really understand anything in a pure text form without its own experience. Is linking from one string of letters to another true ''understanding''? Can the system use text properly in, for example, language translation? Language discourse and language translation are good tests for language understanding. However, tremendous difficulties persist in these areas. Like speech recognition, handwritten character recognition without understanding the meaning of the context cannot go very far.

Recently, several alternative methods have been actively studied, including various artificial neural networks and genetic algorithms. Many encouraging studies have demonstrated that these alternative methods can achieve impressive results that are difficult to achieve using traditional text-based or knowledge-based methods. More recently, behavior-based robotics research puts emphasis on embodiment, situatedness, and behavior generation through interactive learning (12), challenging the traditional disembodied methods. A more flexible learning mode, reinforcement learning, has also been actively studied for various problems (13), However, it is not clear how these alternative methods can be scaled up for dealing with the challenging tasks discussed above.

1.2. The Task-Specific Paradigm

The current major paradigm for developing a system, either intelligent or not, can be characterized by the following steps. (1) Start with a given task. (2) A human being tries his (or her) best to analyze the task. (3) The human derives a task space representation, which may depend on the tool chosen. (4) The human chooses a computational tool and maps the task space representation to the tool. (5) The parameters of the tool are determined using one or a combination of the following methods: (a) They are manually specified using hand-crafted domain knowledge (e.g., the knowledge-based methods in CYC (9,10) and WordNet (11)). (b) They are decomposed manually into system behavior modules (e.g., behavior-based methods in the subsumption architecture (14) and active vision (15)). (c) They are estimated using a training procedure (e.g., learning-based methods in Q-learning (16), eigenfaces (17), and SHOSLIF (18)). (d) They are searched based on a task-specific objective function (e.g., genetic or artificial-life methods in Animate (19), SAGA (20), and AutonoMouse (21)). It is typical to use a combination of the various methods in step 5 (e.g., AutonoMouse (21) used both learning and a genetic algorithm). In a typical research endeavor, steps 2–5 may be repeated many times, which might lead to a modification of the given task in step 1. We call it a task-specific paradigm because it starts with a task and all of the remaining steps depend on the task.

This task-specific paradigm has produced impressive results for those tasks whose space is relatively small, relatively clean (or exact), and relatively easy to model, such as chess or printed character recognition. However, it faces tremendous difficulties for tasks whose space is huge, vague, difficult to fully understand, and difficult to model adequately by hand. Two major restrictions are direct consequences from such a task-specific paradigm:

1. The task space and representation defined manually by humans cannot deal with the full complexity of the real world.
2. Low quantity and low quality of information fed into the system for training.

Due to the limitation of humans in modeling tasks, especially in uncontrolled real-world environments, the task space defined by humans are of a limited scope. Such a limited scope is the major reason for the high brittleness of the existing systems. They cannot extend their capabilities to deal with more complex cases without humans to re-model the task and its representation again.

The brittleness of the system is also attributed to the quantity and quality of the training data. In terms of quantity, the computers are allowed to observe far less environmental variation, context variation, and content domain variation than they really need for the tasks. It is difficult to conduct extensive system training due to the large amount of manual labor that is required in preparing the training data. In terms of quality, these compiled training data are very much *disconnected* from the environment from which the data arise. The rich meaning of the live sensory experience is degenerated into isolated segments, each being tied to a class label that is meaningless to the system. The lack of environmental context in manually fed training data makes it impossible for machines to learn beyond what is possible from the "spoon-fed" data.

Some other systems do not use learning. Instead, the required knowledge is directly programmed into the system. Unfortunately, what is modeled by such knowledge-based programming is typically insufficient for the challenging tasks at hand, due to human limitation in understanding and modeling the complex mechanisms of the required cognitive process. For example, recognition of human faces must cope with a wide variety of variation factors, such as lighting, viewing angle, viewing distance, and facial changes (e.g., expressions, hair styles, eye wear, etc.). A similar situation is true in speech recognition (e.g., variation in time warping, coarticulation, intonation, age, gender, etc.) and language understanding (ambiguity without context, ambiguity without understanding, cultural differences, language differences).

In summary, it is extremely difficult, if not impossible for human designers to adequately represent and model many factors in a challenging task, to design effective knowledge-level rules for them, to collect sufficient training data to cover the variations, and to keep the system up to date. A certain degree of automation is very desirable for these challenging problems.

Therefore, we need to automate not just only the learning phase but also the design of the task space (or problem space).

2. THE DEVELOPMENTAL APPROACH

How does a human being establish his or her cognitive capability? Studies in developmental psychology may shed light on this important question.

2.1. Human Cognitive Development

Jean Piaget (22–24), a well-known developmental psychologist, proposed dividing human cognitive development into four major stages, as summarized in Table 1. There is no doubt that these four stages have a lot to do with neural development in the brain. Furthermore, more recent studies have demonstrated that the progress into each stage depends very much on the learning experience of each individual and, thus, biological age is not an absolute measure for cognitive stages. For example, Bryant and Trabasso (25) showed

Table 1 Piaget's Four Stages in Human Cognitive Development

Stage	Rough ages	Characteristics
Sensorimotor	Birth to age 2	Not capable of symbolic representation
Preoperational	Age 2 to 6	Egocentric; unable to distinguish appearance from reality; incapable of certain types of logical inference
Concrete operational	Age 6 to 12	Capable of the logic of classification and linear ordering
Formal operational	Age 12 and beyond	Capable of formal, deductive, logic reasoning

that given enough drill with the premises, 3- and 4-year old children could do some tasks to construct linear orderings, a deviation from the stage partition proposed by Piaget.

However, it is known that the development of the cognitive capability of each human individual requires many years of learning during which he or she interacts with the environments and learns, while the brain develops to store, accumulate, enrich, generalize, and verify the knowledge learned. The biological learning algorithm that is determined at the birth of a human being enables the human individual to learn more and more tasks through his lifetime without a need of reprogramming.

2.2. AA-Learning

In order to bring out the stark contrast between the developmental learning by an animal and the current task-specific paradigm, we introduce some basic concepts. We introduce the concept of AA-learning (named after *a*utomated, *a*nimallike learning *without claiming to be complete*) for a machine agent.* AA-learning has the following characteristics: (1) domain extensibility; (2) closedness of the brain; (3) simultaneity of learning and performing; (4) integration of different learning types; (5) automatic level building; (6) real time while scaling up.

2.2.1. Domain Extensibility

Domain extensible means that the system is applicable to an open number of task domains and can learn new task domains without a need for reprogramming. A domain-expandable system, like a human, can continuously switch among deferent domains. For example, while reading a magazine, a human can switch among recognizing human faces, recognizing human genders, recognizing written characters, and recognizing other objects. For a domain-extensible system, the system designer cannot use a task-specific objective function (e.g., in training a robot leg hopper, the objective is fixed—keeping balance). He cannot define a specific task space for programming either. Existing learning systems that use a general-purpose learning method typically require the system designer to map a task into the program internal representation (e.g., to map all the possible situations of a task into a set of manually defined system states, such as in Robo-Soar (27)). It is worth noting

---

* An agent is something that perceives and acts (26).

that domain extensibility is not inconsistent with brain modularity. Each domain task may be realized by a particular set of modules in the brain.*

2.2.2. Closedness of the Brain

The internal states of the "brain"† cannot be directly set by external probes for teaching purposes, although they are accessible for research purpose. This is an important characteristic of automated animal learning. Otherwise, a mother must understand the internal brain representation of her daughter before she can teach her. The environment, including human teachers, can only affect the learner's brain through his or her sensors (showing examples, or giving appetitive or aversive stimulus) and effectors (forced actions). Many existing learning methods require humans to directly link internal representation to actual meaning of the content to be learned. The closed brain requirement relieved humans from manually performing this extremely difficult task.

2.2.3. Simultaneity of Learning and Performing

The learning phase is also the performance phase. The human trainers are not in the loop of collecting training data. They are a part of the environment with which the system *continuously* interacts. The system will learn from the real-world environment directly and continuously using its sensors and effectors, without requiring humans to serve as a feature detector or sensory-input-to-symbol converter. This is in contrast with symbolic methods (which works in a symbolic domain) and softbots (which work only in a simulation world). The real world excludes the simulated computer world, such as that used by Shen in his study of autonomous learning from simulated symbolic world (28).

2.2.4. Integration of Different Learning Types

Reinforcement learning (using reward and punishment) and supervised learning (guided actions) are combined in the AA-learning. Some basic behaviors that are typically innate in animals, such as moving forward and simple turning, are taught and memorized through interactive effector guidance (imposition). This allows the system to start with some basic simple behaviors which will facilitate further learning.

2.2.5. Automatic Level Building

Capabilities for learning stimulus–response association, reasoning, and prediction are automatically built up from a low level to higher levels. The system must automatically build a representation for concepts at different scales.‡ This is in contrast with existing single-level reinforcement learning methods such as Q-learning (16) and R-learning (25), manually specifying and building behavior layers in behavior-based approaches (12), and

---

* For example, a tree can be used to partition the input space into regions, each may be considered as a module. For example, human faces may be assigned to a set of such regions represented by a set of notes in a tree.
† For simplicity, we will drop the double quotes for the term "brain" with an understanding that an artificial brain is very different from a real biological brain.
‡ For example, due to the need for understanding visual language (e.g., American sign language) or spoken language (e.g., English, Chinese, or French), the system must be able to automatically learn and recognize words, phrases, and sentences from an environment where one or several languages are used in teaching, without the need to be explicitly programmed for any particular language.

human content-level design in level building for speech recognition (e.g., Refs. 30 and 31).

2.2.6. Real Time While Scaling Up

The system must learn while performing in real time, because it must observe the consequence of its actions as they happen so that it can learn from the events. During the entire developmental life of the system, this real-time requirement must be satisfied while the number of cases the system has learned increases without bound. This is a highly challenging requirement for high-dimensional sensory inputs, such as visual input, and for large-size tasks.

2.3. Living Machines

A machine agent M may have several sensors. At the time of "birth," its sensors fall into one of the two categories: biased and unbiased.* If the agent has a predefined (innate) preference for the signal from a sensor, this sensor is then called biased. Otherwise, it is an unbiased sensor, although the preference can be developed by the agent later through learning. For example, a human being has an innate preference to sweet and bitter tastes from the taste sensor, but does not have a strong preference to visual images of various furniture items. By definition, an *extroceptive* sensor is one that senses external environment (e.g., visual), a *proprioceptive* sensor senses relative position of internal control (e.g., arm position), and an *interoceptive* sensor is one that senses internal events (e.g., internal clock).

We need to accommodate a class of synthetic sensors that typically are not called sensors, such as keyboard input, numerical input from a graphical user interface, and so on. For effectors, the system may also output text or simply numerical numbers. Therefore, it is convenient to define *N-sensors* and *N-effectors*, where N stands for "numerical." The input from an N-sensor at time t is a vector of a dimensionality defined for the sensor. The output to the N-effector is also a vector of certain dimensionality. The meaning of the input from an N-sensor and that of the output to an N-effector is predefined by the sensor or effector. Now, we are ready to give living machine a more precise definition.

Definition 1. A machine agent M conducts AA-learning at discrete time instances if after its "birth," the following conditions are met for all the time instances $t = 0, 1, 2, \ldots$. (1) M has a number of sensors (biased or unbiased, extroceptive, proprioceptive, or interoceptive), whose signal at time t is collectively denoted by $x(t)$. (2) M has a number of effectors, whose control signal at time t is collectively denoted by $a(t)$. The effectors include extro-effectors (those acting on the external world) and intero-effectors (those acting on internal mechanism, e.g., attention). (3) M has a "brain" denoted by $b(t)$ at time t. (4) At each time t, the state-update function f_t updates the "brain" based on sensory input $x(t)$ and the current "brain" $b(t)$,

$$b(t + 1) = f_t(x(t), b(t)) \tag{1}$$

and the action-generation function g_t generates the effector control signal based on the updated "brain" $b(t + 1)$,

---

* This is an engineering definition. For a biological organism, it is hard to say that any of its sensors is absolutely unbiased.

$$a(t + 1) = g_t(b(t + 1)) \tag{2}$$

where $a(t + 1)$ can be a part of the next sensory input $x(t + 1)$. (5) The "brain" of M is closed, in that after the birth (the first operation), $b(t)$ cannot be altered directly by human teachers for teaching purposes. It can only be updated according to Eq. (1).

A machine agent is called a *living machine* if it can perform AA-learning continuously for a very long time to learn to perform various tasks.

For AA-learning, the learner has three types of channels with its environment: sensors, effectors, and reward receiver, as shown in Fig. 1. The reward receivers can be modeled by biased sensors when the agent is "young." They will be enriched by unbiased sensors after the agent has developed its preference pattern for unbiased sensors.

After its birth, a living machine learns autonomously from the environment by sensing the environment through its sensors and acting on the environment through its effectors. The teacher is a part of the environment. The environment can enforce an action of the effectors on the learner. Human teachers, as a part of the system's environment, affect how the system learns. For example, the human teachers will show different object examples, verbally state the characteristics of the object, and then ask questions immediately about the characteristics of the object. He may encourage the system to act properly using different rewards at the appropriate time. He can also directly impose the desired action to execute by imposing control values on the corresponding effector. Such an imposition of action and delivery of rewards occur also in human learning; for example, manipulating a child's hand to hold a pen when teaching a child how to use a pen. Rewards to a human child can be food, a good test score, etc.

We call this type of general-purpose AA-learning machines "living machines" because they "live" in the human environment and autonomously interact with the environment (including humans) on a daily basis. The emphasis of the term is not "life" but, rather, the daily autonomous learning activities that are associated with a biological living thing, especially humans, such as playing, communicating with humans, and learning to perform tasks. Such machines are fundamentally different from a nonliving regular machine, such as an automobile or a computer, because they do not operate autonomously.

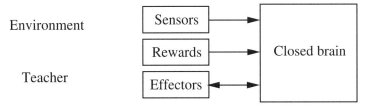

Fig. 1 A living machine has three channels to interact with the environment: sensors, effectors, and reward receiver. The single-headed arrows mean that a human teacher or the environment have access to the corresponding component. The double-headed arrow for the effectors means that actions imposed by humans or the environments on some effectors can be sensed by the brain (through *proprioceptive* sensors).

2.4. Why Living Machines?

Can the major problems for computer vision be solved by dealing with the visual modality only? Can the current task-specific paradigm lead us to solve the challenging problems in artificial intelligence? Why living machines? This chapter will not be able to discuss these issues fully. Here, we examine some major points.

2.4.1. Each Modality Must Be Learned

The cognitive knowledge that is required to communicate with humans in single or multiple modalities in a general setting is too vast in amount and too complicated in nature to be manually modeled adequately and manually spoon-fed sufficiently into a program. Probably few will question the fact that language is learned. Therefore, vision, the modality that many human individuals do not feel requires much learning (i.e., learned subconsciously), is appropriate to demonstrate the importance of learning. How complete is a child's vision system when he or she is born? In fact, as early as the late 19th century, German psychiatrist Paul Emil Flechsig had shown that certain regions of the brain, among them V1, have a mature appearance at birth, whereas other cortical areas, including V2, V3, V4, and V5 regions, continue to develop as though their maturation depended on the acquisition of experience (32). Many studies have been done since then. For example, a cat (kitten) that has only seen vertical lines ever since its birth cannot see horizontal lines (33). It is known that learning plays a central role in the development of human versatile visual capabilities and it takes place over a long period (e.g., Refs. 24 and 34–36). Human vision appears to be more a process of learning and recalling than one that relies on understanding of the physical processes of image formation and object modeling (e.g., the ''Thatcher's illusion'' (37) and the overhead-light-source assumption in shape from shading (38)). Neurologist Oliver Sacks' report (39) indicated that a biologically healthy, adult human vision system that has not learned cannot function, as we take for granted.

Furthermore, recognition by humans takes into account information sources that are not confined to vision. Sinha and Poggio's Clinton–Gore example (40) indicates that humans integrate different sensing modalities and contextual information for face recognition.* With humans, visual learning takes place while the recognizer is *continuously* sensing the visual world around it and interacting with the environment through human actions. The biological brain is so much determined by learning that the normal biological visual cortex is reassigned to tactile sensory functions in the case of the blind.

Many researchers in the field have realized that the visual knowledge required by the human-level performance is certainly too vast and too complex to be adequately modeled by hand. Letting a machine learn autonomously by itself is probably the only way to meet the challenges in vision, speech, and language. In other words, we should move from ''manual labor'' to ''automation.'' Intuitively, manual modeling is hard and costly, whereae automation is more effective in productivity and less costly than human labor. Furthermore, it is extremely difficult, if not impossible, to build an adult human brain that has learned. It appears more reasonable to build a machine ''infant brain'' that can simulate, to some degree, the brain's learning after birth.

---

* The Clinton–Gore example shows that you recognize these two well-known individuals in a picture even when their facial areas are made to be identical to each other.

2.4.2. The Machine Must Sense and Act

The question is then one of how to automate this learning process. In the field of traditional artificial intelligence (AI), the main emphasis of the establishment has been symbolic problem solving (see Minsky's annotated bibliography (41)). Later, due to the need of commonsense knowledge in reasoning systems, some grandiose projects have been launched to manually feed reasoning rules with symbolic, commonsense knowledge via computer keyboards (e.g., the CYC project (9)). The hope is that these rules and commonsense knowledge are complete enough to derive all the needed knowledge. Learning in traditional AI is conducted at a symbolic level, even when autonomy is a goal (e.g., the autonomous learning with the LIVE system (28)). The machine does not have its own sensors and effectors. This has three fundamental problems:

1. The machine cannot deal with all the knowledge that is directly related to sensing and action, such as how to recognize a scene and move around it using vision.
2. Because a large amount of human symbolic knowledge, both low level and high level, is rooted deeply in sensing and action, a sensor-free machine can neither really understand nor properly use it if it is input manually. For example, even for seemingly symbolic problems such as language translation, no reasonable translation is possible without understanding the meaning of what is said (a lot of which is about sensing and action), except for simple cases.
3. The cost and time requirement is extremely high for humans to figure out what knowledge is required for a challenging problem, to build models for content-level knowledge, to input all the knowledge required, and to keep them up to date. It is even more uncertain how to make sure that all this effort, cost, and time will be sufficient to perform the tasks at the end of budgeted period.

Brooks emphasized the need of embodiment for developing an intelligent machine (12). In his view, an intelligent machine must have a body to be situated in the world to sense and act. He advocated that intelligence emerges from the robot's interaction with the world and from sometimes indirect interactions between its components (12). A recent book by Hendricks-Jansen provides perspectives for embodiment and situatedness from psychology, ethology, philosophy, and artificial intelligence (42). Although embodiment, situatedness, sensing, and action have been common practices in robotics and vision communities for many years (43,44), Brooks' work made more researchers in the related fields aware of the importance of embodiment and situatedness. In fact, it is increasingly clear that the emphasis on these important points is useful not only for symbolic AI but also for other fields related to machine intelligence and natural intelligence.

2.4.3. Behaviors Must Also Be Learned

Hand-crafting knowledge-level rules is about modeling the world. If this is not desirable for challenging problems, how about modeling the system itself? Is it more tractable? Unfortunately, it does not seem so.

Aloimonos (15) and others advocated that modeling three-dimensional (3-D) scenes is not always necessary. Each vision problem should be investigated according to its purpose. Brooks (12) attributed the difficulties in vision and mobile robots to the so-called SMPA (*sense–model–plan–act*) framework which started in the late 1960s (see Nilsson's account (45) for a collection of the original reports). However, the behavior-based methods of Aloimonos (15) and Brooks (12) require the human programmer to hand-craft system behaviors. More recent work on robot shaping (21) also requires humans to manually specify behavior modules inside the system. In fact, the pattern-recognition community and the machine learning community have had a long history of recognizing patterns without fully modeling it. Features have been used instead to classify patterns (e.g., Refs. 46–51). The basic difference between a pattern-recognition problem and a typical computer vision problem is that the former has a controlled domain with a limited number of classes and the latter faces a much less controlled environment.

The fundamental problem of the existing SMPA approaches is the *practice of hand-crafting knowledge-level rules for one of two entities: the world and the system.* Avoiding modeling a 3-D surface or abandoning the SMPA framework is not enough. For open-ended applications with unpredictable inputs, hand-crafted behaviors embedded into a programming representation must also be abandoned. No hand-crafted behavior seems to be generally applicable to the open world. Faced with tremendous difficulties in hand-crafting a large number of behaviors, it is not clear how a behavior-based system can meet the challenges in vision, speech, language, and so forth, where the sensing dimensionality and the complexity of the understanding task are much higher than that of sonar sensors (52,53). As we explained in Sec. 2.4.1, not only are hand-crafted knowledge models questionable, but the approach to hand-crafting system behaviors also has a similar problem.

One may say that the human brain is modularized. However, we should not confuse modules grouped according to sensor and effector anatomy with modules grouped according to behaviors. The former concerns the mechanism that guides learning and the latter concerns the content of what is learned. In an artificial system, behavior modules can be represented by branches of the state index tree, state clusters, and action clusters. The decomposition of these behaviors is too complex to be handled manually. The highly complex nature of the information stored in a mature human brain is termed a *society of mind* by Minsky (54). This high complexity is the story of ''what''—what in a mature brain. However, the developmental learning addresses the issue of ''how'' (the mechanism of that guides the developmental learning), which should be more systematic and more fundamental than the story of ''what.''

2.4.4. The Machine Must Learn Autonomously

Totally spoon-fed learning has a fundamental problem of scalability because (1) a large amount of knowledge and behavior must be learned, (2) the system must experience an astronomical number of instances, (3) the system must learn continuously while performing (no human being can practically handle this type of spoon-feeding work on a daily basis), and (4) high-level decisions are based on so much contextual information that only the machine itself can handle it (automatically).

During autonomous learning, a human teacher serves very much like a babysitter (robot-sitter in this case), sending occasional feedback signals, depending on how the living machine is doing. Later, once the robot has learned basic communication skills through normal communication channels (such as speech and visual gesture), the robot-sitter is replaced by school teachers. It appears that only the relatively low cost associated with autonomous learning is practical for meeting the requirement of many challenging problems.

2.4.5. The Machine Must Perform Multimodal Learning in Order to Understand

Studies on humans who are born blind and deaf have demonstrated tremendous difficulties in learning very basic knowledge (55,56). Learning basic skills become virtually impossible with those few who are born blind, deaf, and without arms and legs. For example, a system that cannot see cannot really understand concepts related to vision (e.g., pictures and video, film, color, mirror, etc.) and those concepts that are understood mainly from visual sensing (e.g., trees, mountains, birds, streets, signs, facial expressions, etc.). A system that cannot hear is not able to really understand sentences related to speech and sound (e.g., the sound of music instruments, bird chirps, characteristics of a person's voice). The proverb "a picture is worth a thousand words" vividly points out the deficiency of text (words) in describing information that is best conveyed visually.

Furthermore, understanding any single sensing modality, including vision, speech, language, and text, requires knowledge about other sensing modalities too. A system that does not live and interact with humans cannot really understand the concepts related to human emotions, characteristics, and relationships (e.g., angry, happy, sympathy, care, cruel, friends, colleagues, enemy, spies, etc.). In fact, a system that cannot see, cannot hear, and cannot touch is deprived of the three most important sensing modalities through which a human acquires knowledge. Therefore, such a system has a fundamental limit in understanding any human knowledge and in using such a knowledge even if it is manually fed. Sensor-free systems like CYC have met tremendous difficulties toward applications that require understanding (such as language translation) and they are also very difficult to use due to the lack of any sensing modality for retrieval. Lack of multimodal sensing and action is a major reason why existing knowledge-base systems do not really understand the knowledge they store.

2.5. Comparison with Major Approaches

Existing approaches to artificial intelligence fall into the following four categories.

The *knowledge-based approaches* typically require a predefined task space or world space. Human programmers manually model knowledge and spoon-feed knowledge (MMKSK). Researchers in each subfield have been manually developing knowledge-level theories and methods and using them to write programs or build hardware. Then, they manually "spoon-feed" knowledge into the systems at the programming level (e.g., CART (57), CYC (9,10), WordNet (11)). Such an approach may produce a system that appears to produce some intelligent results. However, the limitation of such systems have been recognized (12,58). Such a methodology requires a huge amount of human labor and it faces a fundamental limit of humans to fully model and to specify the cognitive process required by challenging robotic tasks.

The *behavior-based approaches* avoid modeling the world; instead, they model robot behavior. The subsumption architecture was proposed by Brooks to allow a more sophisticated behavior layer to be added to each existing primitive behavior layers (53). Each layer is a finite-state machine, with states defined and named by the programmer. The programmer is also responsible for programming the state machine in each layer for a desired behavior. Thus, this approach can be characterized by the terms "manually modeling behavior and hand-coding behavior." Aloimonos (15) and others also advocated behavior-based approaches with active vision.

The *evolutionary approaches* are motivated by the evolution of biological species. The law of survival of the fittest is used to select advantageous genotypes which code the structure and/or behavior of simulated robots (20,59). So far, the selection processes have been simulated mostly by computers using a simulated environment, due to the obvious high cost in carrying on the evolutionary process with a large number of robots and performing a long-time physical evolution (21). The simulation method is attractive due to the low-cost benefit. Also, the approach does not require the programmer to code knowledge or behavior rules. However, evolutionary approaches leave the hard task of intelligent system design to the process of random trials and environment selection, an extremely slow and costly process. Three major issues stand out: (1) The chromosome representation for a sophisticated system. The more sophisticated the system, the more sophisticated the chromosome. (2) The extremely high cost of real-robot evolution when high-dimensional perception and sophisticated behaviors are required, such as vision, speech, and language. Simulation is not sufficient for these challenging functionalities. Each system must experience the real world. (3) The time, required to find a good chromosome, is on the order of a large number of generations. So far, genetic algorithms are typically used for a simple environment (e.g., symbolic) and simple behaviors (e.g., symbolic) with carefully designed environment-specific and behavior-specific chromosome representation (see, e.g., Animate (19) and AutonoMouse (21)).

The *learning-based approaches* include all the existing learning methods, such as supervised learning and reinforcement learning. Learning approaches are typically more efficient than the corresponding evolutionary approaches because the learning mechanism of the system is hand-coded by the programmer with the former but is either absent or has to emerge from the trial-and-selection process with the latter. For perception of high-dimensional inputs, learning seems the only viable method. Various learning methods have produced impressive results for challenging cognition tasks involving complex modalities, such as visual recognition (e.g., Refs. 17, 60, and 61), speech recognition (e.g., using HMM (30,31)), vision-guided robot manipulation (e.g., Ref. 62) and vision-guided navigation (e.g., Ref. 18). Supervised learning is typically more efficient than reinforcement learning. However, learning-based methods have not yet produced systems that truly understand anything. The major reasons include the following: (1) All the existing learning methods can only be used for a specific task at a time. For example, a neural network is trained for mapping from every image in a set of face images to a name label. In reinforcement learning using, e.g., the Q-learning algorithm, a task space must be given first. Then, the human designer must translate the task state to the internal representation (e.g., states) of the system model (e.g., Q-learning model). However, a system for a particular task only cannot truly understand anything. (2) A huge amount of manual labor is required in translating a specific task to a learning tool. (3) A huge amount of manual labor is required in training a system, including collecting data and testing the system. These reasons hinder further scaling up to more general, larger-size tasks.

In actuality, a particular system may use a combination of several approaches. For example, Robot-Soar (27) combines a world-knowledge-based approach and a learning approach. It requires the human to feed knowledge about the environment and to define the task space. Then, the system learns to perform predefined tasks. The learning classifier system (21) uses a combination of reinforcement learning and genetic algorithm.

Table 2 summarizes the four existing approaches and the new developmental approach. Among the five approaches in the table, the developmental approach appears to require the least amount of human labor in terms of system design; however, it requires

Table 2 Comparison of Approaches

Approach	Species architecture	World knowledge	System behavior	Task-specific
Knowledge based	Programming	Manual modeling	Manual modeling	Yes
Behavior based	Programming	Avoid modeling	Manual modeling	Yes
Learning based	Programming	Treatment varies	Special-purpose learning	Yes
Evolutionary[a]	Genetic search	Treatment varies	Genetic search	Yes
Developmental	Programming	Avoid modeling	General-purpose learning	No

[a] The engineering definition is used.

a large number of population individuals and a huge amount of genetic search time, which makes physical evolution for complex systems impractical. On the other hand, the knowledge-based, behavior-based, and conventional learning-based approaches all require extensive human labor in task-specific or behavior-specific design, which makes a general-purpose system impractical. The developmental approach seems to be in the middle between the human designer's effort and the system developmental cost. It requires human designers to properly design a general-purpose learning mechanism, but they do not need to explicitly program for the content of what is to be learned—neither for the world knowledge nor for the system behavior. It requires only one or a few physical systems to be built to learn. These physical systems take advantage of the richness of intelligence in the human environment. For example, suppose that in reinforcement learning in a computer simulation environment, a punishment is given when a sequence of system actions eventually leads to a failure. The system does not know which action in the action sequence is wrong. In the developmental approach, however, the human teacher can analyze the observed system action sequence and can tell the system which action is wrong by (a) first bringing the system to the right context (e.g., lead it to the location of its bad action) and then (b) giving it a punishment. Such a very powerful tell-you-what-it-is-for mechanism can speed up learning tremendously. This type of information-rich learning environment is more powerful than the time-discounted average reward model in Q-learning and the time-average reward model in R-learning (63) in dealing with ubiquitous delayed-reward situations in learning.

3. WORK TOWARD LIVING MACHINES

The task of developing living machines consists of two integral aspects. First, develop the physical system, including theory, algorithm, hardware, and software, for autonomous multimodal learning. In some sense, our goal is to build a machine counterpart of an animal newborn, although an exact duplicate is neither possible nor necessary. Second, teach the living machines to do things. In a sense, human beings try to "raise" and teach the machine "babies" properly so that every one of them will become successful in the task field assigned to each.

Although the first aspect is very fundamental, the success depends also very much on the second—teaching. The cognitive development of a human individual relies not only on the learning mechanism in that individual but also how he is taught. It is expected that the methodology for constructing a living machine and the methodology for teaching a living machine will both improve through close interactions between the two aspects.

3.1. Three Phases

The success of this approach requires the multidisciplinary collaboration of academic researchers and the related industries. Therefore, it is beneficial to outline a rough sketch in terms of what we did and where we are heading. Our work at Michigan State University (MSU) has followed a three-phase plan: Phase 1: comprehensive learning; Phase 2: AA-learning; Phase 3: daily living. This plan needs to be refined as we proceed.

3.1.1. Phase 1: Comprehensive Learning

In Phase 1, the task is to develop a framework for basic brain functionalities such as memory store, automatic feature derivation,* self-organization, and fast associative recall. The major goal is *generality* and *scalability*. The generality means that the framework must be applicable to various domains of sensor–effector tasks. The scalability means that it must have a very low time complexity† to allow real-time learning and performance (i.e., scalable to the number of learned cases). A wide variety of sensing and action tasks must be performed to verify the generality and scalability. However, learning at this phase is "spoon-fed," meaning that the learning process is not autonomous.

A lot of results in the fields of pattern recognition, computer vision, and machine learning have contributed to the idea of comprehensive learning. At MSU, the comprehensive learning is represented by the work around what is called SHOSLIF to be discussed in Sec. 4. Phase 1 has been completed.

3.1.2. Phase 2: AA-Learning

This phase is to study the theory and the methodology for AA-learning and to build one or more prototypes of living machines. Experiments using simulation may be conducted before using a real machine. In a virtual-time simulation experiment, the system behaviors can be fully measured in exact time steps, but it is not a replacement for real machines.

In the real machine tests, the living machines are trained to perform certain tasks autonomously, such as moving around, grabbing things, saying simple words, and responding to spoken words, all in a fully autonomous mode and in unrestricted general settings. The tasks used in the training and testing are those that correspond to the *sensorimotor* stage of a human child (from birth to age 2), as described by Jean Piaget (22–24). However, machine development does not necessarily duplicate exactly human cognitive development stages. During the development, the theory and methodology for developing living machines will be modified, depending on how well our living machines can learn and develop.

---

* We do not use the term feature selection here because it means to select from several predetermined feature types, such as edge or area. The term feature extraction has been used for computation of selected feature type from a given image. Feature derivation means automatic derivation of the actual features (e.g., eigenfeatures) to be used based on learned samples.
† For the living machines, the time complexity is logarithmic in the number of cases learned. Thus, the exact term should be "logarithmic scalability"—logarithmic to the scale of the task.

Table 3 Benchmarks of Phase 2 for the Living Machines in General Settings

Task group	Benchmark
Visual recognition	Say hello with correct names of 5 human teachers when they enter the scene. Say the name of 5 toys when being asked.
Speech recognition	Understand sentences: Come. Call me. Hello! Goodbye! Wave your hand. Say hello to . . . . Say goodbye. What's it? Pick this. Put down. Pour water into this. Yes. No. Follow me. Watch this. Stop. Go home. I'm home. I got lost.
Speech synthesis	Respond using sentences: Hello! Goodbye! Yes. No. I'm home. I got lost. Call the names of 5 toys and 5 teachers.
Navigation	Autonomous indoor navigation without running into anything. Outdoor navigation following campus walkways and crossing streets. Follow a teacher to go, via an elevator, from a lab on the third floor of a building to the parking lot outside the building. Return from the parking lot to the lab alone.
Hand action	Correctly pick 1 of the 5 toys. Put a toy down. Wave its hand when saying hello or goodbye. Place one toy on top of another. Pour a cup of water into another cup.

Table 3 lists five integrated task groups which can be the target benchmarks for Phase 2 prototypes. These tasks to be performed by the living machines must be taught and tested in the AA-learning mode. If the benchmark test is successful, this will be a case where a machine really understands something.* By the time it has passed the test, the living machine actually has learned much more because the setting is unrestricted—much more has been seen, much more has been heard, much more has been tried, and much more has been learned. Section 5 discusses the first prototype of the living machine called SAIL being developed at MSU.

It is expected that the progress of this phase will be accelerated when more and more research groups from related disciplines collaborate on this direction of developmental approach. Right now, it is difficult to predict when Phase 2 will be complete and when the next phase begins.

3.1.3. Phase 3: Daily Living

This phase will start when AA-learning algorithms are well developed and the hardware prototypes are operational and reliable. Starting from this phase, the living machine enters what is similar to Piaget's preoperational stage (age 2 to 6). Again, exact duplication of human developmental stages are not necessary and are impossible. Significant improvements of the living machine algorithm and software will continue throughout this phase, similar to the way computer operating systems are improved and upgraded now. Computers move to new levels of storage and speed while their cost continues to fall. The new demands from living machines will stimulate the robotics industry to produce new generations of lightweight, reliable, highly integrated hardware platforms for living machines.

A new emphasis in this phase is to investigate how to teach the living machines to learn things that are taught in human preschools. Because a machine computes fast and

---

* A link from one text string to another, or a mapping from a camera input to a label is probably not understanding, as the machine does not understand the text or label.

is never tired of learning, potentially it can learn faster than a human child, at least in some subjects. At this stage, communications between human teachers and the living machines become mostly visual or vocal, whichever is more convenient. Breakthroughs in vision, speech recognition, speech synthesis, language understanding, robotics, intelligent control, and artificial intelligence are simultaneous at this stage. The benchmark to measure success is a standard entrance test for human preschoolers.

At this time, a new industry will appear. Living machines will be manufactured and delivered to research institutions as experimental machines, to federal agencies for special tasks, to schools as educational material, to amusement parks as interactive attractions, to the media industry as machine personalities, and to homes for those who are physically challenged or just need a friend. The early popular version among all types of living machines is a low-cost version, a software program that can be loaded onto a personal multimedia computer that has a video camera, a microphone, and a speaker. At the end of this phase, the bright future of living machines will be well known to general public.

3.2. Brain Size and Speed of the Living Machine

A question is naturally raised here: How much memory space does the living machine need? How fast can it recall from a large brain? These two important questions cannot be clearly discussed until the methods are outlined. See Sec. 7.2 for a discussion about the brain size and Sec. 7.3 for the speed issue. With the logarithmic time complexity of the living machine and the steady advance of computer storage technology, we predict that, before very long, real-time living machines may have a storage size comparable to that of the human brain at a reasonable cost, although in many applications, we probably do not need as much space as the human brain.

The future of the living machines will be discussed in Sec. 8. Next, we discuss our work in Phase 1, represented by SHOSLIF.

4. SHOSLIF

In Phase 1, we developed the Self-organizing Hierarchical Optimal Subspace Learning and Inference Framework (SHOSLIF) (64–68). SHOSLIF by itself is not domain-extensible. It is meant to be tested on challenging tasks in several domains individually. SHOSLIF is the predecessor of SAIL (to be discussed in Sec. 5).

4.1. Motivations

First, we discuss the motivations behind the work of SHOSLIF.

4.1.1. Comprehensive Learning

The major new concept introduced by the author during Phase 1 was the concept of *comprehensive learning* which the author first presented in an NSF/ARPA sponsored workshop in 1994 (64,66). As illustrated in Fig. 2, the *comprehensive learning* concept consists of two basic ideas:

1. Learning must comprehensively cover the sensed world (visual, auditory, etc.).

Learning in Computer Vision and Beyond

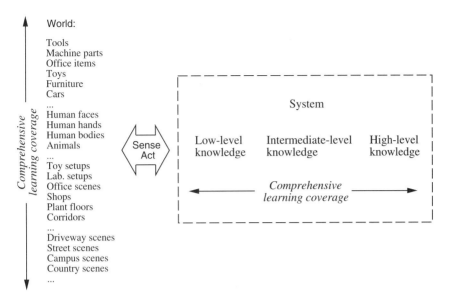

Fig. 2 The concept of comprehensive learning implies that learning must comprehensively cover (1) the sensed world and (2) the understanding system (or called agent). This implies that one needs to avoid hand-crafted content-level rules for the world and the behavior rules for the system.

2. Learning must comprehensively cover the entire understanding system (visual, auditory, etc.).

This concept was motivated from what was discussed in Sec. 2.4. Let us briefly explain its meaning.

The first comprehensive coverage implies that we do not manually model the world. In computer vision, for example, this means that we do not model object shape. Unless we have a very much controlled world environment, no manually built knowledge-level model is general enough for AA-learning. The learning method should not assume which type of scene condition is acceptable by the system. The method must be able to learn from any scene sensed by the sensor, because there exists no automatic condition checker which can tell, given an arbitrary situation, whether the scene condition is acceptable to the method. A system is not able to operate autonomously in the real world if it assumes conditions that it cannot verify by itself. Of course, learning does not mean perfect. The performance of a system is restricted by the sensors and actuators with which it is equipped.

The second comprehensive coverage implies that we should avoid hand-crafting system behaviors. In other words, hand-crafted knowledge-level rules (such as shape from shading rules, shape from contour rules, edge linking rules, collision-avoidance rules, planning rules, reasoning rules, etc.) should be avoided for the programming level as much as possible. There are simply too many behaviors to be hand-crafted effectively by humans. The astronomical number of possible combinations and coordinations of these behaviors further make hand-crafting impractical. Hand-crafted behavior rules result in brittle systems in the open-ended real-world environment. Even if it is necessary that some low-level behaviors are fed into the system as ''innate'' behaviors, they should be well integrated into the learning mechanism of the entire system.

4.1.2. Generality and Scalability

The generality (i.e., the real-world applicability) results from the comprehensive learning requirement, as stated above. The remaining issue is the efficiency. The scalability means that the method must work in real time even when the system has learned a huge number of cases. However, it is difficult to achieve both generality and scalability. A general method does not use preimposed special-purpose constraints and thus tends to be less efficient.

In Phase 1, we applied SHOSLIF to a variety of tasks to test its generality and we used a large number of cases for each task to test its scalability.

4.2. Technical Methods

SHOSLIF is designed for studying the conflicting goals of generality and scalability.

4.2.1. The Core–Shell Structure

A level of SHOSLIF consists of a core and a shell, as shown in Fig. 3. The core is task-independent. It has a network N as its memory, a leaner L for memory updating, and retriever R for memory recall. The core serves the basic function of memory storage, recall, and inference. The shell is task-dependent. It is an interface between the generic core and the actual sensors and effectors. For a particular task with a particular set of sensors and effectors, a shell needs to be designed which converts input data into a vector in space S, to be dealt with by the core. The output of the core is fed into the corresponding effectors by the shell. Mathematically, the SHOSLIF core approximates a high-dimensional function $f: S \mapsto C$ that maps from sensor input space S to the desired output space C. This very general representation is the key to enable the core to be task-independent.

4.2.2. The SHOSLIF Tree

SHOSLIF was developed for the two goals discussed earlier: the generality and the scalability. It must accept the high-dimensional sensory input directly and it must be fast. Specifically, it is a framework for automatically building a tree that is used to approximate the function $\mathbf{Y} = f(\mathbf{X})$ that produces the desired output $\mathbf{Y}$ given the input $\mathbf{X}$, from a set of training samples $\mathbf{L} = \{(\mathbf{X}_i, \mathbf{Y}_i) | i = 1, 2, \ldots, s\}$, where

$$\mathbf{Y}_i = f(\mathbf{X}_i).$$

The dimensionality of $\mathbf{X}$ is typically larger than the number of training samples s in the training set L. Because the high-dimensional sensory input is directly used instead of a

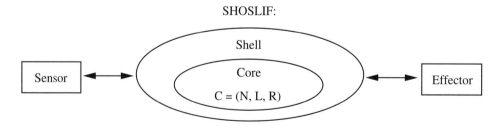

Fig. 3 The SHOSLIF's core and shell, with sensor and effector. Such a core–shell structure can be nested.

Learning in Computer Vision and Beyond

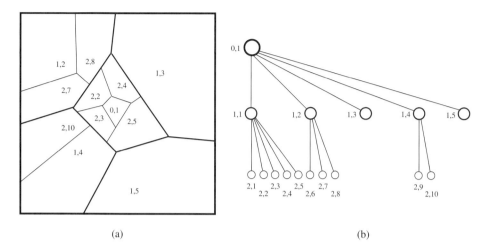

Fig. 4 (a) Samples in the input space marked i, j where i is the level at which its position marks the center of the partition cell and j is the index among the brothers in the SHOSLIF tree (b). The leaves of the tree represent the finest partition of the space. All the samples in each leaf belong to the same class. A class is typically represented by more than one leaf. Linear boundary segments (i.e., corresponding to linear features) at the finest level are sufficient because any smooth shape can be approximated to a desired accuracy by piecewise linear boundaries.

low-dimensional vector of feature measurements extracted by a feature extraction algorithm, typically the noise in sensory input $\mathbf{X}$ is relatively low.

Fig. 4 shows a hierarchical space partition in the input space of $\mathbf{X}$ and its corresponding SHOSLIF tree. The SHOSLIF tree is a classification and regression tree. Therefore, it shares many common characteristics with the well-known tree classifiers and the regression trees in the mathematics community (50), the hierarchical clustering techniques in the pattern recognition community (47,48), and the decision trees or induction trees in the machine learning community (49). The major differences between the SHOSLIF tree and those traditional trees are as follows:

1. The SHOSLIF automatically generates features directly from training images, whereas the traditional trees work on a human preselected set of features. This point is very crucial for the completeness of our representation.
2. Use of PCA (principal component analysis) combined with LDA (linear discriminant analysis) (46,70) for tree generation. The traditional trees have been popularly univariate; that is, they use splits based on a single component of the input vector at each internal node. For example, (a) at each internal node, they search for a partition of the corresponding samples to minimize a cost function (e.g., ID3 (49) and clustering trees (48) or (b) they simply select one of the remaining unused features as the splitter (e.g., the k-d tree). Option (a) results in an exponential complexity that is way too computationally expensive for learning from high-dimensional input-like images. Option (b) implies selecting each pixel as a feature, which does not work for image inputs (in the statistics literature, it generates what is called a dishonest tree (50). There have been several studies

that use multivariate splits. An early study that uses a linear discriminant splitter is the work of Friedman (71), which produces a binary decision tree for every class. A recent excellent survey on these studies is written by Murthy (72). The SHOSLIF is unique among these multivariate split decision trees in that it must deal with such a high-dimensional input space that the number of samples is typically smaller than the dimensionality. Thus, it combines PCA with LDA at each internal node and it uses interpolation among the top-matched nodes in high-dimensional space to address both the classification problem (class label as output) and the regression problem (numerical vector as output).

4.2.3. Automatic Derivation of Features

In each internal node of the SHOSLIF tree, several feature vectors are automatically generated to further partition the training samples. The most discriminant feature (MDF) subspace is such that in that subspace, the ratio of the *between-class scatter* over the *within-class scatter* is maximized. Computationally, the MDF vectors are the eigenvector of $W^{-1}B$ associated with the largest eigenvalues, where W and B are the within- and between-class scatter matrices, respectively. In the case where class information is not available, SHOSLIF uses PCA to compute the principal components of the sample population [which we call the most expressive features (MEF) in order to bring up a contrast with the MDF].

In the training phase, as soon as all the samples that come to a node belong to a single class, the node becomes a leaf node. Fig. 5 shows an example of SHOSLIF binary tree, for which only one feature vector is computed at each node, resulting in a binary tree, which is very fast in retrieval because only one projection needs to be computed at each internal node. As shown, the MDF gives a much smaller tree than the MEF, as it can find good directions to separate classes. In reality, we explore $k > 1$ paths down the

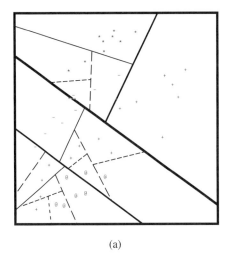
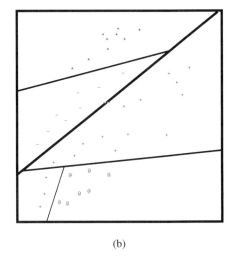

(a)   (b)

Fig. 5 (a) Hierarchical partition of the MEF binary tree. (b) Hierarchical partition of the MDF binary tree, which corresponds to a smaller tree. The symbols of the same type indicate samples of the same class.

tree to get the top k matches. Then, the confidence is estimated from a distance-based confidence interpolation scheme using the top k matches.

4.3. Some Properties of SHOSLIF

Here, we briefly describe some properties that have been established for SHOSLIF. Suppose that the learning task is to approximate a function f from its high-dimensional domain to its range. We avoid a lot of mathematical equations here. In the following, the first two properties address the *generality* and the later two properties deal with the *scalability*.

Point-to-point correct convergence: Loosely stated, as long as an image $\mathbf{X}$ has a positive probability of occurring, the approximating function $\hat{f}$ represented by the SHOSLIF tree approaches the correct $f(\mathbf{X})$ as the number of learning samples increases without bound.

Functionwise correct convergence: Loosely stated, if the training samples are drawn according to the real application and the function to be approximated has a bounded derivative, then the approximate function $\hat{f}$ represented by the SHOSLIF tree approaches the correct f *in the mean square sense** as the number of learning samples increases without bound. The above condition does not mean that the samples must be uniformly drawn from all the possible cases, but, rather, it means that one just takes the samples roughly in the way the system is used in the actual application. This theoretical result means that the system will never get stuck into a local minimum and thus fails to approach the function wanted. This is a property which the artificial neural networks lack.

Rate of convergence: Let R_n be the error risk of the SHOSLIF tree and R^* be the corresponding Bayes risk (which is the smallest possible risk based on a given training set). We have proved the following upper bound on R_n:

$$R_n \leq 2R^* + A(2\beta)^2 k^{2/d} \left(\frac{1}{n}\right)^{2/d} \tag{3}$$

where A is the upper bound on the Jacobian of the function f to be approximated, β is the radius of the bounded domain in which f is to be approximated, k is the number of neighbors used for interpolation employed in SHOSLIF, d is the dimensionality of the feature space, and n is the number of learning samples. This result is consistent with the intuition that the k-sample-based interpolation is useful only for a smooth function, but it slows down the approximation when the function surface is rough. When n goes to infinity, the inequality gives $\lim_{n \to \infty} R_n \leq 2R^*$, which is a well-known result proved by Cover and Hart (24) for the classification task and is later extended to the function approximation by Cover (75). This result gives a theoretical foundation for using the k-nearest-neighbor rule because its resulting error rate is not too far from that of the best possible Bayes estimator. The Bayes estimator

---

* Also in probability 1, which is stronger than the mean square convergence, as discussed in Ref. 73.

is impractical in our case because we do not know the actual distribution function and the estimation of the distribution function in a high-dimensional space is computationally very expensive, even if we impose some artificial distribution models.

Logarithmic complexity: The time complexity for retrieval from the recursive partition tree (RPT) used by SHOSLIF is $O(\log(n))$, where n is the number of samples stored as leaf nodes in the tree, which is typically smaller than the size of the training set. This result is true not only for a balanced SHOSLIF tree (guaranteed by a binary tree version of the SHOSLIF) but also for Bounded Unbalanced Tree typically generated by a general version of the SHOSLIF.

As we know, the above theoretical results gave only some insights into the nature of the task. Evaluating actual error rates, some of which were quoted in this chapter, is a more practical way for evaluating actual algorithms.

4.4. Functionalities Tested for the SHOSLIF

The demonstration of generality and scalability is a very challenging task. It requires us to test SHOSLIF method for a wide array of tasks. We selected several domains of challenging high dimensional tasks to test SHOSLIF theory and performance. The selection of the vision tasks was determined in such a way that they cover major functionalities that must be implemented in Phase 1. Because the living machine requires speech-recognition capability also, we have selected the speech-recognition domain. Table 4 lists the representative tasks that we have selected as test domains for SHOSLIF and the related functionalities that have been tested if applicable.

In the following, a summary of each SHOSLIF subproject is presented. Due to the unified core–shell structure of SHOSLIF, the programming for each subproject was systematic. Basically, an interface needs to be developed as a shell for each subproject.

4.5. SHOSLIF-O: Face and Object Recognition

This project is to use the SHOSLIF method to recognize a large number of objects from their appearance. The SHOSLIF approach is different from other conventional approaches to recognition in that it deals with real-world images without imposing shape rules or

Table 4 Some Tasks Tested Using SHOSLIF

SHOSLIF subproject	SHOSLIF-O (recognition)	SHOSLIF-M (spatiotemporal)	SHOSLIF-N (mobile robot)	SHOSLIF-R (robot arm)	SHOSLIF-S (speech)
Spatial recognition	X	X	X	X	X
Temporal recognition		X	X	X	X
Image segmentation		X		X	
Prediction		X	X	X	
Visual attention		X		X	
Sensorimotor			X	X	
Incremental learning			X		X
On-line learning			X		X

shape models on the scene environment. Rather, it uses a general learning method to make the system learn how to recognize a large number of objects under complex variations. It copes with critical issues associated with such a challenging task, including automatic feature derivation, automatic visual information, self-organization, generalization for object shape variation (including size, position, and orientation), decision optimality, representation efficiency, and efficient indexing into a large database.

In order to test the system using a database that is as large as possible, we combined several different databases for training and testing: (1) MSU face database (38 individuals); (2) FERET face database (303 individuals); (3) MIT face database (16 individuals); (4) Weizmman face database (29 individuals); (5) MSU general object database (526 classes). Fig. 6 shows some examples of face and object images used for recognition. Table 5 summarizes the results obtained. The training images were drawn at random from the pool of available images, with the remaining images serving as a disjoint set of test images.

For training efficiency, SHOSLIF is trained to recognize canonical face views only. In a canonical face view, the position, size, and orientation of each face are all roughly equal to a set of predefined values. In order to deal with some variation beyond the canonical views, the system was also trained to handle a limited amount of variation in size, position, and 3-D orientation within a certain range. We trained the system using samples generated from the original training samples to randomly vary in (a) 30% of size, (b) positional shift of 20% of size; (c) 3-D face orientation by about 45° and testing with 22.5°. The training and test data sizes are similar to that in Table 5. The top 1 and top 10 correct recognition rates were respectively (a) 93.3% and 98.9%, (b) 93.1% and 96.6%, and (c) 78.9% and 89.4%. The data reported here were extracted from Ref. 76. More details are available in Refs. 60 and 77.

Directly treating an image as an appearance vector and recognizing the object by applying statistical methods to such a vector space is called the appearance-based approach. Research for such an approach has become very active. Examples of such an appearance-based approach for face and object recognition include Refs. 17, 60, 61, 78, and 79. The SHOSLIF-O is one of the few appearance-based face and object recognition algorithms that use a tree structure for a logarithmic time complexity, and better performance.

It is worth noting that a successful recognition by the appearance-based approach requires that a well-framed image view be given in which the object of interest is centered reasonably well with an appropriate size, as shown in Fig. 6. It appears that general-purpose goal-directed attention selection is a very challenging task that cannot be effectively addressed without truly understanding the scenes by machines. Currently, mechanical fixed-order image scanning has been attempted. For example, in Refs. 64 and 80, many attention images are obtained by systematically scanning the image using different window sizes along a grid pattern of fixation points. Within the attention image in which the object is recognized, the object is segmented by Cresceptron (64,80) from the background using back-projected edges that have contributed to the recognition. Exhaustive, pixel-to-pixel scanning using an attention window has also been used, especially for the tasks of face detection. Some recent examples for face detection include Refs. 81–84. Motion information and prediction from partial view to global contour can also assist segmentation, as explained in the following sub-section.

4.6. SHOSLIF-M: Motion Event Recognition

For spatio-temporal recognition, we selected a challenge task: recognition of hand signs from American Sign Language. Recently, there has been a significant amount of research

Fig. 6 Some examples of face and object images used for recognition.

Table 5 Experimental Results for Face and Object Recognition from Canonical Views

Type	Training set	Training classes	Test set	Top 1 correct	Top 15 correct
Face	1042 images	384 individuals	246 images	95.5%	97.6%
General	1316 images	526 classes	298 images	95.0%	99.0%

on vision-based hand-sign recognition from images. Compared with other recent studies on hand-sign recognition from image sequences (85–90), the SHOSLIF-M work (91,92) has the following characteristics:

1. The capability to segment a detailed hand (a complex articulated object) from a very complex background without relying on handwear or skin tones, as shown in Fig. 7. With this capability, we can significantly reduce the constraint on what kind of clothes that the signer can wear. This is accomplished through (a) using motion information to reduce the area of attention, (b) using a learning-based prediction-and-verification scheme which predicts the global contour from a local view that partially covers the object of interest but does not contain

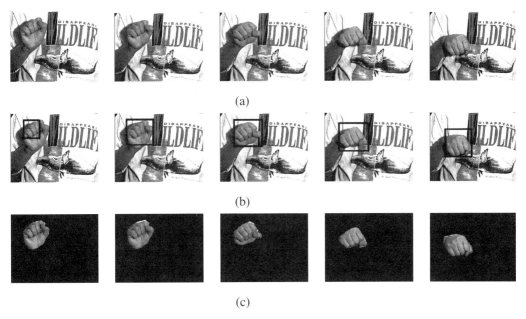

Fig. 7 (a) An example of hand-sign sequences. It means "yes." (b) The results of a motion-based attention mask found, shown with a bounding (dark) rectangular window. Note that the motion-based segmentation alone is not sufficient for hand-sign recognition. Without detailed shape information of the hand, reliable hand-sign recognition is not possible with a large number of hand-sign classes. (c) The result of the final segmentation is shown with the background automatically masked. Such a detailed and accurate segmentation is crucial to the success of hand-sign recognition with a large number of classes and vice versa.

background pixels. Thus, the recognition result is completely independent of the background. In contrast, if a fixed-shape attention window is used, it is required that the background covered by the window does not affect the matching significantly, such as the jets scheme by Malsburg et al. (90).
2. The logarithmic sign-retrieval time complexity $O(\log(n))$.
3. The system distinguishes a large number (over 140) of hand-shape classes, the largest so far among the existing works on static hand-shape recognition from images (a list can be found from Huang and Pavlovic's recent survey (93) and the workshop proceedings in which that survey was published).
4. The relatively large number of hand-signs among the existing works on *hand-wear-free* moving hand-sign recognition (see also the survey article by Huang and Pavlovic (93)).

4.7. SHOSLIF-N: Autonomous Navigation

Autonomous vision-guided navigation is an example for vision-guided effector control. The input to the system is the current image from the camera and the output from the system is the next control vector that specifies the heading direction and the speed. SHOSLIF-N faces the challenging task of performing real-time indoor navigation using only a single video camera without using any other sensors. For learning, it must learn on-line, incrementally, in real time. SHOSLIF-N uses the incremental version of the SHOSLIF core (18), which builds a SHOSLIF tree incrementally by updating the tree after receiving each training sample. It is a good example for studying how real-time on-line learning might be achieved without the need of special image processing hardware, thanks to the extremely low (logarithmic) time complexity of the SHOSLIF.

During the learning, using a joystick, the human teacher controls the robot on-line to navigate along the desired path by updating the control signals in terms of speed and heading direction. The system digitizes the current image frame and links it with the current control signal to form a training sample, which is used to train the SHOSLIF tree. To construct a lean, condensed tree without overlearning, the SHOSLIF tree rejects a training sample when the current tree outputs a control signal vector that is the same (according to the accuracy required) as the human's control signal vector, without learning the current training sample. When almost all of the recent training samples are rejected, the system is almost fully trained and is ready to perform. The on-line learning with both tree retrieval and tree update runs at 5 Hz. The real-time performance, with a tree-retrieval speed of 7 Hz, is accomplished by an on-board Sun SPARC-1 workstation and a SunVideo

Fig. 8 The mobile robot running SHOSLIF navigates autonomously at a walking speed, along hallways, turning at corners, and passing through a hallway door. The real-time, on-line, incremental learning and the real-time performance is accomplished by an on-board Sun SPARC-1 workstation and a SunVideo image digitizer, without any other special-purpose image processing hardware.

Learning in Computer Vision and Beyond

Table 6 Computer Time Difference Between the Flat and the Tree Versions

Time per retrieval	MEF tree	Flat version	
		In MEF space	In image space
Time per retrieval (in seconds)	0.028	0.74	2.9
Slow down wrt MEF tree version	—	26.7 times	103.0 times

image digitizer, without any other special-purpose image-processing hardware. The trained system successfully navigated along the hallways of our Engineering Building, making turns and going through hallway doors. (See Fig. 8.) It was not confused by passers-by in the hallway because it looks at the entire image and uses the most useful features (MEF or MDF), instead of tracking floor edges, which can be easily occluded by human traffic.

A significant amount of research on vision-based outdoor road following has been conducted (e.g., Dickmanns' group (94), CMU Navlab (95), Martin Marida ALV (43), CMU ALVINN (96), and the work at Maryland University (97)). Outdoor navigation must deal with a large degree of light change and weather change. Indoor invigation, in contrast, must deal with the lack of large high-contrast regions in typical indoor scenes (e.g., Refs. 18 and 98), which pose a severe local minima problem for techniques that use iterative minimization in training.

4.7.1. The Speedup of the SHOSLIF Tree

To indicate the scalability performance of the SHOSLIF and the effect of the hierarchical tree, Fig. 6 shows the computer times for SHOSLIF (tree version) and the corresponding two flat versions. A flat version goes through all the training samples one by one to find the nearest neighbor. The MEF flat version does such a linear search in the MEF space constructed from all of the training samples. The image-space version does so in the original image Euclidean space. A total of 2850 images from various hallway sections in the Engineering Building of MSU were used for training. The data in Table 6 show that the use of MEF tree can greatly speed up the retrieval and that real-time navigation can be achieved. The computation times were recorded when the programs ran on a SUN SPARC-10 computer.

4.7.2. MDF Results in a Smaller Tree

For comparison purpose, two types of trees have been experimented with, MEF RPT and MDF RPT. The former uses MEF and the latter uses MDF in each internal node of the respective tree. Both trees used the same 318 training images, 210 from the straight hallway and 108 from the corner. As presented in Table 7, the MDF tree has a total of 69 nodes

Table 7 MDF Results in a Smaller Tree

Tree type	MDF	MEF
Total number of nodes	69	635

only, with only 35 leaf nodes, whereas the MEF tree has a total of 635 nodes, with 318 leaf nodes. Figure 5 explained why MDF can give a smaller tree. Note that the timing data shown in Table 6 is for the MEF tree, which is larger than the MDF tree. An MDF tree is typically much faster. The results quoted here are extracted from Refs. 18 and 99, where more detail is available.

4.8. SHOSLIF-R: Vision-Guided Robot Manipulator

A SHOSLIF-R is for sensorimotor coordination and task-sequence learning. It also uses SHOSLIF-O as the object locator and recognizer. A SHOSLIF-R tree was automatically built from training temporal sensor-guided control sequences (62). The input to the SHOSLIF-R tree is the image position of the objects (from SHOSLIF-O), the current joint angles of robot and the index of the task from the human teacher. The output of the SHOSLIF-R is the incremental values of the six joint angles of the robot manipulator.

As reported in Ref. 100, five actions were learned interactively at several places in the work space. Tests were done randomly in any place in the work space. Fig. 9 shows a sequence of learned actions. The success rate was 100% for all the actions, except that the liquid was partially spilled 20% of the time during the pouring action. More training can improve the pouring accuracy. This is an example of learning by doing, instead of explicit modeling. Modeling the dynamics of poured liquid is not possible because there are too many unknown and unobservable parameters in fluid dynamics. The reader is referred to Ref. 100 for more detail.

Several robotics groups (e.g., Refs. 101 and 102) have recently published works in which a robot manipulator can repeat the action sequence from the human's demonstration using a data glove. Case-specific features and decision rules were written into their programs which the algorithm will use to identify the sequence of actions (such as ''when the speed of the hand is smaller than certain number, do the following...''). SHOSLIF-R is different from those works, in that SHOSLIF-R does not contain any pre-written case-specific rules (hand-crafted knowledge-level rules). Specifically, the SHOSLIF core does not contain any knowledge-level rules and the SHOSLIF-R shell contains only the arm hardware specification [i.e., the anatomy (e.g., degree of freedom of the hand) instead of knowledge-level rules]. Thus, it is, in principle, applicable to any robot manipulator

Fig. 9 A demonstration of various actions learned: approaching the handle of cup A, picking up the cup, moving to top of cup B, pouring, and putting on table.

task. SHOSLIF-R is a robot manipulator learning system that can learn to perform tasks through interactive learning without hand-crafting any knowledge-level rules.

4.9. SHOSLIF-S: Speech Recognition

The objective of this study (103) was to test the feasibility of using SHOSLIF for recognition of isolated spoken-words. The work was performed as a class project in a graduate class. A spectrum feature vector of each isolated world is used as input to the SHOSLIF. The silence is used to segment each isolatedly pronounced word. The output of the SHOSLIF is the word label. An incremental learning version (18) of the SHOSLIF was used for the learning task. Due to the limited available time during the single-semester class, only a small number of training samples were collected. The preliminary experiment was performed with 10 spoken words from "zero" to "nine." The speaker-dependent testing reached 90% accuracy among 20 speakers; each word was trained with only one training sample and tested with 4 different instances.

The fully dynamic speech recognition requires the recurrent version SAIL. The recurrent version is expected to be able to learn to handle time warping, coarticulation, temporal acceleration, and pause, which are very common in the real-world speech.

4.10. Cresceptron: The Predecessor of the SHOSLIF

Our work along this line can be traced back to the early 1990s when Cresceptron was conceptualized for general, open-ended, sensing-based learning. Cresceptron (64,80) is the predecessor of the SHOSLIF. Cresceptron is a system that is capable of learning directly from natural images and performing the task of general recognition *and* segmentation from images of the complex real world, virtually without limiting the type of objects with which the system can deal. It has been tested for recognizing and segmenting human faces and other objects from complex backgrounds. It addressed the issue of self-organizing dynamically by growing a network on-line according to inputs. Although Cresceptron has a very high generality, it does not attempt to solve scalability. Its successor, SHOSLIF, addresses both generality and scalability.

5. SAIL

SAIL (Self-organizing, Autonomous, Incremental Learner) is a living machine under construction at Michigan State University. Its goal is to realize AA-learning. This section discusses some basic issues of AA-learning and some design issues of SAIL. As SAIL is an ongoing project, its design is expected to be modified, refined, and improved in the future.

5.1. System Overview

As discussed previously, the program-level representation should not be constrained by, or embedded with, hand-crafted knowledge-level world models or system behaviors. Thus, the system design of SAIL should be conducted at signal level, instead of knowledge

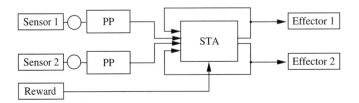

Fig. 10 A schematic illustration of the coarse architecture of the presented system SAIL. A circle represents an attention selector. It is also an effector. PP: preprocessor; STA: spatio-temporal associator.

level. Fig. 10 gives a schematic illustration of the system architecture. The preprocessor (PP) performs some transformations from input, such as intensity normalization, automatic gain (contrast) control, filtering, and so forth. The spatio-temporal associator (STA) is the "brain" of the system. It consists of automatically generated levels. The number of levels depends on the maturity of the system.

5.2. Single-Level Formulation

At each level, the system operates in the following way. At each time instant t, its current internal mental status is represented by its state $s(t) \in S$. It accepts sensory input $x(t) \in X$, which contains both exteroceptive sensors (e.g., visual and auditory), proprioceptive sensors (e.g., effector position), and interoceptive sensors (e.g., internal clock). As shown in Fig. 10, input $x(t)$ at the lowest level includes both sensory input and effector position, and possibly the internal time counts. First, assume that the system is deterministic. At each time instant t, the system accepts an input $x(t) \in X$ while it is at state $s(t) \in S$. Then, it outputs action $a(t) \in A$ and enters a new state $s(t + 1) \in S$. Without loss of generality, we assume that the time is discrete; hence, $t + 1$ is used as the next time instant. In other words, the machine refreshing cycle time is considered as a unit here. With such a deterministic system, the system state transition is represented by a time-varying function $f_t: X \times S \mapsto S$:

$$s(t + 1) = f_t(x(t), s(t)) \tag{4}$$

The state $s(t + 1)$ represents what the system understands from the current input and the current mental status at this level. Then, it figures out what it should do, based on what it understands represented by $s(t + 1)$. Symbolically, the action function is represented by the function $g_t: S \mapsto A$,

$$a(t + 1) = g_t(s(t + 1)) \tag{5}$$

where A is the control signal space of all its effectors.

In order to model the uncertainty, our system is, in fact, nondeterministic. Thus, the system transition function is modeled by the conditional probability:

$$P(s(t+1) = s' \mid x(t) = x, s(t) = s) \tag{6}$$

where $x \in X$, $s \in S$, $s' \in S$, $a \in A$, and P denotes the probability. The action function is modeled by the conditional probability:

$$P(a(t+1) = a \mid s(t+1) = s) \tag{7}$$

The above probability depends on the experience and the reward associated with the experience.

We can see immediately from this notation that it resembles a Markov decision process (63). However, it is more general. For this, we need a more careful examination.

First, A is an Euclidean space here, which contains infinitely many (uncountably many) elements. For a typical Markov decision processes, however, A is a finite set of legal actions. In other words, we have some distance metric defined for the elements in A. We definitely want to use this property to enable the machine to try something it has never tried (e.g., sensorimotor refinement).

Another important issue here is the space S. It should contain certain information about the history. How should we define the state space S? In Markov decision processes (MDP), S is a predefined, finite set of elements, which accounts for a major reason why general-purpose learning fails with MDP.* The human-defined set S for states incorporates a lot of human knowledge. These symbolic states so defined are very artificial and no appropriate distance metric is defined on the finite set S. The definition of a state is not as trivial as it appears. For example, a state can be defined to record the history. Suppose that $s(1)$ and $s(2)$ are the states at time $t = 1$ and $t = 2$, respectively. One may define $s'(2) = (s(1), s(2))$ so that $s'(2)$ contains not just the current state $s(2)$ but also the history $s(1)$! Using this technique, can we use first-order Markov process notation to define Markov process of any order? Unfortunately, this is not that simple because the resulting S would become an infinite set of discrete symbols,† which a practical Markov process does not allow. More complications arise from this too, such as the difficulties in estimating the transition probability.

In SAIL, the system states in S must be defined automatically. We define S to have the same dimensionality as $X \times R(S)$, where $R(S)$ is a dimension reduction operator. For example, if each $s \in S$ is a 128 × 128-pixel vector, the dimension reduction reduces s to a 64 × 64-pixel vector s' through neighboring pixel averaging. Given sensory input $x(t)$ when the current system is at state $s(t)$, the resulting next state, at the lowest level 0, can be simply to record the new situation: $s(t + 1) = (x(t), s'(t))$, where $s'(t)$ is a dimension-reduced version of $s(t)$. For example, if X has a dimensionality of 9000, we set the dimensionality of S to be 18,000. Thus, the dimensionality reduction is to reduce from the 18,000 dimension to the 9000 dimension. With a zero-vector initial state $s'(0)$ at time $t = 0$, we can define the first state at time $t = 1$ to be $s(1) = (x(1), s'(0))$. This explains how the state can be defined automatically based the current state and current

---

* It takes a human being to understand a particular task, using his knowledge to translate the task into the Markov decision process, including defining states, legal actions, and so forth. This is a machine whose "brain" is not closed. It requires humans to dictate its brain states, given each particular task. In this way, the machine is not able to learn on its own.
† This is because such compounding $s'(2) = (s(1), s(2))$ becomes endless with the progress of time. Each compounding defines a new symbol.

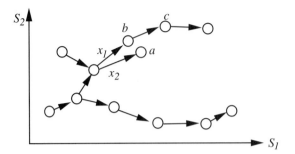

Fig. 11 Representation of a state that enables a generalization across states. Here, the high-dimensional space S is illustrated by a 2-D space only. a is a newly generated state, which does not have any transition experience. The existing state b is the nearest neighbor of a in the state space S. The transition path, from b to c, that is learned by b can be used for figuring out how to act at the newly generated state a in the absence of guidance.

sensory input. Thus, potentially, the number of possible states is infinite. We will discuss how to automatically self-organize the states.

Because S is a high-dimensional Euclidean space, we can naturally define the distance between states as their Euclidean distance. This enables our SAIL model to generalize across states, which a conventional Markov model cannot accomplish. The idea is illustrated in Fig. 11. When a new state a is automatically generated, the current system is at this state. There is no transition probability learned for this state since it is new. With the Markov model, the system will not be able to perform from this new state a.* However, our distance definition (e.g., Euclidean distance) in the state space S allows our model to find the nearest matched existing state b that has a following state c. Thus, the predicted state from a is c, as specified by $f_t(x, b)$ in Eq. (4). Similarly, the appropriate action at this new state a can also be generated by $g_t(b)$ in Eq. (5). In other words, generalization of transition pattern and behavior can be realized based on a distance metric in S without requiring a very large amount of data for training every possible transition among the states.

Due to the dimension reduction in the representation of state $s \in S$, the new state has information about the previous state information. Thus, the single-level model of SAIL is not strictly Markov.

For generality, the system must perform for any input $x \in X$ and any possible $s \in S$. Thus, the sets of possible input x and possible state s are both infinite. However, the number of possible states at each time is finite, and it changes dynamically in the single-level model. When $x(t)$ changes with time, $s(t)$ changes accordingly with time, generating a trajectory in S space, as shown in Fig 11. When SAIL is nondeterministic, SAIL keeps track of $k > 1$ most probable trajectories in S.

---

* It is worth noting that there are a lot of problems with MDP when the number of training samples is small.

We define a notion called *state dictatability*.

Definition 2. A system is state-nondictatable if the trainer of the system is not allowed to dictate the value of state in the system at any time during training.

In all the MDP applications we know, the states are dictatable because humans define the meanings of states. Because the state is very much related to the content of what the system learned, state nondictatability means that the trainer is relieved from the intractable task of coping with content-level representation. He or she can shape the system behavior through its sensors (e.g., presenting examples) and effectors (e.g., imposing actions), but he or she cannot directly set the "brain" of system. This is a very important concept in automating learning.

A subspace $H \subset X$ is called *hard-wired reward space* if the system has predefined preference for the elements in H. For example, we can define $X = X_1 \times H$, where $H = \{x \mid -1 \leq x \leq 1\}$. The system is such that it "likes" positive values in H (appetitive stimulus) and "hates" negative values in H (aversive stimulus). Thus, extreme pleasure, pleasure, neutral, pain, and extreme pain can be represented by 1.0, 0.5, 0, -0.5, and -1.0, respectively. The *hard-wired reward space* is used to enable reinforcement learning by allowing the system to explore on its own while occasionally giving it some appropriate reward or punishment according to its performance.

5.3. Forgetting

Due to a finite memory space, the system cannot remember all the spatio-temporal events that it has come across. In fact, for generalization, many associations must be forgotten. The utilization of association is indicated by *co-occurrence frequency* and *occurrence intervals*.

Let us consider a memory element, a node, or a link, which will be used in our system for memorizing an association it represents. Each element has a memory-residual register whose updating curve is shown in Fig. 12, which resembles what is known about human memory characteristics (36,104).

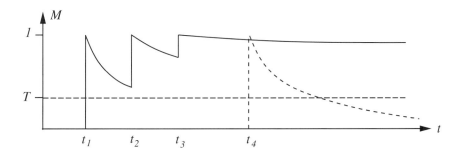

Fig. 12 Update of memory strength M through time t. The solid curve represents an element which is visited often enough to be kept. The dashed curve indicates an element that is not visited often enough; thus, it falls below the threshold T before being visited again.

Each visit to the same element makes the strength reset to 1 and then the curve declines using the next slower speed. For example, we can define a series of *memory fade factors* $\alpha_1 < \alpha_2 < \cdots < \alpha_m \approx 1$. α_i is used for an element that has been visited i times, The memory trace r can be updated by $r \leftarrow r\alpha_i^t$ where t is the number of system cycles (refresh) elapsed since the last visit to the element. Thus, we do not need to update the memory strength for all the elements at every system cycle. When an element is visited, its memory strength is updated first from that which remains from the last visit. If the memory strength falls below the designated threshold, the element should be deleted and so it is marked as to be deleted. If what is deleted is more than a single element (i.e., a subtree), the deleting process will not delete it right away to avoid consuming too much computer time in a real-time process. Instead, it puts the subtree in a garbage buffer which is to be cleaned when the learner is "sleeping."

The memory fade factor is also affected by the score level it receives. An extreme score, near 0 or 1, will result in smaller memory fade factors than a score 0.5, which means "doing OK and keep going." Thus, when the score is 0.5, the memory fade will be faster, so that many details associated with normal routine operations are forgotten quickly.

The memory fade factors control the relative speed of learning and forgetting so that they can keep pace while containing the memory space over time.

5.4. Spatio-Temporal Clustering

The state space S has a finite dimensionality, but the number of elements generated in S through time is unbounded. The objective of spatio-temporal clustering is to form primitive clusters (P-clusters) in S using not only spatial information but also temporal information. A P-cluster consists of a number of prototypes, each being represented by a leaf node. The P-cluster will be used as the state at every level of the system.

At level 0, our goal is to generate clusters of inputs only. Thus, the state $s(t)$ at level 0 depends only on input $x(t)$ but not $s(t)$, as a special case of Eq. (4).

Fig. 13 shows a schematic illustration of the temporal-adjacency cluster. The cluster represents a mapping $h: X \mapsto Y$ which maps from input space X to output space Y. h is either f_t in Eq. (4) or g_t in Eq. (5). Given x, $y = h(x)$ is the image of x, representing the corresponding P-cluster. In order to preserve the necessary topology in X, Y has the same dimensionality as X or has a reduced dimensionality. If X has a dimensionality $n \times n$ corresponding to an image space of $n \times n$ pixels, then $Y = R^{m \times m}$, where R is the set of all real numbers. The ratio of dimensionality difference $(n \times n)/(m \times m)$ is called the factor of dimensionality reduction (FDR) (e.g., FDR = 4). This reduction of dimensionality is reasonable because all we need is to have the Y space to roughly keep the topology of the X space. When a δ-prototype is first created, it is represented by a single training sample x_i. We let its image y be the same as x_i, $y_i = g(x_i) = x_i$ (with FDR taken into account if FDR > 1. This is a short-term memory effect. Later, successful matching with x_i will cause y_i to be pulled toward the center of the cluster formed.

Consider an input stream $x = \{x(0), x(2), x(3), \ldots, x(t), \ldots\}$, where each frame $x(i)$ is in X and i is the time index. The stream x forms a trajectory in $X \times T$ space, where $T = \{0, 1, 2, \ldots, t, \ldots\}$ represents the set of time indexes. We need to chunk the trajectory into segments of some appropriate length of events. Each segment corresponds to a P-cluster. Given each frame $x(t)$ at time t, its top k matched prototypes are considered. The

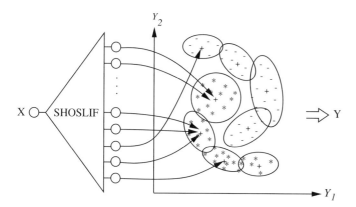

Fig. 13 A schematic illustration of the temporal adjacency cluster. Each region in Y space represents a P-cluster. A dashed curve is used to roughly indicate the samples that fall into the cluster. The circles to the right of the SHOSLIF tree are the leaf nodes of the tree. Each $-$, $+$, or $*$ sign indicates the relative position of a leaf node (δ-prototype) in the input space to the SHOSLIF tree. Two different signs representing two different input sequences. Each leaf node in the SHOSLIF tree has a pointer to the Y space, which points to the center of the corresponding P-cluster represented by the mean and the covariance matrix. Y is shown here as 2-D for visualization, but it is typically of the same dimensionality as X or with a reduced dimension by a factor of dimension reduction.

top match gives its image y. Among top k-matched δ prototypes with a sufficient matching confidence, if two δ-prototypes are less than σ ($\sigma > \delta$) distance apart in X space and they are temporally adjacent, the images of these two δ-prototypes are pulled together using a simulated force, where the number of visits of each prototypes is the "mass."* Thus, among the temporally adjacent δ-prototypes with a σ radius in X, those most often visited prototypes tend to become the center of the cluster. When the y images of two prototypes are very close, their y images become one. Merged y vectors form the center of the P-cluster. Merged centers of y vectors have a higher mass. To avoid two neighboring P-clusters from being slowly pulled together, a maturity schedule is applied so that P-clusters that are mature enough will no longer be moved—an effect of long-term memory. Fig. 14 shows that the pointers emitting from leaf nodes gradually point to the center of the P-cluster in Y space.

When the spatio-temporal clustering is applied to input directly, it produces the lowest-level discrete state: level 0. The technique of spatio-temporal clustering is also used for forming states in the higher levels, which will be discussed in the following sections.

The techniques used here have a close relationship with the learning vector quantizer (LVQ) (105,106) and k-mean clustering (48). The differences are (1) the state concept in

---

* In fact, a probabilistic model is applied here. Pulling is modelled as updating the mean and variance of a cluster using a new sample. For simplicity, we describe the scheme intuitively, instead.

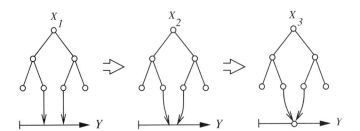

Fig. 14 The effect of pulling in spatio-temporal clustering. The SHOSLIF tree is repeatedly visited by $x_1 \approx x_2 \approx x_3$ that are nearby in the spatio-temporal domain. The two leaf nodes are among the top k matched leaf nodes, although they are separated early in the tree. Consequently, their Y vectors are merged when they are sufficiently close and the corresponding prototypes of the two leaf nodes belong to the same P-cluster.

the case of f_t, (2) the unknown number of clusters, (3) the nearest neighbors must be found quickly, thus the SHOSLIF tree is used, and (4) learning is incremental.

5.5. Automatic Spatio-Temporal Level Building

Each state at level i, $i = 0, 1, 2, 3, \ldots$, is a cluster at level i, Such a cluster is called level-i state, which is formed using the spatiotemporal clustering explained above.

The hidden Markov model (HMM, also called the partially observable Markov model, POMM) has been successfully used in speech recognition (31,107) and in computer vision (e.g., Refs. 87 and 108) when what is to be recognized is a temporal event. In the temporal domain, the system must deal with time warping and segmentation. Typically, the HMM is used in the following way. A HMM is trained to recognize a predefined sequence class, such as a word or an action. A different sequence class uses a different HMM. Given an input sequence, multiple HMMs are being run in parallel. At each time instant, each HMM reports its accumulated probability for the current partial sequence to belong to the class the HMM represents. A high probability from HMM indicates that a complete sequence has passed through that HMM. Due to the probabilistic model of state transition, the HMM can deal with time warping and segmentation effectively. Level building has been used to build short units into long units, such as linking digits to form numbers and linking words to form phrases (30,31). The task of level building involves a great amount of manual data preparation but uses basically the same HMM structure. Little attention has been paid to automatic level building.

Fig. 15 gives an illustration of time warping and the basic idea of level building. The probabilistic model at each level is an extension of HMM. The model is hidden because the system is not completely sure which current state is the best state in the long run to generate actions that receive the best reward. A higher-level HMM accepts inputs from the preceding lower-level HMM and takes care of a sequence that covers longer state transitions.

In manual level building in the speech community, each possible combination of concatenation of lower-level HMMs is given a new HMM. This results in a large number of HMMs typically. In the presented work, the state generation and survival all depend

Learning in Computer Vision and Beyond

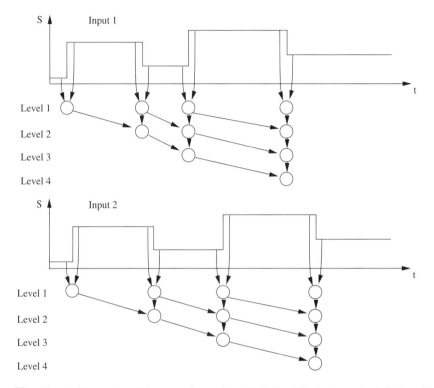

Fig. 15 An illustration of time warping and level building. The horizontal axis is time. The vertical axis represents the state value at level 0. The vertical axis indicates the Y image of P-clusters. Each circle indicates the arrival at a new state, at that level, with a high probability. At each time index, a state can transit to itself (i.e., staying at the same state).

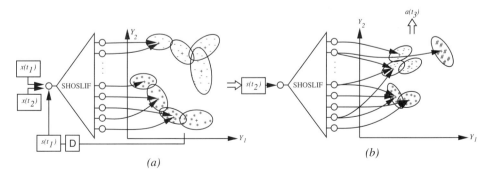

Fig. 16 Level building element (LBE). Each LBE contains two components: (a) the transition associator for f_t in Eq. (4), and (b) the action generator for g_t in Eq. (5). D denotes a delay register.

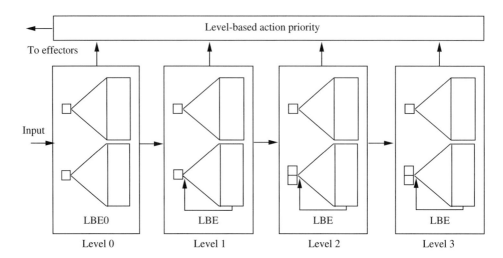

Fig. 17 Automatically building levels. The number of levels is grown according to the maturity of existing levels.

on our forgetting model discussed in Sec. 5.3. To fully utilize the time warping capability of a short state sequence with HMM, we define a concatenation of two or more adjacent states at a low level as the input to the state at the next higher level. However, the number of generated and survived states at the higher level depends on the actual co-occurrence and clustering at the higher level, not the number of all possible state combinations (i.e., transitions) at the lower level, which could be very large. This advantage is difficult to achieve with the manual level building approach.

The level building element (LBE) is shown in Fig. 16. Each level has a single LBE, it dynamically generates new states and forgets states. At each system cycle, the LBE accepts states from the lower level. It has two components: the transition associator and the action generator. The purpose of the transition associator in Fig. 16 is to realize f_t in Eq. (4) and that for action generator is to realize g_t in Eq. (5).

When the system is run for the first time, the system automatically creates a LBE. With a continuous input stream, the number of nodes of the SHOSLIF tree in the lower-level LBE will increase or decrease (forgetting) dynamically. When the number of nodes in the SHOSLIF tree is sufficiently large, a new LBE is automatically created at the next higher level, which starts to collect inputs to form clusters. The resulting architecture is shown in Fig. 17. We call it the spatio-temporal associator (STA). The effect of learning mainly causes the number of nodes and clusters to increase within each LBE. At level m, each state corresponds roughly to an input sequence of length 2^m in terms of P-clusters. A sequence of length of power-of-2 is expected to appear at a lower level. Due to the time warping effect, the actual time length of the input sequence at each level can vary tremendously.

Fig. 18 is a flowchart of the SAIL algorithm. The system is designed to be turned on for an extended period every day. As soon as the system is on, it updates itself according to the flowchart. Each loop corresponds to an input refreshing period called the machine cycle. The human teacher interactively plays with the system using two modes: the super-

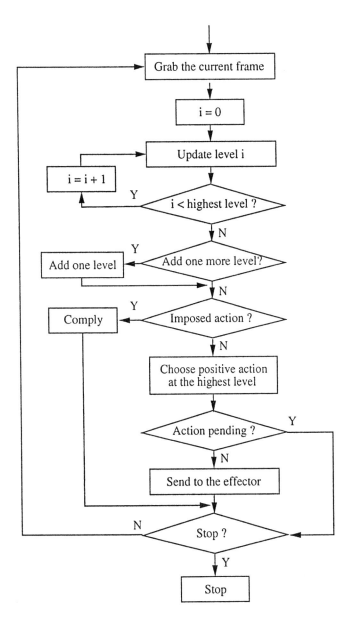

Fig. 18 The flowchart of SAIL algorithm.

vised mode and the reinforcement mode. Whenever there is an action imposed, the teaching is at the supervised mode. As indicated by the flowchart, the system always complies with the action imposed. Otherwise, the teaching is at the reinforcement mode. The system is allowed to try on its own, receiving a reward, occasionally, from the teacher.

5.6. Chunking

One very important phenomenon that will be explored is termed *selective chunking*. For example, let us consider an input sequence `this word`, assuming that a letter corresponds to a P-cluster at level 1. All the possible sequences of length 4 will be `this`, `his w`, `is w`, `s wo`, and so forth. However, our automatic level building method will not generate all such states at level 2 if the string `this word` is learned in a structured way. For example, a teacher may present the words `this` and `word` individually to learn first. Then, when learning the phrase `this word`, the successful response of `this` and `word` at level 2 means that the clusters for `this` and `word` have already been formed. The transition of two states is recorded at level 3, which forms a single cluster for `this word`. Such a selective chunking will discourage other chunks to be formed at level 2 because of the effect of the memory fade factor for a score near 0.5, as discussed in Sec. 5.3.

A well-recognized property of human short-term memory is called "short-term memory bottleneck"* (109,110). It is easier to remember if the sequence is grouped into richer, more complex items called chunks (or units).† The level building scheme intends to allow the system to learn and memorize short chunks at a lower level and then remember longer sequences at a higher level from the learned chunks at the lower level.

5.7. Recognition Through Spatio-Temporal Association

In the above discussion, we concentrated primarily on a temporal sequence. In fact, recognition of a spatial object is performed in time, as the learning process of recognition is also dynamic in the autonomous learning mode.

The recognition result is given by the system as an action through the N-effector discussed in Sec. 2.3. For example, we can teach the system to give a code which indicates the name of the object. A question can be coded as a number in the N-sensor. Thus, we need to teach the system to give a correct code whenever it recognized an object.

The presentation of an object is continuous. For example, the object is rotated continuously. The temporal sequence of the input is learned as a temporal event. We can impose a correct action (i.e., giving the correct code) as soon as a "question" is asked through the N-sensor, during the presentation of the object. Temporal frames will cause the system to cluster their trajectory into states. Because the views from different angles are presented

---

* For example, if you hear a string of about 10 single digits, read at a constant and fairly rapid rate, and are then asked to reproduce the string, you generally cannot recall more than about 7 or so of the digits.

† Try to remember the following 40 letters:

            BYGROUPINGITEMSINTOUNITSWEREMEMBERBETTER

No one can remember these 40 letters correctly if they are treated as 40 separate, unrelated letters in a string.

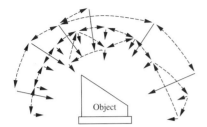

Fig. 19 A graphic explanation of the multilevel priming which allows the generalization of association of different views to the same class name. A solid short arrow indicates a view represented by a P-cluster. A solid long arrow indicates a subsequence of views represented at a higher level. A dashed arrow indicates the link of priming, from the prime to the target. A double-direction link indicates that priming in both directions exist. The forgetting process will merge some of the clusters, thus resulting in some generalization and forgetting some detail.

without a particular order, a view may be followed by many other similar views in the presentation. This implies that a view can prime other similar views, represented by P-clusters. At a higher level, such a priming action covers larger P-cluster sequences. During the learning, the desired action may be associated with several views (P-clusters) at different levels. For the test, the recognizer is shown with an object to be recognized and a question is asked through the N-sensor; the STA runs to find primed P-clusters and to find the best associated reward among the primed P-clusters. If the result is positive, the action is executed through the N-effector. This action is then associated with the current view, allowing it to be used as the target of later priming activities. Fig. 19 graphically explains how temporal association at different levels is applied to the object, which allows the object to be recognized from different angles. The state distribution pattern shown in Fig. 19 reflects the learning experience of the recognizer, although the behavior of recognition does not necessarily clearly show the distribution details of this pattern.

What if an object is only viewed from a limited number of viewing angles? For example, if the front view of the object is presented but not the back view. Then, the recognizer will not be able to recognize the object from the back view if the front view is very different from the back view and the SHOSLIF top k matches will not give a matched leaf node with sufficient confidence. If the back view is presented individually, the system still cannot link the front view with the back view. Only when the front view and the back view are presented consecutively or the object is turned around slowly can the system link the front view with the back view through the temporal association.

5.8. The Effect of Forgetting

As we discussed in Sec. 5.3, the forgetting process is performed according to the memory strength shown in Fig. 12. When a leaf node in the SHOSLIF is forgotten, the leaf node is merged into its parent and the corresponding state disappears. Later visits to the forgotten node will be directed to its nearest-neighbor cluster. This means that a cluster will correspond to a larger region. Thus, the effect is that details are forgotten and generalization results. For example, in Fig. 19, the dashed links indicate the path of state transition during learning. After a certain amount of time, some of the states will be merged, making some sections along the path of learning be forgotten and thus resulting in some generalization. This is because in an HMM with fewer states along a temporal trajectory, each state represents larger regions in the input space.

5.9. Incremental, On-line Learning Algorithm Using Markov Models

The HMM learning algorithms in the speech community are for a batch, off-line, supervised learning mode. Furthermore, a separate HMM is needed for every sequence (e.g., word) and each word typically has only a small number of states (e.g., five to six). In the operations research community and the AI community, the Q-learning and R-learning algorithms for reinforcement learning of POMM require a world model. There are simple algorithms for on-line learning, provided that a world model is available.

The presented method is intrinsically incremental and fully takes advantage of SHOSLIF incremental, on-line, learning-performance unification capability. Each level works on a different level of representation. A low level represents events of short time duration, whereas a higher level represents those of long duration. The system keeps the top k most probable states (MPS) $s(t_1)$ at each time t_1. An observation $x(t_2)$ is made from the input, which indicates the state transition from $x(t_1)$ to $x(t_2)$ at the lower level. It is worth noting that the input at level 1 contains not only the sensory input but also the previous action just executed, as shown in Fig. 10. For each of the current MPS $s(t_1)$ at each level, the vector $(s(t_1), x(t_2))$ is used by the transition associator to determine the next top k MPS $s(t_2)$. The k MPSs at time t_1, each predicting k MPSs for $s(t_2)$, results in k^2 MPSs for $s(t_2)$. The next states predicted at time t_1 for $s(t_2)$ are used to increase the confidence of those predicted. Only top k of those $s(t_2)$ are kept as the MPSs for time t_2 at this level. The action that records the best reward value is executed. The predictor will predict the set of the next state $s(t_3)$, and the top k of them will be kept.

In summary, we use (a) an internal high-dimensional representation for states, which addresses the cross-state generalization drawback of Q-learning, (b) a predictor for priming generation, and (c) the automatic level building to address the delayed reward problem and thus avoid the drawback of the simple discounted (or average) reward model used by the Q-learning and R-learning algorithms. The forgetting mechanism is central for controlling the size of the STA. The memory fade factors will be used to investigate their effect in the learning speed and the speed of the network growth.

5.10. Reinforcing Desired Behavior

The above discussion about perception is also applicable to the association of sensory input to action. Once a set of sensory inputs have activated the associated action, the action should be checked for the recalled reward experience before execution.

At an early learning stage, the expected reward is recalled directly as the reward value associated with similar cases. At each state, among the learned actions associated with the current state, the one with the highest reward record is executed.

At high levels, most actions are not associated with an extreme value of the score (-1 or 1). For example, many things that a human adult does everyday do not directly relate to physical pain or pleasure. The selection of alternative actions are based on the best-matched past behavior pattern that has been learned and reinforced. The physical pain or pleasure are responsible for establishing those behavior patterns directly at a low level but only indirectly at high levels. A hypothetical example is helpful for explaining this point. Suppose the system was taught verbally by its human teacher (a) not to say a rude words to humans. It also had learned (b) that if it does not listen to the human teacher, it will feel "pain" (a -1 reward). It has also a wired-in mechanism (c) that a recall of "pain" will suppress the action. Suppose that, in some scenario, it recalled some rude words. If it recalls (a) and (b), it will recall "pain" and the mechanism (c) will make it suppress using the rude words. Therefore, after basic behavior has been established through a process of using the physical reward representing "pain" and "pleasure," the system is expected to be taught by being told, without need of resorting to the physical reward very often. Of course, such a high-level behavior cannot be established without a significant period of learning.

6. HOW THE LIVING MACHINES WORK

This section discusses how the living machines work through interactions with its environments.

6.1. Early Learning

As in the case with a human infant, robot-sitting is needed during the early stage of the cognitive development. Initially, the human teacher uses the action imposition to feed actual control signals to the system via a graphical user interface, a joystick, touch sensors, or a robot-arm teaching pendant. This is to simulate the way a human caretaker teaches a human baby by, for example, holding the baby's hands. The trainer may use good actions with a good reward value. After a sufficient period of hand-in-hand training, the living machine can be set free to try by itself, first in the same environment and then gradually in slightly different environments. During this stage, some mistakes may be made, depending on how extensive hand-in-hand training was and how different the new environment is from the training one. The human teacher enters an appropriate reward value in real time to discourage actions that may potentially lead to a failure and encourage actions that can lead to a good result. At this stage, the human teacher may want to speak and use gestures in addition to feeding a reward value so that the system can associate the visual and auditory signal from the human trainer with the corresponding reward values. Later, spoken words and gestures will have the similar effects as the reward values.

6.2. Concept Learning and Behavior Learning

Our task is to investigate how a machine can develop high-level knowledge and motives from low-level "physical feedback." The ideas presented here were inspired by studies in child developmental psychology.

SAIL associates each sensory input with the visual and auditory signals from the human to learn the concepts taught by the human. It will learn good behavior from very early stages, such as "do not run toward a wall," "do not run too fast when there are things nearby," "handle it if you see something like that," and so forth.

The understanding of concepts will be built up through interactions with human, such as the concepts of left, right, up, down, fast, slow, good, bad, right, and wrong. Many experiences and occasions are associated with the concept that is used. The representation of these concepts is implicit in the system. It represents a pattern of response in the network.

Later, many instances will allow the living machine to associate good reward with its actions that follow what the teacher wants. It also will gradually accumulate the information about what the human teacher likes and expects from it. It will then link the concepts of "good and bad" and "right and wrong" with the many instances it experienced.

Generalization of the concepts occurs naturally. For example, the living machine first associates the human teacher's words with the reward value. Some words appear together with a bad reward and they become aversive stimuli; other words appear with a good reward and they become become appetitive stimuli. The living machine gradually establishes a behavior to choose actions that will produce appetitive stimuli (to make the human teacher happy) and avoid actions that will produce aversive stimuli. This is called second-order conditioning in psychology (see, e.g., Ref. 111). At that time, the importance of the physical reward value gradually diminishes. A default physical reward value is often enough during this stage. A more mature system will receive most of its human feedback from normal sensory input, via speech, gestures, and so on.

Further on, the system will link the sense of good and bad to other more complicated activities, such as schooling, understanding the evaluation system in its school, knowing what to do in order to receive a good evaluation from its school, and so forth. In summary, as long as the teacher uses the reward value correctly, he or she will be able to train the living machine to do what he or she wants, from simple to complex, even during later stages when physical reward is used very rarely.

6.3. Cognitive Maturity

In an autonomous learning process, the environment provides an endless sequence of stimuli coupled with the actions from the living machines. Certainly, the living machine should not just remember the sequence as, for example, a single 8-h sensing-action sequence everyday, because a lot of events do not have close relationships. The machine should only remember things that are important at its current cognitive developmental stage.

The cognitive maturity determines what a machine can learn. For example, a child in the sensorimotor stage will not be able to learn formal reasoning, even if a teacher tries that. Before a sufficient amount of low-level knowledge and skills has been learned, it is not possible to learn higher-level knowledge and skills. In the living machine, the maturity scheduling is realized automatically by the process of state self-organization, the forgetting process, and the level building process.

When a set of stimuli representing a high-level concept is sensed by an immature machine, the configuration of the current SHOSLIF tree is not sufficient to learn the new concept associated with the stimuli. This is reflected by the fact that the features generated by the SHOSLIF tree constructed so far is not able to successfully recall a set of learned concepts. Thus, the corresponding stimuli looks meaningless to the living machine and are just simply memorized temporarily. Without extracting recallable features, stimuli containing the same high-level concepts in different occasions look very different. Later on, that temporary memory is quickly forgotten (deleted) by the forgetting process because there is no recall to this memorized stimuli within a certain time period.

If the living machine has learned a sufficient number of low-level concepts and skills, indicated by the corresponding nodes in the SHOSLIF tree, the same stimuli mentioned above will result in a very significant amount of feature recall from the mature SHOSLIF tree. Thus, the corresponding stimuli are memorized at a higher level of the STA hierarchy. Within a certain period of time, if the same concept appears in the stimuli on another occasion, the mature SHOSLIF tree can recall a sufficient number of features and retrieve the newly memorized concept. Such a successful retrieval indicates a memory reinforcement. The forgetting process will record this reinforcement at the corresponding node and apply a much slower memory decay curve to this concept. This concept is then successfully learned by the living machine.

In this way, the living machine is able to learn autonomously and to go through various cognitive developmental stages which are probably similar to those characterized by Jean Piaget (see Table 1). During each day, it learns what it can learn and forgets what it has to forget. As is the case with humans, entering a new cognitive stage by a living machine is natural and gradual, depending on the learning experience with each machine individual. There is no need for the living machine designer to enforce each development stage into the program.

6.4. Thinking

A living machine must be able to think autonomously. How does a machine think? This is a question that has fascinated scientists and nonscientists alike (112). Not many computer researchers like to regard the computational process implemented by a computer as thinking. Otherwise, any program is doing thinking. The thinking process seems to be autonomous. However, autonomy is not sufficient for qualifying thinking. Otherwise, an autonomous road-following vehicle is doing thinking.

6.4.1. We Must Not Program Logic Rules into the Living Machines

A fundamental characteristic of thinking is that the contents and the rules of reasoning cannot be prearranged (i.e., programmed in). Let us consider an example. A commonsense knowledge "all human are mortal" can be represented by compound proposition (tautology) $\forall x[p(x) \rightarrow q(x)]$, where $p(x) = $ "x is a human" and $q(x) = $ "x is mortal." Then, with input $p(\text{Tom}) = $ True, a program can prove $q(\text{Tom}) = $ True using logic deduction rules. When doing the above reasoning, the program is not thinking, because the logic deduction rules are preprogrammed by the human. In order to make a machine think, its representation must be knowledge-free. In our example, the knowledge is formal logic. The representation for $\forall x[p(x) \rightarrow q(x)]$ is not knowledge-free.

It is worth pointing out that symbolic reasoning is not as difficult as the other parts of cognition. In the above case, for example, it is much more difficult to figure out that Tom is a human from visual sensing than to perform the symbolic reasoning.

6.4.2. What Is Thinking?

There is no widely accepted definition of thinking (113). We probably should not attempt to define it here. The following characterization of a thinking process may be considered: (1) A thinking process is autonomous. (2) The representation of the thinking program is free of knowledge (i.e., content independent). (3) The thinking process is conducted according to knowledge that has been autonomously learned. The first two points are meant to distinguish a preprogrammed computation from an autonomous thinking process. The last point is to make sure that thinking is not a useless process or directly directed by humans.

6.4.3. Thinking at the Mental Cycle Level

With the SAIL living machine, a thinking process might result if the attention selection for external sensors stays off while state vectors are fed into each level as we saw earlier. A mental cycle of the thinking process is to go through STA once for each input frame, as shown in Fig 10. A long thinking process corresponds to many consecutive mental cycles going through STA, each with a new state vector from the former mental cycle at each level. Thus, a long thinking process may allow prediction of many chained events. Naturally, all the thinking processes originate from the stimuli in the environment. During a thinking process, the stimuli input from the sensors may be temporarily turned off to allow prediction and planning to be performed internally. When to think and how to think are determined by the system's learned behavior.

6.5. Reasoning and Planning

One might think that in order to conduct reasoning and planning, there must be a controller which controls what to think about and organizes various stages of reasoning and planning. However, no such a controller can work.

6.5.1. Must Not Have a Thinking Controller

There is a fundamental dilemma with this "controller thinking." This controller, as a supervisor, must be smarter than what it controls. In other words, it must know the global situation, understand it, and figure out what to do and how to do it. Furthermore, this controller must be general purpose because the living machine is a general-purpose creature. We know that no controller can perform a reasonable control task without understanding what is occurring. However, understanding is exactly our original problem. Thus, we have a chicken-and-egg problem—one cannot be solved without first having the other.

6.5.2. Programming Behaviors into the System Cannot Work Either

Another alternative is that we do not use any controller. This is the case with some behavior-based methods (e.g., see Ref. 12). However, the behavior-based methods cannot go beyond the very limited number of behaviors that have been explicitly modeled and programmed. None of the existing behavior-based approaches really try to understand the world. Human beings, on the other hand, can gradually learn and understand more and more complex concepts in the world and their high-level behaviors are based on understanding instead of pure wired-in reflexes.

Reasoning and planning are special types of thinking process in the living machines. The thinking process described above is used for reasoning and planning as soon as the machine is thinking about the sequence of events where one event is likely to follow the next event. Humans may teach the living machine how to reason by demonstrating visual examples or by stating a story or theory after the living machine has established a certain language capability.

7. FURTHER THOUGHTS

7.1. Generality

Can a living machine potentially do anything that a human can? This is an open question. The problem here is not just a static function approximation which can be performed by, for example, a three-layer neural network with back-propagation. The critical issues include the dynamic generation of concepts from sensor–effector-based autonomous learning, concept generalization through experience, the capability to automatically generate feature space, and self-organizing the dynamically changing network. Autonomous learning through real-time sensing and action without hand-crafted knowledge-level rules or behaviors is a totally new subject of study. The upcoming study will answer some very important and fundamental questions.

7.2. Space Complexity

Space complexity is directly related to the amount of information learned. The human brain has about 10^{11} neurons, each being connected by roughly 10^3 synapses on average (114–116). If each synaptic link is considered a number, the human brain can store about 10^{14} numbers. This amount is now within reach by hard disks as far as the cost is concerned, thanks to the fast advance in computer storage technology. It is worth noting that the nature of the SHOSLIF tree data allows a moderate compression that is typical in video compression.

It is known that a large part of the neurons in the human brain is not activated. Furthermore, the living machine does not need its brain to control heart beating, breathing, and digestion, which are served mainly by the medulla and the pons in the human brain. It does not need much service from the somatosensory system. The taste system and the olfactory system are not needed either, except for certain special applications. It probably does not need its brain to serve for sexual drive either, which is taken care of partially by the amygdala in the human brain. Further, computers have effective high-level computational mechanisms, such as the fast and effective algorithms for computing eigenvector and eigenvalues of a huge matrix, which is a major part of the computations in SHOSLIF. The known methods for computing eigenvectors by artificial neural networks are slow, iterative, and not as accurate (117–119). Thus, the living machine might be able to reach a good performance with a disk space that is significantly smaller than the absolute storage size of the human brain. Of course, this also depends on the scope of the domain to be learned and the required resolution of the sensors. Because the cost of magnetic hard disks and rewritable optical disks is decreasing fast and consistently, it is now possible to equip a system with a disk system of 1000 GB (about 20% of the absolute storage size of the human brain with a moderate compression) at a cost of about $20,000. It is not clear what kind of performance SAIL can reach with a storage size ranging from 10 GB to 1000 GB. This is one of the major questions to be answered in our project.

If the size of the brain were not an important issue, perhaps a monkey could serve as a living machine—a seemingly cheaper one. However, the size is probably just one of the problems with a monkey's brain. Although a monkey can perform sensing and action tasks that no machine can do so far, the genetic coding of the monkey brain might not enable it to work up to a high level comparable to that of humans. We cannot control the self-organization scheme of the monkey's brain, at least not now. With a living machine, we do not have such a limitation.

7.3. Time Complexity

Time complexity should not be addressed in a conventional way. Here, the time needed to train the system to reach a certain level of performance is on the order of months or years. It is expected that a machine can learn faster than human beings because it does not feel tired and it computes faster. The speed of learning with the living machines seems more controllable than biological systems, such as humans.

The critical type of complexity is the time complexity for each mental cycle when the size of the network becomes very large. Because of our on-line learning and the subspace method, the time complexity for each mental cycle is $O(\log(n))$, which is an inverse function of exponential explosion. In other words, when the processing speed increases, the number of cases it can handle in a fixed 100-ms time interval grows exponentially. To see how low the logarithmic complexity is, assume a system with a time complexity of $O(\log_b(n))$, regardless of how large the constant coefficient is in the time complexity. If this system can complete a mental cycle in 100 ms with 1000 stored cases (which is about the speed SHOSLIF has reached), the same program running on another computer whose speed is 4.7 times faster can finish a mental cycle in the same time interval for a network that has stored 10^{14} cases (the absolute size of the human brain)! This displays a very high potential for the logarithmic time complexity.

7.4. Knowledge Base

The subject of knowledge representation and knowledge base has been studied for many years without serious consideration about sensing and action. A huge amount of human power has been spent in modeling human knowledge, its representation, and input of data into knowledge bases. However, the resulting systems are hard to use, hard to maintain, hard to keep up to date, and are brittle because of a lack of understanding of what has been stored.

What the field of knowledge representation and knowledge base has experienced is a natural stage that we humans must go through on our way toward understanding ourselves, machines, our environments, and their relationships. However, it is about time that we tried a fundamentally different approach, an approach that is much closer to the way a human acquires knowledge. Not too surprisingly, the living machine approach seems to require much less manpower and costs less than many conventional knowledge-base projects, because the human is relieved from the tremendous task of building rules for human knowledge and spoon-feeding human knowledge. For knowledge-base construction, we want to move from manual labor to automation.

7.5. Is It a Formidable Endeavor?

To consider whether the proposed living machine project is formidable, it is helpful to compare the nature of the proposed domain-independent approach with that of task-specific ones.

> **Task-specific approach.** With a task-specific approach, humans manually model knowledge for each task. Thus, each task is very hard because the knowledge required in each task is too vast in amount and too complicated in nature. Accustomed to task-specific approaches, it is hard for us to believe that any algorithm can handle general-purpose learning because each task is already too hard.
>
> **The living machine approach.** With the proposed living machine approach, humans are relieved of the tasks for developing knowledge-level representation, man-

ually modeling knowledge, and programming knowledge-level rules into programs. Instead, we develop a systematic, unified method to model the system's self-organization scheme at the signal level (instead of the knowledge level). Thus, the development task tends to be less labor-intensive because the algorithm is very systematic and domain independent. No knowledge needs to be programmed into the program and one program is meant for various sensing and action modalities.

Developing living machines is certainly not easy. Many unknowns need to be explored and studied. It is naive to think that challenging tasks such as vision, speech, and language are easy to meet once we have an AA-learning framework that probably will eventually work. However, the development of AA-learning systems appears more tractable than many task-specific projects which use intensive task-level knowledges in each area, such as vision, speech, handwritten and mix-printed document recognition, autonomous robots, knowledge-base development, and language understanding.

8. THE FUTURE OF THE LIVING MACHINES

The success of Phase 2 may require coordinated efforts from many research groups in various areas. It is expected that the learning algorithm developed in Phase 2 must undergo many versions of improvement before the goals of Phase 2 can be realized. Such improvements would continue even in Phase 3. Although it is now too early to predict the time table for Phase 2, the implication of the developmental approach and the living machine concepts have opened a new way of thinking about intelligent machines and a new array of future possibilities. Here, we discuss some of those possibilities.

8.1. Smart Toys

Before the living machines have reached a mature level, the first possible mass market application is probably a new generation of smart living toys. They are trainable, smart toys that live with human beings. Such a toy has its sensors (visual, audio, ultrasound, tactile, text-input, etc.) and its effectors (speaker, hand, steerable wheels, text output, etc.), depending on the chosen sophistication level for the toy. The most basic, low-cost version of these smart living toys does not need a mechanical robot body. It is a piece of software in a multimedia computer. A joystick connected to the computer feeds human trainer reward or punishment into the computer [e.g., 1.0 representing a reward (like food) and -1.0 representing a punishment (like hunger)]. The human trainer can also guide actions using a joystick or a graphical interface, very much like the case in which a human teacher trains a dog how to reach a ball by using the dog's legs to touch the ball. Each living toy is loaded with a learning program which controls the toy to learn and to act to maximize the expected reward. Because the trainer trains the toy by playing with it, instead of programming knowledge rules into it, the learning of the living toy is automatic. Thus, any person is able to train such a living toy for fun. Various toy competitions at various sophistication levels can be held once such "living toys" have reached the mass market, which, in turn, further promotes wide ownership and upgrading of such toys.

8.2. A Revolutionary Way of Developing Intelligent Software

After completion of Phase 3 in the above plan, continued education for living machines is more or less like teaching a human child. A large number of living machines will be made. Each newly made machine is loaded with the Phase 3 brain (software and network as a database) as the starting point. Computer experts work with educational experts to teach professional skills to living machines, very much like the way human students are taught. Each professional field or subject has a living machine school, in which several living machine robots are taught to master the professional knowledge. What each school does is to train a generic living machine to become a professional expert. A living machine can be trained to become a spacecraft pilot, a fighter-plane pilot, a deep-sea diver, a waste-site cleaner, a security-zone monitor, an entertainer, a nursing-home caretaker, a tutor, or a personal assistant. Each living machine is a software factory. This represents a revolutionary way of making intelligent software: Hand programming is no longer a primary mode for developing intelligent software, but school teaching is. Teaching and learning are carried out through visual; auditory, and tactile communications. Two types of products will be sold: expert software brains and expert living machines.

8.3. Expert Software Brains as Product

The expert brain (software and network) of each specially trained living machine is downloaded and many copies are sold as software brain. The software-only brain is useful and cheap, but it does not have a full learning capability because of its loss of most sensors and effectors. For example, a software-only brain running on the Internet can only learn from those available on the Internet, based on the knowledge it learned when it had a full array of sensors and effectors. Depending on the richness of the information available from the Internet and how his owner instructs its learning from the Internet, this software brain's knowledge may gradually become obsolete. This is very similar to the case in which a human's knowledge becomes obsolete, once he or she does not keep learning. However, software brain buyers can always get upgrades regularly from software brain makers, very much like the way commercial software gets upgraded now.

8.4. Expert Living Machines as Product

Each living machine school also sells expert "brains" with a physical body as a full living machine with sensors and effectors. Such complete living machines can continue to learn at their application sites to adapt to the new environment and to learn new skills. This type of full living machines are used to extend three types of human capabilities: (1) physical capability, such as operating a machine or driving a car, (2) thinking capability, such as finding the most related literature from a library or replying piled-up electronic mails, (3) learning capability, such as learning to use computers for the handicapped or learning to predict economic ups and downs better. Humans can do things that they enjoy doing, leaving living machines to do other things. Furthermore, the living machines can work round-the-clock without fatigue and loss of concentration, which is something that no human being can match.

8.5. The Longevity of Living Machine

The living machines eventually *outlive* human individuals because once their hardware is broken or worn, the brain can be downloaded and then uploaded to a new hardware. They can also learn faster and learn more hours every day than any human individual. This longevity may lead to tremendous creativities. For example, Albert Einstein passed away and humanity lost his creativity. The creativity and the knowledge that each living machine learned can be preserved for future human generations for more innovations and better service.

8.6. Social Issues with Living Machines

Just as the case with human children or pets, each human teacher is responsible for his or her living machine. If a living machine happens to have learned something that we do not want it to learn, we can always replace its brain with the backed-up copy of, say, yesterday, to take it back to yesterday's status. We do not have such a convenience with a human child or a pet.

8.7. A New Industry

Today, the automobile industry is huge because every household needs at least one automobile. The computer industry is enormous today because the computer is general purpose in extending human's computational needs (and all the functionalities that accompany the computation). Unlike all the special-purpose robots that have been built so far, the living machine is general purpose. Therefore, the market for the living machine is huge. With the decreasing cost that accompanies advances in robotic technology and computer technology and the ever-increasing volume of mass market sales, the living machines will not only enter every business and industrial sector but also millions of households to serve as intelligent personal assistants.

9. CONCLUSIONS

A new approach—the developmental approach—has been introduced here with an algorithmic description.* The developmental approach is motivated by human cognitive development from infancy to adulthood. The key for success of this approach is to realize the automation of general-purpose learning (AA-learning) and to make it effective. The developmental approach does not just mean growing a network from small to large, from simple to complex (80,121), although these concepts are related to cognitive development. More importantly, it means AA-learning: free of manual design for task space, free of manual design for behavior-level modules or decomposition, closed brain, autonomous learning, and learning while performing. Because learning by the living machines is automated and the training process is similar to the way the human teaches children, training for living machines is natural and widely practicable by nonprogrammers. The future

---

* Our earlier account of the concept and motivation of the living machines appeared in a technical report dated December 1996 (120). This chapter is a slightly edited version of that technical report.

challenge is to further improve the basic framework described here and realize the design by real systems (see Refs. 122–124 for some preliminary results of AA-learning using the algorithmic design outlined here). The research on the living machines may help us to further understand sensorimotor learning, visual and speech learning, language acquisition, and thinking, as well as how the mind works. Although it is now too early to promise a success, the future of this direction seems very exciting.

ACKNOWLEDGMENTS

The SHOSLIF work was supported in part by NSF Grant IRI 9410741 and ONR grant No. N00014-95-1-0637. The author would like to thank Daniel L. Swets, Yuntao Cui, Shaoyun Chen, Sally Howden and Wey-Shiuan Hwang, and Jason Brotherton for discussions and development of SHOSLIF-O, SHOSLIF-M, SHOSLIF-N, SHOSLIF-R, and SHOSLIF-S, respectively. Thanks also go to an anonymous reviewer who made constructive comments that were very useful for the improvement of the presentation here.

REFERENCES

1. AM Turing. On computable numbers with an application to the Entscheidungsproblem. Proc London Math Soc 2(42):230–265, 1936.
2. T Pavlidis. Why progress in machine vision is so slow. Pattern Recogn Lett 13:221–225, April 1992.
3. WEL Grimson, JL Mundy. Computer vision applications. Commun ACM 37(3):45–51, 1994.
4. IR Alexander, GH Alexander, KF Lee. Survey of current speech technology. Commun ACM 37(3):52–57, 1994.
5. A Waibel, K Lee. Readings in Speech Recognition. San Mateo, CA: Morgan Kaufmann, 1990.
6. F Jelinek. Self-organized language modeling for speech recognition. In: A Waibel, K Lee, eds. Readings in Speech Recognition. San Mateo, CA: Morgan Kaufmann, 1990, pp 450–506.
7. SR Young, AG Hauptmann, WH Ward, ET Smith, P Werner. High-level knowledge sources in usable speech recognition systems. Commun ACM 32(2):183–194, 1989.
8. KW Church, LF Rau. Commercial applications of natural language processing. Commun ACM 38(11):71–79, 1995.
9. DB Lenat. CYC: A large-scale investment in knowledge infrastructure. Comm ACM 38(11):33–38, 1995.
10. DB Lenat, G Miller, TT Yokoi. CYC, WordNet, and EDR: Critiques and responses. Commun ACM 38(11):45–48, 1995.
11. GA Miller. Worknet: A lexical database for English. Commun ACM 38(11):39–41, 1995.
12. R Brooks. Intelligence without reason. In Proceedings of the International Joint Conference on Artificial Intelligence, Sydney, 1991, pp 569–595.
13. LP Kaelbling, ML Littman, AW Moore. Reinforcement learning: A survey. J Artif Intell Res 4:237–285, 1996.
14. RA Brooks. A robots layered control system for a mobile robot. IEEE J Robot Automat 2(1):14–23, 1986.
15. J Aloimonos. Purposive and qualitative active vision. Proceedings of the 10th International Conference on Pattern Recognition, Atlantic City, NJ, 1990, pp 346–360.
16. C Watkins. Learning from delayed rewards. PhD thesis, King's College, Cambridge, 1989.
17. M Turk, A Pentland. Eigenfaces for recognition. J Cogn Neurosci 3(1):71–86, 1991.
18. J Weng, S Chen. Incremental learning for vision-based navigation. Proceedings of the International Conference on Pattern Recognition, Vienna, 1996, vol IV, pp 45–49.

19. S Wilson. Classifier systems and the Animat problem. Machine Learning 2(3):199–228, 1987.
20. I Harvey. Evolutionary robotic and SAGA: The case for hill crawling and tournament selection. Technical Report CSRP 222, University of Sussex, Brighton, UK, 1992.
21. M Dorigo, M Colombetti. Robot shaping: Developing autonomous agents through learning. Artif Intell 71(2):321–370, 1994.
22. HE Gruber, JJ Voneche. The Essential Piaget. New York: Basic Books, 1977.
23. S Carey. Cognitive development. In: DN Osherson, EE Smith, eds. Thinking. Cambridge, MA: The MIT Press, 1990, pp 147–172.
24. S Carey. Conceptual Change in Childhood. Chambridge, MA: The MIT Press, 1985.
25. PE Bryant, T Trabasso. Transitive inferences and memory in young children. Nature 232:456–458, 1971.
26. S Russell, P Norvig. Artificial Intelligence: A Mordern Approach. Enqlewood Cliffs, NJ: Prentice-Hall, 1995.
27. JE Laird, ES Yager, M Hucka, CM Tuck. Robo-Soar: An integration of external interaction, planning, and learning using Soar. Robot Auton Syst 8:113–129, 1991.
28. W-M Shen. Autonomous Learning from the Environment. New York: Computer Science Press, 1994.
29. A Schwartz. A reinforcement learning method for maximizing undiscounted rewards. Proceedings of the international Joint Conference on Artificial Intelligence, Chambery, France, 1993, pp 289–305.
30. LR Rabiner, LG Wilpon, FK Soong. High performance connected digit recognition using hidden Markov models. IEEE Trans Acoustics Speech Signal Proces ASSP-37:1214–1225, 1989.
31. JR Deller Jr, JG Proakis, JHL Hansen. Discrete-Time Processing of Speech Signals. New York: Macmillan, 1993.
32. S Zeki. The visual image in mind and brain. Sci Am 267(3):69–76, 1992.
33. C Blakemore, GF Cooper. Development of the brain depends on the visual environment. Nature 228:477–478, October 1970.
34. DH Hubel. Eye, Brain, and Vision. New York: Scientific American Library, 1998.
35. JR Anderson. Cognitive Psychology and Its Implications. 3rd ed. New York: Freeman, 1990.
36. JL Martinez Jr, RP Kessner (eds.). Learning & Memory: A Biological View. 2nd ed. San Diego, CA: Academic Press, 1991.
37. P Thompson. Margaret Thatcher: A new illusion. Perception 9:483–484, 1980.
38. VS Ramachandran. Perceiving shape from shading. In I Rock, ed. The Perceptual World. San Francisco: Freeman, 1990, pp 127–138.
39. O Sacks. To see and not see. The New Yorker, 59–73, May 1993.
40. P Sinha, T Poggio. I think I know that face . . . . Nature 384:404, December 1996.
41. M Minsky. A selected descriptor-indexed bibliography to the literature on artificial intelligence. In: EA Feigenbaum, J Feldman, eds. Computers and Thought. New York: McGraw-Hill, 1963, pp 453–523.
42. H Hendriks-Jansen. Catching Ourselves in the Act: Situated Activity, Interactive Emergence, Evolution and Human Thought. Cambridge, MA: The MIT Press, 1996.
43. MA Turk, DG Morgenthaler, KD Gremban, M Marra. VITS—A vision system for autonomous land vehicle navigation. IEEE Trans Pattern Anal Machine Intell PAMI-10(3):342–361, 1988.
44. C Thorpe, MH Hebert, T Kanade, S Shafer. Vision and navigation for the Carnegie-Mellon Navlab. IEEE Trans Pattern Anal Machine Intell PAMI-10(3):362–373, 1988.
45. N Nilsson. SRI A.I. Center Technical Note. Technical Report 323, Stanford Research Institute, April 1984.
46. K Fukunaga. Introduction to Statistical Pattern Recognition. 2nd ed. New York: Academic Press, 1990.
47. R Duda, P Hart. Pattern Classification and Scene Analysis. New York: Wiley, 1973.

48. AK Jain, RC Dubes. Algorithms for Clustering Data. Englewood Cliffs, NJ: Prentice-Hall, 1988.
49. J Quinlan. Introduction of decision trees. Machine Learning 1(1):81–106, 1986.
50. L Breiman, J Friedman, R Olshen, C Stone. Classification and Regression Trees. New York: Chapman & Hall, 1993.
51. R Michalski, I Mozetic, J Hong, N Lavrac. The multi-purpose incremental learning system AQ15 and its testing application to three medical domains. Proceedings of the Fifth Annual National Conference on Artificial Intelligence, Philadelphia, 1986, pp 1041–1045.
52. MJ Mataric. Integration of representation into goal-driven behavior-based robots. IEEE Trans Robot Autom RA-8(3):304–312, 1992.
53. R Brooks, LA Stein. Building brains for bodies. Technical Report 1439, MIT AI Lab Memo, Chambridge, MA, August 1993.
54. M Minsky. The Society of Mind. New York: Simon and Schuster, 1986.
55. C Yoken. Living with Deaf-Blindness: Nine Profiles. Washington, DC: National Academy of Gallaudet College, 1979.
56. JM McInnes, JA Treffry. Deaf–Blind Infants and Children. Toronto: University of Toronto Press, 1982.
57. HP Moravec. The Stanford Cart and the CMU Rover. Proc IEEE 71(7):872–884, 1982.
58. S Franklin. Artificial Minds. Cambridge, MA: The MIT Press, 1997.
59. MJ Mataric. A distributed model for mobile robot environment—Learning and navigation. Technical Report AI-TR-1228, MIT Artificial Intelligence Laboratory, Cambridge, MA, 1990.
60. DL Swets, J Weng. Using discriminant eigenfeatures for image retrieval. IEEE Trans Pattern Anal Machine Intell PAMI-18(8):831–836, 1996.
61. H Murase, SK Nayar. Visual learning and recognition of 3-D objects from appearance. Int J Computer Vision 14(1):5–24, 1995.
62. W Hwang, SJ Howden, J Weng. Performing temporal action with a hand–eye system using the SHOSLIF approach. Proceedings of the International Conference on Pattern Recognition, Vienna, 1996.
63. ML Puterman. Markov Decision Processes. New York: Wiley, 1994.
64. J Weng. Cresceptron and SHOSLIF: Toward comprehensive visual learning. In: SK Nayar, T Poggio, eds. Early Visual Learning. New York: Oxford University Press, 1996.
65. J Weng. SHOSLIF: A framework for sensor-based learning for high-dimensional complex systems. Proceedings, IEEE Workshop on Architectures for Semiotic Modeling and Situation Analysis in Large Complex Systems, Monterey, CA, 1995, pp 303–313.
66. J Weng. On comprehensive visual learning. Proceedings of the NSF/ARPA Workshop on Performance vs. Methodology in Computer Vision, Seattle, WA, 1994, pp 152–166.
67. J Weng. SHOSLIF: A framework for object recognition from images. Proceedings, IEEE International Conference on Neural Networks, Orlando, FL, 1994, pp 4204–4209.
68. J Weng. SHOSLIF: A learning system for vision and control. Proceedings, IEEE Annual Workshop on Architectures for Intelligent Control Systems, Columbus, OH, 1994.
69. K Karhunen. Uber lineare methoden in der wahrscheinlichkeitsrechnung. Ann Acad Sci Fennicae, ser A1, Math Phys, 37, 1946.
70. IT Jolliffe. Principal Component Analysis. New York: Springer-Verlag, 1986.
71. JH Friedman. A recursive partition decision rule for nonparametric classification. IEEE Trans Computers C-26:404–408, April 1977.
72. SK Murthy. Automatic construction of decision trees from data: A multidisciplinary survey. Data Mining Knowledge Discovery, 1998.
73. Q Li, J Weng. SHOSLIF convergence properties. Technical Report CPS 96-26, Department of Computer Science, Michigan State University, East Lansing, MI, May 1996.
74. TM Cover, PE Hart. Nearest neighbor pattern classification. IEEE Trans Inform Theory IT-13:21–27, 1967.

75. TM Cover. Estimation by the nearest neighbor rule. IEEE Trans Inform Theory IT-14:50–55, 1968.
76. D Swets, J Weng. Discriminant analysis and eigenspace partition tree for face and object recognition from views. Proceedings of the International Conference on Automatic Face- and Gesture-Recognition, Killington, VT, 1996, pp 192–197.
77. DL Swets, J Weng. Hierarchical discriminant analysis for image retreival. IEEE Trans Pattern Anal Machine Intell, accepted to appear, 1999.
78. K Etemad, R Chellappa. Discriminant analysis for recognition of human face images. Proceedings of the International Conference on Acoustics, Speech, Signal Processing, Atlanta, GA, 1994, pp 2148–2151.
79. PN Belhumeur, JP Hespanha, DJ Kriegman. Eigenfaces vs fisherfaces: Recognition using class specific linear projection. IEEE Trans Pattern Anal Machine Intell PAMI-19(7): 711–720, 1997.
80. J Weng, N Ahuja, TS Huang. Learning recognition using the Cresceptron. Int J Computer Vision 25(2):109–143, 1997.
81. KK Sung, T Poggio. Example-based learning for view-based human face detection. IEEE Trans Pattern Anal Machine Intell PAMI-20(1):39–51, 1998.
82. AJ Colmenarez, TS Huang. Face detection with information-based maximum discrimination. Proceedings, IEEE Conference on Computer Vision Pattern Recognition, 1997, pp 782–787.
83. HA Rowley, S Baluja, T Kanade. Neural network-based face detection. IEEE Trans Pattern Anal Machine Intell PAMI-20(1):23–38, 1998.
84. B Moghaddam, A Pentland. Probabilistic visual learning for object detection. Proceedings of the International Conference on Computer Vision, Cambridge, MA, 1995, pp 84–91.
85. T Darrell, A Pentland. Space-time gesture. Proceedings, IEEE Conference on Computer Vision Pattern Recognition, New York, 1993, pp 335–340.
86. A Bobick, A Wilson. A state-based technique for the summarization and recognition of gesture. Proceedings of the 5th International Conference on Computer Vision, Boston, 1995, pp 382–388.
87. T Starner, A Pentland. Visual recognition of American sign language using hidden Markov models. Proceedings of the International Workshop on Automatic Face- and Gesture-Recognition, Zurich, 1995, pp 189–194.
88. A Lanitis, CJ Taylor, TF Cootes, T Ahmed. Automatic interpretation of human faces and hand gestures using flexible models. Proceedings of the International Workshop on Automatic Face- and Gesture-Recognition, Zurich, 1995, pp 98–103.
89. R Kjeldsen, J Kender. Visual hand gesture recognition for window system control. Proceedings of the International Workshop on Automatic Face- and Gesture-Recognition, Zurich, 1995, pp 184–188.
90. J Triesch, C von der Malsburg. Robust classification of hand posture against complex background. Proceedings of the International Conference on Automatic Face- and Gesture-Recognition, Killington, VT, 1996, pp 170–175.
91. Y Cui, D Swets, J Weng. Learning-based hand sign recognition using SHOSLIF-M. Proceedings, IEEE International Conference on Computer Vision, Cambridge, MA, 1995, pp 631–636.
92. Y Cui, J Weng. Hand segmentation using learning-based prediction and verification for hand-sign recognition. Proceedings, IEEE Conference on Computer Vision Pattern Recognition, 1996, pp 88–93.
93. TS Huang, VI Pavlovic. Hand gesture modeling, analysis, and synthesis. Proceedings of the International Workshop on Automatic Face- and Gesture-Recognition, 1995, pp 73–79.
94. ED Dickmanns, A Zapp. A curvature-based scheme for improving road vehicle guidance by computer vision. Proceedings, SPIE Mobile Robot Conference, Cambridge, MA, 1986, pp 161–168.

95. C Thorpe, M Herbert, T Kanade, S Shafer. Toward autonomous driving: The CMU Navlab. IEEE Expert 6(4):31–42, 1991.
96. DA Pomerleau. Efficient training of artificial neural networks for autonomous navigation. Neural Comput 3(1):88–97, 1991.
97. M Rosenblum, S Davis. An improved radial basis function network for visual autonomous road following. IEEE Trans Neural Networks NN-7(5):1111–1120, 1996.
98. M Meng, AC Kak. Mobile robot navigation using neural networks and nonmetrical environment models. IEEE Control Syst 31–42, August 1993.
99. J Weng, S Chen. Vision-guided navigation using SHOSLIF. Neural Networks, II, pp. 1511–1529, 1998.
100. W Hwang, J Weng. Vision-guilded robot manipulator control as learning and recall using SHOSLIF. Proceedings, IEEE International Conference on Robotics and Automation, Albuquerque, NM, 1997.
101. SB Kang, K Ikeuchi. A robot system that observes and replicates grasping tasks. Proceedings of the International Conference on Computer Vision, Cambridge MA, 1995, pp 1093–1099.
102. T Kuniyoshi, M Inaba, H Inoue. Learning by watching: Extracting reusable task knowledge from visual observation of human performance. IEEE Trans Robot Automat RA-10(6):799–822, 1994.
103. J Brotherton, J Weng. HEARME: Speaker independent word recognition using SHOSLIF. Technical Report CPS 96-33, Department of Computer Science, Michigan State University, East Lansing, MI, October 1996.
104. MH Ashcraft. Human Memory and Cognition. New York: Harper Collins College Publishers, 1994.
105. T Kohonen. Self-Organization and Associative Memory. 2nd ed. Berlin: Springer-Verlag, 1988.
106. T Kohonen. Self-Organizing Maps. 2nd ed. Berlin: Springer-Verlag, 1997.
107. LR Rabiner. A tutorial on hidden Markov models and selected applications in speech recognition. Proc IEEE 77(2):257–286, 1989.
108. J Yamato, J Ohya, K Ishii. Recognizing human action in time-sequential images using hidden Markov model. IEEE Conf. Computer Vision and Pattern Recognition. 379–385, 1992.
109. GA Miller. The magical number seven, plus or minus two: Some limits on our capability for prcessing information. Psychol Rev 63:81–97, 1956.
110. AD Baddeley, N Thompson, M Buchanan. Word length and the structure of short-term memory. J Verbal Learning Verbal Behav 14:575–589, 1975.
111. M Domjan. The Principles of Learning and Behavior. 4th ed. Belmont, CA: Brooks/Cole, 1998.
112. R Wright. Can machines think? Time, 50–56, March 25, 1996.
113. A Newell. Unified Theories of Cognition. Cambridge, MA: Harvard University Press, 1990.
114. MI Posner, ME Raichle. Images of Mind. New York: Scientific American Library, 1994.
115. B Kolb, IQ Whishaw. Fundamentals of Human Neuropsychology. 3rd ed. New York: Freeman, 1990.
116. JR Anderson. Cognitive and psychological computation with neural models. IEEE Trans Syst Man Cybern SMC-13(5):799–815, 1983.
117. J Mao, AK Jain. Artificial neural networks for feature extraction and multivariate data projection. IEEE Trans Neural Networks NN-6(2):296–317, 1995.
118. J Rubner, K Schulten. Development of feature detectors by self-organization. Biol Cybern 62:193–199, 1990.
119. J Rubner, P Tavan. A self-organizing network for principal component analysis. Europhys Lett 10:693–698, 1989.
120. J Weng. The living machine initiative. Technical Report CPS 96-60, Department of Computer Science, Michigan State University, East Lansing, MI, December 1996.

121. JL Elman. Learning and development in neural networks: The importance of starting small. Cognition 48:71–99, 1993.
122. J Weng, W Hwang. Toward automation of learning: The state self-organization problem for a face recognizer. Proceedings of the 3rd International Conference on Automatic Face- and Gesture-Recognition, Nara, Japan, 1998.
123. J Weng, W Hwang. Sensorimotor action sequence learning with application to face recognition under discourse. Proceedings of the International Conference on Pattern Recognition, Brisbane, 1998.
124. J Weng, W Hwang. A sensorimotor system for spatiotemporal learning with application to face recognition under discourse. Technical Report CPS 98-9, Department of Computer Science, Michigan State University, East Lansing, MI, March 1998.

16
Principles of Halftoning with Stochastic Screens

Qing Yu* and Kevin J. Parker
University of Rochester, Rochester, New York

1. INTRODUCTION TO DIGITAL HALFTONING

Digital halftoning is the process used to reduce the number of quantization levels per pixel in a digital image while maintaining the gray or color appearance of the image at normal viewing distance. This technique is widely employed in the printing and display industries, where many devices are binary in nature or have a limited number of output levels. It has been widely accepted that the human visual system (HVS) has low-pass characteristics, which implies that the eye tends to average a region around a pixel. This averaging process can create the illusion of many gray and color levels even though the actual image is rendered with only black and white dots, or cyan, magenta, yellow and black dots for the color case. The halftoning process governs the distribution of these dots with the constraint that the perceived gray or color level is preserved.

In this chapter, we briefly review some basic concepts in halftoning, and we focus on active research areas in stochastic screening. The term stochastic screening has been applied to techniques that exhibit relatively unstructured but visually pleasing halftones.

1.1. Ordered Dither

The history of halftoning technology can be dated back to the last century, when physical screens and gauzes were used to generate halftone images. These techniques have been translated into digital halftoning directly. Some excellent comprehensive reviews have been published, including Refs. 1–4. A brief orientation is given here. Ordered dither is the natural digital solution, where a two-dimensional threshold array is designed and the halftoning process is accomplished by a simple pixelwise comparison of the gray-scale image against the array (Fig. 1). This method is straightforward and requires little computation; thus, ordered dither is the most popular and widely used technique. Depending on the progressive ordering of how halftone dots in a cell are turned on/off, ordered dither can be classified into clustered-dot and dispersed-dot classes.

---

* *Current affiliation*: Eastman Kodak Company, Rochester, New York.

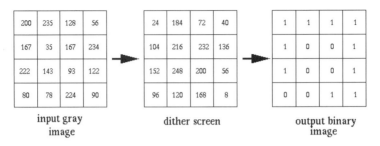

Fig. 1 Schematic of ordered dithering.

In clustered-dot ordered dither, adjacent pixels in a cell are turned "on" sequentially so as to form a cluster in the halftone cell. Clustered-dot dither is primarily used for printing devices that have difficulty producing isolated pixels. Obviously, this congregation of pixels will result in significant low-frequency structure in the output image. On the other hand, in dispersed-dot ordered dither, halftone dots in a cell are turned "on" individually without grouping into clusters. Therefore, sharp edges can be better rendered than in clustered-dot dither. However, dispersed-dot techniques are more susceptible to dot gain, a problem that is considered in a later section. Figure 2 compares clustered-dot and dispersed-dot patterns.

1.2. Stochastic Processes

Many problems of ordered dither, including susceptibility to Moiré patterns and highly visible texture, can be traced back to the rigid regular structures of the halftone cell. To break these regular structures, researchers searched for less obtrusive halftone patterns. Blue-noise halftoning, also called stochastic screening or frequency-modulation (FM) screening, has been the most active research field in digital halftoning in recent years. These terms have been loosely applied to both algorithm approaches and the screen ap-

Fig. 2 Image obtained with ordered dithering: (a) clustered dot; (b) dispersed dot.

proach. Error diffusion (5) is the algorithm approach that has been studied most extensively, whereas the blue Noise Mask (6,7) is the term first applied to a screen, or threshold array, that produces an unstructured, visually appealing halftone pattern. In order to follow a precise definition from this point, the term "stochastic screening" is used if a threshold array is actually employed. Also, "mask" and "screen" will be used interchangeably when both will refer to a threshold array.

1.2.1. Error Diffusion

Error diffusion is an adaptive algorithm that produces patterns with different spatial-frequency content depending on the gray value of the input image. It forces the average gray value to remain the same and attempts to localize the distribution of tone levels. Figure 3 is the flowchart for error diffusion. This approach was first presented by Floyd and Steinberg in the 1970s (5). Subsequently, many modifications and derivations have been proposed to the error filter design (8), the choice of threshold value (9), the feedback loop (10), as well as the sequence in which the pixels are processed (11). Although all algorithms that have been listed involve intensive computation and exhibit some artifacts, the quality of the halftone image is generally considered excellent (12), particularly the sharp edges and many image details. Most of the success of error diffusion lies in the fact that, under many conditions, it is a "good blue-noise generator," as pointed out by Ulichney (1). In the academic literature, the nature of various types of noise is often described by a reference to a color (i.e., white noise is so named because of its "flat" power spectrum so that all frequencies have approximately equal power). On the other hand, blue noise has most of its energy located at higher spatial frequencies, with very little low-frequency content. Typical white-noise and blue-noise radial-average power spectra (RAPS) are shown in Fig. 4. The RAPS plot is a convenient one-dimensional plot derived from the two-dimensional power spectrum of a binary pattern (1). Patterns with blue-noise characteristics generally exhibit aperiodic uncorrelated dot patterns without low-frequency graininess.

1.2.2. Stochastic Screens

Stochastic screen halftoning is the subject of active research. It combines the simplicity of ordered dither with the blue-noise quality of error diffusion. Stochastic screen halftoning

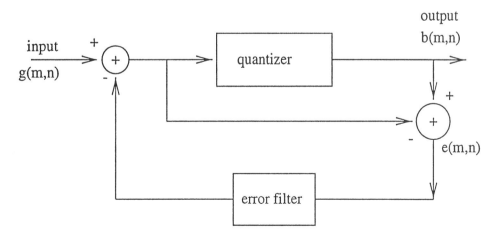

Fig. 3 Computational algorithm for standard error diffusion.

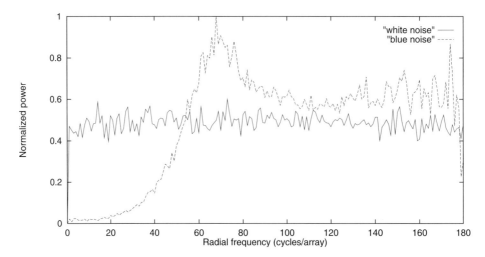

Fig. 4 Typical white-noise and blue-noise RAPS.

is a point-comparison process, so it is easy to implement in devices currently using the ordered dither technique. The halftone image from a stochastic screen will have the visually pleasing blue-noise characteristics, which is guaranteed when the screen is generated from blue-noise dot patterns of individual gray levels. The Blue Noise Mask proposed by Mitsa and Parker is the first stochastic screen to realize the above scheme (6,7,13). (See Fig. 5).

The following sections will consider the design of stochastic screens and their application in black-and-white halftoning, multilevel halftoning, and color halftoning. Our review focuses on the scientific literature published in peer-reviewed forums. In Sec. 2, the construction of the prototypical stochastic screen, the Blue Noise Mask, is outlined. Section 3 will detail the common filter approaches in screen construction, and various filter design

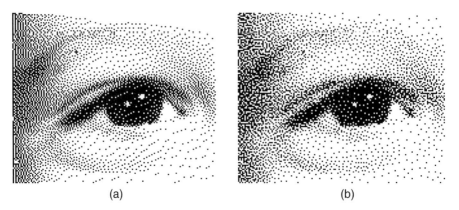

Fig. 5 Images obtained with stochastic techniques: (a) error diffusion (serpentine); (b) the Blue Noise Mask.

techniques will be examined. In Sec. 4, the optimality of blue-noise binary patterns in terms of screen design is pursued. In Sec. 5, various modifications of stochastic screens to meet special application requirements are introduced, such as screens with dot-gain compensation, screens for fax encoding, and screens for multilevel output devices. In Sec. 6, different schemes for stochastic color halftoning are presented, followed by an evaluation based on a human visual model. Section 7 is a summary, and current problems with stochastic screen halftoning will be identified and future research will be proposed.

2. CONSTRUCTION OF A STOCHASTIC SCREEN—BLUE NOISE MASK

In this section, the algorithm to generate a Blue Noise Mask (BNM) is presented (6,7,13,14). First, a good initial binary pattern $b[i, j, g]$ [a two-dimensional (2-D) binary pattern at gray level g) is required for some intermediate level g ($0 < g < 255$, assuming an 8-bit mask). With the filtering technique which will be presented in Sec. 3, such a pattern with blue-noise characteristic is obtained and used as the initial pattern. From this initial pattern, an initial mask $m[i, j]$ is generated, which produces the initial binary pattern $b[i, j, g]$ when used to halftone the constant gray image of level g.

Once the binary pattern for level g is completed, the binary pattern for the next brighter level, $g + 1$, is processed as shown in Fig. 6. The blue-noise pattern is created by converting the appropriate number (K) of "0"s to "1"s in the previous pattern g, with K given by

$$K = \frac{N^2}{L} \tag{1}$$

where N is the size of the mask and L is the total number of levels. After the binary pattern

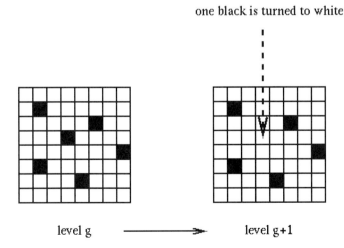

Fig. 6 Build a blue-noise binary pattern for level $g + 1$ from a blue-noise binary pattern at level g.

for level $g + 1$ is done, the mask $m[i, j]$ is updated as follows:

$$m[i, j] = m[i, j] + \text{NOT}(b[i, j, g + 1]) \tag{2}$$

where NOT is the logical NOT operation, $m[i, j]$ is the mask file, and $b[i, j, g + 1]$ is the binary pattern for level $g + 1$. This process is repeated until the mask has been updated for all the levels above g to level 255. To construct the mask for levels below the initial level g, the status of "1"s and "0"s are reversed to the upward case and the mask updating is given by

$$m[i, j] = m[i, j] - b[i, j, g - 1] \tag{3}$$

Otherwise, the downward construction is basically identical to the upward construction. The resulting single-valued 8-bit function $m[i, j]$ will be the final BNM.

The following algorithm (6,7,13,14) has been proposed as an generic but efficient way to generate blue-noise binary patterns for any individual gray level. The algorithm employs a filter to determine suitable pixel locations for the placement of black and/or white elements into an existing binary pattern:

1. Set the initial number M of pairs of "1"s and "0"s to be swapped in each iteration.
2. Specify a 2-D low-pass filter with optional anisotropy (approximating eye-sensitivity characteristics) appropriate for the gray level. The filter can be specified in either image or frequency domain.
3. Filter the binary pattern with the 2-D low-pass filter in the appropriate domain.
4. Form an error array by computing the difference between the filtered pattern and the gray level the pattern represents.
5. Sort the errors into two cases: For all pixels containing a "1" in the binary pattern, sort the positive error; for all pixels containing a "0" in the binary pattern, sort the negative error.
6. Swap the M pairs of "1"s and "0"s that have the highest (absolute value) positive and negative errors.
7. Compute the frequency-weighted mean square error (FWMSE) of the filtered pattern with respect to the grey level. If the FWMSE drops, go to Step 3 and proceed to the next iteration; if FWMSE increases but $M > 1$, reduce M by half, go to Step 3; otherwise, stop.

To apply this algorithm to mask construction, only K [defined in Eq. (1)] new pixels are converted from "1" to "0" or vice versa for each binary pattern. (See Fig. 7).

There is a significant constraint on the converting and swapping operation in this mask construction. In making a mask, the binary patterns at different levels are not independent. For example, in the upward construction process, all the "1"s in the binary pattern for level g are contained in the binary pattern for level $g + 1$; thus, when swapping "1"s and "0"s, these common "1"s shared by the two neighboring levels cannot be changed. This is a requirement for the creation of the mask as a single-valued function.

Another important part of this swapping scheme is its computational efficiency. As shown above, the process for each level begins by swapping a fairly large proportion (initial M value) of the K converted "1"s with an equal number of filter-selected "0"s. When the FWMSE stabilizes, M is reduced by half to further decrease the FWMSE. This process is repeated until M reaches 1 and the FWMSE stops decreasing. This scheme

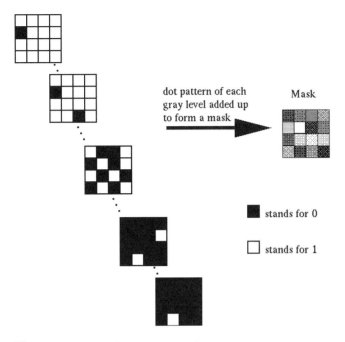

Fig. 7 Schematic of BNM construction.

would achieve the effect of simulated annealing (10), which is a computationally intensive process, with much less time.

The construction technique outlined above is quite general and has enabled the generation of masks with different properties such as 8-bit depth (level 0–255) and 12-bit depth (level 0–4095), small size (64 × 64) and large size (256 × 256), and isotropic and anisotropic (explained at the end of next section).

There are two critical issues in designing a good mask: the digital filters and the optimization. These topics will be covered in Secs. 3 and 4, respectively.

3. COMMON FILTER APPROACHES

As described in the previous section, the algorithm for generating the BNM constructs the binary pattern at any gray level $g + 1$ from the binary pattern at level g. Furthermore, the BNM algorithm filters the binary pattern for level g to select the location of pixels that would be the "best candidates" for adding to the majority pixels required for level $g + 1$. Once binary patterns for all gray levels have been sequentially produced from some "seed" level in this way, the binary patterns can be summed to produce a threshold array or Blue Noise Mask. The role of filters in this procedure warrants close attention, and a summary of some important results is given in this section.

Mitsa and Parker (6,7,13) selected the location of pixels that should be changed from minority to majority values by finding the extremes of an error function $e(i, j)$. This error function was generated by directly filtering the binary pattern of level g and then subtracting the original binary pattern from the filtered pattern. Specifically, the error function in the image domain is

$$e(i, j) = [h_{hp_g}(i, j) * b_g(i, j)] - b_g(i, j) \qquad (4)$$

where $h_{hp_g}(i, j)$ is a high-pass filter selected for level g, $b_g(i, j)$ is the initial binary pattern, and $*$ denotes circular convolution. In the spatial frequency domain, the corresponding equation is

$$E(k, l) = B(k, l)[H_{hp_g}(k, l) - 1] \qquad (5)$$

Because $H_{hp_g}(k, l)$ was described as a blue-noise (high-pass) filter, the overall filter $[H_{hp_g}(k, l) - 1]$ is a low-pass filter. Thus, an essential feature of this process is that low-frequency "clumps" of binary patterns can be located directly by filtering the binary pattern. Also, the high-pass region (or corresponding low-pass region) can be related to the principal frequency f_g, which is a function of the gray level g. Mitsa and Parker also suggested that the filter H_{hp_g} could be adapted to directly shape the RAPS of the binary pattern for level $g + 1$. This filter is derived from

$$H_{hp_g}(\rho) = \sqrt{\frac{D(\rho)}{B_g(\rho)}} \qquad (6)$$

where $D(\rho)$ is the desired blue-noise RAPS for level $g + 1$ and $B_g(\rho)$ is the known RAPS for level g. As with any filtering approach, care should be taken to avoid unwanted discontinuities in the spatial-frequency domain that produce "ringing" in the image domain. This approach is depicted in Figs. 8 and 9. Note that this specific filter is computationally more complicated than simple low-pass filters, but the approach demonstrates the central requirement of providing a low-pass filter with a cutoff frequency linked to gray

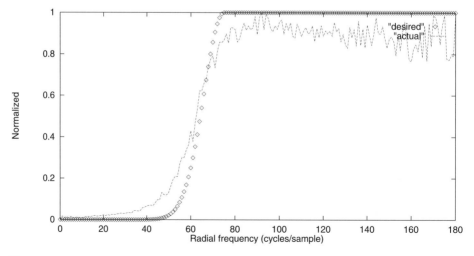

Fig. 8 Desired blue-noise RAPS (filter $D(\rho)$) and actual blue-noise RAPS $B_g(\rho)$ for a blue-noise pattern at gray level 210.

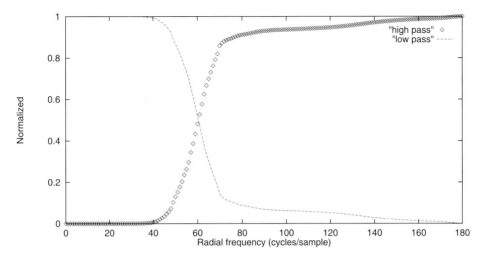

Fig. 9 High-pass filter generated with Eq. (6) and corresponding low-pass filter according to Eq. (5).

level g by the principal frequency (15):

$$f_g = \begin{cases} \dfrac{\sqrt{g}}{R} & \text{for } g \leq \tfrac{1}{2} \\ \dfrac{\sqrt{1-g}}{R} & \text{for } g > \tfrac{1}{2}, \end{cases} \qquad (7)$$

where R is the distance between the addressable points of the printing or display device. This relationship is plotted in Fig. 10. The dashed line in Fig. 10 corresponds to the factor of $1/\sqrt{2}$ in principal frequency (or $\sqrt{2}$ in average separation) that was discussed by Mitsa and Parker (16) as an empirical choice of the transition cutoff frequency for some filters.

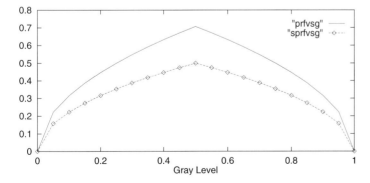

Fig. 10 Principal frequency (solid line) for each gray level (gray levels are normalized to the range [0:1]).

In a later paper, Parker et al. (17) demonstrated how changes in the filter cutoff frequency could produce different types of final halftone patterns at a single gray level. Specifically, if f_c represents the cutoff frequency of a filter, let $f_c = Kf_g$, where K is an adjustable scaling factor. For $K = 0.5$ to 1.0, binary patterns of different "textures" were produced. K was set to $1/\sqrt{2}$ for use with the filters employed in Ref. 17.

Ulichney (18) further explored the filter issue, choosing a 2-D Gaussian filter implemented in the image domain for direct operation on the binary pattern. He demonstrated that for arrays of size 32×32 and smaller, even a single Gaussian filter (not adjusted for gray levels as in the previous work) could produce a useful BNM for some applications. Of course, the "seed" pattern and the Gaussian width of the filter had to be chosen carefully to produce a desirable result. In general, it is beneficial to make the cutoff frequency (or Gaussian width parameter) adapt to gray levels (14,16). Yao and Parker (14) also demonstrated that a variety of low-pass filter shapes could produce desirable halftone patterns, so long as the filter parameters were adjusted for appropriate cutoff with respect to the principal frequency. Mitsa and Brathwarte (19) further developed the concept of the filter bank for different gray levels as a wavelet concept. Low-pass filters were each adjusted for a specific cutoff below the principal frequency ($K = \frac{1}{2}$, as suggested) and related to a wavelet-type filter bank. Dalton (20) described the use of bandpass filters to produce textured binary patterns. Thus, the general lessons from this work are that direct filtering of binary patterns can be useful for selecting pixels to be changed to produce a desired result in the image domain and, thus, a correspondingly approximate power spectrum in the spatial frequency domain. To illustrate the varieties of filters that have been used, Figs. 11 and 12 depict the shape of the filter function in the spatial-frequency domain and image domain from Mitsa and Parker (6,16), Ulichney (18), and Yao and Parker (14). In each case, the filter represented is the one for level 210 out of 256. Only isotropic filters are shown, but anisotropic filters have also been used (14).

Note that the specification of a filter as a low-pass filter in the image domain provides perhaps the easiest way to explain the algorithm to new designers. However, the benefit of specifying the filter in the spatial-frequency domain, and as an initial high-pass operation, is that the final RAPS of the binary pattern will generally approximate the shape of the high-pass filter that is specified in Eq. (4) or (5) (14,16,17). Thus, a halftone designer considering

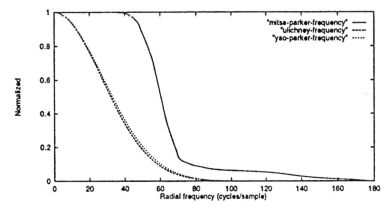

Fig. 11 Three different filters in the spatial frequency domain.

Halftoning with Stochastic Screens

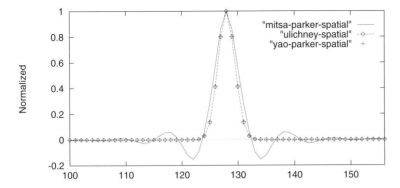

Fig. 12 Three different filters in the image domain.

the final RAPS of the binary pattern can envision changes resulting from different filter shapes by considering the filter initially as high pass as taught in the early references (6,7,13,16).

Modifications of the filters may be further explored for specific applications, such as the manipulation of the spectral peak at the principal frequency. In filtering the binary pattern with a simple low-pass filter, such as a Gaussian, the low-frequency contents of the binary pattern are minimized by changing certain identified pixels. By careful choice of a bandpass filter, one can enhance selective frequencies in the spatial-frequency domain, through the proper identification of certain pixels in the image domain. For example, consider the binary pattern at level 210 shown in Fig. 13 as

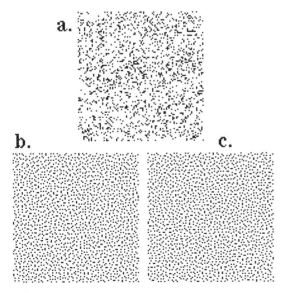

Fig. 13 (a) White-noise "seed" pattern; (b) final pattern from simple Gaussian filter; (c) final pattern from Gaussian filter with band-reject component.

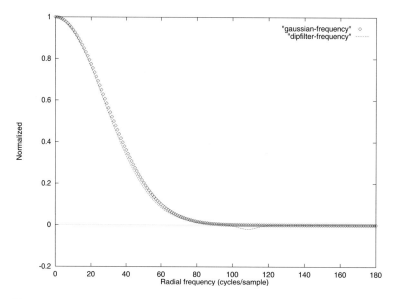

Fig. 14 Radial profiles of the two filters in frequency domain. A small "dip" is seen near frequency sample 110 in one curve.

pattern a. Now, apply two different filters with the algorithm for identifying and swapping pixel values:

1. Gaussian:
$$F(u, v) = e^{-(u^2+v^2)/2\sigma^2} \tag{8}$$

2. Gaussian with a band-reject component:
$$F(u, v) = e^{-(u^2+v^2)/2\sigma^2} - ae^{-(\sqrt{u^2+u^2}-f_g)^2/2\sigma'^2} \tag{9}$$

Figure 14 shows both filters in the spatial-frequency domain, where a is set as 0.02 and f_g is approximately 110. Each of these two filters was repeatedly applied to the white-noise "seed" pattern with pixel swapping until the FWMSE of the binary pattern stopped decreasing. The resulting patterns are given in Figs. 13b and 13c and their corresponding RAPS in Fig. 15.

Thus, selective enhancement or suppression of portions of the power spectrum can be implemented by the proper selection of filters. Note also that although the difference in low-pass filters (Fig. 14) is subtle, the resulting binary patterns have a recognizably different RAPS. The examples given here demonstrate the wide utility of designing binary patterns with particular characteristics.

The characteristic response of the human visual system (HVS), as reported by Sullivan et al. (10), can also be incorporated into the filter. The HVS model they employed is shown in Fig. 16. Their approach generated unstructured binary patterns, each one 32 × 32 in size and "tileable." An independent binary pattern was generated for each gray level by using a cost function with HVS weighting to guide a Monte Carlo approach with simulated annealing (10) in the creation of individual binary patterns.

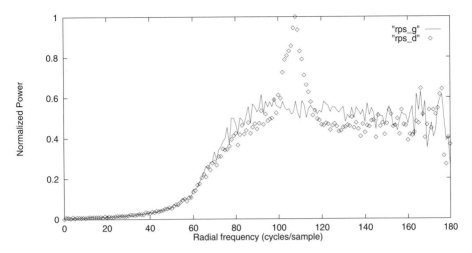

Fig. 15 RAPS for the two final blue-noise patterns from two different filters.

As a general rule, the human eye has maximum sensitivity in the horizontal and vertical directions and minimum sensitivity in the diagonal direction. Therefore, an anisotropic filter may be designed so that the resulting halftone pattern will have more energy in the less visible diagonal direction than in the horizontal and vertical directions. For example, the 2-D Gaussian filter $F(u, v)$ in Eq. (8) may be modified in the following way:

$$F'(u, v) = [1 + c \cos(4\theta)] F(u, v) \qquad (10)$$

where θ is the central angle around the DC point and c is a constant between 0 and 1 which controls the energy distribution (14).

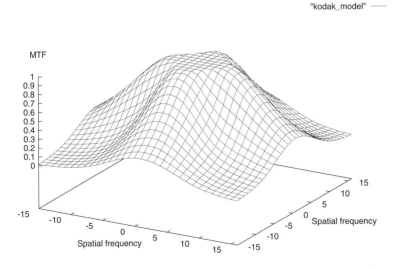

Fig. 16 Human visual model by Sullivan et al.; c is set to 0 in this case. (From Ref. 10.)

4. OPTIMALITY OF BLUE-NOISE BINARY PATTERN AND STOCHASTIC SCREEN

The filtering techniques presented in last section produce visually pleasing blue-noise binary patterns. This section addresses the question of whether the binary patterns are optimal for their corresponding gray levels and for the construction of a mask.

Yao et al. (21) has given a detailed mathematical analysis of the construction of the Blue Noise Mask based on a human visual model. This analysis provides insights to the filtering process and also prescribes the locations of dots that result in a binary pattern with minimum perceived MSE (or FWMSE) when swapped. The analysis puts a lower bound on the lowest FWMSE that is achievable assuming that a filter based on the HVS model is used to measure FWMSE. As Yao et al. pointed out, the difference between the local filtered output of the largest white "clump" and the largest black "clump" (defined as DWB) must be greater than a certain value T in order for the FWMSE to be further reduced:

$$T = \frac{1}{4\pi\sigma^2} \tag{11}$$

where σ is the standard deviation of the adaptive filter based on a human visual model. Therefore, if the filtering technique is employed and the DWB value keeps decreasing during the iterations, a nonzero lower limit of FWMSE will be reached. An example will be given later.

To surpass this limit, a postfiltering algorithm may be used. By locally enforcing a vector process after filtering, the FWMSE of a binary pattern is further reduced, and more visually pleasing binary patterns are obtained. This new algorithm (22) will be presented first. Then, a series of binary patterns for a certain gray level are generated, ranging from a white-noise pattern to a very structured pattern. From the study of these patterns in both the spatial-frequency domain and image domain, the question of optimization is explored.

4.1. Electrostatic Force Algorithm

This new algorithm is based on the model of electrostatic force between charges. As illustrated in Fig. 17, point charges of same polarity repel each other and charges of different polarity attract. The force is proportional to $1/r^2$, where r is the distance between the charges. In the case of a binary pattern where the pixel values are either "1" or "0", if all the minority pixels are treated as point charges with " + " polarity and all the minority pixels are treated as point charges with " − " polarity, then there will be interactive forces between all the pixels. Furthermore, if those minority pixels are allowed to move freely under the net electrostatic force from pixel charges in a neighborhood, a nearly homogeneous distribution of those minority pixels should be expected after a period of time. As every pixel has some amount of force acting on it, a threshold value should be set such that only when the force surpasses the threshold value will the pixel move to minimize the net force. Remember that the starting pattern for this new algorithm is obtained from the filtering process, which is already free of "clumps." Therefore, the histogram of the net force on every minority pixel should be highly peaked near the value of 0. In this case, it is reasonable to assume the mean value of net force on every minority pixel to be 0 and the threshold value (T) can be related to the standard deviation (SD) of force distribution as

Halftoning with Stochastic Screens

attract each other

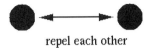
repel each other

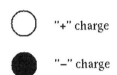
"+" charge

"−" charge

Fig. 17 The electrostatic model.

$$T = V(\text{SD}) \tag{12}$$

V is a variable that will adapt to the gray level and iteration number. As binary patterns are two dimensional, the force calculation and pixel movement should be done in the horizontal and vertical directions (or X and Y directions), respectively.

The steps of this new algorithm are as follows:

1. Set the neighborhood size according to the gray level of the pattern and initial value of V (normally around 1). The maximum value and increment of V should also be set in this step.
2. Make a copy of the starting pattern 1 and denote this as pattern 2.
3. Calculate the net force on each minority pixel in pattern 1 from its neighborhood in the X and Y directions.
4. Calculate the standard deviation of these two groups of force and set the initial threshold T_x and T_y using Eq. (12).
5. For every minority pixel j of pattern 1, compare the absolute value of $(f_x)_j$ against T_x. If the force exceeds T_x, move the corresponding pixel of pattern 2 one pixel in the direction determined by the sign of $(f_x)_j$. Otherwise, make no movement. The same procedure is done in the Y direction for that same pixel.
6. After this process is completed for every minority pixel in pattern 1, we obtain a new pattern 2. The FWMSE (between the binary pattern and corresponding uniform gray pattern) for pattern 1 and for pattern 2 are compared. If the FWMSE of pattern 2 is less than that of pattern 1, pattern 2 is accepted and used to update pattern 1, and go back to Step 2; otherwise, V is increased by a certain amount. If V is less than the maximum value set in Step 1, go back to Step 2. Otherwise, the process is terminated.

A different force-relaxation model for adaptive halftoning of images was proposed by Eschbach and Hauck (23).

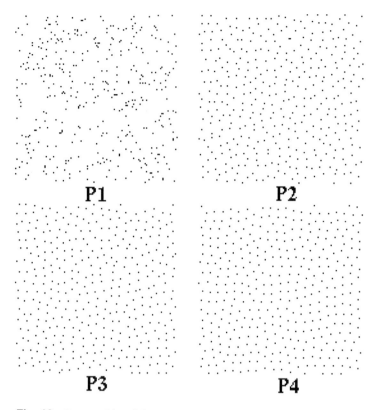

Fig. 18 Patterns P1 to P4.

4.2. A Progressive Series of Binary Patterns

To illustrate the procedure, a white-noise pattern (P1 in Fig. 18) at level 245 is used as an initial pattern, then the frequency-domain filtering is applied, where the standard deviation for the filter is empirically set as 2.4 (out of a 128 × 128 frequency-domain Gaussian filter) in this case. Figure 19 shows the FWMSE drop versus iteration number and Fig. 20 shows the DWB value for each iteration. Because the initial pattern is white noise, the DWB is quite large; therefore, the FWMSE of the binary pattern keeps going down in each iteration. After a certain number of iterations, the DWB approaches the lower bound set by Eq. (11), in this case approximately 0.0137; at this point, filtering can no longer improve the binary pattern in terms of FWMSE. Figure 18 shows the binary pattern (P2) obtained from the filtering process with a FWMSE value of 0.263. From this pattern, the force algorithm is implemented. The neighborhood size is set as 13 × 13 and the starting value of V is set as 1.5. Figure 18 shows the binary pattern (P3) after just five iterations with FWMSE of 0.165. It is quite obvious that by locally enforcing the vector force process, FWMSE could be furthered reduced and more uniform patterns could be generated.

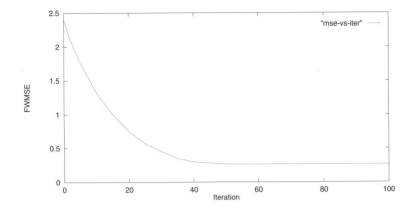

Fig. 19 FWMSE versus iteration in the filtering process.

4.3. Optimality Issue

Without strict proof here, it is noted that the force algorithm converges after many iterations. Figure 18 shows the final pattern (P4) obtained when the algorithm converges after 75 iterations. The final FWMSE is 0.087. This pattern has a relatively ordered structure.

Because all the binary patterns in a mask are constructed from a "seed" pattern, the decision to choose a "seed" pattern should be based on its suitability for mask generation. In other words, an optimal binary pattern at level g should not degrade the quality of its neighbor levels, such as levels $g + 1$, $g - 1$, and so on.

Obviously, a white-noise pattern cannot be the optimal one because of the existence of low-frequency structures. However, a highly structured pattern such as pattern P4 in Fig. 18 is not optimal either. If this highly structured pattern is used as a "seed" pattern and neighboring levels are constructed, the binary patterns of these neighboring levels are

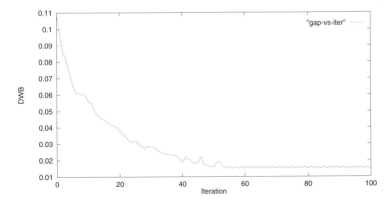

Fig. 20 Difference between the filter response to the largest white clump and the largest black clump in the filtering process.

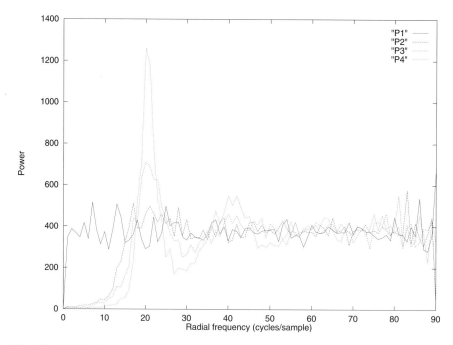

Fig. 21 RAPS of patterns P1 to P4.

generally visually annoying due to the noticeable disruption of the semiregular patterns established by that "seed" pattern. This leads to the question of what pattern constitutes an optimal blue-noise pattern for BNM construction.

Figure 21 shows the RAPS of the patterns presented in Fig. 18. P1 has the typical flat white-noise characteristics, P2 and P3 have the typical blue-noise characteristics, and P4 has a very high peak at the principal frequency along with an emerging "second harmonic" peak. The trend is very obvious that the spectrum starts from the white-noise shape, gradually switches into the blue-noise shape, and ends up with a concentration of energy around the principal frequency of the corresponding gray level. Therefore, in carrying out the force algorithm, a criterion should be set that will enable the computer to terminate the process once an excessive concentration of energy at corresponding principal frequency is reached.

Another experiment has been carried out to study the effect on neighboring binary patterns by blue-noise patterns at one level. For an intermediate gray level (level 245), another similar series of binary patterns are generated in the same way as above. In this case, pattern 0 corresponds to the output from the filtering technique, pattern 5 is the output after 5 iterations of the force algorithm on pattern 0, and pattern 50 is the output after 50 iterations of the force algorithm on pattern 1. Then, each of these patterns is used as a seed pattern to build three different BNMs (mask0, mask5, and mask50). The plots of FWMSE versus gray level for the three different BNMs are shown in Fig. 22 (only levels 240–250 are shown). The plots for mask0 (corresponding to pattern 0) and mask50 (corresponding to pattern 50) show extreme values around the initial level (level 245).

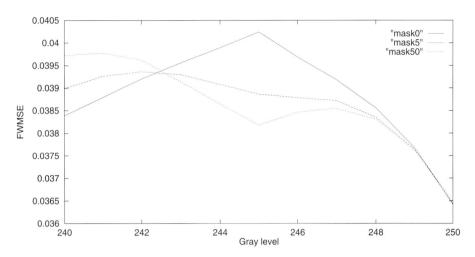

Fig. 22 FWMSE versus gray level for three different masks (mask0, mask5, and mask50). Mask50 had the most ordered seed pattern at level 245, and mask0 had the least ordered blue-noise seed pattern at level 245.

Therefore, the smoothness of a FWMSE transition could serve as a parameter to design an optimal binary pattern.

5. SPECIAL APPLICATIONS

All the discussions so far have considered the design of an ideal stochastic screen for an ideal device. However, because real printers and displays are not ideal, special screens can be designed to meet individual application requirements. Although these requirements could be met with different preprocessing and postprocessing techniques, by incorporating the device characteristics into screen design, both the rendering time and the memory requirement can be reduced. Beyond these, with careful design, stochastic screens can be furthered employed in applications such as fax encoding and watermarking.

5.1. Dot-Gain Compensation

In digital printing, one major concern is dot gain (24). When dots are printed on paper, the actual dot area coverage is normally much higher than the expected value. The dot-gain problem can be attributed to ink spread or dot overlap and usually a combination of both, as shown in Fig. 23. Another way to look at dot gain is that it is related to the area-to-perimeter ratio of printed dots. The area refers to the area of paper covered by a dot, measured in pixels, and the perimeter is the total length of travel around a printed dot. It is easy to see that the smaller the ratio, the bigger the dot gain. For an isolated dot, this ratio is 0.25, assuming a unit diameter for a printed dot. In clustered-dot dither, because the halftone dots in a cell are connected to each other, this ratio is less than 0.25, which is the reason that clustered-dot dither generally shows less dot gain. With stochastic screens,

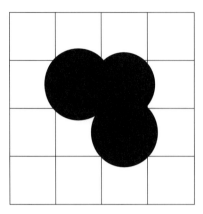

Fig. 23 Dot-gain model.

isolated and dispersed halftone dots are typically generated; therefore, dot-gain compensation will be a necessary step. In practice, dot-gain compensation is usually performed with lookup tables. With stochastic screen halftoning, dot-gain compensation can be actually included in the design process (25). In general, these approaches can be classified into two categories. One requires printing fewer black dots than required in the ideal case. The other is by printing black dots in a preferred way to reduce dot gain without changing the number of black dots for each level.

5.1.1. Printing Fewer Black Dots

The nonlinearity of a specific printer can be directly accounted for in the construction of a mask. Gray patches of certain levels are first printed to measure the printer input–output characteristic curve. Then, a corresponding curve to compensate for this nonlinearity is generated. This curve will show how many dots are actually needed to correctly render a gray level. Thus, instead of converting K pairs of ''1''s and ''0''s to moving up/down one level, a variable number of dot pairs are converted according to the compensation curve. Figure 24 shows the printer curve, an ideal mask curve, and a typical mask curve with dot-gain compensation for a printer.

Sometimes, if the printer characteristic curve is not available during mask design or if several masks must be designed, a mask with 12 bits of dynamic range (levels 0–4095) can be built first instead of the typical 8-bit case. The mask construction is exactly as in Sec. 2, except that 4096 gray levels are considered. Once that 12-bit mask is completed, it will be very easy to map it back to an 8-bit mask when the printer curve is available. Also, many different 8-bit masks might be generated from this 12-bit one with different mapping strategies.

5.1.2. Printing Dots in a Preferred Way

The above two methods reduce dot gain by printing fewer black dots than in an ideal case (i.e., the number of black dots printed corresponds to a lighter level than the desired one, but the desired level is achieved due to ink spread or dot overlap). Another approach to reduce dot gain is to increase the area-to-perimeter ratio. Two special masks called nonsymmetric mask and checkmask have been designed based on this approach.

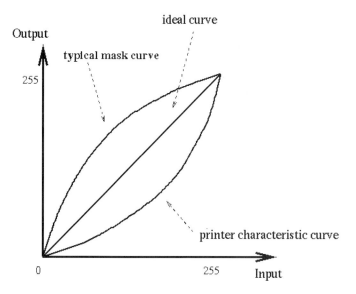

Fig. 24 Printer characteristic curve and lookup table curve.

5.1.3. Nonsymmetric Mask

Generally speaking, dot gain can be severe for dark gray levels because black dots are the majority dots. In the construction of a BNM, to go from level g to level $g - 1$, certain white dots have to be replaced with black ones. Normally, with a low-pass filter to pick up those white-dot candidates (as described in Sec. 2), connected white dots are more likely to be selected than isolated white ones. Therefore, as the total number of white dots is decreasing, the number of isolated white dots is actually increasing, which leads to a decreased area-to-perimeter ratio and a potential dot-gain problem. To avoid this, isolated white dots can be eliminated first as we move to lower levels, while keeping the connected white dots intact until there are no more isolated white ones. With this modified algorithm, dot gain could be reduced to a certain degree. Because the resulting binary patterns at darker levels contain connected white dots in small "clusters," the use of this nonsymmetric mask is analogous to automatically switching the printer to a coarser resolution in dark regions where dot gain is a problem, as illustrated in Fig. 25.

5.1.4. Checkmask

In making the checkmask, a mid-gray checkerboard pattern is generated first, where each pixel is replicated to a 2 × 2 dot. This replicated checkerboard is used as the binary pattern for level 128, hence the name checkmask for the final mask built from this "seed" pattern. There are two reasons to use a replicated checkerboard as a starting pattern. First, the area-to-perimeter ratio is increased. Second, it will be suitable to those printers having difficulty printing fine detail at the largest spatial frequency. Assuming an 8-bit mask, a binary pattern at level 128 will have its principal frequency at high frequency and most of the energy is concentrated around the principal frequency, which is also the highest spatial frequency in this case. A replicated 2 × 2 checkerboard has its energy shifted to lower frequencies, thus avoiding the highest spatial frequency and easing the requirement on a printer. A partial gray ramp of one checkmask is shown in Fig. 26.

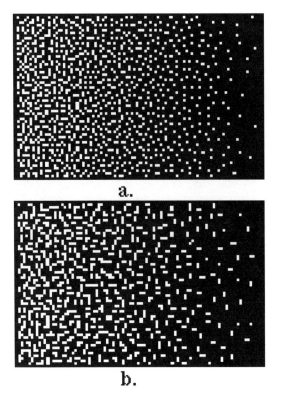

Fig. 25 Comparison between partial ramps of (a) a regular mask and (b) a nonsymmetric mask.

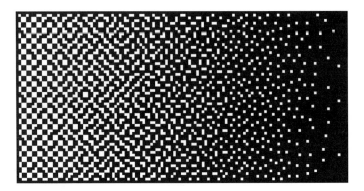

Fig. 26 A checkmask (partial ramp).

5.2. Fax Encoding

The conventional method for sending images via facsimile machines employs run-length coding for compression. For halftone images of stochastic screens where halftone dots are usually dispersed, the run-length coding algorithm is very inefficient. However, if the typical stochastic screen is known to both the transmitter and receiver, then the characteristics of that screen can be exploited in the encoder and decoder design, which leads to higher compression ratios and less transmission time.

With knowledge of the screen, the following processing techniques have been proposed and included in the ToneFac algorithm (26–28):

1. Block segmentation that represents halftone images by mean block values and an error image
2. Quantization and coding of block values
3. Bit switching that transforms the error image into a more compressible image
4. Spurious dot filtering which removes perceptually insignificant dots.

Use of the ToneFac algorithm can reduce transmission time by a factor of 8, depending on image characteristics. Applications to halftone video segments have also been derived (29).

5.3. Stealth Fax and Digital Watermarking

The randomness and correlation statistics of the stochastic screen and individual binary patterns can also be utilized for concealing or encoding information. When two uncorrelated blue-noise binary patterns are overlayed or added, the result yields an unstructured superposition. However, consider the special case where the overlay is either the complement of the blue-noise pattern (black pixels become white and vice versa) or is the original pattern itself. In the former case, the result of superposition is an uniformly black field. In the latter case, the result of superposition is identical to the original pattern. However, if we perturb the binary pattern only within a small preselected region such as within a boldface alphanumeric character, then the result of superposition using the original unperturbed pattern or its complement will reveal the perturbed pattern, thus making the preselected region visible. This concept was first announced by Yao and Parker (30,31) for a fax application.

Fax machines and other information-transfer machines are often shared by several users. This calls for the need to encrypt the message during transmission of private messages while allowing the receiving individual to decode the information. This encryption can be done in the halftoning process at the transmitting end and decoded at the receiving end with the same stochastic screen.

In the following, we will explain a method and apparatus for private binary messages. An input image with text, images, or other information is sent to the system, which generates an unreadable private message and an individual key. The private message contains the original information but is encrypted in such a way that it is not recognizable to the human eye. The individual key is sent in advance to be used by the recipient to decode the information in the private message. A stochastic screen can conveniently generate those private keys. (See Fig. 27.)

The detailed procedures are as follows:

1. From a specific screen, the binary pattern for a certain gray level is generated. This pattern will be used as the private key for an individual.

Fig. 27 Fax encoding and decoding: (a) original image; (b) key image; (c) encoded image; (d) decoded image.

2. As the system is receiving the input image, whenever a background pixel is encountered in the image, the corresponding pixel value of the key is inverted and assigned to the encoded image at that position. In comparison, whenever a pixel with text information (characters) is encountered, a different encoding is employed (e.g., the key is shifted by one pixel in either the horizontal or vertical direction) and the new pixel value at original position is complemented and assigned to the encoded image. In these regions, black pixels become white and vice versa. Thus, the location of the text information is encoded into the halftone binary pattern in the form of shifted halftone pixels. Due to the interaction of the halftone pixels corresponding to text information and their neighboring pixels, the text will not be recognizable by the eye as long as a stochastic screen is used. So far, we have assumed that the key has the same size as the input image; if this is not the case, then the key pattern will be tiled to cover the whole image.
3. When the encrypted document is received, a logical OR operation of the private message and the individual key immediately brings out the original information. Because the background part of the encoded image and the individual key are complementary to each other, the OR operation will blacken the entire back-

ground, leaving only the region with text information in a certain shade of gray due to the shifted pixels, thus making it readable. The logic OR operation can be realized either digitally or optically. The digital operation can be done by a computer, whereas the optical operation can be implemented by overlaying the individual key in the form of a sheet of transparency on the received private message, revealing light text characters on a dark background. As most fax machines print at 200 dpi, (dots per inch), the private message could be transmitted at 100 dpi so that the fax machines can print a reliable "2 × 2" dot.

An almost infinite number of distinct individual keys can be generated by the system because it is possible to produce a nearly infinite set of unique stochastic screens. Therefore, it can be different to decode the process without knowing the exact key pattern.

Knox and Wang (32) also investigated the overlay properties of the stochastic screen and proposed a scheme to incorporate digital watermarks in printed halftone images using the stochastic screen, based the principles outlined in this section.

5.4. Multitoning

Stochastic screen halftoning could easily be generalized for use with devices having a multilevel output. Typically, such techniques are referred to as multitoning. Figure 28 shows a generalization of the stochastic screen technique for application to multilevel devices. It can be seen that this is equivalent to the binary implementation except that the quantization operation replaces the threshold operation.

Assume an input image $I(x, y)$ has p different levels and the device has q possible output levels, also assume the screen $M(x, y)$ is 256 × 256 in dimension and has 256 levels. Then, the multitoning process can be given by

$$O(x, y) = \text{INT}\left[\left(I(x, y) + \frac{M(x_m, y_m)}{256} \frac{p-1}{q-1}\right)\left(\frac{p-1}{q-1}\right)^{-1}\right] \qquad (13)$$

where $O(x, y)$ is the output value, $x_m = x \text{ MOD } 256$, and $y_m = y \text{ MOD } 256$, INT ( ) indicates an integer truncation, and MOD stands for the modulation operation. It can be

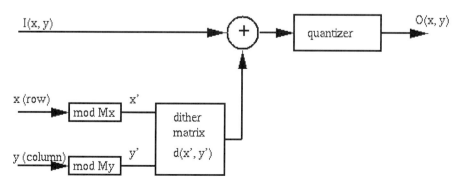

Fig. 28 Schematic of multitoning process.

seen that the screen value is scaled and added to the input image before a simple threshold operation.

Although any stochastic screen designed for black-and-white halftoning could be used in this processing for multitoning, improvement would result by optimizing screens specifically for multitoning. Spaulding and Ray (33) have investigated methods that minimize a visual cost function within a quantization interval; they have also reported using nonuniform quantization functions in the multitoning implementation.

6. COLOR HALFTONING

Color imaging normally requires mixing of three additive primary colors [red, green, and blue (RGB)) for cathode-ray tube (CRT) display or subtractive primary colors [cyan, magenta, and yellow (CMY)] for print. Printing technology can also utilize a fourth primary (K) to provide a better black hue, enlarge the color gamut, and improve image quality. Color halftoning is the process of generating halftone images for the different color planes of a printing device, such as cyan, magenta, yellow, and black. Color-image halftoning is significantly more complicated than halftoning a gray-scale image. All the qualities required of black-and-white halftone images apply to color halftone images that are composed of multiple color planes, but, in addition, the interactions between color planes must be precisely controlled.

6.1. Conventional Color Halftoning Approach

In conventional halftoning, the same clustered-dot screen can be used to halftone the C, M, Y, and K planes separately to obtain four halftone images, which are then used to control the placing of color on paper. One immediate problem of this scheme is the appearance of Moiré patterns, which are caused by the low-frequency components of the interference of different color planes. When combining periodic signals, such as two color halftone screens of vector frequency f_1 and f_2, interference produces a "beat" at the vector difference frequency $f_b = f_1 - f_2$ (2). If the individual color screens were made at the same angle and frequency, any slight spatial-frequency modulation due to misregistration in the printing process forms a low-frequency visually objectionable beat. To reduce the visibility of Moiré patterns, the screens are typically oriented at different angles, usually 30° apart. At 30° separation, the frequency of the Moiré is about half the screen frequency, thereby producing a high-frequency "rosette" pattern. Color errors caused by misregistration only appear at high frequencies and therefore are not easily detected by eye. To achieve the highest frequency and least visible rosette patterns, it is typical practice to orient cyan at 105°, magenta at 75°, yellow at 0°, and black at 45° (Fig. 29). Because yellow and black have the least and most impact on visual sensitivity, respectively, they are oriented at angles where human eyes are most and least sensitive. Although yellow is at 15° relative to the other color planes, its impact on intercolor Moiré is not objectionable because of low contrast. Recently, research on color printing has been directed at introducing more colors to expand the color gamut (4). Rotation angles of these colors must be assigned to prevent visible Moiré patterns. However, there is a limit to the number of angles. Therefore, it can be difficult to apply the conventional color halftoning technique to high-fidelity color printing.

Halftoning with Stochastic Screens

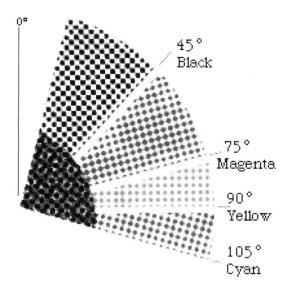

Fig. 29 Screen-angle rotation.

Stochastic halftoning such as error diffusion and stochastic screen eliminate the Moiré concern completely, as the halftone dots created by these processes are relatively unstructured, thus eliminating the constraints of the rotation angle. Therefore, high-fidelity color printing is relatively simple to implement using stochastic halftoning.

6.2. Color Halftoning Using Error Diffusion

One characteristic of the error diffusion technique is that when applied to the color planes separately (the so-called "scalar error diffusion"), the halftone dot patterns for different color planes can be highly correlated. To improve the visual quality of color halftones using error diffusion, Miller and Sullivan (34) processed the color image in a vector color space. Each image pixel is treated as a color vector. This techniques is called vector error diffusion. The color image is first converted to a nonseparable color space such as CIELAB, and the pixel is assigned with the closest halftone color in that color space. The resulting vector halftone error is distributed to neighboring pixels in the same manner as scalar error diffusion.

Klassen et al. (35) also proposed a vector error diffusion technique that minimized the visibility of color halftone noise. It is a well-known property of the human visual system that the contrast sensitivity decreases rapidly with increasing spatial frequency. Thus, the minimum threshold above which patterns are visible rises rapidly with increasing spatial frequency. One approach to increase the spatial frequency of color halftone dots so as to minimize the visibility of color halftone noise is to select low-contrast color combinations wherever possible and to generate the finest possible "mosaic" pattern without large clumps. Therefore, light gray is printed using nonoverlapping cyan, magenta, yellow, and unprinted white pixels, as opposed to printing occasional black clusters on a large white background.

This concept of utilizing the finest possible patterns also serves as a fundamental rule for designing schemes that employ stochastic screens for color halftoning.

6.3. Color Halftoning Using Stochastic Screens

The following schemes have been proposed (36) to apply the BNMs to halftone color planes. These schemes are illustrated in Fig. 30.

6.3.1. The Dot-On-Dot Scheme

The simplest application of the mask utilizes the same mask for each color plane; this is known as the dot-on-dot technique. Although this approach is the easiest to implement, it is rarely used in practice because it results in the highest level of luminance modulation and is most sensitive to misregistration.

Suppose a gray patch is to be rendered with equal levels of cyan, magenta, and yellow. With the dot-on-dot approach, the halftone image for each color plane is identical. Therefore, cyan, magenta, and yellow dots will be printed at the same locations. As a result, the final output patch will be formed with dispersed black dots on the white background. This situation is analogous to using only black and white dots to render a gray even when the output device has multilevel output ability. This will produce a higher level of luminance modulation than in the cases where color dots are not coincident. Therefore, the halftone patterns will be more visible. Another possible outcome under this scenario

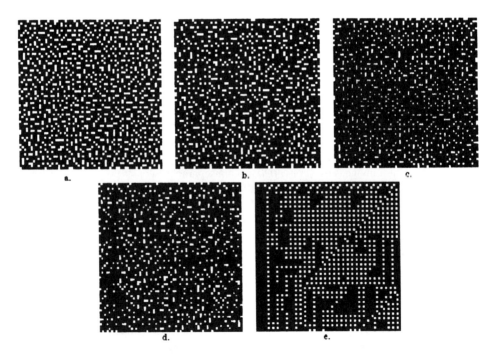

Fig. 30 A color patch halftoned with different schemes: (a) dot on dot; (b) shifted mask; (c) inverted mask; (d) four mask; (e) error diffusion.

is that if one color plane is misregistered relative to others, the annoying color "banding" effect will appear with possible color shifts.

6.3.2. The Shifted-Mask Scheme

To decrease correlation of the color planes, spatially shifted masks are employed for halftoning each color plane. This will also increase the spatial frequency of the printed dots. For example, one mask can be used for the cyan plane, then this mask can be shifted in the horizontal and vertical directions in a wraparound manner and applied to the magenta plane. Similarly, the mask can be shifted by different amounts to be applied to yellow and then black. This technique will tolerate misregistration and, therefore, is more robust for real printing processes. However, if the set of shifts are not chosen carefully, low-frequency structures may appear if a color pattern is overlapped with its shifted version. Generally, the shift values must be tested first by printing some overlapping gray patches halftoned with the shifted masks and deciding if these shifts are acceptable.

6.3.3. The Inverted-Mask Scheme

In this strategy, one mask is applied to one color plane and its complement is applied to another. The 8-bit complement mask is

$$m_i[i, j] = 255 - m[i, j] \tag{14}$$

In light regions, this scheme results in the nonoverlapping arrangement of color dots that exhibits large spatial frequencies. However, this scheme is only applicable for two color planes (typically, cyan and magenta); therefore, some other scheme has to be used for the other color planes.

6.3.4. The Four-Mask Scheme

This scheme is actually an extension of the complement technique. It is based on the same idea of increasing the spatial frequency of the color halftone dots and minimizing the low-frequency energy introduced by the overlapping of color planes. However, this scheme places no limit on the number of masks that can be used, as the inverted scheme does, so it is more appropriate for halftoning high-fidelity color images with multiple color planes.

In the case of four CMYK color planes, four anticorrelated masks are generated from four mutually exclusive "seed" patterns, as illustrated in Fig. 31. The steps are as follows:

1. Generate one Blue Noise Mask that produces visually pleasing unstructured binary patterns.
2. Assume that the mask values are from 0 to 255; for all the mask locations that have a value in the interval 0–63, a binary pattern is defined by setting the pixels corresponding to these locations to black dots, and the remaining pixels are set to white.
3. Three more binary patterns are made in the same way by picking the location of pixels with values in the range 64–127, 128–191, and 192–255.
4. This construction ensures that these binary patterns exhibit blue-noise characteristics and are mutually exclusive. The filtering and swapping technique can be further used to eliminate any residual periodic structures.
5. These four binary patterns are used as "seed" patterns to generate four masks.

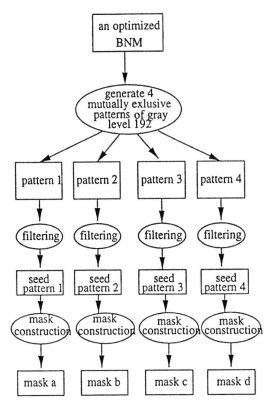

Fig. 31 Schematic of four-mask set construction.

When these four masks are applied to different color planes, they generate color halftone dots that are maximally dispersed, therefore achieving the highest spatial frequency, especially at highlight areas.

6.3.5. Evaluation Using a Human Visual Model in CIELAB Space

In this section, a human visual model is used to evaluate the effect of different stochastic color halftone schemes on the luminance and chrominance of a color test patch (37). Figure 32 shows the block diagram of this evaluation; the HVS model by Sullivan et al. (10) is used in this case. Both the halftoned and the original color patch are passed through the human visual model and then converted to the CIELAB space. The color difference in the CIELAB space is given by

$$\Delta E = \sqrt[2]{(\Delta L^*)^2 + (\Delta a^*)^2 + (\Delta b^*)^2} \tag{15}$$

where ΔL^*, Δa^*, and Δb^* are corresponding differences between two colors. This color difference can be further broken up into components of luminance error ΔL^* and chrominance error ΔC^*, with the later given by

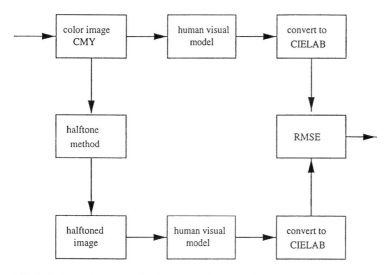

Fig. 32 Flowchart for color halftone scheme evaluation.

$$\Delta C^* = \sqrt[2]{(\Delta a^*)^2 + (\Delta b^*)^2} \tag{16}$$

Our analysis (36,37) shows that different perceived errors are produced by different schemes. In general, the dot-on-dot scheme results in minimum chrominance error but maximum luminance error, whereas the four-mask scheme results in minimum luminance error but maximum chrominance error. The shift scheme falls between these two errors.

6.3.6. Adaptive Color Halftoning

Beyond the previous methods, one solution to reduce perceived colorimetric error is to apply two mutually exclusive masks on two color planes first and then to apply an adaptive scheme on other planes. Another advantage of this adaptive scheme is that color reproduction could be taken into account (38). Figures 33 and 34 show the flowchart of this new scheme.

For this method, it is necessary to know the $L^*a^*b^*$ values of the eight primary colors of the destination printer. Thus, solid patches of the primary colors are first printed, then measured, and a lookup table (LUT) of the $L^*a^*b^*$ values for each primary color is generated.

Next, one mask is applied to the cyan plane and its inverted version on the magenta plane. In this way, the highest spatial frequency is achieved. At each image pixel, there are only two possible values for the yellow plane, either 255 or 0. Therefore, only one from two possible primary colors (c_1 and c_2) for that image pixel should be selected. To do that, the corresponding $L^*a^*b^*$ values for c_1 and c_2 as well as the $L^*a^*b^*$ values for the original pixel (c_0) have to be identified. Then the distances in CIELAB space between c_1 and c_0, and c_2 and c_0 are calculated, and the primary color which gives the smaller distance is selected. Finally, the luminance and chrominance errors between the chosen halftone color and original color are calculated and passed to neighboring pixels in the error diffusion sense.

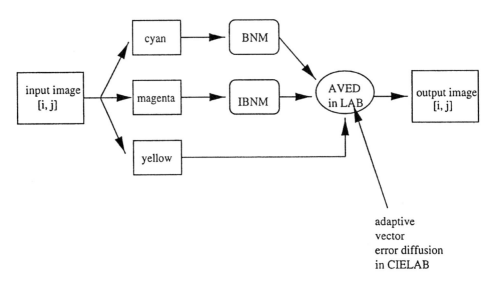

Fig. 33 Overview of adaptive color halftone scheme.

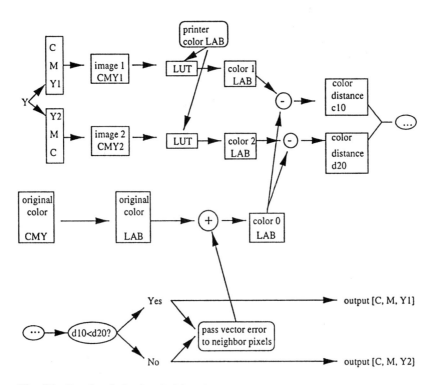

Fig. 34 Details of adaptive decision step.

To compare the performance of this adaptive scheme with other schemes mentioned earlier, a real image is halftoned with all the schemes presented so far and the perceived luminance and chrominance errors of the resulting halftone images are evaluated (38).

Assuming the viewing distance at 10 in. and printer resolution at 300 dpi, the FWMSE between the original image and each color halftone image for the luminance channel and chrominance channel are calculated. As the results show (38), the lowest colorimetric error (especially luminance error) is achieved by using the adaptive scheme. The trade-off is computational complexity. However, by having one color channel be adaptive, increased flexibility is obtained to manipulate the output so as to reduce colorimetric error while permitting customization to specific printing hardware. It can be seen that the approach is easily extended to black and other high-fidelity color inks.

7. CONCLUSION

The introduction of stochastic screens in recent years has opened new possibilities for rendering images on display and printing devices with limited output states. The Blue Noise Mask, the prototypical stochastic screen, combines the speed of threshold arrays with the desirable unstructured patterns of error diffusion. Critical topics for analysis and research include the questions of filtering, optimality, color strategies, and application-dependent design. These topics have been discussed in the previous sections. Further research will need to consider refinements to the concept of optimal design for color rendering, and more advanced models of the human visual perception, particularly for printers with multilevel output capability.

REFERENCES

1. R Ulichney. Digital Halftoning. Cambridge, MA: MIT Press, 1987.
2. PG Roetling, RP Loce. In: ER Dougherty, ed. Digital Image Processing Methods New York: Marcel Dekker, 1994, pp 363–413.
3. PR Jones. J Electron Imaging 3(3):257–275, 1994.
4. H Kang. Color Technology for Electrionic Imaging Devices. Bellingham, WA: SPIE, 1997.
5. RW Floyd, L Steinberg. Proc Soc Inform Display 17(2):75–77, 1976.
6. T Mitsa, KJ Parker. In: ICASSP 91: 1991 International Conference on Acoustics, Speech, and Signal Processing. New York: IEEE, 1991, vol 2, pp 2809–2812.
7. T Mitsa, KJ Parker. Opt Soc Am A 9, 1920–1929, 1992.
8. PW Wong. In: ICASSP 94: 1994 IEEE International Conference on Acoustics, Speech, and Signal Processing. New York: IEEE, 1994, vol 5, pp 113–116.
9. KT Knox. In: BE Rogowitz, JP Allebach, eds. Proceedings, SPIE—The International Society for Optical Engineering: Human Vision, Visual Processing, and Digital Display V, San Jose, CA: SPIE, 1994, vol. 2179, pp. 159–169.
10. J Sullivan, L Ray, R Miller. IEEE Trans Syst Man Cybernet SMC-21:33–38, 1991.
11. G Marcu, S Abe. In: Proceedings, NIP12: International Conference on Digital Printing Technologies. San Antonio, TX: IS&T, 1996, pp 132–135.
12. KT Knox. In: JP Allebach, BE Rogowitz, eds. Proceedings, SPIE—The International Society for Optical Engineering: Human Vision, Visual Processing, and Digital Display IV, San Jose, CA: SPIE, 1993, vol 1913, pp. 326–331.
13. T Mitsa, KJ Parker. Image Process Algorithms and Techniques III, Toronto, Canada: SPIE, 1991, vol 1452, pp 47–56.

14. M Yao, KJ Parker. J Electron Imaging 3(1):92–97, 1994.
15. RA Ulichney. Proc IEEE 76:56–79, 1988.
16. T Mitsa, KJ Parker. In: ICASSP 92: 1992 IEEE International Conference on Acoustics, Speech, and Signal Processing. New York: IEEE, 1992, vol 3, pp 193–196.
17. KJ Parker, T Mitsa, R Ulichney. A new algorithm for manipulating the power spectrum of halftone patterns. SPSE's 7th Int. Congress on Non-Impact Printing, 1991, pp 471–475.
18. R Ulichney. In: JP Allebach, BE Rogowitz, eds. Proceedings, SPIE—The International Society for Optical Engineering: Human Vision, Visual Processing, and Digital Display IV, San Jose, CA: SPIE, 1993, vol 1913, pp 332–343.
19. T Mitsa, P Brathwaite. In: BE Rogowitz, JP Allebach, eds. Proceedings, SPIE—The International Society for Optical Engineering: Human Vision, Visual Processing, and Digital Display VI, New York: SPIE, 1995, vol 2411, pp 228–238.
20. J Dalton, In: BE Rogowitz, JP Allebach, eds. Proceedings, SPIE—The International Society for Optical Engineering: Human Vision, Visual Processing, and Digital Display VI, New York: SPIE, 1995, vol 2411, pp 207–220.
21. M Yao, L Gao, KJ Parker. In: Proceedings, NIP12: International Conference on Digital Printing Technologies. San Antonio, TX: IS&T, 1996.
22. Q Yu, K J Parker, M Yao. In: Proceedings, NIP12: International Conference on Digital Printing Technologies. San Antonio, TX: IS&T, 1996, pp 66–69.
23. R Eschbach, R Hauck. Opt Commun 62:300–304, 1987.
24. M Rodriguez. In: BE Rogowitz, JP Allebach, eds. Proceedings, SPIE—The International Society for Optical Engineering: Human Vision, Visual Processing, and Digital Display V, San Jose, CA: SPIE, 1994, vol. 2179, pp 144–149.
25. M Yao, KJ Parker, In: BE Rogowitz, JP Allebach, eds. Proceedings, SPIE—The International Society for Optical Engineering: Human Vision, Visual Processing, and Digital Display VI, New York: SPIE, 1995, vol 2411, pp 221–227.
26. AC Cheung, KJ Parker. J Electron Imaging. 1:203–208, 1992.
27. HT Fung, KJ Parker. In: M Rabbani, EJ Delp, SA Rajala, eds. Proceedings, SPIE—The International Society for Optical Engineering: Still-Image Compression. San Jose, CA: SPIE, 1995, vol 2418, pp 221–228.
28. HT Fung, KJ Parker. J Electron Imaging 5:496–506, 1996.
29. HT Fung, KJ Parker. Electron Imaging 1:388–395, 1992.
30. M Yao, KJ Parker. Electron Imaging Rev 2(3):4, 1995.
31. M Yao, KJ Parker. Popular Sci 248:38, June 1996.
32. KT Knox, S Wang. In GB Beretta, R Eschbach, eds. Proceedings, SPIE—The International Society for Optical Engineering: Color Imaging: Device-Independent Color, Color Hard Copy, and Graphic Arts II, San Jose, CA: SPIE, 1997, vol 3018, pp 316–322.
33. K Spaulding, LA Ray. U.S. Patent application No. 08/131,801 assigned to Eastman Kodak Company.
34. R Miller, J Sullivan. Color halftoning using error diffusion and a human visual system model, SPSE's 43rd Annual Conference, 1990, pp 149–152.
35. RV Klassen, R Eschbach, K Bharat. In: Proceedings, IS&T's 47th Annual Conference/ICP, Rochester, NY: IS&T, 1994, pp 489–491.
36. M Yao, KJ Parker. In: Proceedings, NIP11: International Conference on Digital Printing Technologies. IS&T, 1995.
37. Q Yu, KJ Parker, M Yao. In: Proceedings, Fourth Color Imaging Conference: Color Science, Systems, and Applications, Scottsdale, AZ: IS&T/SID, 1996, pp 77–80.
38. Q Yu, KJ Parker. In: GB Beretta, R Eschbach, eds. Proceedings, SPIE—The International Society for Optical Engineering: Color Imaging: Device-Independent Color, Color Hard Copy, and Graphic Arts II, San Jose, CA: SPIE, 1997, vol 3018, pp 272–277.

17
An Image-Algebra-Based SIMD Image-Processing Environment

Joseph N. Wilson, E. Jason Riedy and Gerhard X. Ritter
University of Florida, Gainesville, Florida

Hongchi Shi
University of Missouri–Columbia, Columbia, Missouri

1. INTRODUCTION

The advent of VLSI technology led image-processing researchers to use such SIMD, mesh-connected computers as the ILLIAC (1), the CLIP series (2–4), the DAP (5), the MPP (6), the GAPP (7,8), and the Hughes 3D Computer (9) for improving their codes' performance. Both the performance and cost-effectiveness of SIMD machines have improved steadily. However, the current software development systems are still comparable to assembly language programming for traditional sequential systems (10,11). Each parallel computer has its own language which runs efficiently on its architecture only. Various approaches striving for architecture independence have been proposed (12), ranging from the use special-purpose languages (13) to special implementations of general-purpose languages (14). A unified software development environment for these parallel systems has yet to appear.

Unified frameworks for image-related applications have interested many researchers. Many efforts have been devoted to searching for such a unified framework that can serve as a model for algorithms dealing with image objects and fit well into the theory and practice of parallel computing (15). Mathematical morphology provides a mathematical framework for expressing a large number of algorithms for image processing and analysis (16–19) through image filtering and structuring elements. Morphology-based systems ignore important operations like transformations between different domains and between different value sets. The image algebra developed by Ritter and his colleagues (20,21) provides a more general framework for image-related applications. Image algebra incorporates and extends mathematical morphology, providing more general image-template operations that support the elements missing from morphology. It defines images in the broadest sense and is widely applicable. Image algebra provides a common algebraic framework for algorithm development, optimization, comparison, coding, and evaluation.

In this chapter, we present a parallel environment based on image algebra suitable for SIMD machines. The environment keeps developers at a comfortable level of abstraction, specifying algorithms symbolically and algebraically while automatically partitioning the image data and scheduling operations to achieve optimal performance. Specifically, we discuss the use of the retargetable Image Algebra C++ object library iac++ (22) for image processing on the Lockheed Martin PAL-I computer, a fine-grained, SIMD-parallel computer. Modifying the iac++ library to provide efficient SIMD execution on the PAL system requires the development of a new image representation class, implementation of a client–server system, development of a strategy to reduce data transfers, and creation of a cost measure to control that strategy.

The next section briefly describes the properties of the PAL-I IOC/SA SIMD machine. The image-algebra environment structure and the operations and operands provided by the iac++ library are described in Sec. 3. Section 4 discusses retargeting iac++ to the PAL. The cost function used to direct the evaluation of programs on the SIMD array to reduce the required processing time is presented in Sec. 5. The heuristics applied to optimize the cost function are described in Sec. 6. Section 7 contains possible improvements to the cost model and heuristics. Section 8 follows with a few examples of a simple cost model and evaluation heuristic. Finally, Sec. 9 closes with a direction for future work and notes on the implementation in progress.

2. THE PAL-I SIMD PROCESSING SYSTEM

The PAL-I IOC/SA system is a workstation-based image-processing system. The PAL processor array itself is an attached image-processing accelerator. The system's current workstation platform is a Sun SparcStation 4. The PAL processor array is attached to such a workstation with an EDT SCD-40 configurable DMA interface (23). This provides a nominal 40-megabyte per second (MB/s) connection between the host workstation and the PAL processor array. The PAL processor array is a multiple-board 6U VME system with one controller board and one or more processor array boards. Each processor board contains 4608 one-bit processing elements (PE) arranged in a 72 × 64 grid. The typical configuration of a PAL-I system has two processor array boards and contains 9216 PE. Clocked at 40 MH, this system can execute over 368 billion bit operations per second. For 32-bit integer operations, this translates into execution speeds on the order of up to billions of operations per second. Because the processors are connected in a two-dimensional mesh network, performance of this system on local neighborhood operations can far outstrip any sequential computer.

Efficiently programming such a SIMD processor system presents challenges. The sustained throughput of the SCD-40 card in real situations is somewhat less that 10 MB/s. Thus, transfer of a 9216-pixel image with 8-bit pixel values will typically consume over 30,000 SIMD clock cycles. Even with its single-bit architecture, the PAL system can implement most 32-bit floating-point operations in just hundreds of clock cycles. The system is capable of executing between dozens and hundreds of operations in the time it takes to transfer an operand from host memory onto the processor array.

Two primary mechanisms are used to overcome such problems: designing the system to simultaneously transfer data and execute operations, or employing a cache memory hierarchy to place data near the processors. Both of these activities can improve performance of a SIMD system such as the PAL-I, but they cannot completely overcome the severe imbalance between transfer rate and computation speed presented by the IOC/SA. Although it can halve the processing time spent on any computation, simultaneous transfer fails because the processor speed in our setting is still much faster than the doubled transfer speed. Providing a cache memory hierarchy will improve performance but will still face the problem that image operands frequently occupy megabytes of storage. Even if one can store a few operands of this sort directly in the processing elements, the next level of the cache will incur some transfer penalty. A large frame buffer can dramatically improve many algorithms' performance, however.

The user interface provided by Lockheed Martin for programming the PAL system is the PAL Workstation Development Library (PAL_WS). This library supports parallel computation on SIMD-array-sized chunks of images and provides rudimentary support for creating larger images and directing operations to be carried out upon these images. Images are stored on the host workstation and transferred to the PAL processor array as necessary to carry out the specified operations.

3. SOFTWARE ENVIRONMENT BASED ON IMAGE ALGEBRA

The ideal environment for application software development would rise from combining a simple, reasonable model with compilers that bridge the gap between the model and specific architecture. Sequential computing has such systems, but parallel computing does not. Although there are more parallel computer models than parallel computer vendors, no suitable, all-encompassing model exists (24); neither do compilers capable of exploiting every extant parallel architecture. Indeed, difficulties in formulating a single, useful model make it appear unlikely that we will every have a single, useful environment for all parallel development. Thus, a practical methodology for application software development on parallel computers is to write algorithms using abstract libraries of fundamental data operations, implementations of which are optimized for specific computers. We adopt this approach and develop an image-algebra-based parallel environment that augments the C++ language with image-algebra operations.

3.1. Operands and Operations of the iac++ Library

The iac++ library provides classes of objects and related operations that are well suited to specifying image processing and computer vision algorithms. It was developed at a high level of abstraction, thus it is independent of any specific computer architecture. Furthermore, its software architecture allows it to be retargeted to exploit the capabilities of special-purpose computer architectures and devices. This class structure is shown in Fig. 1. In this figure, we use the notation of the Unified Modeling Language of Booch, Rumbaugh, and Jacobson (25). Boxes represent classes. The word abstract indicates that there are no direct object instances of a given class. The dashed boxes denote any parameters of the class. Solid lines imply that the class at the diamond end contains a pointer to the class at the box end. A class at the tail of a dashed directed line derives from the class at the head of that line.

The operands of iac++ are drawn from the image algebra developed at the University of Florida by Ritter and Wilson (26) and fall into the following categories:

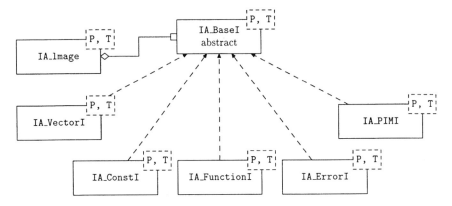

Fig. 1 The image classes in the `iac++` library are specialized for efficiency.

1. Points and sets of points
2. Values and sets of values
3. Images (functions from points to values)
4. Neighborhoods (images with values that are sets of points)
5. Templates (images with values that are images)

The `iac++` class library provides these operands via a collection of C++ template classes (27). The groups of operands listed above are represented as follows:

1. `IA_Point⟨int⟩` and `IA_Point⟨double⟩` represents points with integral and floating-point coordinates, respectively, and `IA_Set⟨IA_Point⟨int⟩⟩` and `IA_Set⟨IA_Point⟨double⟩⟩` represent sets containing these points.
2. Values are represented by the built-in C++ types `bool`, `unsigned char`, `int`, and `float` and the additional types `IA_RGB` and `IA_Complex`. Sets containing elements of type T are provided by the type `IA_Set⟨T⟩`.
3. The image classes have C++ template arguments specifying the kind of points mapped and the type, T, of elements. Most images are discretely sampled and of the form `IA_Image⟨IA_Point⟨int⟩, T⟩`. One continuous image type, `IA_Image⟨IA_Point⟨double⟩, float⟩`, is provided.
4. The neighborhood classes have C++ template arguments specifying the coordinate type of the domain points and the range points. The only presently implemented class is `IA_Neighborhood⟨int, int⟩`.
5. The presently implemented templates map discrete point sets to discrete images. The C++ template argument, T, to the class `IA_DDTemplate⟨T⟩` tells the type of image to which the template maps integral coordinate points. Templates on images with integer points and value types `bool`, `unsigned char`, `int`, `float`, and `IA_Complex` are supported in the library.

Operations upon images fall into several general categories:

1. Binary and unary pointwise operations
2. Global reductions
3. Neighborhood and template reductions
4. Composition with point-to-point or value-to-value functions

Image-Algebra-Based Environment

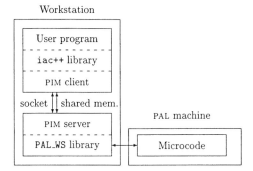

Fig. 2 The user's program uses the PIM client through the `iac++` library to talk to the PIM server. The server, in turn, uses the `PAL_WS` library to communicate with the PAL device.

These operations are provided, as appropriate, by overloaded operations on the image classes, by class member functions, and by overloaded functions.

3.2. The Image-Algebra Environment Structure

The structure of the image-algebra environment on the PAL-I system is depicted in Fig. 2. The current PAL image representation class, `IA_PIMI⟨P,T⟩`, implements its operations through a client–server system. The `iac++` client application directs the PAL Image Manager (PIM) through socket-sent operations. Image data are placed in shared-memory segments, eliminating costly, needless copying. This client–server architecture allows us to mediate PAL access between multiple clients.

To effectively support image-algebra-based programming on the PAL-I, we needed to retarget the `iac++` library to the PAL-I and to implement data-partitioning and operation-scheduling strategies to efficiently schedule image-algebra operations on the PAL-I under a cost model.

4. RETARGETING THE `iac++` LIBRARY TO THE PAL

The retargeting of the library is supported by the separation of the user interface from the representation employed in the class hierarchy. As shown in Fig. 1, the image classes are all instances of a single C++ template class, `IA_Image⟨P,T⟩`. These classes provide the user interface to such images. Each of these user-interface classes contains a reference (or handle) to an object of type `IA_BaseI⟨P,T⟩`, the base class for image representations. A variety of specific representation classes are derived from `IA_BaseI⟨P,T⟩` to effect efficient sequential implementation of the library, such as `IA_VectorI⟨P,T⟩`, `IA_ConstI⟨P,T⟩`, and `IA_FunctionI⟨P,T⟩`, which represent an image as a vector of values, a constant value, and a function mapping points to values, respectively. This arrangement of classes ensures that the behavior of any image in response to user-issued operations is dependent on its representation class. To retarget the library to support the PAL IOC/SA system, we have developed an image representation class, `IA_PIMI⟨P,T⟩`, whose operations are implemented on the PAL.

If we are to use a SIMD computer to operate upon images having many more pixels than there are SIMD processing elements, then we must serialize our computation to some extent. One straightforward way is to break image operands up into smaller image subframes or blocks. This breaks the operation into a sequence of array-sized, parallel computations.

Suppose we wish to evaluate the expression $R = (A + B)C$. The value of each pixel in R is the sum of the corresponding pixels in A and B times the value of corresponding pixel in C. Assume that A, B, and C are so large that they occupy more memory than the entire PE array contains. Any single-image operation cannot be computed on the PE array without being serialized. (This assumption is, in fact, warranted for many applications of the PAL and other SIMD systems.) Suppose we divide the images into k corresponding blocks $A_1, \ldots, A_k, B_1, \ldots, B_k,$ and $C_1, \ldots, C_k$. There are two fundamentally different, yet correct strategies to evaluate these operations.

In the first approach, we carry out operations on images in the order specified by the expression, fully evaluating each image operation and yielding an entire image result. In our example, for each i in 1 to k, we calculate $T_i = A_i + B_i$, then for each i in 1 to k, we calculate $R_i = T_i C_i$. This approach produces temporary results that must be stored in the host.

In the second approach, we carry out operations on the entire expression for each subframe block, generating an image composite after completing the work on each of the subframes. Thus, for each i in 1 to k, we calculate $R_i = (A_i + B_i)C_i$.

Temporary results should be avoided when they must be transferred over a slow path. In the second approach, these temporary operands need never be transferred; hence, computations are completed more quickly. The PIM server exploits this fact, using lazy evaluation (28) of image operations. When a client program operates upon images, the server constructs an expression tree that represents the computation to be performed. When the client program attempts to use the pixel values of an expression's result, we evaluate the associated tree and carry out as many operations as possible on each subframe.

It might appear that the second evaluation approach is the best and should be used as the only serialization rule. This ignores neighborhood and template operations' requirements of data from neighboring pixels. Longer sequences of neighborhood or template operations require larger neighboring regions in order to calculate a pixel's value. We cannot calculate valid results for neighborhood and template operations at every processing element, but only those whose neighbors have valid values. This leads us to divide image operands into overlapping subframes. As the number of such subframes increases, the time to communicate the original image grows, eventually overcoming the savings we achieve by avoiding temporary results. If we can express the cost incurred during a sequence of operations, we can attempt to analyze and minimize the cost dynamically.

5. COST MODEL

We assume a two-dimensional system. Most restrictions to one dimension and extensions to higher dimensions are straightforward. The total cost breaks into two subcosts: the cost of transferring data and the cost of computing results. Because the goal is to minimize execution time, the cost is expressed in units of time.

5.1. Cost of Transfers

Assume the image of interest is larger than the SIMD array. Then, any computation must be broken into array-sized blocks or subframes. Each input subframe needs to be transferred to the array, and each result subframe needs to be shipped back out. The total transfer cost is the number of array-sized subframes transferred times the cost of transferring a block.

5.1.1. Number of Subframes

The straightforward method of breaking images into exactly array-sized blocks produces erroneous results. Template and neighborhood operations along image borders produce boundary effects by sampling outside the image. The boundary effects also occur along the induced, internal borders. Figure 3 shows these effects.

The common solution for external boundary effects is to extend the image with a known value. This solution works for many operations, but the padding around the image must still be transferred to the array. For the current discussion, assume the image's padding is loaded as a part of the image. In practice, initializing a transfer buffer with the padding value and assembling the image into the buffer also solves problems associated with images having nonrectangular point sets. Some platforms may have an operation that extends a subframe on the SIMD array more efficiently. The padding term in Eq. (2) below is still necessary to calculate the number of blocks.

Some operations, such as an image-template product which combines through addition and reduces by minimization, do not have a single, convenient padding value. For additive minimum operations, the reduction identity is not preserved by the combination operator. For example, combining an 8-bit, unsigned padding value of 255 with a template value of 1 results in an 8-bit value of 0. That zero will be the minimum value and the result of the template. Fixing the padding values requires finding the correct minimum. That circularity implies that another method is needed.

The min operator uses the sum's result only if that point is in the original image; otherwise, it uses its identity. The correct value from either the sum or the identity is chosen according to a mask image. The mask has a value of 1 at each point in the original image's point set and 0 elsewhere. This mask image must be transferred to the array as well. The mask should be sent only if necessary on a per-block basis. This is dependent on the point set of the original image; a sparse point set such as $\{(i^2, j) | 0 \leq i < 16, 0 \leq j < 256\}$ would require a mask on every subframe. Again, some architectures may provide

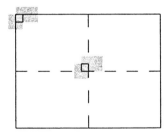

Fig. 3 Blindly applying templates leads to boundary effects (lightly shaded) along both image (solid) and array-induced (dashed) borders.

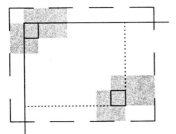

Fig. 4 Each subframe (dashed) yields a smaller result (dotted).

more efficient mechanisms for creating masks on the processor array. The results below assume a mask is transferred with every block and provide an upper bound on the number of transfers.

The internal boundary effects also need variable-valued padding. The necessary padding values, however, are already available in the image and are loaded with the block. Avoiding internal boundary effects reduces the region of the array containing valid results as shown in Fig. 4. This valid region tiles the image and determines the number of subframes to be loaded.

For two-dimensional images on a two-dimensional processor array, the number of block boundaries along each image axis is the length of the padded image divided by the length of the valid region. The total number of subframes is the product of these across all the image's dimensions. If the dimensionality of the image does not match the dimensionality of the processor array, more complicated subframing methods must be applied.

Let t be a sequence of templates and neighborhoods applied to an $\mathcal{L}_x \times \mathcal{L}_y$ image A. The SIMD array contains $L_x \times L_y$ processing elements. Each template can be fit inside a bounding box. The sequence t has a corresponding sequence of bounding boxes, **a**. On these boxes, define the functions $l_d(b)$ and $r_d(b)$ as shown in Fig. 5. The functions determine the extent of the bounding box on either side of the template's origin along dimension $d \in \{x, y\}$.

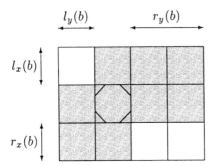

Fig. 5 The functions l_d and r_d denote the extent of a template's bounding box on either side of the origin.

Define the function pad_d on sequences of bounding boxes to be

$$\mathrm{pad}_d(\mathbf{a}) = \max\{l_d(b)|b \in \mathbf{a}\} + \max\{r_d(b)|b \in \mathbf{a}\}. \tag{1}$$

This is the total length of padding needed along dimension d.

The total number of subframes of A to be transferred is

$$B = \prod_{d \in \{x,y\}} \left\lceil \frac{\mathcal{L}_d + \mathrm{pad}_d(\mathbf{a})}{L_d - \Sigma_{b \in \mathbf{a}}(l_d(b) + r_d(b))} \right\rceil. \tag{2}$$

The ceiling operator takes care of image sizes which are not exact multiples of the array's size. The resultant image has the same point set as A. Any extra computed values are ignored. Note that each neighborhood or template operation decreases the size of the subframe's valid region.

Building an expression tree for a sequence of template and neighborhood operations requires four state variables per dimension, the sum and maximum of the l_d and r_d values. The method for keeping track of the valid region's size is clear. Adding another operation to an existing expression tree needs only the current root's state variables. The number of blocks to be transferred is compositional with these operations. In general, B is a function of the immediate history of the computation.

Equation (2) only gives the number of subframes necessary for a sequence of template and neighborhood operations. Unary and image-scalar operations do not affect the valid region's size, so they can be introduced freely. Operations between two images do affect the valid region, as shown in Fig. 6. The result's valid region is the intersection of the arguments' valid regions. The new state variables are the maximum values of the arguments' state variables. Note that the regions must be aligned consistently. In general, results of template operations should be shifted back to their original points. This adds to the computational cost. The merges from image–image operations also add dependencies on the order of operations and introduce a limited form of shared subexpressions.

5.1.2. Transfer Time

The final cost of transferring the images depends on both the number of subframes and the time to transfer each subframe. Blocks are transferred for both operand and result images.

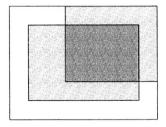

Fig. 6 The valid region of image–image operations is the intersection of the operands' valid regions.

```
1     IA_Set< IA_Point<int> > domain =
2        IA_boxy_pset (IA_Point<int> (0, 0),
3                     IA_Point<int> (255, 255));
4
5     // Images A and B are to use the SIMD array, C and D
6     // will inherit that property...
7     IA_Image< IA_Set< IA_Point<int> >, int >
8        A = IA_PIMImage
9              (IA_Image< IA_Point<int>, int > (domain, 1)),
10       B = IA_PIMImage
11             (IA_Image< IA_Point<int>, int > (domain, 2)),
12       C, D;
13
14    C = A + B;
15    D = C * A;
16    cout << sum (D);
```

Fig. 7 Because C is still in scope, evaluation of D also evaluates and stores C.

Not all computations return a single result. Figure 7 shows one such case. The global reduction in line 15 triggers the evaluation of D, which in turn evaluates C. The results for C may be used again later, so they must be transferred back to the host. Had line 15 been written as D = (A + B) * A, the temporary result would be out of scope by line 16, and the subresult of A + B would not be returned. The example is trivial, but real code often contains unused temporaries for readability. Conservative dependency assumptions are a significant limitation of systems that do not modify the original source.

For a general sequence of operations, the total transfer time into the array, T_{in}, and the total transfer time out of the array, T_{out}, are

$$T_{\text{in}} = B\left(\text{setup} + \sum_{A \in \{\text{operands and masks}\}} T(\text{rep-size}(A))\right)$$
$$T_{\text{out}} = B\left(\text{setup} + \sum_{A \in \{\text{results}\}} T(\text{rep-size}(A))\right) \quad (3)$$

where rep-size(A) is the size of the machine representation of A's values, and $T(s)$ is the time needed to load the block of representation size s onto the processor array. The setup time is the small time required to initiate the transfer.

Many architectures are designed for streaming data and instructions. The general, transparent method attempted here must deal with the staccato bursts caused by intermediate return values and by delays in the controlling program. These bursts will often produce nonlinear times from low-level setup costs. Hence, the values of T should be determined experimentally to counter possibly nonlinear transfer times.

Note that the set of temporary results may change during execution. Results are not returned to host-side variables which have passed out of scope. At any given point in the tree, the T_{out} term determined during creation is an upper bound. Updating the T_{out} values and propagating the new values up the tree is similar to the shared-subexpression problem mentioned later.

5.2. Cost of Computations

The cost of computation may be small compared to the cost of transferring the images, yet it must be considered as a portion of the total cost. With future PAL or other SIMD systems, these costs may be much closer. Each operation has costs associated with each step of the computation. In this context, consider the same mathematical operation on different representation sizes to be different operations.

5.2.1. Image–Template Operations

For general image–template products, $A \circledgamma t$, a template is treated as a sequence of points in the support and associated values. Assume the support of all templates and neighborhoods fit on the array. The ordering of the sequence is given by the convolution paths. The first operation also initializes the accumulator image. The computation proceeds by shifting a pixel value to the target point, combining the entire subframe with each template point's value and reducing the intermediate result into the result subframe.

The per-subframe cost of computing the general template product is the sum of the time to shift each pixel value to the target point and the time to combine and reduce at the target point. This process is illustrated in Fig. 8. Other convolution paths are possible. If there is an extra area of processor memory available, the path can be split into two separate paths starting from different template locations. The cost of moving processing element memory into a buffer negates the savings from a few shifts, so this optimization is only useful when temporary space already exists.

The cost C for computing an image–template product $A \circledGamma t$ with B subframes is

$$C = B(|t|(\text{op-time}(\circ) + \text{op-time}(\gamma)) + \text{op-time}(\text{shift})|\text{path}|) \qquad (4)$$

Section 5.1.1 mentions the need for a masking image with certain image–template operations. The masking adds to the per-template-point operation time.

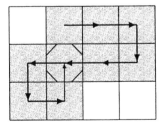

Fig. 8 This convolution path involves nine shift operations–eight for calculations and one for returning to the origin.

5.2.2. Neighborhood Operations

Neighborhood operations are simply local reductions. They are computed as templates without the combination operation. The cost C for a neighborhood product $A \textcircled{\Gamma} n$ with B subframes is

$$C = B(|n| \text{op-time}(\gamma) + \text{op-time(shift)}|\text{path}|) \tag{5}$$

5.2.3. Unary, Image–Scalar, Image–Image Operations

Once the necessary data reside on the processor array, unary, image–scalar, and image–image operations are calculated through a single operation per block. The cost C for unary operations (op A), image–scalar operations (s op A), and image–image operations (A_1 op A_2) is

$$C = B \cdot \text{op-time(op)} \tag{6}$$

5.2.4. Global Reductions

Global reductions produce a scalar result. Most languages, including C++, have no capacity for delaying the computation and storage of their basic types. Thus, all global reductions in the iac++ system must be evaluated immediately. The cost is immaterial for that purpose. Note that some common, important operations such as equality testing are essentially global reductions.

If there were a method for delaying scalar results, global reductions could be considered neighborhood reductions over an image's entire support. This breaks the assumption that supports are smaller than the processor array, requiring decomposition of the support.

5.2.5. Functional Composition

Compositions between functions and images are currently not supported on the PAL machine. Extending support to general functions would place more restrictions on the source code. General functions will contain many sequential operations on types native to the client. With a stock C++ system, these operations cannot be supported transparently.

For instance, composing an integer-valued image with a function that adds one would require parallelizing the application of the function. If the function worked with the standard int type, a stock compiler would produce standard sequential code. The library would need a way to execute that specific code on each PE. No such facility exists for the PAL, which has opcodes and data sizes different from the host processor.

6. SCHEDULING EVALUATION OF TREES

The costs discussed so far are accumulated as an expression tree is built. At some point, the delayed operations must be evaluated. The system only knows the history of the computation so far, not the entire span of computation. A new operation is potentially the optimum place to stop building the tree, so some heuristic must guide the system from the incomplete information available.

One simple heuristic is to examine the average cost per operation in a tree. As long as adding a new operation decreases the average cost, it can be delayed. If the new operation increases the average cost, the tree should be evaluated before the new operation is added.

Template and neighborhood operations decrease the valid region and initially increase the number of blocks to be transferred. Multiple operations can be performed per block, however, obviating the need for intermediate, temporary results. The two effects are balanced according to the time constants on a particular system. Unary and image–scalar operations will always decrease the average cost, so they will never trigger an evaluation.

An image–image operation can increase the average cost by reducing the valid region. If it does, only one of the two subtrees needs to be evaluated. One possible choice is to evaluate the subtree with the smallest valid region. The smaller region cannot be delayed for many more image–template products, so this is a good choice when those operations predominate. Evaluating the subtree with the lower average cost is also attractive. The higher cost is likely to be decreased further than the lower. The lower average cost might correspond to the larger valid region, however. A sampling of real examples will be necessary to determine which is more effective.

Other possible heuristics include hysteresis extensions on the average cost to support small bumps or waiting until a result is required or the valid region is completely destroyed before partitioning the computation. The average-cost heuristic has the advantages of being somewhat simple and relatively intuitive.

7. POSSIBLE IMPROVEMENTS

Some of the assumptions made so far are not always suitable for real systems. Templates and neighborhoods may be larger than the processor array. One delayed computation may be shared by multiple expression trees. A mostly transparent system cannot detect these and modify the source program; it must deal with them as they occur.

The processor array and the sequential host only have finite resources to devote to temporary results, as well. Most hosts have limited shared memory for transferring images, and most processor arrays have limited on-board memory. Some algorithms use very little space on the host but need many temporaries on the processor array. The resource needs of others are reversed. Using nodes from a preallocated node pool controls the resource use on both ends, but particular allocation strategies may be suboptimal for certain algorithms.

7.1. Weak Template Decomposition

Not all templates and neighborhoods will fit within a single subframe. The supports need to be decomposed, and the decomposed pieces need to be calculated separately and reduced together (29). Image algebra already requires the reduction operator to be associative, so no ordering problems arise.

If template t decomposes into templates t_1 and t_2, the template product becomes

$$A\textcircled{\gamma}t \Rightarrow (A\textcircled{\gamma}t_1)\gamma\text{translate}(A\textcircled{\gamma}t'_2) \tag{7}$$

The translate function is a general notation for lining up the appropriate subframes, and t'_2 is t_2 with the origin relocated to lie within the same subframe.

If the subexpressions $A\textcircled{\gamma}t_1$ and $A\textcircled{\gamma}t'_2$ can be broken into the same blocks, the calculation can be pipelined up to the number of temporary stores available in the array. Subframes could be loaded as needed and used multiple times before being replaced.

7.2. Shared Subexpressions

Shared subexpressions seem to offer potential for further optimizations. However, they destroy the compositional nature of the cost function. Expressions are no longer disjoint; they join to form a directed, acyclic graph. We will loosely refer to the expression graphs as trees for the remainder of the chapter.

The evaluation of any subtree in an expression will invalidate the state information carried at the root of the tree. Because subtrees can be shared between trees, the evaluation of one expression tree can trigger the evaluation of a subtree within another expression tree. Figure 9 shows how even a simple sequence of operations can invalidate state information.

This is similar to the earlier problem of overestimating the number of intermediate results. The current solutions to both problems are to ignore them. The cost function composition method provides an upper bound. One extension which could tighten the bound would be to keep track of a limited number of parents per node and update their state information. Future tests will show how many updates will be both useful and feasible for practical algorithms.

Subexpressions shared only within one tree do not suffer from early evaluation. They could provide extra loading optimizations, as seen in the support decomposition discussion.

7.3. Finer Optimizations

The cost estimate is an upper bound. Many fine-tuning optimizations exist beyond the general framework presented here. Some, such as always loading array-sized blocks of the image and shifting unused pieces into the position for the next block, may save more time at considerable programming expense. Others, such as tracking the origin of result blocks and shifting them to agree rather than always returning the result to its origin, are of marginal savings and little programming expense. Any implementation must find an acceptable balance between the different aspects of efficiency (30).

```
1    // Say A, B, C, and D are defined as before,
2    // and t1 and t2 are simple, small templates
3    // of the appropriate type.
4
5    C = linear_product (A, t1) + B;
6    D = C * linear_product (A, t2);
7    cout << sum (C);
```

Fig. 9 The global reduction of C will invalidate the state information for the tree rooted at D.

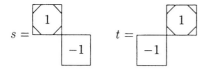

Fig. 10 The templates s and t for the Roberts edge detector.

8. EXAMPLES

Any cost model/heuristic pair must be evaluated in the context of numerous examples. Here, we present two fairly simple ones, a Roberts edge detector and a three-level wavelet transform. We take the simple cost model proposed above and combine it with a heuristic that dictates evaluation of a stored tree when adding a new operation would increase the average cost per operation. We also assume no shared subexpressions, so the implementation cost of this heuristic is very low. These algorithms are to run on a single-board PAL system, so the array dimensions are 72×64.

8.1. Roberts Edge Detector

The Roberts edge detector (26) consists of a simple, typical expression. The edge image E is derived from an input image A by $e = \sqrt{(A \oplus s)^2 + (A \oplus t)^2}$; Fig. 10 displays the templates s and t. The table in Fig. 11 shows how the estimated cost per operation Q varies as the tree grows when applying the algorithm to a 1024×1024 image. Because

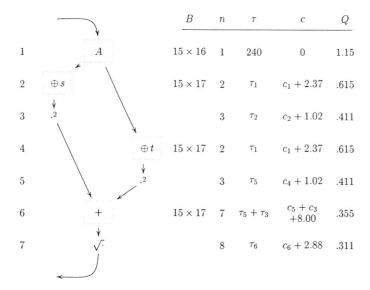

	B	n	τ	c	Q
1	15×16	1	240	0	1.15
2	15×17	2	τ_1	$c_1 + 2.37$	.615
3		3	τ_2	$c_2 + 1.02$	.411
4	15×17	2	τ_1	$c_1 + 2.37$	.615
5		3	τ_5	$c_4 + 1.02$	.411
6	15×17	7	$\tau_5 + \tau_3$	$c_5 + c_3$ +8.00	.355
7		8	τ_6	$c_6 + 2.88$	.311

Fig. 11 The cost Q associated with the expression tree for $\sqrt{(A \oplus s)^2 + (A \oplus t)^2}$ strictly decreases. The times for τ and c are in units of 10^{-5} s, and the time for Q is in seconds.

the reduction identity, zero, is preserved by the combination operator, multiplication, there are no mask images in this example. The loading times, τ_i, and computation times, c_i, are per block. The number of operations, n_i, is tallied as if there are no shared subtrees. For example, n_6 is $n_5 + n_3 + 1$, ignoring the fact that the subtrees share their initial operation. The approximate average cost per operation at step i is then $Q_i = B_i(2\tau_i + c_i)/n_i$.

The costs are approximately correct for a PAL-I IOC/SA system. Two multiplications, two additions, and two shifts comprise the 2.37×10^{-5} s for each template. The extra 8.00×10^{-5} s included in the computation time of step 6 is the time required to move the result of Step 3 into each PE's memory before continuing. The 2.40×10^{-3} s for loading a block assumes a throughput of around 8 MB/s to the PAL unit.

Note that both the time to load a block and the number of operations are overestimated in Step 6 of Fig. 11. Each subtree is conservatively assumed not to share nodes. The actual average cost per operation is 0.210 s. Each node could maintain a dependency list and image–image operations could determine their costs more intelligently, but this imposes an extremely large overhead on computations that form long chains.

Another important observation is that the cost of communication outweighs the cost of computation for a small expression by nearly two orders of magnitude. If every expression in an application is fairly small, as when each tree consists of fewer than 100 operations, updating the computational cost is needless overhead.

One of the major factors attributing to the severe imbalance between computation and communication costs is the repeated transmission of data in the overlap regions. A simple cache hierarchy, like the 64-MB frame buffer in the forthcoming IOC/FB, will let the system transfer the whole image once. This saves time not only by sending each value once but also by better using the link's available bandwidth. The total time to load will then be a function of only the image size plus a per-block constant on the order of 5×10^{-5} s.

8.2. Three-Level Wavelet Transform

Next, examine a three-level wavelet transform on a 512×512 image A. Assume the base wavelet filters have six taps like Daubechies' $\mathcal{W}_6$ wavelet (31). The supports of the high-pass filters h_i and low-pass filters g_i are shown Fig. 12. Here, we ignore the negligible cost of computation and focus on the number of blocks to be transmitted between the array and the host. Also, we only apply the transform along the x dimension for simplicity.

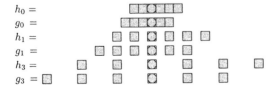

Fig. 12 The high-pass (h_i) and low-pass (g_i) filters sample from successively farther locations. The shaded boxes form the nonzero support of each template.

Image-Algebra-Based Environment

```
1    S = IA_PIMImage (A);
2    for (i = 0; i < 4; i++) {
3      D[i] = linear_product (S, g[i]);
4      S = linear_product (S, h[i]);
5    }
```

Fig. 13 A short, simple wavelet decomposition along one dimension.

Figure 13 holds the basic wavelet decomposition code. The arrays hold the similarly named filters for each level i. As the server builds the tree in Fig. 14, it maintains the maximum and sum of l_x and r_x over the templates to be applied to the image. The total number of blocks B is calculated as in Eq. (2).

When the server adds the $\oplus g_2$ operation to the tree, the total number of blocks transferred per operation increases. The simple heuristic we use indicates that the delayed expression $A \oplus h_0 \oplus h_1$ should be evaluated. As Fig. 14 demonstrates, this evaluation decreases the number of blocks, but it also decreases the number of operations. The average number of blocks transferred actually increases more when $A \oplus h_0 \oplus h_1$ is fully evaluated. This suggests another possible heuristic: Evaluate when evaluation does not increase the cost more than another delay does.

In the previous example, we counted the transfer time for both loading and returning the image data. The transfer cost out is substantial in the current example. A frame buffer can hold the results and delay returning them to the host. The potential savings for such algorithms as wavelet analysis, where the images will be filtered and recombined before being returned, are great. A limited frame buffer presents significant challanges for the cost model. The cost of managing the memory allocation should be included, especially with access from multiple clients. Currently, we ignore the return time and assume the result remains in the frame buffer. This should be a fair approximation for algorithms that

Step		$\max l_x$	$\max r_x$	$\sum l_x$	$\sum r_x$	B	n	B/n
1	A	0	0	0	0	8×8	1	64
2	$\oplus g_0$	3	2	3	2	8×8	2	32
3	$\oplus h_0$	2	3	2	3	8×8	2	32
4	$\oplus g_1$	6	4	8	7	10×8	3	26.6
5	$\oplus h_1$	4	6	6	9	10×8	3	26.6
6	$\oplus g_2$	12	8	18	17	15×8	4	30
		12	8	12	8	11×8	2	44
7	$\oplus h_2$	8	12	8	12	11×8	2	44

Fig. 14 Assuming the transfer cost overwhelms the computation cost, this expression tree's cost per operation will increase after Step 6, forcing evaluation of $A \oplus h_0 \oplus h_1$.

produce few temporary results and for environments where few clients use the same IOC/FB unit.

9. SUMMARY

We have presented a general, cost-based model for optimizing image-processing operations for the PAL computer. We have shown how the operations of image algebra affect both the cost of computation of results on the PAL and the cost of communication of data to the PAL. The model balances these costs with a set of heuristics, scheduling expression evaluation on the fly.

Clearly, there are many possible choices for heuristics and even for parameters within the heuristics. A systematic study of image-processing algorithms should reveal which parameters work well for sets of algorithms. Such a study requires specifications of algorithms in image algebra, `iac++` implementations of those algorithms, and a statistic-collecting PIM server. Catalogs like Ref. 26 provide a useful starting point for the former, and most of the algorithms in it have already been implemented. Only minor modifications are necessary to begin a study, but we are still determining which parameters to examine.

We are also building a real system with this general, cost-based model. In the implementation, expression node objects contain cost objects through an aggregation system. The cost object classes share a common interface to the node objects, but each class can maintain a different internal interface. The cost of a new node is determined by its cost object's examination of the childrens' cost objects. Thus, different cost objects can be used within different trees. Initially, we assume that each client can select a single cost class provided by the server. We may find it useful to allow clients to have multiple contexts; expressions within one context may use a different cost class than expressions within another context. Moving an image to a different context may be implemented by fully evaluating the image on a context switch or by providing a common, inter-cost-class interface to allow the classes to determine if they are compatible.

ACKNOWLEDGMENTS

This work was sponsored in part by the PAL Consortium, funded by the Wright Laboratory Armament Directorate at Eglin Air Force Base, Lockheed Martin Electronics and Missiles in Orlando, the University of Florida, and the University of Missouri–Columbia. We would like to thank the consortium members for their efforts in support of this and other related work.

REFERENCES

1. BH McCormick. The Illinois pattern recognition computer—ILLIAC III. IEEE Trans Electron Computers EC-12:791–813, 1963.
2. MJB Duff, DM Watson, TJ Fountain, GK Shaw. A cellular logic array for image processing. Pattern Recogn 5(3):229–247, 1973.
3. MJB Duff. Clip4. In: KS Fu, T Ichikawa, eds. Special Computer Architectures for Pattern Processing. Boca Raton, FL: CRC Press, 1982, pp 65–86.

4. TJ Fountain, KN Matthews, MJB Duff. The CLIP7A image processor. IEEE Trans Pattern Anal Machine Intell PAMI-10(3):310–319, 1988.
5. S Reddaway. The DAP approach. In: C Jesshope, R Hockney, eds. Infotech State of the Art Report: Supercomputers, Vols. 1 & 2. Maidenhead, England: Infotech Int. Ltd., 1979, pp 309–329.
6. KE Batcher. Design of a massively parallel processor. IEEE Trans Computers C-29(9): 836–840, 1980.
7. EL Cloud. Geometric arithmetic parallel processor: Architecture and implementation. In: VK Prasanna, ed. Parallel Architectures and Algorithms for Image Understanding. San Diego, CA: Academic Press, 1991, pp 279–305.
8. MS Tomassi, RD Jackson. An evolving SIMD architecture approach for a changing image processing environment. DSP Multimedia Technol 5:1–7, October 1994.
9. MJ Little, J Grinberg. The 3-D computer: An integrated stack of WSI wafers. In: EE Swartzlander, ed. Wafer Scale Integration. Boston: Kluwer Academic Publishers, 1989.
10. JA Webb. Steps toward architecture-independent image processing. Computer 2:21–31, February 1992.
11. R Davoli, LA Giachini, O Babaoglu, A Amoroso, L Alvisi. Parallel computing in networks of workstations with Paralex. IEEE Trans Parallel Distrib Syst PDS-7(4):371–384, 1996.
12. GE Blelloch, JC Hardwick, J Sipelstein, M Zagha. Implementation of a portable nested data-parallel language. J Parallel Distrib Comput 21:4–14, 1994.
13. J Rose, G Steele. C*: An extended C language for data parallel programming. Technical Report, Thinking Machines Corporation, Cambridge, MA, 1987.
14. J Adams, W Brainerd, J Martin, B Smith, J Wagener. Fortran 90 Handbook. New York: McGraw-Hill, 1992.
15. H Shi, GX Ritter, JN Wilson. Parallel image processing with image algebra on SIMD mesh-connected computers. In: PW Hawkes, ed. Advances in Imaging and Electron Physics, Volume 90. San Diego, CA: Academic Press, 1995, chap 4.
16. J Serra. Image Analysis and Mathematical Morphology. London: Academic Press, 1982.
17. SR Sternberg. An overview of image algebra and related architectures. In: S Leviaidi, ed. Integrated Technology for Parallel Image Processing (Polignano, Italy, June 1–3, 1983). London: Academic Press, 1985, pp. 79–100.
18. P Maragos. A Unified Theory of Translation-Invariant Systems with Applications to Morphological Analysis and Coding of Images. PhD thesis, Georgia Institute of Technology, 1985.
19. ER Dougherty, CR Giardina. Image algebra—induced operators and induced subalgebras. In: SPIE Proceedings of Visual Communications and Image Processing II, Bellingham, WA: SPIE, 1987, vol 845, pp 270–275.
20. GX Ritter, JN Wilson, JL Davidson. Image algebra: An overview. Computer Vision Graphics Image Process 49(3):297–331, 1990.
21. GX Ritter. Recent developments in image algebra. In: P. Hawkes, ed. Advances in Electronics and Electron Physics, *Volume 80*. New York: Academic Press, 1991.
22. JN Wilson. Supporting image algebra in the C++ language. In: Image Algebra and Morphological Image Processing IV. Bellingham, WA: SPIE, 1993, vol 2030, pp 315–326.
23. Engineering Design Team, Inc. SBus Configurable DMA Interface User's Guide. Document #00-00419-04 edition, 1997.
24. S Sahni, V Thanvantri. Parallel computing: Metrics and models. Unpublished data.
25. G Booch, J Rumbaugh, I Jacobson. The Unified Modeling Language User Guide. Reading MA: Addison-Wesley, 1999.
26. GX Ritter, JN Wilson. Handbook of Computer Vision Algorithms in Image Algebra. Boca Raton, FL: CRC Press, 1996.
27. MA Ellis, B Stroustrup. The Annotated C++ Reference Manual. Reading, MA: Addison-Wesley, 1990.
28. RJM Hughes. Why functional programming matters. Computer J 32(2):98–107, 1989.

29. GX Ritter. Image algebra with applications. Available via anonymous ftp from `ftp://ftp.cise.ufl.edu/pub/arc/ia/documents/ia.*.ps.gz`, 1994.
30. L Wall, T Christiansen, RL Schwartz. Programming Perl. 2nd. ed. O'Reilly & Associates, Inc., Sebastopol, CA: 1996, pp 537–546.
31. I Daubechies. Ten Lectures on Wavelets. CBMS–NSF Regional Conference Series in Applied Mathematics. Philadelphia: SIAM, 1992.

Index

AA-learning, 436
Abstract muscle action procedures (AMAP), 105
AC prediction, 281, 282
Access unit layer, 304
Adaptive arithmetic coding, 83
    adaptive context, 84
    context, 83, 84
    conditional probability, 83, 84
    local probability distribution, 83
Adaptive quantization
    scene adaptive, 50
    signal adaptive, 50
Additive primary colors, 514
Affine transform, 132, 141, 153
Algorithm complexity, 78
AM-FM models, 345, 347, 348, 351–352, 361–367, 370–381
AM-FM representations, 345, 361–367, 373–381
American sign language, 453
Analysis, 98, 100
Analytic image, 353
Animation, 100, 104, 130, 133, 135, 152, 160, 161
    realistic animations, 107
Area-to-perimeter ratio, 507, 509
Arithmetic coding, 330, 331
Augmented reality, 135, 150, 152, 160
Autonomous navigation, 456
Auxilliary function, 348–350, 368, 369, 380

Bayes risk, 451
Bayesian estimation, 41, 51
Behavior-based approach, 441, 443
Between-class scatter, 450
Biased sensor, 436
Bilinear interpolation, 396
Binary shape, 277, 278
Bit-plane, 83
    magnitude coding, 83

Bitstream syntax, 317, 318, 319
Black-and-white halftoning, 492
Block, 321
Blue noise, 491, 506
Blue noise mask (BNM), 492
    binary pattern design, 493
    mask construction, 493, 494
    optimality, 505–507
Breadth-first traversal, 148

CART, 432
CELP coder, 296
Centroid condition, 13, 20, 21
Channelized components, 362, 363, 374–377
Checkmask, 508, 509
Chess, 431
Chroma key, 339
Chrominance, 320
Chunking, 470
CIELAB, 515, 518
Classification tree, 449
CLIP, 523
Clustered-dot dithering, 489, 507, 514
Clustering, 41, 51, 55, 79, 464
    area threshold, 81
    cluster elimination, 81
    clustering analysis, 421
    conditioned dilation, 79, 80
    connected component, 80
    correlation, 77
    finite iteration, 80
    geometric connectivity, 80
    insignificant-to-significant ratio, 81
    intersection, 80
    irregular-shaped cluster, 78, 79
    marker, 80
    quantization, 41, 50
    significant cluster, 81
    structuring element, 79, 80

Codebook search techniques, 18
  exhaustive search, 18
  M-search, 19
  sequential search, 18
Coded-domain combining, 208
Coding
  entropy coding, 103
  fixed-length coding (FLC), 47
  intermode coding, 103
  intramode coding, 103
  predictive coding, 103
  pyramid progressive coding, 103
  redundancies, 69
  strategies, 69
  variable-length coding (VLC), 47
Coding categories, 92
  interframe mode, 92
  intraframe mode, 92
Color gamut, 514
Color halftoning
  adaptive, 519–521
  conventional, 514
  dot-on-dot, 516
  four-mask, 517
  inverted-mask, 517
  scalar error diffusion, 515
  shifted-mask, 517
  vector error diffusion, 515
Common intermediate format (CIF), 320
Complexity
  blocking effect, 73
  compression algorithms, 74, 78, 85, 86, 90, 92
  DCT, 92
  EZW, 74, 78, 85, 86
  image, 74
  JPEG, 92
  MRWD, 74, 78, 79, 86
  natural image, 86
  scalable compression schemes, 103
  SFQ, 90
  SLCCA, 74, 79
  space complexity, 477
  SPIHT, 74, 78, 79, 85, 86
  texture image, 86
  time complexity, 478
  video, 7, 92
Compression, 73
Computer vision algorithms, 415
Connected component analysis, 80
  clustering, 79
Contaminated Gaussian noise, 415

Content-based mesh, 134
  representation, 129
  video compression, 129, 250
  video manipulation, 149, 161
Continuity
  higher-order continuity, 101
  zero-order continuity, 101
Contour fitting, 101
Convergence, 451
Cost function, 416
Cutoff frequency, 496, 497, 498
Cyberware
  Cyberware head scan, 114, 121
  Cyberware range data, 115
  Cyberware scanner, 101, 118, 120
CYC, 432, 439, 441

Data organization and representation, 74, 78
  boundary zero, 79, 81
  cluster, 78
  embedded zerotree wavelet (EZW), 74, 78, 85, 86
  isolated zero, 79
  nonchild descendents, 79
  significance-link, 82
  space-frequency quantization (SFQ), 90
  tree structure, 78
  tree-root, 79
  zerotree, 78, 79
Data partitioning, 286
Daubechies wavelet, 538
DC prediction, 281
Dead zone, 55
Decomposition
  spatio-temporal, 42, 43
Delaunay, 137, 138, 143, 146, 156
Delaunay triangulation, 102
Developmental approach, 443
Digital halftoning, 489–522
  black-and-white halftoning, 492
  color halftoning, 514–521
  multilevel halftoning, 513–514
Discrete cosine transform (DCT), 23, 24, 322, 323, 329, 330, 336
Dispersed-dot dithering, 489
Distribution
  Gaussian, 53, 54
  Generalized Gaussian, 53, 54
  Laplacian, 41, 53, 54
Dithering (*see* Digital halftoning)

Index

Dominant component analysis, 351, 359–361, 370–375, 380
Dot gain, 490, 507
   dot gain compensation, 507–510
Dynamic mesh coding, 143

Edge region, 75
   high frequency coefficients, 75
   low frequency coefficients, 75
   object boundary, 75, 77
Electrostatic force, 502
Elementary function, 349, 350, 367
Elementary signal, 343, 344, 347–349
Elementary streams, 303
Embodiment, 432
Emergent frequencies, 360, 370
Enhancement, 57
Entropy coding, 83
   adaptive arithmetic coding, 83
   run-length coding, 81
Error, 163
   erasure, 163
   random bit, 163
   residual, 198
Error concealment, 163, 287
   forward, 168
   interactive, 190
   postprocessing, 179
   selective, 191
Error control, 163
Error detection, 167
Error diffusion, 490, 491, 515
   Floyd and Steinberg, 491
Error resilience, 164
Error robustness, 266, 285
Euclidean space, 11
Euler method, 105
Evolutionary approach, 442, 443
Exhaustive search, 18
Expression
   expression synthesis, 102
   facial expression, 105, 109
Expression tree, 531
   evaluation scheduling, 534
   shared subexpressions, 534
Extroceptive sensor, 436

Face recognition, 121, 452
Facial action coding (FAC), 104

Facial action parameters (FAP), 104, 105
Facial animation, 288
Facial definition, 290
Fax encoding, 511
Features, 98
   facial features, 109, 110
   feature points, 109
Feature tracking, 392
FEC (forward error correction), 177
Finite-element method, 104, 105
First-in-first-out (FIFO) queue, 85
   scan list, 85
   scan order, 85
Forgetting, 463, 471
Frequency-modulation (FM) screening, 490
Frequency-weighted mean square error (FWMSE), 494, 500, 502, 503

Gabor, 343, 347, 348, 352
   coefficients, 349–351, 368, 380
   expansion, 345, 348–351, 367, 380
   filter, 346, 354–359, 362, 363, 366
   function, 343–345, 349, 368
   transform, 344, 345, 347–351, 367–370, 380
GAPP, 523
Gateway, 198
Gaussian component MF-estimator, 419
Gaussian mixture density decomposition, 422
Generality, 448, 477
Generalized Lloyd algorithm, 14
Geometric connectivity, 80
   4-connected, 80
   6-connected, 80
   8-connected, 80
Gibbs Random Field (GRF), 41, 51
   Huber-Markov random field model, 57
   neighborhood system, 52
   potential function, 51
Gisting, 121
GOB combiner, 217
Group of blocks (GOB), 321, 322, 329

H.261, 317
H.263, 29, 32, 257, 317
H.263 Decoding, 259
H.223 Multiplex, 260
Hand sign recognition, 455
Hilbert transform, 347, 353
Homogeneous region, 75, 77

Hughes 3D computer, 523
Human-computer interaction, 100
Human-computer interface, 99
HVS (human visual system), 48, 67, 68, 489, 500, 515
  scalar quantization, 49
  vector quantization, 46

iac++ library, 525
  operands, 526
  operations, 526
  retargeting, 527
ICM (iterative conditional mode), 55, 56, 57
  clustering, 55
  enhancement, 57
  spatial constraint, 55
ILIAC, 523
Image
  neighborhood operation, 534
  operations, 534
  padding, 529
  rep-size, 532
  subframe transfer time, 531
  template product, 533
  valid region, 530
Image algebra, 523
Image stabilization, 387
  algorithm, 390
  composition, 395
  evaluation, 400
  fidelity, 401
  performance, 403
  real-time, 389
Image warping, 132
Information diagram, 343
Interactive, 190
International Organization for Standardization (ISO), 317, 318
International Telecommunication Union—Telecommunication Standardization Sector (ITU-T), 317, 318
Interpolation, 101, 185
  B-spline interpolation, 101
  bi-cubic interpolation, 101
  frequency, 186, 187
  interpolation methods, 103
  linear interpolation, 107
  polynomial interpolation, 102
  scattered data interpolation, 101, 102
  spatial, 186
Intra update, 287

Introceptive sensor, 436
Inverse distance methods, 102
ISDN, 318, 328

K-means algorithm, 14
K-nearest-neighbor rule, 451
Kalman filter, 112, 121, 364–366, 386
Karhunen-Loeve transform (KLT), 106
Knowledge, 98, 102, 117
Knowledge base, 478
Knowledge-based approach, 441, 443
Kohonen's self-organizing feature map, 14
Kolmogorov-Smirnov normality test, 427

Lagrangian multiplier, 36
Laplacian pyramid, 394
Layered coding, 169
LDA, 449
Learning
  AA-learning, 436, 444
  automated animal-like learning, 436
  comprehensive learning, 444, 446
  developmental learning, 436
  Q-learning, 432, 435, 472
  R-learning, 472
  reinforcement learning, 432, 472
Learning-based approach, 442, 443
Least square method, 108
Level-building, 466, 467
Living machine, 437
Lloyd-Max conditions, 14
Local coherency, 347, 353
Logarithmic complexity, 452
Longevity, 481
Loop filter, 327
Luminance, 320

M-search, 19
Macroblock, 321
Magnetic resonance imaging, 101
MAP (maximum a posteriori), 51, 55, 57
  clustering, 55
  enhancement, 57
Mapping, 102
  mapping functions, 107
Markov decision process, 461
Markov random field, 73, 78
Markov source, 83
Masking function, 23, 24, 25

Index

Mathematical morphology, 79, 523
  conditioned dilation, 79, 80
  connected component, 80
  intersection, 80
  structuring element, 79, 80
Maximum likelihood estimate, 416
MDC (multiple description coding), 171
MDF, 450, 457
Media analysis and recognition, 229
  comparison of DAR and DVMAR, 230–231
  DAR (Document Image Analysis and Recognition), 229
  DVMAR (Digital Video Media Analysis and Recognition), 227, 229
Media representations, 229
  analog media, 230
  content recognized media, 230
  digitized media, 230
MEF, 450, 457
Memory fade factors, 463
Mesh
  animation, 291
  articulation model, 100, 104, 105
  boundary, 137, 142, 146, 149, 158
  design, 130, 134, 135, 137, 143, 153, 154
  ellipsoidal model, 109
  face model, 101, 103, 104, 107, 114
  facial articulation model, 105
  generic face model, 101
  geometric articulation model, 104
  geometric model, 100
  geometry, 130, 133, 137, 143, 146
  head model, 114, 119, 125
  hierarchical meshes, 101
  local parametric models, 109
Mesh-based modeling, 130, 135, 153
Meshes, 101
MF (model fitting)-estimator, 417
Microblocks, 321
Minimum-error Bayesian classification, 417
Min-max estimator, 416
Model, 98
  model deformation, 101
  model fitting, 101, 102
  model parameters, 99
  motion, 132, 133, 143, 146
  muscle model, 105
  parametric models, 116
  physical-based models, 104, 105
  polygonal meshes, 100, 101
  polygonal model, 101

[Model]
  quadrilateral meshes, 103
  rectangular meshes, 101
  surface meshes, 100
  surface models, 102
  tracking, 130, 133, 135, 137, 152, 155
  triangular meshes, 100
Model-based
  model-based approach, 97, 98, 108, 114
  model-based coding, 97, 99
  model-based paradigm, 125
  model-based techniques, 125
  model-based video compression, 100
Moire, 490, 514
Morlet wavelet, 344
Mosaicking, 387, 399
  composition, 389, 396
Motion, 98
  global motion, 123
  motion prediction, 109
  nonrigid facial motion, 108, 109, 113, 116, 123, 125
  nonrigid motion estimation, 114
  rigid head motion, 108
Motion coder, 275, 279
Motion compensation, 10, 319, 322, 325, 327, 329, 332, 333, 339, 394
Motion estimation, 27, 131, 134, 143, 325, 328, 392
  confidence measures, 393
Motion modeling, 131, 133
Motion vectors, 10
MPEG-2, 255
MPEG-2 Systems, 263, 271
MPEG-4, 129, 133, 251, 255
MPEG-4 Audio, 263, 294–297
MPEG-4 Functionality, 264
MPEG-4 System, 272
MPEG-4 Terminal, 256
MPEG-4 Video, 256, 269, 273, 274
MPEG-7, 311
MSE (mean square error), 40
  HVS-weighted MSE, 46
Multilevel dithering (multitoning), 492, 513
Multiplexing, 256, 304
Multipoint control unit, (MCU), 206
Multipoint controller (MC), 206
Multipoint processor (MP), 206
Multipoint videoconferencing, 205
  centralized, 210
  continuous presence, 207
  distributed, 210
  switched presence, 207

Multiresolution, 394
Multistage predictor, 195

Natural and synthetic, 129, 150, 160
Nearest-neighbor condition, 13, 21
N-effector, 436
Neighborhood system, 52, 65
Neural net, 109
News video parsing, indexing and browsing, 245
Node point coordinate coding, 146
Non-symmetric mask, 508, 509
N-sensor, 436

Object-based
  coding, 130, 156
  mesh, 130, 134, 135, 137
  representation, 129
  tracking, 135, 136
Object mesh, 289
Object recognition, 452
Optic flow, 108
  optic-flow-based methods, 109
  optic-flow equations, 109
Optimality
  causal stages optimality, 21
  overall optimality, 20
Ordered dither, 489
  clustered-dot dithering, 489
  dispersed-dot dithering, 489
Orthographic projection, 110, 115
Outlier process, 415

Packet loss, 167
Packetization, 179
PAL Image Manager (PIM), 527
PAL-I, 524
Parametric coder, 296
Parke head model, 125
Partial modeling, 419
PCA, 449
Perceptual
  grouping, 50, 68
  scalar quantization, 48
  vector quantization, 48
Performance comparison, 86
  coding efficiency, 78
  Peak Signal-To-Noise Ratio (PSNR), 86
Personal presence system, 218

Pixel Aspect Ratios (PAR), 335
Pixel-domain combining, 208
POCS (projection onto convex sets), 184
Poisson distribution, 79
Polytopal vector quantizer, 12
Pose, 98, 123
  global head pose, 113
  pose estimation, 110
Positional information, 81
  overhead, 81
Postfiltering, 366, 367, 374, 375
Postfiltering process, 502
Postprocessing, 179
Potential net, 109
Prediction, 181
  motion-compensated, 181
  temporal, 181
Predictive residual vector quantizer, 32
Principal component analysis (PCA), 105, 107
Principal frequency, 496–498
Printer curve, 508
Prioritization, 169
Probability density estimation, 424
Probability density function, 17
Profiles, 307
  audio profile, 308
  systems profile, 309
  video profile, 307
Proprioceptive sensor, 436
Pruning techniques
  optimal pruning, 23
  top-down pruning, 23
PSNR (peak signal-to-noise ratio), 62, 402
Public service telephone network (PSTN), 318, 328, 333

Quadtree, 34, 36
  decomposition, 36
  optimal quadtree, 36
  quadtree-based VQ, 34, 37
  quadtree-VQ encoder, 37
Quantization, 9, 85, 323, 325, 328, 329, 336, 340
  optimal bit allocation, 73, 86
  quantization error, 12, 17
  residual quantization, 17
  scalar quantization, 9, 85
  uniform, 85
  vector quantization, 11
Quarter common intermediate format (QCIF), 320

Radial
    radial basis function, 102
    radial basis methods, 103
    radial functions, 103
Radial-average power spectrum (RAPS), 491
Rate-distortion optimization, 23, 36
Real time, 100, 125, 126
    real-time applications, 107
Reconstruction, 164
Recovery, 188, 189
    coding mode, 189
    motion vector, 188
Redundancies, 69, 165
    crossband, 69
    intraband, 69
    psychovisual, 69
    symbol coding, 69
Region
    of interest, 68
    salient feature, 68
Regression tree, 449
Render, 100, 105, 120, 123
Residual vector quantization, 17
    codebook design, 22
    optimality conditions, 20
    pruned variable-block-size RVQ, 23
    search techniques, 18
    structure of RVQ, 19
    variable-block-size RVQ, 22
Residual vector quantizers (RVQ), 18, 19, 22, 23
Resynchronization, 286
Retransmission, 193
    multicopy, 197
    without wait, 194
Reversible VLC, 287
Roberts edge detector, 537
Robot manipulator, 458
Robot sitter, 440
Robust entropy coding, 178
Robust estimators, 415
Robust video, 285, 287
Rosette, 514
Rotation matrix, 112

SAIL, 459, 469
Scalability, 170, 266, 283, 284, 285, 294, 448, 451
Scalable video, 283, 287
Scalar quantization, 45, 47, 48
    Lloyd-Max, 56

Scene description, 256, 301, 303
Sequential search, 18
Shape coder, 275, 277
SHOSLIF, 451, 446
Shot, 231
    shot attributes, 231
    shot detection, 234
    shot length distribution, 232
Shot detection algorithms, 234–243
    edge pixel comparison, 236
    histogram comparison, 236
    pixel value comparison, 235
Shot detection for gradual transition, 237
    edge model fitting, 239
    edge pixel comparison, 239
    twin-comparison, 238
Shot detection for MPEG compressed video, 239
    DCT coefficient-based shot detection, 240
    hybrid algorithm, 242
    motion-vector-based shot detection, 242
Significance field, 78
Significance-link, 74, 82
    child, 74
    parent, 74
Significance map, 80
    marker, 80
Significant coefficient, 85
SIMD, 523
    PE array, 528
Situatedness, 432
Smart toys, 479
Smoothness, 182
SMPA framework, 439
Social issues, 481
Sophisticated signal, 347, 351, 353, 360, 363, 370
Space complexity, 477
Spatial constraints, 51, 52
    homogeneity, 68
Spatial scalability, 285
Speech recognition, 431, 459
Sprite, 283
SSD, 393
State dictatability, 462
State-nondictatable, 462
Statistical properties, 73, 75
    clustering, 73, 75, 77
    cross-subband similarity, 73, 77
    decaying, 73, 78
    energy compaction, 73, 75
    localization, 73, 75

Stereo vision, 113
Still texture, 292
Stochastic screening, 490
Stream map table, 303
Streaming, 197
  Internet, 197
Structure, 40
  edge, 40, 42, 56
Structured audio, 298
Subband, 75
  analysis, 41
  base, 44
  coding, 73
  coefficients, 41
  high-frequency highpass subband, 75
  HPT (highpass temporal) low-frequency, lowpass subband, 75
  LPT (lowpass temporal) orientation, 41
  resolution, 41
  synthesis, 41
Subband decomposition, 73
  finite impulse response (FIR), 75
  infinite impulse response (IIR), 75
  short time Fourier transform (STFT), 75
Subpixel tracking, 393
Subtractive primary colors, 514
Surfaces
  Bezier surfaces, 100
  B-spline surfaces, 100, 107
  free-form surface deformation, 105
  parametric surfaces, 100
  surface models, 102
Syntactic description, 306
Synthesis, 100, 125, 126
Systems decoder, 299

Talking head, 104
Task space, 432
Task-specific paradigm, 432
Teager-Kaiser operator, 352, 370
Teleconferencing, 104
Template, 98
  template matching, 108, 121
  template decomposition, 535
Temporal scalability, 284
Text to speech, 297
Texture, 100, 344–346, 359
  texture coder, 275, 280
  texture mapping, 100, 105, 107, 120, 123, 130, 150, 158

[Texture]
  texture region, 75, 77
  texture segmentation, 370–375, 380
  texture updates, 119
Thinking, 475
Threshold array, 489
Tracked multicomponent analysis, 363–367, 375–379
Tracking, 97
  feature tracking, 108
  head tracking, 108, 121, 126
Tracking algorithms, 105, 109
Transform coding, 322
Transform vector quantization, 23
Transformations, 391
  affine, 391
  Euclidean, 391
  similarity, 392
Triangular mesh, 130, 138, 141, 146, 147, 160
Triangulation, 130, 137, 138, 143, 146

Unbiased sensor, 436
Uncertainty principle, 343, 354
Uniform mesh, 137
Unsupervised learning, 421

Variable block size, 22, 23
Variable-length-coding (VLC), 146, 149, 324, 326, 329, 331, 336
Variable rate, 26
Vector quantization, 11
  codebook design, 14
  entropy-constrained VQ, 14
  GVQ (Geometric Vector Quantization), 40
  HVS-weighted MSE, 46
  LBG (Linde-Buzo-Gray) algorithm, 46
  MSE, 46
  optimality conditions, 13
  quadtree-based VQ, 34
  quantization error, 12
  residual vector quantization, 17
Vector quantizers (VQ), 12, 13, 14
Vectors
  normal vectors, 101
Verification tests, 310
Veronoi
  Veronoi cells, 103
  Veronoi diagram, 103

Index

Vertices, 101
Video bridge, 206
   coded-domain, 214
   pixel-domain, 218
Video browsing, 248
Video coding, 130, 131, 158
Video compression, 9, 129, 130, 131, 133, 156, 161
   quadtree-based vector quantization, 37
   residual vector quantization, 24
   vector quantization, 15
Video manipulation, 130, 134, 150
Video object (VO), 273, 275
Video object plane (VOP), 273, 274, 275
Video session, 276
View dependence, 293
Virtual agents, 99, 100
   intelligent human agents, 104
Virtual Reality Modeling Language, 129, 130, 301
Vision
   coding, 66
   computer, 68
   high-level, 66
   human, 68
   low-level, 66
   mid-level, 66
   sub high-level, 66
   sub mid-level, 66

Warping
   image warping, 102, 103
Watermark, 511
Wavelet
   coding, 40
   filterbanks, 43
   transform, 40, 538
Wavelet coefficients, 75
   ancestor, 77
   children, 77, 82
   descendents, 77
   insignificant, 75, 79
   parent, 82
   significant, 75, 79
   zero coefficients, 75
Wavelet transform, 73
   coarse resolution, 75
   filter bank, 73
   fine resolution, 75
   iterative decomposition, 75
   multiresolution representation, 73
   parent-child dependency, 82
   pyramid decomposition, 73
   uniform compression, 73
White noise, 491, 506
Wire-frames
   adaptation of wire-frame model, 109
   object-oriented wire frames, 103
   wire-frame head model, 108, 123
   wire-frame structure, 103
With-class scatter, 450
WordNet, 432, 439, 441

Zak transform, 348–350
Zerotree, 62
   embedded zerotree wavelet (EZW) coding, 62